D1524701

STANDARD SOIL METHODS

FOR

LONG-TERM ECOLOGICAL RESEARCH

LONG-TERM ECOLOGICAL RESEARCH NETWORK SERIES

1. Grassland Dynamics: Long-Term Ecological Research in Tallgrass Prairie
Edited by
Alan K. Knapp, John M. Briggs, David C. Hartnett, and Scott L. Collins

2. Standard Soil Methods for Long-Term Ecological Research
Edited by
G. Philip Robertson, David C. Coleman, Caroline S. Bledsoe, and Phillip Sollins

STANDARD SOIL METHODS

FOR

LONG-TERM ECOLOGICAL RESEARCH

Edited by

G. Philip Robertson

David C. Coleman

Caroline S. Bledsoe

Phillip Sollins

LTER

New York Oxford • Oxford University Press 1999

Oxford University Press

Oxford New York
Athens Auckland Bangkok Bogotá Buenos Aires Calcutta
Cape Town Chennai Dar es Salaam Delhi Florence Hong Kong Istanbul
Karachi Kuala Lumpur Madrid Melbourne Mexico City Mumbai
Nairobi Paris São Paulo Singapore Taipei Tokyo Toronto Warsaw

and associated companies in
Berlin Ibadan

Published by Oxford University Press, Inc.
198 Madison Avenue, New York, New York 10016

Library of Congress Cataloging-in-Publication Data
Standard soil methods for long-term ecological research / edited by
G. Philip Robertson . . . [et al.].
p. cm.—(Long-term ecological research network series ; 2)
Includes bibliographical references and index.
ISBN 0-19-512083-3
1. Soil ecology—Methodology—Congresses. 2. Soil ecology—Research—Congresses. I. Robertson, G. P. II. Series.
QH541.5.S6S688 1999
577.5'7'072—dc21 98-28545

9 8 7 6 5 4 3

Printed in the United States of America
on acid-free paper

Series Preface

The U.S. LTER Network is a collaborative ecological research effort that promotes synthesis and comparative research across disparate ecosystems and ecological research programs. The Network provides an important addition to the information, infrastructure, and culture underlying environmental science in the United States. From an initial six sites selected in 1980, the Network has grown to twenty-one sites, ranging from arctic tundra to hot desert, from tropical rainforest to suburban watersheds, and it represents the joint efforts of more than 1000 scientists. This community is performing research that will carry into the future a legacy of well-designed, long-term experiments and data sets that should provide unprecedented opportunities for the evaluation and synthesis of important ecological questions.

The value of community effort is demonstrated in many ways as individual scientists realize results, through collaborations, that are not possible via traditional, individually based research programs. However, effective collaborations do not occur without strong commitments to cooperation and consensus decisions, and the development of standard methods, the endeavor of this volume, is a good example of the effort required for significant long-term payoff.

The LTER Network is proud of this second volume in the LTER Series. We anticipate many additional volumes that will contribute to our long-term understanding of important ecological questions and that will promote the synthesis of knowledge from many sites.

James R. Gosz
Chair, LTER Network

Preface

One of the great opportunities in long-term ecological research is the ability to compare and synthesize patterns and processes across decades and sites that might otherwise stand alone as isolated snapshots. There are many examples of the value of such comparisons, ranging from formal efforts such as the International Biological Programme (IBP) of the 1960s (e.g., Bliss 1977; Edmonds 1982; Mueller-Dombois et al. 1981) and the current Tropical Soil Biology and Fertility Programme (Anderson and Ingram 1993; Woomer and Swift 1994) to less formal, a posteriori efforts such as comparisons of soils under long-term agronomic management (e.g., Jenkinson 1991; Paul et al. 1997; Rasmussen et al. 1998). But such comparisons work best only when the measurements made are comparable; without appropriate standardization it can be very difficult if not impossible to separate an environmental cause of cross-site or through-time differences from simple differences in methodology. Standardization is thus key to addressing many of the most exciting questions in long-term ecological research.

Soils present special problems with respect to standardization. First, it is common to have three or more methods available for measuring a particular soil property, and many such methods are equally appropriate so long as they are applied across some restricted range of soil types. For example, available soil phosphorus is commonly measured by extraction with HCl-NH_4F, $NaHCO_3$, or anion-exchange resin. Results from any one of these techniques can correlate well with plant response to phosphorus nutrient additions over a range of roughly similar soils, but none is the optimum choice across all soil types worldwide. Agreement on a common protocol that will work in most situations—with alternative procedures identified for the odd soil—could greatly help to simplify the interpretation of both subtle and major differences that invariably emerge in long-term or cross-site studies.

For properties known to be strongly affected by methodology, e.g., soil microbial biomass, a priori agreement can be even more advantageous.

The present volume is the result of the need perceived by soil scientists working within the National Science Foundation's Long-Term Ecological Research (LTER) Network for standard soil methods. Many scientists have been frustrated by a limited ability to make cross-site comparisons with other long-term research sites, and in 1996 Network and other soil scientists agreed to participate in an effort to develop a set of common protocols that could be used to characterize the physical, chemical, and biological properties of soil from disparate Network sites. Soils at these sites range from tundra permafrost to desert aridosols, with land use ranging from annual cropping systems to old-growth forest. Over 40 scientists participated in a spirited workshop held at the Sevilleta LTER Site in central New Mexico to discuss standardized methods for these soils, and many more were involved in the review and revision of the resulting papers. The editors greatly appreciate the sense of cooperation and commitment that drove most discussions toward the consensus noted herein.

In each of the papers that follow we have attempted to lay out a specific protocol for a number of soil properties known to be ecologically useful. The reader should be aware, however, that (1) not all protocols will be appropriate under all circumstances, and (2) several protocols are under active development. Protocols under active development include in particular the techniques based on molecular biology, the fine root production protocols, and the measurements of microbial biomass. For circumstances under which the protocol of choice is inappropriate, we have attempted to provide alternate, near-equivalent methods. For those protocols under active development, we have attempted to flag potential pitfalls. For all protocols, we expect to provide added detail as appropriate and to publish errata at a Web site available via the LTER Network (www.lternet.edu).

We thank many individuals who contributed in different ways to the success of this volume. In particular we thank fellow Network scientists who encouraged and supported this endeavor; we thank participants at the Sevilleta Workshop and our local hosts, Robert Parmenter and James Gosz; we thank over 100 ad hoc reviewers who provided many valuable comments to authors and editors; and we thank Barbara Fox for countless hours of coordinating reviewer responses and copy editing. Finally, we thank the authors, who in many cases suspended their own specific methodological prejudices in favor of a broader group consensus.

We are hopeful that this volume will find a place on many laboratory benches, and that the data resulting from its use will indeed be useful in efforts to identify cross-site and cross-decade patterns and processes. It remains a work in progress; please contact us through the Web address noted above to provide and obtain feedback and comments on specific protocols.

February 1998

G. P. R.
C. S. B.
D. C. C.
P. S.

References

Anderson, J. M., and Ingram, J. S. I. 1993. *Tropical Soil Biology and Fertility: A Handbook of Methods.* 2d edition. CAB International, Wallingford, UK.

Bliss, L. C. 1977. *Truelove Lowland, Devon Island, Canada: A High Arctic Ecosystem.* University Alberta Press, Edmonton, Alberta, Canada.

Cole, J., G. Lovett, and S. Findlay, editors. 1991. *Comparative Analyses of Ecosystems: Patterns, Mechanisms and Theories.* Springer-Verlag, New York, New York, USA.

Edmonds, R. L., editor. 1982. *Analysis of Coniferous Forest Ecosystems in the Western United States.* Hutchinson Ross, Stroudsburg, Pennsylvania, USA.

Jenkinson, D. S. 1991. The Rothamsted long-term experiments: are they still of use? *Agronomy Journal* 83:2–10.

Likens, G. E., editor. 1989. *Long-Term Studies in Ecology: Approaches and Alternatives.* Springer-Verlag, New York, New York, USA.

Mueller-Dombois, D., K. W. Bridges, and H. L. Carson, editors. 1981. *Island Ecosystems.* Hutchinson Ross, Stroudsburg, Pennsylvania, USA.

Paul, E. A., K. A. Paustian, E. T. Elliot, and C. V. Cole, editors. 1997. *Soil Organic Matter in Temperate Ecosystems: Long-Term Experiments in North America.* Lewis Publishers, Boca Raton, Florida, USA.

Rasmussen, P. E., K. W. T. Goulding, J. R. Brown, P. R. Grace, H. H. Janzen, and M. Körschens. 1998. Long-term agroecosystem experiments: assessing agricultural sustainability and global change. *Science* 282:893–896.

Risser, P. G. 1991. Long-Term Ecological Research: An International Perspective. SCOPE No. 47. Wiley, New York, New York, USA.

Sollins, P. 1998. Factors influencing rain-forest species composition: does soil matter? *Ecology* 79:23–30.

Woomer, P. L., and M. J. Swift, editors. 1994. *The Biological Management of Tropical Soil Fertility.* Wiley, Chichester, UK.

Contents

Contributors

Robert J. Ahrens
Natural Resources Conservation Service
Lincoln, NE 68508

David E. Armstrong
Water Chemistry Program
University of Wisconsin
Madison, WI 53706

James C. Bell
Department of Soil, Water, and Climate
University of Minnesota
St. Paul, MN 55108

John M. Blair
Division of Biology
Kansas State University
Manhattan, KS 66506

Caroline S. Bledsoe
Department of Land, Air, and Water Resources
University of California
Davis, CA 95616

Richard D. Boone
Institute of Arctic Biology
University of Alaska
Fairbanks, AK 99775

David C. Coleman
Institute of Ecology
University of Georgia
Athens, GA 30602

Harold P. Collins
W. K. Kellogg Biological Station
Michigan State University
Hickory Corners, MI 49060

Frank P. Day
Department of Biological Sciences
Old Dominion University
Norfolk, VA 23529

Charles T. Driscoll
Department of Civil and Environmental Engineering
Syracuse University
Syracuse, NY 13244

Edward T. Elliott
Natural Resource Ecology Laboratory
Colorado State University
Ft. Collins, CO 80523

Boyd G. Ellis
Department of Crop and Soil Sciences
Michigan State University
East Lansing, MI 48824

Timothy J. Fahey
Department of Natural Resources
Cornell University
Ithaca, NY 14853

Carol Glassman
Forest Science Department
Oregon State University
Corvallis, OR 97331

James R. Gosz
Biology Department
University of New Mexico
Albuquerque, NM 87131

James Greenberg
Atmospheric Chemistry Division
National Center for Atmospheric Research
Boulder, CO 80307

David F. Grigal
Department of Soil, Water, and Climate
University of Minnesota
St. Paul, MN 55108

Peter M. Groffman
Institute of Ecosystem Studies
Millbrook, NY 12545

Mark E. Harmon
Department of Forest Science
Oregon State University
Corvallis, OR 97331

David Harris
Stable Isotope Facility
University of California
Davis, CA 95616

Justin W. Heil
3920 Dewey Avenue
Omaha, NE 68105

Elisabeth A. Holland
Atmospheric Chemistry Division
National Center for Atmospheric Research
Boulder, CO 80307
[and]
MPI für Biogeochemie
Tatzendpromenade 1a
07745 Jena
Germany

Wesley M. Jarrell
Department of Environmental Science
and Engineering
Oregon Graduate Institute
Portland, OR 97291

Dale W. Johnson
Biological Sciences Center
Desert Research Institute
Reno, NV 89506
[and]
Environmental and Resource Science
University of Nevada
Reno, NV 89512

Nancy C. Johnson
Department of Environmental
and Biological Sciences
University of Northern Arizona
Flagstaff, AZ 86011

Eugene F. Kelly
Department of Soil and Crop Sciences
Colorado State University
Fort Collins, CO 80523

Michael J. Klug
Department of Microbiology and
W. K. Kellogg Biological Station
Michigan State University
Hickory Corners, MI 49060

Kate Lajtha
Department of Botany and Plant Pathology
Oregon State University
Corvallis, OR 97331

H. Curtis Monger
Department of Agronomy and Horticulture
New Mexico State University
Las Cruces, NM 88003

David D. Myrold
Department of Crop and Soil Science
Oregon State University
Corvallis, OR 97331

Knute J. Nadelhoffer
The Ecosystems Center
Marine Biological Laboratory
Woods Hole, MA 02543

Thomas E. O'Dell
USDA Forest Service
Forest Sciences Laboratory
Corvallis, OR 97331

Eldor A. Paul
Department of Crop and Soil Sciences
Michigan State University
East Lansing, MI 48824

Søren O. Petersen
Danish Institute of Agricultural Sciences
Department of Plant Physiology
and Soil Science
Tjele, DK 8830
Denmark

G. Philip Robertson
Department of Crop and Soil Sciences and
W. K. Kellogg Biological Station
Michigan State University
Hickory Corners, MI 49060

Roger W. Ruess
Institute of Arctic Biology
University of Alaska
Fairbanks, AK 99775

Robert L. Sinsabaugh
Biology Department
University of Toledo
Toledo, OH 43606-3390

Alvin J. M. Smucker
Department of Crop and Soil Sciences
Michigan State University
East Lansing, MI 48824

Phillip Sollins
Forest Science Department
Oregon State University
Corvallis, OR 97331

Christopher W. Swanston
Forest Science Department
Oregon State University
Corvallis, OR 97331

Diana H. Wall
Natural Resource Ecology Laboratory
Colorado State University
Ft. Collins, CO 80523

David A. Wedin
School of Natural Resource Sciences
University of Nebraska
Lincoln, NE 68583

Phillip E. Yeager
Biology Department
University of Toledo
Toledo, OH 43606

Xiaoming Zou
Institute for Tropical Ecosystems Studies
University of Puerto Rico
San Juan, PR 00936

STANDARD SOIL METHODS

FOR

LONG-TERM ECOLOGICAL RESEARCH

1

Soil Sampling, Preparation, Archiving, and Quality Control

Richard D. Boone
David F. Grigal
Phillip Sollins
Robert J. Ahrens
David E. Armstrong

The saying "The devil is in the details" appropriately describes the many decisions that must be made when conducting soils research, particularly long-term investigations where consistency over time and among scientific staff is essential. In addition to technical and often sophisticated analytical methods, soils research also includes the basic sampling and processing of samples as well as adherence to general standards for quality control in the laboratory. Though many decisions about the more nonanalytical steps of soils research may seem inconsequential at the time, they often largely determine the quality of data and their general utility later on.

In this chapter we recommend general protocols for the sampling and general laboratory processing of soils for long-term studies. We discuss soil variability and make practical suggestions for determining sample numbers, collecting soils, and preparing soils for analysis. We propose three levels of soil sampling intensity for long-term research, the lowest level being a minimum standard and the highest level the most comprehensive. We also outline general quality control procedures for the laboratory, including the use of replicates, blanks, spiked samples, and reference materials. Finally, we recommend protocols for the archiving of soil samples and specify the elements of metadata that are essential for the soils database.

Soil Variability

The variability of soil properties in space and time presents a challenge for site assessment and the detection of changes within or among sites. Spatial variation includes horizontal variation across a landscape and vertical variation with horizon depth. In nonagricultural systems this variability is due to numerous factors, including microrelief, animal activity, windthrow, litter and wood inputs, any human activity, and the effect of individual plants on soil microclimate and precipitation chemistry. Variability in agricultural systems is caused by amendments (e.g., fertilizers and lime), tillage, cropping sequences, animal dung and urine, and compaction from grazers and farm equipment. The spatial aspect of disturbance history is a key factor for many systems. Failure to appreciate and adjust for site variability can compromise an otherwise well-designed and carefully conducted investigation.

The literature on the spatial variability of soil properties (e.g., Zinke and Crocker 1962; Mader 1963; Beckett and Webster 1971; Biggar 1978; Riha et al. 1986; Grigal et al. 1991; Robertson and Gross 1994) is reasonably consistent. Generally, variance increases with size of area sampled, even for areas regarded as the same sampling unit; forest soils tend to be more variable than agricultural soils (though see Beckett and Webster 1971; Robertson et al. 1988, 1997); some properties (e.g., extractable cations) are more variable than others; and data often are not normally distributed. Also, horizontal soil variability in both natural and managed ecosystems can be spatially complex and may vary in scale from 1 m to over 100 m. Many investigators (e.g., Warrick et al. 1986; Robertson et al. 1988; Cambardella et al. 1994; Chien et al. 1997) have applied geostatistics as a means of evaluating spatial correlation in soils, although this technique generally has been limited to agricultural sites.

There are no general rules regarding spatial variance of soil properties within horizons and by depth. Although lower variances within horizons and with greater soil depth are often assumed (Petersen and Calvin 1986; Crêpin and Johnson 1993), the few data on this topic are ambiguous. Several investigators have found that variability may be high within horizons and may actually increase with depth. In forest soils the spatial heterogeneity caused by windthrows may contribute to variability of chemical properties within horizons. Cline (1944), who outlined some of the original principles of soil sampling, noted that physical heterogeneity vertically does not guarantee chemical heterogeneity. Mader (1963), in an analysis of several soil properties in Massachusetts red pine plantations, reported a wide range in variance by horizon. Coefficients of variation ranged from 7% for bulk density in the A horizon to 83% for exchangeable calcium and magnesium in the B horizon. Most properties showed no difference in variation between the A and B horizons. Mader (1963) concluded that variability in soil properties did not decrease with depth and suggested that variance may be higher when nutrient concentrations are low. Clearly, whenever possible, the variance of soil properties within horizons and by depth should not be assumed but rather should be determined directly for site-level work.

Temporal variation (from days to years) within a soil horizon or depth interval can be substantial for many soil properties. Temporal changes may reflect seasonal and annual variations in climate and microclimate as well as management regime (e.g., plowing and manuring, fertilization, liming, forest cutting) and alteration of the amounts and chemical quality of organic matter inputs. Many biological soil processes (e.g., microbial respiration and nitrogen mineralization) are strongly controlled by temperature and moisture and often have seasonal patterns specific to a particular site or ecosystem. The amounts, timing, and chemical quality of organic matter inputs (leaf litter, dead wood, crop residues) also may influence temporal changes in soils. For both forest and agricultural systems there is a large collective literature variously documenting seasonal and interannual changes for nitrogen mineralization potential, total organic matter, active organic matter, microbial biomass, light-fraction organic matter, soluble carbon and nitrogen, and extractable cations (Gupta and Rorison 1975; Spycher et al. 1983; Bonde and Rosswall 1987; Boone 1992; Collins et al. 1992; DeLuca and Keeney 1994; Maxwell and Coleman 1995; Sollins 1998). If any characteristics subject to seasonal changes are compared over years or among sites, it is imperative that samples be collected at roughly the same time or under similar climatic and site management conditions.

Field Sampling

Preliminary Assessment

Site assessment prior to the establishment of experimental plots or the adoption of a sampling design should include exploratory soil sampling. For many purposes, soil in a potential field site can be rapidly assessed with hand probes or augers, or by exposing the upper part of the soil with a spade. Soil pits, if interpreted by knowledgeable pedologists, provide the most information on the pedogenic processes at a site and potentially on properties relating to site productivity (e.g., redox conditions, texture, rooting depth). Profiles can reveal evidence of deposition, erosion, and previous land use, and can sometimes serve as a rough gauge of time since previous major soil disturbances. Some profile features, such as plow (Ap) layers, may persist in soils for centuries after plowing has ceased. Indeed, determining past human impacts on the soil may be critical to understanding current soil conditions, how a soil will change in response to disturbance, and how a soil's physical and biological features will change with time. Further recommendations on site and landscape-level assessment are provided in Chapter 2, this volume.

Number of Samples

The objective in measuring a soil property is most often to precisely estimate its mean; for example, to produce an estimate that has a 90% confidence of being within 10% of the mean. Many physical, chemical, and biological soil properties (e.g., aggregate sizes, exchangeable bases, soil gases, bacterial numbers) are not normally distributed but are more nearly lognormally distributed, so this objective requires careful consideration. If the properties are lognormally distributed, then fewer samples are usually required to achieve similar precision of their estimated mean than if they are normally distributed (Grigal et al. 1991). What level of accuracy is necessary? Do we believe that soil-dependent processes are markedly different at two sites whose mean values differ by 10% (e.g., exchangeable Ca^{+2} of 4.5 versus 4.1 cmol (+) kg^{-1})? At what level of differences in properties do we expect differences in processes: at 20% of the mean, or 50%, or even at differences of an order of magnitude (e.g., from 4.5 to 0.4 cmol (+) kg^{-1})? Acceptable levels of precision will vary by study and soil property. For most of the soil protocols described in this volume, reasonable care will keep analytical variability lower than field variability. In general, as a consequence of high variability in the field, the presence of lognormal distributions, the triviality of small differences, and the limited resources with which to process many samples, acceptable laboratory procedures need not be as precise as those presented in many methods manuals, and the investigator will need to strike a reasonable balance between precision and accuracy.

Testing for Normality

The first step in calculating the number of samples to collect is to ascertain whether the frequency distributions for the soil properties of interest are normally or lognormally distributed. This is best done by examining sample data from previous analyses. Good sources are data sets from a pilot study or from other investigators. In this regard, databases on the World Wide Web may be useful if values for samples (versus means alone) are included. Generally, frequency distribution information is limited or unavailable from data in the published literature.

Normality (or lognormality) of data can be assessed visually by graphs and more rigorously by statistical tests. One simple approach is to construct a histogram, which will reveal obvious skewness. If data are lognormally distributed, they often obtain a normal distribution after a natural log–transformation (Parkin and Robinson 1994), although this may not always be the case (Grigal et al. 1991). An alternative and more diagnostic graphical method for identifying normality is a probit plot (Miller 1986; Parkin and Robinson 1992). Statistical methods that can be used to test for normality (or lognormality) include the *W* or Shapiro-Wilk test for sample sizes of up to 50 (Shapiro and Wilk 1965; Parkin and Robinson 1994) and the D'Agostino test (Gilbert 1987; Parkin and Robinson 1992) for sample sizes greater than 50 but less than 1000. If only data summary statistics are available, asymmetry is indicated by a high coefficient of variation (CV), a wide difference between the mean and median (or geometric mean), and a high coefficient of skewness (Parkin and Robinson 1994).

Sample Number Calculation

If data are distributed normally, then the number of samples that are necessary for a given level of accuracy can be found relatively simply by using the relationship

$$n = t^2C^2/E^2$$

where

n = the number of samples to be collected
t = Student's t statistic that is appropriate for the level of confidence and number of samples being collected
C = the coefficient of variation (standard deviation divided by the mean)
E = the acceptable error as a proportion of the mean

For example, to collect sufficient samples for the sample mean to be within 10% of the true population mean with a 95% probability, the t statistic is approximately 2 (1.96 for a 95% confidence interval for a sample of infinitely large size) and E = 0.1. Prior data from similar samples, either collected on site or from the literature, can be used to estimate C. Several studies and reviews (e.g., Mader 1963; Beckett and Webster 1971; Blyth and MacLeod 1978; Grigal et al. 1991) provide useful tables with CVs for numerous chemical and physical soil properties.

If data are distributed lognormally, the necessary number of samples can be calculated by using log-transformed data in the preceding equation. This will give lower sample numbers than would be obtained if a distribution mistakenly were assumed to be symmetrical. For many soil data, scaling by multiplying by 100 or 1000 eliminates values less than 1, making the use of logarithms more straightforward.

Composite Sampling

Compositing or combining sampling units into a single sample for analysis is an effective method for obtaining an accurate estimate of the population mean while reducing cost and analytical time. The requirements for compositing samples are (1) the sample volume represents a homogeneous sample, (2) each sample contributes an equal amount to the composite, and (3) there are no interactions between the sample units within the composite that would significantly affect the composite value. When these conditions are met, values from composites agree well with means obtained from single sampling units (Jackson 1958; Cline 1944). However, compositing does not provide a direct estimate of the population variance, which may be no less important than the mean. For hypothesis testing, at least two composites must be collected from a population to obtain a measure of the variance of the estimated mean. In that case, the estimated mean is the average of the two composites, and the standard deviation of the two composite values can be considered an approximation of the standard error. Field and laboratory costs, the desired accuracy of the estimate, and the expected error of laboratory measurements ultimately should determine the optimum number of field composites after the required number of field samples has been determined (see Mroz and Reed 1991).

Sampling Time and Frequency

Sampling time and frequency are determined by the conditions and objectives of the study. For comparison of flux measurements (e.g., nutrients, gases, water) among sites or across years, we recommend that measurements of all fluxes be carried out on at least a growing season basis, but preferably for a full calendar year. Determining interannual variation in flux measurements certainly is an essential component of a long-term program and is necessary for legitimate site comparisons. Often winter fluxes can be a significant fraction of total annual fluxes and should be measured if at all possible, especially at sites where summer drought limits biological activity. Soil chemistry (e.g., extractable cations, pH, nitrate, active carbon) and soil biotic pools (microbiota and soil animals), which change with season, vegetation phenology, weather, and site conditions, should be determined under common conditions among years whenever possible and ideally during those periods that are most stable or at least repeatable.

Sampling Intensity

Debate continues concerning the costs and benefits of sampling soil by uniform depth increments versus sampling by horizons. Reasons to favor depth-increment sampling are that:

- A large number of samples can be collected relatively easily with augers or similar devices so that variation in soil properties can be adequately captured. Excavating pits for horizon sampling is more costly and time-consuming.
- If several crews are used, lack of uniformity among crews is a valid concern. Differences in horizon descriptions and subsequent analytical results could be attributed to a "lumper" versus "splitter" approach to description and sampling.
- Budgeting often requires a firm estimate of the number of samples to be analyzed. Estimating sample numbers is difficult when sampling is by horizon because of variation in soil profiles and potentially in personnel.
- A major objective of many studies is an inventory of the total elemental or water content of the soil, requiring analysis of the entire soil. There is a risk when sampling by horizon that some thin or discontinuous horizons may not be included when a profile is sampled; if they are combined with another horizon, the concept of sampling by horizon is violated.

Certainly, fixed-depth sampling is not without problems. It may skew results, obscure soil changes, and lead to false conclusions, particularly on sites where erosion, deposition, or compaction has occurred (see later discussion). Horizon-based sampling in some cases effectively reduces both depthwise and horizontal variability (though see Mader 1963) and is fundamental to studies of pedogenesis. In the end, which sampling approach is "best" depends on the site conditions and study objectives.

For long-term research and for cross-site comparisons, we recommend a hierarchical sampling scheme, with depthwise sampling at the lowest intensity level (I), horizon sampling at the highest intensity level (III), and a blend of approaches at the intermediate level (II). The lowest level (I) provides the minimum amount of information acceptable for cross-site or long-term studies, while the highest level (III) is designed to capture at least 90% of the variation in a property at a site (e.g., 90% of total net N mineralization over the full soil profile). Sampling is by fixed depth (20 cm) at Level I and by horizon at Level III; sampling by horizon is encouraged when appropriate at Level II. All three sampling levels provide soil data on at least a 20 cm depth basis. We have deliberately chosen the 0–20 cm depth as a minimum standard because it extends below the plowing depth in most agricultural soils and includes the majority of root biomass. Samples taken from 0–15 cm depth are discouraged because they often do not include the full plow depth in soils. Level III information is recommended as the goal for long-term research sites and for the most meaningful cross-site comparisons.

Level I Sampling (Least Intensive)

The minimum intensity for long-term and cross-site research includes

- sampling of organic horizons;
- sampling of mineral soil from 0 to 20 cm depth;
- description of horizons or distinct soil layers within the sampling zone on a site basis; and
- collection of ancillary soil profile information allowing conversion of data to a 20 cm soil depth basis, if soils are sampled from 0 cm to below 20 cm depth.

Level II Sampling (More Intensive)

An intermediate sampling intensity is preferable for long-term and cross-site research and includes

- sampling of organic horizons;
- sampling of mineral soil at 0–10, 10–20, 20–50, and 50–100 cm depths;
- further subdivision of depth intervals into horizons if there are obvious changes in pertinent soil properties; and
- field description of soil profile, including characterization of all horizons (depth, color, and texture) and determination of rooting depth.

Level III (Most Intensive)

A comprehensive sampling ensures valid site comparisons and includes

- sampling by horizon over the full profile;
- certification (based on previous work) that more than 90% of the soil property's level has been captured by the sampling;
- collection of ancillary data as required to allow determination of properties over 0–20 cm mineral soil depth; and
- full soil profile characterization according to Natural Resources Conservation Service (NRCS) format.

Outside expertise will most likely be required to characterize the soil according to the NRCS format, which is not a trivial task.

"Equivalent depth sampling" (Crépin and Johnson 1993) may be appropriate when depth sampling alone (Level I) is employed and control soils are compared with those that have been physically disturbed. To illustrate, consider a comparison of organic matter contents between a heavily compacted cultivated soil and an uncultivated soil that has not been obviously compacted. If both are sampled to the same depth, the cultivated soil will have a greater "effective depth" (Davidson and Ackerman 1993) and soil mass because of the inclusion of material from deeper horizons. Accordingly, the value for a given property may be skewed. One means of adjusting for this is to sample the less compacted soil more deeply to obtain an equivalent soil mass, though this method is generally feasible only when soil rock content is low and not variable.

Field Procedures

Organic horizons, which consist of undecomposed and partially decomposed litter, are sometimes difficult to sample because they may have variable thickness and poor delineation, may be held together tightly by fine roots, and are not always visually distinct from the mineral soil. The Soil Survey Division Staff (1993) defines an organic horizon as one that

- is never saturated with water for more than a few days and has 20% or more organic carbon (by weight); or
- is saturated for longer periods, or has been artificially drained, and has an organic-carbon content (by weight) of 18% or more if 60% or more of the mineral fraction is clay; 12% or more if the mineral fraction has no clay; or 12 + (clay percentage multiplied by 0.1)% or more if the mineral fraction contains less than 60% clay.

Federer (1982) reviewed the problems associated with identification of organic horizons in forest soils and proposed that 40% organic matter (determined by loss on ignition) is a more convenient definition for organic horizons. The NRCS (1996) has defined standards of "rubbed fiber content" for the Oi, Oe, and Oa organic horizons, though this assay is not commonly used by ecologists. Because most investigators identify organic horizons subjectively, based on composition and color, field separations should be calibrated against the quantitative criteria. This calibration is particularly important when there is a change of field personnel, and when organic horizon data are used in cross-site comparisons or to examine changes over time.

Organic horizons can be collected with a knife and spatula or other flat blade to lift away material from a uniform area marked with a template or frame. Garden clippers or scissors may be useful for separating organic horizons from one another and from the mineral soil if roots make sampling difficult. Surface moss, if present, can be removed by hand if loosely attached to the soil surface, or it can be cut away. Investigators should decide prior to sampling and based on their scientific questions and objectives whether or not to include well-decomposed deadwood (often embedded in the ground) in soil samples (see Chapter 11, this volume).

A bucket or screw auger or a drive-type corer can be used for depthwise sampling of the mineral soil; these corers permit rapid collection of a sample of uniform cross-section area and minimize contamination. A relatively undisturbed sample collected with a corer can be used for measurements of physical properties, including total pore volume, field capacity, and bulk density, which is necessary to convert weight or concentration data to a volume or area basis. A tapered tip on a corer allows the cutting of a sample with a diameter less than the tube, thus facilitating sample removal. Coring devices do not work well in stony soils or dry sandy soils, though augers are available for these conditions. Coring devices obviously exclude rock fragments larger than the core diameter but also minimize the collection of smaller rocks. The degree to which rocks are excluded by the sampling method should be considered when soil concentration data are expressed on a volume or area basis.

Often a soil pit is necessary for sampling by horizon. When loose samples are collected by horizon from a pit, it is usually best to sample from the "bottom up." That is, after horizons have been delineated and described, the deepest horizons should be sampled first. As samples are collected, soil materials slough to the bottom of the pit, and deeper horizons can become contaminated or even buried by material from above. Soil from horizons, which can be collected with a blade (trowel, spade, or knife), should be a composite of samples from more than one face of the pit and should represent the full horizon depth interval equally. Complete sampling of the full horizon depth interval is generally difficult if cores or short augers are used. An alternative to horizontal sampling from the bottom up is to sample a profile from the top down, removing each horizon after sampling to produce a flat surface for the sampling of the next horizon. Determination of both rock content (see Chapters 2 and 4, this volume) and bulk density (i.e., the ratio of the total mass of solids to the total or bulk volume) is necessary for conversion of soil measurements from weight to an area basis if sampling area is not known.

We recommend that soils collected for biological and chemical assays be sealed in plastic bags, stored in coolers with blue ice or ice packs, and returned promptly to the laboratory for analysis. Polyethylene bags (1–2 mil thickness), which are convenient and commonly used, are relatively gas-permeable and somewhat permeable to water vapor. Thicker-gauge polyethylene bags or double bags may be required to reduce the possibility of tears or punctures when samples contain a large number of rocks, sticks, or sharp roots. Thicker bags better retard moisture loss but also can more readily lead to anaerobic conditions. Cooling and rapid processing are not as necessary for samples collected for physical measurements, though maintenance of field moisture may be important for some variates (e.g., aggregation and texture).

Laboratory Processing

Soil Handling

The techniques chosen and the time required for preparation of soil samples returned to the laboratory should be considered carefully to minimize changes in properties of interest and to avoid compromising future analyses. In the absence of information from laboratory trials, the analyst should not assume that soil properties are stable during sample preparation. Changes in properties after soil sampling and during processing can be assessed through the use of field addition samples (see section "Spiked (or Fortified) Samples," below). The processing protocol should be the same for all samples, and field and laboratory replicates should be randomized before processing to minimize systematic errors.

Storage and Drying

The soil variates of interest, the questions asked, and sometimes the soil type determine how samples should be best stored in the laboratory and whether they should be analyzed field-moist or dried. Investigators should recognize that soils can undergo significant changes under any storage condition, whether soils are refrigerated, frozen, or dried before analysis, and that long-term effects of storage on soil properties have not always been examined adequately. Readers are directed to the following chapters for variate-specific recommendations on storage conditions and the advisability of using field-moist versus air-dried samples. Here we discuss generally the more common approaches.

Air drying (at ambient laboratory temperature and humidity) is convenient and often appropriate for measurement of many nonbiological soil properties. Air-dried soil has relatively constant weight and minimal biological activity. Soils after collection can be spread to air-dry in a thin layer in trays or on paper or plastic in a room free of contaminants. If maintaining soil structure is not important for analyses, the soil can be rolled gently with a roller and clods can be broken to facilitate drying. Air drying notably can cause variable and significant changes in soil chemistry (e.g., soluble organic matter, pH, total S, and extractable K^+, NH_4^+, and NO_3^-; Schalscha et al. 1965; Bartlett and James 1980; Kalra and Maynard 1991; Bates 1993; Tan 1996), and changes can continue even after soils have been air-dried. Especially in the case of soils rich in allophane and other amorphous clay, air drying can cause irreversible aggregation and substantial changes in texture (Schalscha et al. 1965; Bartlett and James 1980). Air-dried soils that are remoistened should be allowed sufficient equilibration time before analysis. *Oven drying* at 35 °C to accelerate drying may be acceptable, although oven drying at higher temperatures is strongly discouraged because of large effects on soil properties (Hesse 1971).

Refrigeration or cooling (near 4 °C) of field-moist samples often is justified provided that storage time is minimized if biological or biochemical properties are assayed. Long-term storage of moist samples in a refrigerator (several months or more) is not recommended because of possible major shifts in the microbial community (Stotzky et al. 1962) and the potential development of anaerobic conditions (Gordon 1988). Using field-moist samples is often preferable for biological assays and some physical properties (e.g., water retention) but may be problematic given that field-collected samples can range from air-dry to saturated. Adjusting water content (amendments, or removal by evaporation) may be necessary for biological assays.

Freezing at low temperature (≤ -20 °C) can be suitable for long-term storage, given that microbial activity is effectively minimized, though it too has some drawbacks. Freezing promotes desiccation, lyses microbial cells, and disrupts soil organic matter (SOM) structure, and it may alter exchangeable NH_4^+ and soluble P concentrations (Allen and Grimshaw 1962; Nelson and Bremner 1972; Bartlett and James 1980). Typically there is a flush of biological activity in thawed soils due to the decomposition of soil microbial cells lysed by the freezing.

In summary, no storage condition is perfect, and the absence of a storage effect on soil properties should be checked rather than assumed.

Sieving and Grinding

Conventionally, mineral soil samples for chemical and physical analyses should be passed through a 2 mm sieve to obtain representative subsamples and to exclude larger particles (small surface-to-volume ratio) that are relatively less reactive.

Aggregates, unless otherwise examined and considered, can be forced through the mesh by hand or with a larger rubber stopper. Whenever possible, samples should be sieved or ground in an air-dry condition; moist samples can be sieved and ground but often with difficulty. Subsamples from the sieved fraction subsequently may be ground to pass at least a 0.5 mm sieve (40-mesh) to reduce subsampling error for micro- or semimicro-analyses that require lower sample weight (micrograms to several grams). Organic horizons with macro-organic matter can be passed through a coarse sieve (e.g., 5.6 mm) to remove sticks and stones before grinding for chemical analysis. In some cases, hand picking of larger coarse fragments from organic horizons may be appropriate. Common grinding devices are mortar and pestle, ball or rod mill, Wig-L-Bug shaker, and SPEX mill for mineral soil; a Wiley mill or manual meat grinder can be used for organic material. Stainless steel or nylon sieves and mortar and pestle should be considered to minimize the possibility of sample contamination when trace elements are measured.

Allophanic soils, and others rich in amorphous clays, again deserve special mention because of their tendency to aggregate irreversibly upon air drying. Such soils may need to be sieved fresh, as best possible, or hand-picked to remove rocks and large plant fragments, then ground after air drying to break up aggregates formed during drying. Cautious air drying of the fresh soil prior to sieving may lower the moisture content to a level at which samples can be sieved but aggregates will not yet have formed.

Often soils should be analyzed unsieved or after passage through a coarse sieve only. This is true when maintenance of soil structure or retention of all particle sizes is important for biological and physical assays. Sieving of mineral soil, for example, breaks aggregates and exposes formerly physically protected SOM to soil microbiota. This may not be a trivial consideration for interpretation of nutrient dynamics, microbial respiration, and microbial biomass during laboratory incubations. In the case of organic horizons, using material that passes through a 2 mm sieve often is impractical because of the large fraction left on the screen. Conveniently, organic material can be passed through a coarser sieve (e.g., 5.6 mm) before biological assays.

All sample weights (whether air-dry or field-moist) should be converted to an oven-dry (105 °C) basis, determined by oven drying subsamples (>24 hours) taken at the time of sample analysis. Prior to some analyses (e.g., total nitrogen), it may be appropriate to dry soils at a lower temperature (e.g., 70 °C) to avert volatilization losses. In this case, an additional conversion to 105 °C weight equivalent may be necessary. When a sieved fraction is used for analysis, the material larger than mesh diameter should be oven-dried (105 °C) and weighed to provide a correction factor between sieved oven-dry weight and unsieved oven-dry weight.

Analytical Issues

Here we focus on general aspects of analytical methods that are critical to the quality of long-term data. Problems associated with accuracy and analytical bias must be avoided in assessing long-term trends or making cross-site comparisons. Thus the analytical data set must contain information reflecting and verifying method accuracy. The most important approach for assessing and documenting accuracy is the use of reference materials or external standards that can be analyzed regularly by each laboratory. Analytical methods for specific soil constituents are described by Sparks (1996) and in other standard soil methods volumes.

The analyst must be prepared to make a major commitment to quality control and good laboratory procedures (Taylor 1987; Association of Official Analytical Chemists-International 1990; American Society for Testing and Materials 1991; American Public Health Association 1992). In addition to the use of reference materials, each analytical laboratory involved in producing long-term data sets must develop well-documented, standard operating procedures and incorporate regular analyses of blanks (field and laboratory) and spiked or fortified samples to enable detection of contamination or loss during storage and analysis and to document method accuracy. Blanks, spikes, reference materials, and replicates will constitute a substantial part of the sample load, analytical time, and cost.

When implementing a new method, the analyst should establish its precision and accuracy by analysis of replicate field samples and reference materials, respectively. Problems associated with the method will be reflected by these "trial runs." Control charts with various action levels established (e.g., warning versus shutdown) must be used to follow, and to react to, changes in method performance. A formal system of data quality flags and codes should be established to maintain uniformity across all work groups.

Blanks

Blanks are a critical component of the analytical scheme for detecting contamination from sample containers, analytical reagents, and sample handling. For blanks to be meaningful, cleaning, sampling, and sample handling techniques must be evaluated and standardized for all sample containers, sampling apparatus, and laboratory glassware and plastic ware used in the procedure. Sample containers that minimize blank values should be chosen whenever possible. For example, glass containers should not be used for samples to be analyzed for silica or trace metals, while plastic materials should not be used for organic components. Sample bottles manufactured from many other polymers, including low-density polyethylene (LDPE) and especially fluorinated LDPE, after detergent and acid washing, are suitable for dissolved organic carbon (DOC) sample collection, though DOC storage in these bottles has not been adequately tested. Appropriately cleaned Teflon (PTFE, PFA, FEP) is compatible with analysis of total organic carbon, DOC, many trace organic compounds, and nearly all inorganic compounds.

Two types of blanks should be incorporated into the sampling and analytical protocol. *Field blanks* should be used to detect possible contamination associated with sampling and storage. To the extent possible, field blanks (sample containers) should be exposed to all steps in the sampling and analysis scheme. These blanks are generally more critical for aqueous soil solution samples and soil extracts than for solid soil samples. For soil solution sampling, pure water should be added to the sampling container and carried through all field and laboratory steps such as sample transfers, filtrations, digestions, and reagent additions. Field blanks should represent at least

2% of the field samples, but a higher frequency may be appropriate for procedures unavoidably subject to contamination.

Laboratory blanks or reagent blanks should be incorporated at the beginning of the laboratory analysis scheme. Laboratory blanks monitor reagent purity and the overall laboratory procedure. Again, laboratory blanks should represent at least 2% of the samples. More frequent blanks should be used when contamination is problematic, when new reagents are incorporated into the analysis, and when sample and blank values are not so different (trace level analyses). Replicate spikes should be used to establish the limit of detection and limit of quantification for the method in use. Results (including those for blanks) below these values should be flagged as "below the limit of detection" or "below the limit of quantification." Preconcentrating the analyte may be necessary to achieve concentrations above blank values for accurate quantitation. If quantitation at this level is essential, the availability of more sensitive and less variable alternative methods should be explored.

Spiked (or Fortified) Samples

Spiked (or fortified) samples are necessary for monitoring analyte losses during sample storage and analysis and for determining the accuracy or bias of the analytical procedure. Spikes should be included at two points in the analytical scheme. Field spikes should be used for soil solution samples, which involves adding standards to the field sample container and to a subset of the field samples, then carrying the spiked samples through the remainder of the analytical scheme. These samples serve to detect analyte adsorption by the sample container. Laboratory spikes should be used for both soil and soil solution samples to monitor analyte recovery and matrix effects (either positive or negative interferences). For soil samples, the standard should be added to the soil extract. For soil solution samples, the standard should be added when the sample is transferred from the field sample container to the first container used in the laboratory analysis. Spikes should be made at concentrations approximately twice that expected in the sample, and spike frequency should correspond to at least 10% of the field samples. Spikes should not be used to correct analyte values for the regular samples because the spike may be more easily recovered or more readily detectable than that occurring in the samples naturally, and the spike may be reduced or otherwise altered by the sample and the sample matrix. Although any recovery standard is somewhat subjective, spike recovery of 80–120% is acceptable for most analyses (Klesta and Bartz 1996).

Laboratory Replicates

The degree of replication must be sufficient to document the precision of the sampling and analysis scheme. Field replicates monitor the precision of the overall procedure and overall field variability. Laboratory replicates or split samples (i.e., splits from a homogeneous sample in the laboratory) monitor the precision of the lab method. Precision of laboratory replicates can be assessed by calculating a CV for analytical replicates, referred to in the quality assurance/quality control (QA/QC) literature as the relative standard deviation

$$RSD = (s/x) \times 100$$

where

s = standard deviation
x = mean

RSD values $\leq 10\%$ for laboratory replicates indicate that analyses are sufficiently precise. The number of replicates required will vary by protocol and should be evaluated carefully on a laboratory-by-laboratory basis. For example, some analyses will require even triplicate replicates to bring analytical CVs to $\leq 10\%$.

Quality Control Check Samples

Quality control check samples (distinct from calibration standards) are certified reference materials or in-house standards used to determine analytical precision and accuracy. Quality control check samples should be matched with respect to matrix and the analyte concentration range of the routine samples, and should be handled the same as routine samples in the processing and analytical stream. In-house standards and calibration standards should come from different sources. Quality control samples should be used only during their known shelf life (i.e., the period of stability for the parameter of interest). If the shelf life is not known, the integrity of quality control check samples can be examined by comparison with fresh certified reference standards or by participation in an interlaboratory sample exchange program (see later discussion). Mean values (and standard deviations) for in-house standards should be determined upon repeat analysis and should be traceable to certified reference standards whenever possible. Analyses are regarded as sufficiently precise if the value for quality control check samples is within two standard deviations of the mean value (Klesta and Bartz 1996). Otherwise a problem with the analysis (method or instrumentation or both) is indicated.

Reference Materials

Incorporation of reference materials or external standards in a quality control scheme is necessary to determine analytical bias caused by the measurement protocol and the analyst. *Reference material* is defined as "a material or substance, one or more properties of which are sufficiently well established to be used for the calibration of an apparatus, the assessment of a measurement method, or for assigning values to materials" (National Institute of Standards and Technology 1995). Primary or certified reference materials are those with properties certified by a national standards laboratory or other organization with appropriate legal authority and are accompanied by a certificate from the issuer (Ihnat 1993). The National Institute of Standards and Technology (NIST), formerly the U.S. National Bureau of Standards, uses the term *standard reference material* (SRM) for its certified reference materials. The philosophy of certification is that the value for a given measured property is independent of method.

Soil reference materials with certified values for chemical constituents are uncommon, and the assays (reflecting predominantly the needs of geochemists, geologists, and reclamation specialists) are biased toward metals and cations. There are few reference soils with certified values for total nitrogen, total carbon, organic carbon, and extractable cations, or for any physical properties. NIST reference soils include Peruvian Soil (SRM 4355), Buffalo River Sediment (SRM 2704), San Joaquin Soil (SRM 2709), and Montana Soil (SRMs 2710 and 2711). Although all have certified values for numerous metals, only Montana Soil and Buffalo River Sediment have certified values for carbon, and none have certified values for nitrogen. Mineral soils available from several other agencies outside the United States (e.g., the Community Bureau of Reference, or BCR, program of the European Commission) also tend to have certified values for metals and cations only. The Canadian Centre for Mineral and Energy Technology (CANMET) offers several reference soils and sediments with consensus values (means based on participating labs, each using methods of its choice) for major elements and some metal oxides. CANMET soil samples SO-2, SO-3, and SO-4 have consensus values for carbon, nitrogen, and loss-on-ignition, but the precision of their nitrogen values is low. Notably, mineral soils with NIST-traceable values for carbon, nitrogen, and sulfur may be obtained from some companies that manufacture soil nutrient analyzers (e.g., LECO Corporation).

We know of no certified material for soil organic material. However, several types of leaves (tomato, apple, peach, pine) are available from NIST with certified values for carbon and nitrogen. Similarly, 10 different plant reference materials (including beech and spruce leaves) are available from the BCR Reference Materials program of the European Commission, with consensus values for metals and cations (all materials) and for carbon and nitrogen (leaves only). Simulated rainwater reference materials with certified values for major cations and anions also are available from the BCR program. Several companies (e.g., SPEX CertiPrep, Inc.) offer certified standards for trace metals, minerals (e.g., calcium, magnesium, sodium, potassium), anions, and nutrients (e.g., ammonium, nitrate, total nitrogen and phosphorus) in a water matrix.

Certified reference materials should be used to calibrate in-house standards and to determine analytical bias. Although many labs commonly use in-house standards for quality control purposes and to measure precision, these do not provide a measure of accuracy or analytical error unless compared against certified standards. Certified reference materials are particularly critical for intersite work and for measurement of long-term changes in soil properties. NIST is now initiating inclusion of certified values for carbon and nitrogen in all of its soil SRMs (B. MacDonald, NIST, personal communication), although the process is likely to take 3 years. Discussions within the LTER Network are now under way to develop a library of certified soil standards designed to cover properties of common interest to ecologists. Table 1.1 lists a selection of current soil reference material providers.

Interlaboratory Exchanges

An interlaboratory exchange program is one means of judging values for in-house standards and promoting comparability of results among laboratories. The largest and most extensive sample exchange program is the International Soil-Analytical

Table 1.1. Selected Providers of Soil Reference Materials

Provider	Address
NIST	Standard Reference Materials Program National Institute of Standards and Technology Gaithersburg, MD 20899-0001 Phone: 310-975-6776 FAX: 301-948-3730 email: SRMINFO@enh.nist.gov http://www.nist.gov
CANMET	Canadian Certified Reference Materials Project (CCRMP) Natural Resources Canada 555 Booth Street Ottawa, Ontario Canada K1A 0G1 Phone: 613-995-4738 FAX: 613-943-0573 email: ccrmp@nrcan.gc.ca http://www.nrcan.gc.ca/mets/ccrmp
BCR	BCR Reference Materials European Commission, Joint Research Centre Institute for Reference Materials and Measurements (IRMM) Management of Reference Materials (MRM) Unit Retieseweg B-2440 Geel Belgium Phone: +32-14-571211 FAX: +32-14-590406 http://www.irmm.jrc.be/mrm.html
BAM	Bundesanstalt für Materialforschung und-prüfung [Federal Institute for Materials Research and Testing] Section 1.01, Quality Assurance and Methodology in Chemical Analyses Rudower Chaussee 5 D-12489 Berlin Phone: +49-30-63 92 58 47 FAX: +49-30-67 77 06 10 http://www.bam-berlin.de/e3org.html

Notes. See text for a description of available materials.
A more comprehensive listing can be found in Ihnat (1993).

Exchange Programme operated by the Wageningen Evaluating Program (WEPAL) of Wageningen Agricultural University in the Netherlands. For a subscription fee, participating laboratories receive in common four soil samples every 3 months for analysis of parameters (only those they decide are useful to them) by methods of their choice. Consensus values (with statistics) for measured parameters subsequently are compiled by WEPAL and distributed back to the member laboratories. Laboratories additionally may submit their own samples and in-house standards to WEPAL for interlaboratory measurements. Currently nearly 300 laboratories worldwide participate in the WEPAL program, which is described in more detail at www.benp.wau.nl/wepal.

Calculations

For cross-site purposes we recommend that soils data be expressed on an areal basis, whenever possible and practical, and that soil sampling depth always be specified. In some cases it may be useful and appropriate to express soils data additionally by soil mass, volume, or horizons. However, these units alone do not lend themselves as well to cross-site comparisons. The disadvantage to comparing soil data among sites on a soil mass basis, for example, is that differences in coarse fragment content and bulk density are commonly ignored. Adopting the standard of expressing soils data on an areal basis whenever possible should facilitate cross-site comparisons and synthesis, and better ensure the comparability of long-term data sets.

The steps required to express soil data on an areal basis depend on whether samples are collected from a fixed surface area. If soil samples are collected quantitatively within a fixed surface area (e.g., defined by a sampling frame or coring device), soil nutrient concentration data are easily converted to an area basis by the equation

$$y = a \times b$$

where

y = variate mass/m^2
a = soil nutrient concentration (mass, equivalents, molar quantities)/soil mass
b = mass of soil/m^2

The mass of the sieved fraction must be used for b in the equation, assuming that soils are sieved before analysis. Conventionally, soil mass is defined as oven-dry (105 °C) mass.

If samples are not collected from a fixed area, the equation must include bulk density, rock content, and soil volume m^{-2} for the given sampling depth, and becomes

$$y = a \times b \times c \times d$$

where

y = variate mass/m^2
a = soil nutrient concentration (mass, equivalents, molar quantities)/soil mass
b = bulk density of sieved material
c = soil volume for 1 m^2 at given sampling depth
$d = (1 - [\%\text{ rock volume}/100])$

Bulk density of sieved material can be determined by subtracting the rock fragment volume (best measured by water displacement) from a bulk density sample of unsieved soil. Note that the equation becomes increasingly sensitive to rock content as rock content increases, and correspondingly less sensitive to bulk density. Certainly a major impediment to comparing sites is the scarcity of reliable data on rock content (see Homann et al. 1995). Methods for determining rock volume are given in Chapters 2 and 4, this volume.

Sample Archiving

The archiving of soil samples is an essential component of a long-term soils research program. Archived soils are invaluable "time capsules" for assessing temporal changes in soil properties, particularly as new analytical tools become available. Examples of the profitable use of archived soils include detection of reduced soil lead levels following the banning of leaded gasoline (Siccama et al. 1980; Friedland et al. 1992) and refined measurements of soil organic matter turnover based on changes in the ^{14}C bomb signal (Trumbore 1993). Certainly the creation and maintenance of an archive for soil or other physical samples (e.g., leaf litter, tree cores, sediments) is not trivial and requires continued financial support and institutional commitment. One option for reducing the costs of an archive facility, particularly for research sites with a smaller sample inventory, may be to share existing archives or to establish a national-level facility or network.

The following are commonsense recommendations for long-term storage of soil samples, based in part on the experience of scientists who have developed the Sample Archive Building (a library of nearly 10,000 physical samples) at the Hubbard Brook LTER site:

- Samples should be kept air-dry at room temperature in a secure location with low probability of water damage (e.g., broken pipes, flooding from weather or storm events), chemical contamination, fire, or other catastrophes. Temperature fluctuations in the archive should be minimized because of the potential for condensation inside containers. Dehumidification may be necessary during warmer months. Long-term storage of field-moist samples in refrigerators or freezers generally is not recommended because of inevitable power failures and the cost of backup power units.
- Containers should have secure lids and should be made of long-lasting materials (plastic or glass) with low potential to contaminate the sample and alter soil chemical properties.
- Container labels should be carefully evaluated for permanence. Labels should always be placed on the container itself rather than placed only on the lid. If there is significant risk that labels will be defaced or that they will not be permanently affixed, a copy of the label on plastic or similar material should be placed inside the container with the sample.
- Labels should include both sample number and the degree of fineness of the sample (e.g.,$<$2 mm).
- Records of sample collection (investigators, location, method, sampling time), processing (e.g., prior storage conditions, sieving, grinding), and available data, including analytical methods, should be readily accessible and main-

tained by personnel responsible for the archive. A copy of the records should be kept in a location near the archive if convenient. The location of the records can be written on the sample container. Ideally the soil archive inventory and sample data should be electronically cataloged and made accessible electronically both on-site and remotely.

- Each archive should have a written policy regarding use and access, and a log of activities and users should be maintained. The original investigator should have free and easy access to samples.
- Subsampling of archived soil is wasteful. People often take more material than they need. It is better for users to take the complete sample, use only the amount necessary, and return the sample. To protect against loss of a sample, archives can maintain a subsample for use only in the event that the "working" sample is lost.
- Changes in properties will occur during sample storage and should be monitored by periodic analysis of archived soil reference materials or in-house standards.

The amount of soil for archiving cannot be easily fixed because it depends on the projected number of future users, the amount required for analyses, and the cost and logistics of soil storage. A minimum of several hundred grams may be appropriate. Destructive sampling of archived material should be minimized.

Metadata

Metadata, or the supporting documentation necessary to interpret a data set, are essential for data sufficiently valuable to preserve for potential reuse. Without such documentation the value of data depreciates rapidly due to human and institutional memory loss; the loss of field and laboratory notes; and career changes, retirements, and deaths of the data originators. Metadata are particularly critical for data from long-term studies given the high reuse value of such data and because long-term research projects generally have a changing group of investigators. Without adequate accompanying metadata, which can be tedious, time-consuming, and expensive to assemble, soils data may have limited value beyond the original study.

We strongly encourage researchers conducting long-term soils research to include sufficient metadata with data that have future potential value. What metadata are sufficient? Michener et al. (1997), in a thorough review of metadata (costs, benefits, and implementation), propose a metadata standard for nongeospatial ecological data similar to that already established for geospatial data (e.g., National Institute of Standards and Technology 1992; Federal Geographic Data Committee 1994). We adapted their version to accommodate long-term soils work (Tab. 1.2). Data from long-term studies should be able to meet the "20-year test" (Webster 1991; Strebel et al. 1994; Michener et al. 1997), meaning that someone unfamiliar with a study should be able to readily utilize data 20 years after their collection. The monetary costs associated with developing and maintaining metadata are not trivial but must be considered obligatory for those conducting and funding long-term research.

Table 1.2. Metadata Descriptors

Descriptor	Explanation
Class I. Data Set Descriptors	
Data set identity	Title or theme of data set
Data set identification code	Accession number or code specified by the data set originator or data management personnel to identify a data set
Data set description	Summary of research objectives, data set contents (including temporal and spatial context), and potential uses of the data
Originators	Name(s) and address(es) of principal investigator(s) associated with data set
Abstract	Summary of research objectives, data set contents (including temporal and spatial context), and potential uses of the data
Keywords	Theme and contents, ecosystem type, location
Class II. Research Origin Descriptors	
Overall project description	(Note: this section may be essential if the data set represents a component of a larger or more comprehensive database; otherwise relevant items may be incorporated into the subproject description, below.)
Identity	Project title or theme
Originator(s)	Name(s) and address(es) of principal investigator(s) associated with the project
Study period	Start date and end date or expected duration
Objectives	Scope and purpose of research project
Abstract	Summary of the broader scope of the overall research project
Funding source(s)	Name(s) and address(es) of funding sources, grant and contract numbers, and funding period, if available
Subproject description	
Site description	
Location	Latitude and longitude, political geography, permanent landmarks or reference points
Physiographic region	Ecoregion, physiographic province, major land resource area
Landform component	Backslope, summit, floodplain, stream terrace
Watershed(s)	Size, boundaries, receiving waterways (streams, rivers)
Terrain attributes	Slope, aspect, slope curvature, elevation, microtopography, catchment area, catena position
Soils	Taxonomic unit (order and series if available), depth, texture, thaw depth, pans, presence of upper organic horizons or organic debris
Predominant soil parent material type	Residuum, alluvium, glacial drift, colluvium, lacustrine deposits, etc.
Lithology of predominant soil parent material	Sandstone, shale, limestone, granite, gneiss, etc.
Geomorphic history and approximate age of geomorphic features	Erosional and depositional events, slumps, etc.
Predominant vegetation communities	Tall-grass prairie, eastern temperate forest, tilled agricultural field, etc.
History of land-use and natural disturbances	Management activity (e.g., plowing, fertilization, liming, grazing, cutting, clearing, scarification), wildfires, drainage, depositional and erosional events, pest outbreaks, severe storms, severe climatic events, and other "acts of God" (e.g., volcanoes and earthquakes)

(continued)

Table 1.2 *(continued)*

Descriptor	Explanation
Climate	Summary of climate statistics
Experimental or sampling design	
Design characteristics	Experimental design, field replication, subsampling and compositing, decisions regarding inclusion or exclusion of heterogeneous features (e.g., deadwood, rocks, furrows)
Permanent plots	Dimension, location, vegetation characteristics
Data collection period and frequency	
Sampling area, depth, horizons	Area, depth, and horizons sampled for each analysis
Research methods	
Field and laboratory	Description of protocols, including references to standard methods
Sample processing	Sieving, storage time and conditions (e.g., refrigeration, freezing, air drying), removal or inclusion or roots, grinding
Instrumentation	Type, manufacturer, and model
Standards	Use and frequency of standards (certified and other)
Project personnel	Principal and associated investigator(s), technicians, and students
Class III. Data Set Status and Accessibility	
Status	
Latest update	Date of last modification
Latest archive date	Date of last data set backup
Metadata status	Date of last metadata update and current status
Data verification	Status of data quality assurance checking
Accessibility	
Storage location and medium	Pointers to where data reside (including redundant archival sites)
Contact person(s)	Name and address, phone, fax, electronic mail address, and web home page
Copyright restrictions	Whether copyright restrictions prohibit use of all or portions of data set
Proprietary restrictions	Any other restrictions that may prevent use of all or portions of data set
Release date	Date when proprietary restrictions expire
Citation	How data may be appropriately cited
Disclaimer	Any disclaimer that should be acknowledged by secondary users
Costs	Costs associated with acquiring data (may vary by size of data request, desired medium)
Class IV. Data Structural Descriptors	
Status	
Identity	Unique file names or codes
Size	Number of records, record length, number of bytes
Format and storage code	File type (e.g., ASCII, binary), any compression schemes used
Header information	Description of any header data or information attached to file (Note: may include elements related to "Variable information" [below]; if so could be linked to appropriate section[s])
Alphanumeric attributes	Mixed, uppercase, or lowercase
Special characters/fields	Methods used to denote comments or to flag modified or questionable data

(continued)

Table 1.2 (*continued*)

Descriptor	Explanation
Authentication procedure	Digital signature, checksum, actual subset(s) of data, and other techniques for assuring accurate transmission of data to secondary users.
Variable information	
Variable identity	Unique variable or code
Variable definition	Precise definition of variables in data set
Units of measure	SI units of measurement associated with each variable
Data type	
Storage type	Integer, floating point, character, string, etc.
List and definition of variable codes	Description of any codes associated with variables
Range for numeric values	Minimum and maximum
Missing value codes	Description of how missing values are represented in data set
Precision	Number of significant digits
Data format	
Fixed, variable length	
Columns	Start and end columns
Optional number of decimal places	
Data anomalies	Description of missing data, anomalous data, calibration errors
Class V. Supplemental Descriptors	
Data acquisition	
Data forms or acquisition methods	Description or examples of data forms, automated data loggers, digitizing procedures, etc.
Location of completed data forms	Physical location (address, building name, room, or office number)
Data entry verification procedures	Methods employed to identify and correct errors during data entry
Quality assurance/quality control procedures	Identification and treatment of outliers, description of quality assessments
Supplemental materials	References and location of maps, photographs, slides, videos, GIS data layers, physical specimens, field note-books, comments, etc.
Computer programs and data-processing algorithms	Description or listing of any algorithm or software package used in deriving, processing, or transforming data
Archiving	
Archival procedures	Description of how data are archived for long-term storage and access
Redundant archival sites	Locations and procedures followed to provide redundant copies as a security measure
Publications and results	List of publications resulting from or related to the study, graphical and statistical data representations, primary Web site(s) for the data and the study
History of data set usage	
Data request history	Log of who requested data and for what purpose
Date set update history	Description of any updates performed on the data set
Questions and comments from secondary users	Questionable or unusual data discovered by secondary users, limitations or problems encountered in specific applications of data, unresolved questions or comments

Sources Adapted from Michener et al. (1997) and Chapter 2, this volume.

Acknowledgments We thank E. T. Elliott and E. F. Kelly for highly useful contributions to our discussion of soil sampling issues; T. Siccama for recommendations on soil archiving; A. Doyle, J. Canary, and three anonymous referees for critical review of the manuscript; and G. P. Robertson for many helpful editorial suggestions.

References

Allen, S. E., and H. M. Grimshaw. 1962. Effect of low temperature storage on the extractable nutrient ions in soils. *Journal of Science, Food and Agriculture* 13:525–529.

American Public Health Association. 1992. *Standard Methods for the Estimation of Water and Wastewater.* 18th edition. American Public Health Association, American Water Works Association, Water Environment Federation, Washington, DC, USA.

American Society for Testing and Materials (ASTM). 1991. *Annual Book of ASTM Standards.* ASTM, Philadelphia, Pennsylvania, USA.

Association of Official Analytical Chemists-International (AOAC). 1990. *Official Methods of Analysis.* 15th edition. AOAC-International, Arlington, Virginia, USA.

Bartlett, R., and B. James. 1980. Studying dried stored soil samples: some pitfalls. *Soil Science Society of America Journal* 44:721–724.

Bates, T. E. 1993. Soil handling and preparation. Pages 19–24 *in* M. R. Carter, editor, *Soil Sampling and Methods of Analysis.* Lewis Publishers, Boca Raton, Florida, USA.

Beckett, P. H. T., and R. Webster. 1971. Soil variability: a review. *Soils and Fertilizers* 34: 1–15.

Biggar, J. W. 1978. Spatial variability of nitrogen in soils. Pages 201–222 *in* D. R. Nielsen and J. G. MacDonald, editors, *Nitrogen in the Environment. Volume 1, Nitrogen Behavior in Field Soil.* Academic Press, New York, New York, USA.

Blyth, J. F., and D. A. MacLeod. 1978. The significance of soil variability for forest soil studies in northeast Scotland. *Journal of Soil Science* 29:419–430.

Bonde, T. A., and T. Rosswall. 1987. Seasonal variability of potentially mineralizable nitrogen in four cropping systems. *Soil Science Society of America Journal* 51:1508–1514.

Boone, R. D. 1992. Influence of sampling date and substrate on nitrogen mineralization: comparison of laboratory-incubation and buried-bag methods for two Massachusetts soils. *Canadian Journal of Forest Research* 22:1895–1900.

Cambardella, C. A., T. B. Moorman, J. M. Novak, T. B. Parkin, D. L. Karlen, R. F. Turco, and A. E. Konopka. 1994. Field-scale variability of soil properties in central Iowa soils. *Soil Science Society of America Journal* 58:1501–1511.

Chien, Y-J., D-Y. Lee, H-Y. Guo, and K-H. Houng. 1997. Geostatistical analysis of soil properties of Mid-West Taiwan soils. *Soil Science* 162:291–298.

Cline, M. G. 1944. Principles of soil sampling. *Soil Science* 58:275–288.

Collins, H. P., P. E. Rasmussen, and C. L. Douglas Jr. 1992. Crop rotation and residue management effects on soil carbon and microbial dynamics. *Soil Science Society of America Journal* 56:783–788.

Crépin, J., and R. L. Johnson. 1993. Soil sampling for environmental assessment. Pages 5–24 *in* M. R. Carter, editor, *Soil Sampling and Methods of Analysis.* Lewis Publishers, Boca Raton, Florida, USA.

Davidson, E. A., and I. L. Ackerman. 1993. Changes in soil carbon inventories following cultivation of previously tilled soils. *Biogeochemistry* 20:161–193.

DeLuca, T. H., and D. R. Keeney. 1994. Soluble carbon and nitrogen pools of prairie and cultivated soils: seasonal variation. *Soil Science Society of America Journal* 58:835–840.

Federal Geographic Data Committee. 1994. *Content Standards for Digital Geospatial Metadata* (June 8). Federal Geographic Data Committee, Washington, DC, USA.

Federer, C. A. 1982. Subjectivity in the separation of organic horizons of the forest floor. *Soil Science Society of America Journal* 46:1090–1093.

Friedland, A. J., B. W. Craig, E. K. Miller, G. T. Herrick, T. G. Siccama, and A. H. Johnson. 1992. Decreasing lead levels in the forest floor of the northeastern USA. *Ambio* 21:400–403.

Gilbert, R. O. 1987. *Statistical Methods for Environmental Pollution Monitoring.* Van Nostrand Reinhold, New York, New York, USA.

Gordon, A. M. 1988. Use of polyethylene bags and films in soil incubation studies. *Soil Science Society of America Journal* 52:1519–1520.

Grigal, D. F., R. E. McRoberts, and L. F. Ohmann. 1991. Spatial variability in chemical properties of forest floor and surface mineral soil in the north central United States. *Soil Science* 151:282–290.

Gupta, P. L., and I. H. Rorison. 1975. Seasonal differences in the availability of nutrients down a podzolic profile. *Journal of Ecology* 63:521–534.

Hesse, P. R. 1971. *A Textbook of Soil Chemical Analysis.* Chemical Publishing, New York, New York, USA.

Homann, P. S., P. Sollins, H. N. Chappell, and A. G. Stangenberger. 1995. Soil organic carbon in a mountainous, forested region: relation to site characteristics. *Soil Science Society of America Journal* 59:1468–1475.

Ihnat, M. 1993. Reference materials for data quality. Pages 247–262 *in* M. R. Carter, editor, *Soil Sampling and Methods of Analysis.* Lewis Publishers, Boca Raton, Florida, USA.

Jackson, M. L. 1958. *Soil Chemical Analysis.* Prentice-Hall, Englewood Cliffs, New Jersey, USA.

Kalra, Y. P., and D. G. Maynard. 1991. *Methods Manual for Forest Soil and Plant Analysis.* Information Report NOR-X-319. Forestry Canada, Northwest Region, Northern Forestry Centre, Edmonton, Alberta, Canada.

Klesta, E. J., and J. K. Bartz. 1996. Quality assurance and quality control. Pages 19–48 *in* D. L. Sparks, editor, *Methods of Soil Analysis. Part 3, Chemical Methods.* American Society of Agronomy–Soil Science Society of America, Madison, Wisconsin, USA.

Mader, D. L. 1963. Soil variability: a serious problem in soil-site studies in the Northeast. *Soil Science Society of America Proceedings* 27:707–709.

Maxwell, R. A., and D. C. Coleman. 1995. Seasonal dynamics of nematode and microbial biomass in soils of riparian-zone forests of the southern Appalachians. *Soil Biology and Biochemistry* 27:79–84.

Michener, W. K., J. W. Brunt, J. J. Helly, T. B. Kirchner, and S. G. Stafford. 1997. Nongeospatial metadata for the ecological sciences. *Ecological Applications* 7:330–342.

Miller, R. G. 1986. *Beyond ANOVA: Basics of Applied Statistics.* Wiley, New York, New York, USA.

Mroz, G. D., and D. D. Reed. 1991. Forest soil sampling efficiency: matching laboratory analyses and field sampling procedures. *Soil Science Society of America Journal* 55:1413–1416.

National Institute of Standards and Technology. 1992. *Spatial Data Transfer Standard (Federal Information Processing Standard 173).* National Institute of Standards and Technology, Gaithersburg, Maryland, USA.

National Institute of Standards and Technology (NIST). 1995. *Standard Reference Materials Catalog 1995–96.* NIST Special Publication 260. U.S. Government Printing Office, Washington, DC, USA.

Natural Resource Conservation Service (NRCS). 1996. *Soil Survey Laboratory Methods Manual.* Soil Survey Investigations Report No. 42, Version 3.0. National Soil Survey Center, USDA, Washington, DC, USA.

Nelson, D. W., and J. M. Bremner. 1972. Preservation of soil samples for inorganic nitrogen analysis. *Agronomy Journal* 64:196–199.

Parkin, T. B., and J. A. Robinson. 1992. Analysis of lognormal data. *Advances in Soil Science* 20:193–235.

Parkin, T. B., and J. A. Robinson. 1994. Statistical treatment of microbial data. Pages 15–39 *in* R. W. Weaver, J. S. Angle, and P. S. Bottomley, editors, *Methods of Soil Analysis. Part 2, Microbiological and Biochemical Properties*. American Society of Agronomy–Soil Science Society of America, Madison, Wisconsin, USA.

Petersen, R. G., and L. D. Calvin. 1986. Sampling. Pages 33–51 *in* A. Klute, editor, *Methods of Soil Analysis. Part 1, Physical and Mineralogical Methods. 2d edition*. American Society of Agronomy–Soil Science Society of America, Madison, Wisconsin, USA.

Riha, S. J., G. Senesac, and E. Pallant. 1986. Effects of forest vegetation on spatial variability of surface mineral soil pH, soluble aluminum and carbon. *Water, Air and Soil Pollution* 31:929–940.

Robertson, G. P., and K. L. Gross. 1994. Assessing the heterogeneity of below-ground resources: quantifying pattern and scale. Pages 237–253 *in* M. M. Caldwell and R. W. Pearcy, editors, *Plant Exploitation of Environmental Heterogeneity*. Academic Press, New York, New York, USA.

Robertson, G. P., M. A. Huston, F. C. Evans, and J. M. Tiedje. 1988. Spatial variability in a successional plant community: patterns of nitrogen availability. *Ecology* 69:1517–1524.

Robertson, G. P., K. M. Klingensmith, M. J. Klug, E. A. Paul, J. C. Crum, and B. G. Ellis. 1997. Soil resources, microbial activity, and primary production across an agricultural ecosystem. *Ecological Applications* 7:158–170.

Schalscha, E. B., C. Gonzalez, I. Vergara, G. Galindo, and A. Schatz. 1965. Effect of drying on volcanic ash soils in Chile. *Soil Science Society of America Proceedings* 29:481–482.

Shapiro, S. S., and M. B. Wilk. 1965. An analysis of variance test for normality (complete samples). *Biometrika* 52:591–611.

Siccama, T. G., W. H. Smith, and D. L. Mader. 1980. Changes in lead, zinc, copper, dry weight, and organic matter content of the forest floor of white pine stands in central Massachusetts over 16 years. *Environmental Science and Technology* 14:54–56.

Soil Survey Division Staff. 1993. *Soil Survey Manual*. USDA Handbook No. 18. U.S. Government Printing Office, Washington, DC, USA.

Sollins, P. 1998. Factors influencing species composition in tropical lowland rain forests: does soil matter? *Ecology* 79:23–30.

Sparks, D. L., editor. 1996. *Methods of Soil Analysis. Part 3, Chemical Methods*. American Society of Agronomy, Madison, Wisconsin, USA.

Spycher, P., P. Sollins, and S. Rose. 1983. Carbon and nitrogen in the light fraction of a forest soil: vertical distribution and seasonal patterns. *Soil Science* 135:79–87.

Stotzky, G., R. D. Goos, and M. I. Timonin. 1962. Microbial changes occurring in soil as a result of storage. *Plant and Soil* 16:1–18.

Strebel, D. E., B. W. Meeson, and A. K. Nelson. 1994. Scientific information systems: a conceptual framework. Pages 59–85 *in* W. K. Michener, J. W. Brunt, and S. G. Stafford, editors, *Environmental Information Management and Analysis: Ecosystem to Global Scales*. Taylor and Francis, London, UK.

Tan, K. H. 1996. *Soil Sampling, Preparation, and Analysis*. Marcel Dekker, New York, New York, USA.

Taylor, J. K. 1987. *Quality Assurance of Chemical Measurements*. Lewis Publishers, Chelsea, Michigan, USA.

Troedsson, T., and C. O. Tamm. 1969. Small-scale spatial variation in forest soil properties and its implications for sampling procedures. *Studia Forestalia Suecica* 74:4–30.

Trumbore, S. E. 1993. Comparison of carbon dynamics in tropical and temperate soils using radiocarbon measurements. *Global Biogeochemical Cycles* 7:275–290.

Warrick, A. W., D. E. Myers, and D. E. Nielsen. 1986. Geostatistical methods applied to soil science. Pages 53–82 *in* A. Klute, editor, *Methods of Soil Analysis. Part 1, Physical and Mineralogical Methods.* American Society of Agronomy–Soil Science Society of America, Madison, Wisconsin, USA.

Webster, F. 1991. *Solving the Global Change Puzzle: A U.S. Strategy for Managing Data and Information. Report by the Committee on Geophysical Data Commission on Geosciences, Environment and Resources, National Research Council.* National Academy Press, Washington, DC, USA.

Zinke, P. J., and R. L. Crocker. 1962. The influence of giant Sequoia on soil properties. *Forest Science* 8:2–11.

2

Site and Landscape Characterization for Ecological Studies

David F. Grigal
James C. Bell
Robert J. Ahrens
Richard D. Boone
Eugene F. Kelly
H. Curtis Monger
Phillip Sollins

There is growing awareness of the importance of adequately defining the environment when conducting ecological studies. Our objective in this chapter is to present the rationale and procedures for describing soils and landscapes (sites) for such studies. We use the term *site characterization* to refer to the entire suite of soil biogeophysical descriptors that places a site into an environmental context. A primary goal of any environmental study is to understand a phenomenon or set of linked phenomena, whether attributes or processes. Macro- and microclimate, soil, and landscape properties influence both ecosystem attributes and processes. To understand our own data or those of others, we must understand the environment in which they were collected. In addition, we are increasingly called upon to apply our research results to land management issues. Extrapolation from study sites can be either statistically rigorous or qualitative. Both understanding and extrapolation demand good site characterization.

The importance of soil and landform for affecting and defining the environment has been directly and indirectly assessed, and an understanding of the landscape can clarify ecological relationships. The strong influence of slope and aspect on the productivity of sites for forest growth has been universally recognized (Carmean 1975), and the greater lushness and productivity of cove forests compared with ridgetops (Whittaker 1956) is an ecological axiom. Standing stocks of nitrogen, phosphorus, and carbon vary dramatically but not monotonically along a topographic sequence of sites in arctic Alaska (Giblin et al. 1991). McAuliffe (1994) clearly demonstrated that previous assumptions of a simple gradient model of vegetation across *bajadas* of Arizona was incorrect. A mosaic of distinct landscape patches related to landform age and erosional history affect vegetation patterns and ecosystem processes. Under the gradient assumption, this mosaic is considered to be statistical noise. In fact, knowledge of the causes and consequences of the mosaic leads to insights about the system. Many environmental measurements and soil and landscape features are related and therefore predictable (Hall and Olson 1991), providing strong justification for their characterization.

Site descriptions must consider the spatial and temporal dimensions of ecological studies. Spatial descriptions should define the extent, scale, and specific location of the study. *Spatial extent* refers to the size of the study area, *scale* refers to the minimum size of a spatial feature that is considered or can be represented, and *location* refers to the geographic coordinates of the study area or of specific sampling points. The hierarchical levels of organization for the components of the soil-landscape system can be described by distinct scales (Fig. 2.1). Each scale requires a different set of descriptors and defines a different set of processes. The temporal dimension refers to the frequency of observations that are made at a specific site. For example, some sites may be visited only once to collect data, whereas other sites may be visited repeatedly for intensive study. If a site is visited only once, then the required site description is likely to be less detailed than that for a site undergoing intensive study.

Because the level of detail required either for an individual study or for comparison across sites varies with the property being measured and its spatial and temporal scales, we recognize here three levels of intensity of site description and soil sampling (see also Chapter 1, this volume). At the first level, the primary interest is in properties or processes of the surface soil, and description and sampling may be carried out by personnel who are relatively inexperienced in soil science. At the second level, more detailed information is collected following standard procedures (Soil Survey Division Staff 1993); work is usually carried out by advanced graduate students or experienced field scientists. At the third level, detailed descriptions are made and samples collected by scientists trained in soil pedogenesis.

When is a site adequately characterized? Is there a raison d'être for site characterization, or is data collection simply a rote procedure to ensure that all possible questions are addressed? In our view, the rationale for site characterization is to describe the biophysical soil environment of the site as the basis for a wide range of ecological investigations, but especially with respect to its suitability for organisms. The four operationally defined environmental factors of light, nutrients, heat, and water provide a focus for our discussion.

Site characterization helps to describe the environment for light and for the entire spectrum of solar radiation. Location affects photoperiod and is quantified by latitude and longitude or by other georeferencing systems. Topographic setting may also affect the environment for light through the differential shading of solar radiation. Characterization of the environment for nutrients is complex, and many facets will be discussed elsewhere in this volume, but routine soil characterization includes measurement of a suite of nutrient-related properties. Heat, or the instantaneous measure of temperature, is clearly affected by location. Slope inclination and aspect as they affect incoming solar radiation, simple elevation above a datum, and the possibility of air drainage in response to temperature gradients all affect the thermal regime of a site.

Specifying the operational environment with respect to water provides a further focus for site characterization. Location and elevation define the macroclimate of a

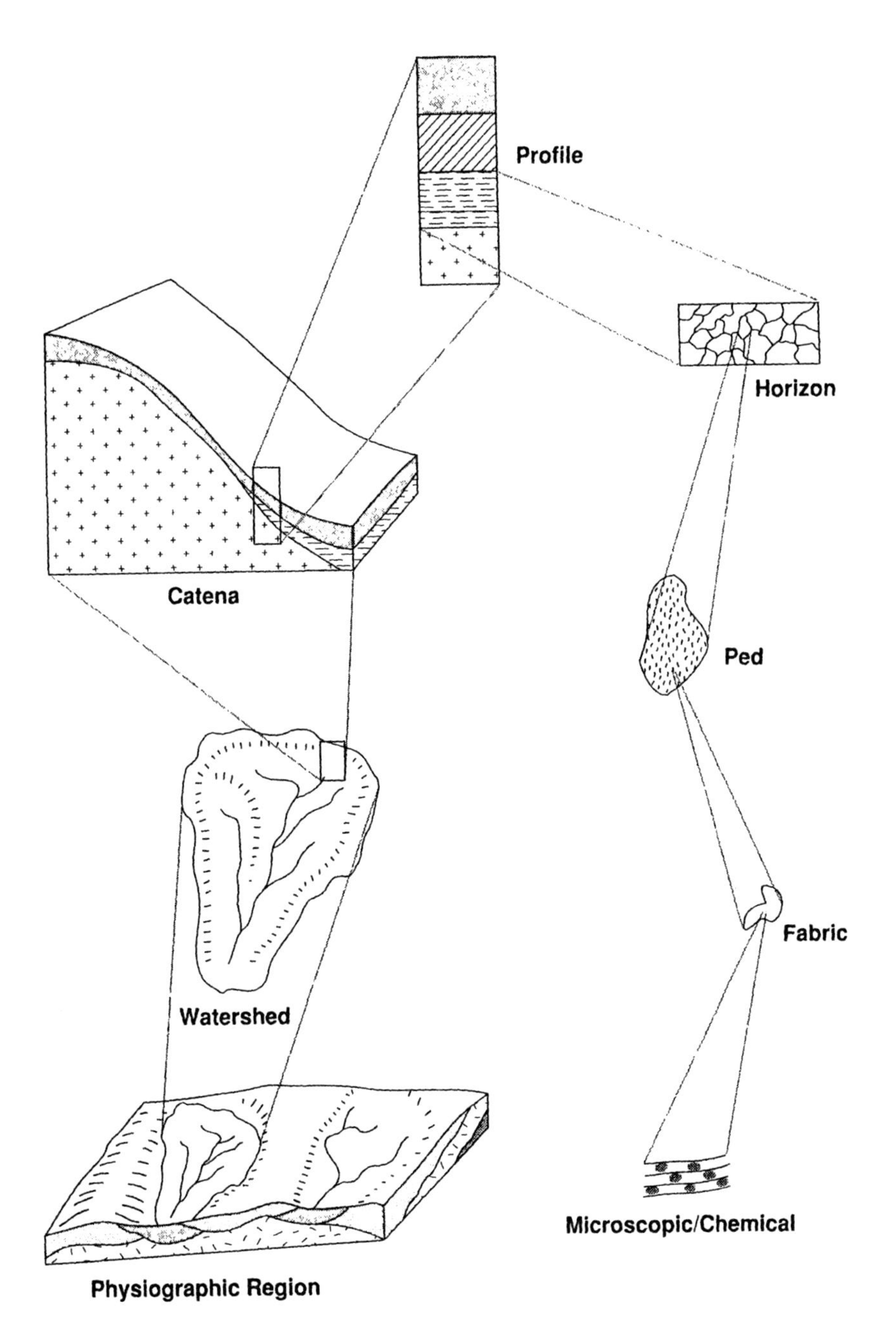

Figure 2.1. Hierarchical levels of organization of the soil-landscape system.

site and profoundly affect its water regime. Within similar macroclimatic zones, this regime is also affected by the surface and subsurface flow paths of water through the system. Is the site steeply sloping or nearly horizontal? Is it underlain by an impermeable soil or rock stratum? Is it seasonally impermeable? Is there an indication of redoximorphic features (mottles) in the soil? At what depth? Is there a local, seasonally defined water table? At what depth is the regional water table? Is evapotranspiration from the site affected by slope and aspect? What is the texture and the water-holding capacity of the soil—both by layer or horizon and summed over all layers? In summary, a good site characterization should provide at least a qualitative description of the environment with respect to light, nutrients, heat, and water, providing a unifying basis and rationale for the characterization.

Structural Organization of Soil Landscapes

Soil-landscape systems are inherently spatial phenomena, and any description of their characteristics is made within the context of spatial scale. A complex mosaic of processes occurs at several spatial scales within the soil-landscape system. Jenny (1941) defined the factors that influence the development of soils in landscapes as

$$s = f(cl, o, r, p, t, \ldots)$$

where

s = a soil property
cl = climate
o = organisms representing the combined effect of both fauna and flora
r = topographic relief
p = the soil parent material or geologic substrate
t = time or relative soil age
. . . = other site-specific factors such as human disturbance

From a simplistic view, these factors define an equilibrium state toward which soil characteristics adjust through time, the underlying assumption being that a unique soil will develop for that equilibrium state. In reality, these factors continually change, and the soil never reaches an equilibrium. Nonetheless, this framework is a useful conceptual model for understanding the environmental factors that influence both past and present ecosystem processes.

Recognizing that the factors influencing soil development operate at a range of spatial scales, we can define a hierarchy that describes the organization of the soil-landscape system (Fig. 2.1). Site characterization requires a description of the soil-landscape system at these multiple scales, and we will use this multiscale model as a framework for describing site characterization. The interaction of biophysical processes occurring from regional to microscopic scales determines the environment of the entire terrestrial ecosystem.

The physiographic region and landform (Fig. 2.1) can be described in broad terms from existing information concerning soils, landscapes, climate, and vegetation. Watersheds provide a means for further subdividing physiographic regions into spatial units linked by hydrologic processes. Repeating soil patterns within watersheds can often be described by catenas. Considerable information at the level of the soil catena complex (Fig. 2.1) can be determined from knowledge of general soil stratigraphy and the configuration of the topographic surface. The spatial organization of soil components at this level is closely related to the age of the surface and to water and particle movement through the soil, with the accumulation of particles in the upper 1–2 meters of the profile. The profile, horizon, and ped levels of organization (Fig. 2.1) are described at specific points in the landscape by standard methods.

While information concerning the fabric and microscopic/chemical levels of organization (Fig. 2.1) can be gleaned from field observations, laboratory analyses are usually required for comprehensive physical, chemical, and mineralogical analysis. We will divide our discussion of this multiscale soil-landscape system into three general categories: (1) we will focus on the physical geography of study sites at the physiographic region, landform, and watershed/catena levels of organization; (2) we will use soil morphology at specific points in the landscape to describe the profile, horizon, and ped levels of organization; and (3) we will refer to soil laboratory analyses to characterize the structural organization of soils at the fabric and microscopic/chemical levels.

Physical Geography

Physiographic Region/Landform

At a broad level of generalization, land masses having similar physical structure have been classified into *physiographic regions* that divide the earth into unique sets of landforms and geologic substrate(s) that have been and are being influenced by similar geomorphic and/or geologic processes, climate, and vegetation. Physiographic regions usually cover large areas, such as the Ridge and Valley Physiographic Province of the eastern United States, extending from Pennsylvania to northern Georgia. While the composition of soil parent material may vary within physiographic regions, the geologic processes responsible for its occurrence at a site are usually similar. Hence, physiographic regions are often delineated by commonality in bedrock lithology, regional landforms, or the depositional environment for transported soil parent materials (glacial tills, outwash, loess). Landforms further divide physiographic regions into more homogeneous subunits. In the Ridge and Valley Province, for example, two obvious landforms are ridges and valleys, but those landforms can be further subdivided.

There are numerous classification systems that consider physiographic regions, each with a different emphasis depending on the purpose of the classification. Such systems include ecoregions (Bailey and Cushwa 1981), major land resource areas (Austin 1972), and physiographic provinces (Fenneman 1928). Physiographic regions are typically delineated on maps at scales of 1:100,000 to 1:1,000,000. Major land resource areas, with national coverage at map scale of 1:10,000,000, provide the basis for defining working regions for the National Cooperative Soil Survey.

Although the emphasis in characterizing a site is on properties at the surface, lithology of the underlying bedrock may also be important. Although in some cases the bedrock is well below the weathered soil horizons, knowledge of bedrock lithology may help one to understand the chemistry and the mineralogy of the surface materials and even the evolution of the landform that is being studied. The history of land use and vegetation change is also important information. The presence of a plow layer—a clearly delineated, constant-depth surface zone of organic-rich mineral soil—indicates former cultivation, as does the presence of stone fences or stone piles. Old fields in the local area also indicate the potential for past agricultural activities at the study site. The successional status of the vegetation may also provide clues to disturbances that have affected the site and the soil-related processes therein such as nitrogen mineralization. For example, processes in a primary successional sequence following flooding, glacial recession, or volcanic activity may be associated with much different soil characteristics than those associated with secondary succession, such as following logging, intense grazing, agriculture, or fire.

The characteristics of a site can also be more fully understood by historical information on a much longer time scale. Paleosols, pollen diagrams, isotope ratios, and other indicators of past environments help place the current site and environment in a temporal context. Landform evolution, soil development, and vegetation composition are linked to past climatic history.

A complete description of the physiographic region and landform of a site should include:

- physiographic classifications of interest (ecoregion, physiographic province, major land resource area);
- major landform (mountains, till plain, basin and range, etc.);
- predominant soil parent material type (residuum, alluvium, glacial drift, loess, colluvium, lacustrine deposit, etc.);
- lithology of the predominant soil parent material (sandstone, shale, limestone, granite, gneiss, etc.);
- approximate age of the geomorphic surfaces;
- geomorphic history;
- predominant vegetation communities (tallgrass prairie, boreal forest, oak savanna, etc.);
- 30-year average annual and monthly climate data, including precipitation and maximum and minimum temperatures;
- freezing depth;
- evapotranspiration potential;
- other physiographic features;
- land-use history; and
- location.

Most of this information can be obtained from existing broad-scale inventories. County-level (2nd order; Soil Survey Division Staff 1993) soil surveys also contain information on the range of soil parent materials, the geomorphic history of an area, and summaries of local climatic data.

Watershed/Soil Catenas

Watersheds are logical subdivisions of physiographic regions that define portions of the landscape linked by hydrologic processes. The hierarchical arrangement of watersheds with respect to stream order provides a method for defining soil and landscape variability at spatial scales intermediate to physiographic regions and hillslope catenas. Where hydrologic processes are a major factor influencing soil formation and subsequent variability within physiographic regions, watersheds are appropriate spatial units to further partition variability. In many cases, however, topographically based watershed boundaries are superimposed over a complex mosaic of soil variability attributable to differences in other soil-forming factors such as parent material and microclimate. As such, watersheds may not always be ideal spatial units to subdivide soil and landscape variability because not all pedogenic, geomorphic, and/or geologic processes occur within the confines of watershed delineations. Eolian (wind) erosion and transport, for example, often transcends watershed boundaries.

Soil variability within watersheds may be further subdivided according to repeating patterns of soil variability occurring along hillslopes. These repeating patterns are described as soil catenas and are often linked to topographic variability. Milne (1935) first described a *catena* as a "chain" of soils hanging between two summits. A catena can be viewed as a hydrologically-linked segment of the landscape. Soil variability along hillslopes is usually related to two factors: (1) changes in soil parent material resulting from geologic or geomorphic processes, and (2) alterations of the soil by hillslope-scale hydrologic, biotic, and biogeochemical processes. Geologic materials often differ in resistance to erosion. As a result, the higher elevations on erosional landscapes are usually composed of resistant materials, and lower elevations are composed of more-erodible materials that may be covered by a mantle of depositional sediment (colluvium or alluvium). Hillslope relationships can be very complex in landscapes where multiple geomorphic and pedogenic cycles have occurred. Soil parent materials on hillslopes may range from homogeneous to very heterogeneous depending on the history of the site.

Superimposed onto this variability in parent material are alterations of the soil mantle due to past and contemporary hydrologic and geomorphic processes. Soil water on hillslopes may follow one or several pathways:

- surface runoff;
- lateral flow in shallow soil horizons above relatively impermeable subsoils;
- vertical flow through the soil profile;
- some combination of the above two flow paths; and
- return to the atmosphere via evapotranspiration.

The precise pathway of water flow depends on the shape of the land surface, surface stratigraphy, the quantities, timing, and frequency of precipitation, and the potential for evapotranspiration. In humid climates the magnitude, direction, frequency, and timing of water movement are primary factors affecting differential soil development on hillslopes and the associated presence of different biological communities. These pathways of water movement in the upper few meters of soil are responsible for migration of soil particles and solutes from uplands to lower hillslope positions. This inextricable linkage of soils and landscapes by hydrologic and geomorphic processes is often ignored, leading to such environmental problems as accelerated soil erosion, constriction or elimination of wetlands, and soil salinization. The imprint of these hydrologic processes can often be found in the morphologic characteristics of soils, and the pathways of movement can often be inferred by an analysis of terrain attributes.

While the catena concept describes soil variability along hillslopes, differences among soils within a landscape can also be related to specific landforms. In many cases these landforms are the result of specific geomorphic processes that redistribute soils and soil parent material by transport and deposition via wind or water. These processes are often episodic, and the current arrangement of soil materials may have been manifested over very long periods by processes that may or may not be active today. Changes in radial patterns of soil texture along the surface of alluvial fans are examples of deposition. Soil variability in sequences of stream terraces formed in response to variations in sea level is an example of erosional processes.

Information on soil catenas and other soil patterns can be found in soil survey reports. The composition of soil map units is defined by soil taxonomic units (usually soil series). Three kinds of soil map units are used: (1) soil associations define areas where two or more soil types follow discernible patterns across the landscape, (2) soil complexes define areas where the pattern of two or more soil types is random or not distinguishable at the mapping scale, and (3) soil consociations are map units that primarily contain a single soil type. Modern reports usually have a series of block diagrams that depict the topographic relationships of the predominant soils in the major landforms of the survey area (e.g., Calus 1996). If a soil survey is not available for the immediate area of interest, surveys of adjacent areas may have similar landform and soil relationships. While soil surveys can provide valuable background information, they are *not* intended for site-specific application. On-site investigations are absolutely necessary to adequately document soil and site conditions for ecological investigations. This requires both a description of the topographic surface and of soil profiles located along a hillslope catena or in other soil patterns.

Terrain Attributes

Attributes of the topographic surface can be either measured or estimated from visual inspection. Local terrain attributes describe the shape and orientation of a small segment or patch of the landscape, and regional attributes consider the patch relative to the overall hillslope and/or catchment. A fundamental set of terrain attributes (slope gradient, aspect direction, plan and profile curvature, and catchment area) define the shape and orientation of the topographic surface in geometric space and the convergence or divergence of water flow and other pedogenic influences.

Local Terrain Attributes

Local terrain attributes refer to the shape and orientation of the land surface within a patch, often called a "window." The window may be on the order of 100–1000 m^2 in highly variable landscapes and 1000–10,000+ m^2 in less variable landscapes. Slope gradient, slope curvature, and aspect direction are local terrain attributes that would be similar within an appropriately sized window. Similarly, a window would occur exclusively on a single hillslope position.

Mathematically, slope gradient is the first derivative of elevation measured in a specific direction. Although the direction is usually that of maximum descent, other directions such as perpendicular to maximum descent may be useful in complex terrain. Slope gradient plays a major role in determining the potential rate of water movement over and within the soil. Soil materials and water are likely to be transported downslope from steeper to lower slope gradients. Slope gradient can be measured in the field using simple hand-held instruments such as clinometers or Abney levels. More precise measurements can be made with stationary surveying devices, but this level of precision is seldom justified in ecological studies. When measuring slope, care should be taken to avoid including a break or change in slope within the distance necessary for the measuring instrument.

The distance over which the slope gradient and other local terrain attributes are measured should always be recorded, and is a function of both topographic variability and expected use of the data. The frequency of variation in the topographic surface varies considerably among landscapes. For example, consider slope gradients in two distinctly different terrains, one a highly undulating glacial moraine landscape and the other a major mountain range. Neglecting microrelief (differences of <1 meter), measured slope gradients from the top of a hillslope to distances of 10, 100, and 1000 m in the glacial landscape would probably all be different. However, measured gradients at equivalent distances from the top of a major ridge in the Appalachian Mountains would probably all be similar. The scale of variation within the local landscape must therefore be considered.

Slope curvature is the rate of change in slope gradient, and mathematically is the second derivative of elevation. It is quantitatively expressed in units of distance per distance squared (m/m^2). Slope curvature is also a directional attribute, usually measured both along the axis of maximum descent (profile curvature) and perpendicular to that direction (plan curvature). In the field, slope curvature is usually described qualitatively by the change in slope gradient from the local window into adjacent windows:

- *concave,* in which the slope gradient decreases in direction of measurement;
- *straight,* in which the slope gradient remains constant in direction of measurement; and
- *convex,* in which the slope gradient increases in direction of measurement.

Based on these three curvature classes, nine possible combinations of plan and profile curvature can be defined. Slope curvature affects the relative dispersion or accumulation of water in the landscape. As a broad generality, convex slopes are water spreading (divergent flow), straight slopes are water transporting (parallel flow), and concave slopes are water accumulating (convergent flow). Water and sediment tend to accumulate in zones of convergent flow and are removed from zones of divergent flow, affecting soil characteristics. Soils in concave areas (convergent flow) often have accumulations of sediment, organic matter, and redoximorphic features indicating prolonged periods of saturation and biochemical reduction. Soils on convex slopes (divergent) often are eroded, with consequently lower levels of organic matter and plant nutrients. Although field descriptions of slope curvature are usually qualitative, quantitative measurements can be calculated from digital elevation models using techniques of digital terrain analysis (Moore et al. 1993).

Aspect refers to the compass direction of the vector describing the direction of maximum slope gradient. In the field, aspect is determined as a compass bearing in the direction that the hillslope is facing (the compass is pointed away from the hillslope). If magnetic declination is appreciable, correction for the deviation between true and magnetic north must be made. For long and/or steep slopes, aspect determines the incident solar radiation and can dramatically influence soil water and the type of vegetation. Simulation models can be used in conjunction with digital elevation models to derive and map estimates of incident solar radiation across the landscape (Moore et al. 1991). Because incident radiation is also affected by other factors, including latitude, shading by adjacent hillslopes, and cloud cover, these models require considerable parameterization to obtain actual as opposed to relative estimates. Shading by adjacent landforms is especially important at higher latitudes.

Regional Terrain Attributes

While hydrologic and geomorphic processes are influenced by the shape and orientation of the immediate landscape, they are also influenced by the location of the window in the larger landscape. The use of the term *regional* in this context refers to a larger region surrounding the local terrain window, usually the entire hillslope or basin. The primary attribute affecting hydrologic and geomorphic processes at this scale is the potential rate and quantity of water that can enter the window from upslope, either over the soil surface as run-on or through the soil as lateral interflow. To assess the potential influence of these processes, information on both soil stratigraphy and landform configuration is needed. In general, highly stratified or low-permeability soils will be more strongly affected by these processes than unstratified, highly permeable soils. Two general approaches can be used to assess the influence of the regional terrain—landscape position descriptions and catchment area measurements.

The regional setting of a local landscape window can be qualitatively described by the hillslope position; these positions have been defined in the profile direction (Fig. 2.2). Cross-sectional hillslope positions can be more completely described by combinations of the slope gradient, profile curvature, and catchment area (Tab. 2.1). These positions are relatively easy to recognize on simple idealized hillslopes. Landscapes are often complex, however, with variation in the topographic surface occurring at several different spatial scales so that local hillslopes may be nested within a larger hillslope complex. In these situations the hillslope position of a spe-

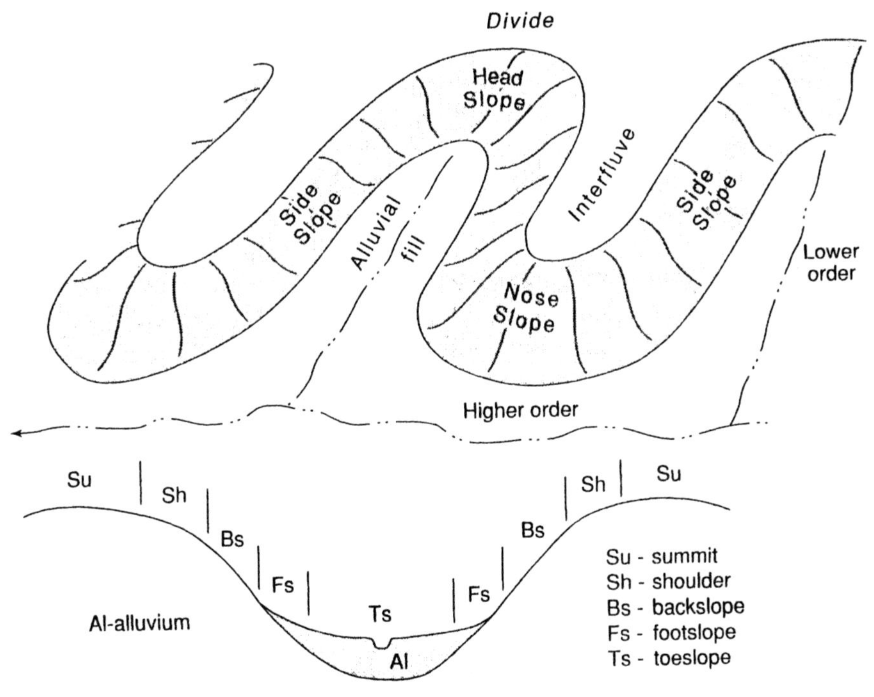

Figure 2.2. The components of an incised valley (above) and of a hillslope profile (below). Modified from Ruhe and Walker (1968).

cific point on the landscape can be described in terms of both the major and minor hillslope components. For example, a site may be located both on the sideslope of a major hillslope component and on the toeslope of a minor hillslope component. The cross-sectional hillslope position describes the convergence or divergence of hillslope processes along only one axis (downslope profile direction). To accurately portray hillslope morphology, a qualitative description of hillslope characteristics in the plan direction (perpendicular to profile) is also necessary. This requires description of slope positions along the contour of the slope; divergent slope positions are interfluves and noseslopes, convergent positions are headslopes and drainageways, and parallel slopes are backslopes (Fig. 2.2).

Table 2.1. General Description of Cross-Sectional Hillslope Positions in Terms of Slope Gradient, Profile Curvature, and Catchment Area

Hillslope Position	Slope Gradiant	Profile Curvature	Catchment Area
Summit	Low	Straight	Lowest
Shoulder	Moderate	Convex	Low
Sideslope	Steep	Straight	Moderate
Footslope	Moderate	Concave	Moderate
Toeslope	Low	Concave	High
Depression/Drainageway	Low	Concave	Highest

The regional terrain can be alternatively and more quantitatively described by catchment area measurements. These measurements determine the upslope watershed area that contributes flow to a particular point in the landscape. Drainage divides have no catchment area, and drainageways or depressions have relatively large catchment areas. Direct measurement of catchment area is difficult in the field but can be estimated from topographic maps. Calculations of catchment area for an entire landscape can be carried out by applying flow-tracing algorithms to a digital terrain model (digitized contours, triangular irregular networks, or digital elevation models; Moore 1992). If a digital terrain model is available, most flow-tracing algorithms are relatively easy to apply, are able to compensate for spurious "pits" in the data, and are applicable for landscapes with both open and deranged surface drainage. Estimates of catchment area are especially useful for defining spatial patterns of soils in landscapes with appreciable lateral flow (Moore et al. 1993; Bell et al. 1994). The vertical proximity of a point on the landscape to the nearest expression of the local free water surface (lake, stream, wetland, etc.) appears to be a useful terrain attribute to explain patterns of soil in landscapes without appreciable lateral flow. Vertical proximity can be estimated in the field, from topographic maps, or calculated from digital terrain models using simple algorithms (Bell et al. 1992).

Soil Morphology

The characteristics of the soil at specific points in the landscape are the focus of the next level of soil-landscape organization. The extrapolation of soil characteristics from specific points to the three-dimensional landscape requires an understanding of the spatial relationships between soil and landscape characteristics. The techniques used by professional soil scientists to create soil maps by spatial extrapolation (Holmgren 1988) combine art and science, and their discussion is beyond the scope of this chapter. Our objective here is to discuss techniques for describing soil characteristics at specific points in the landscape.

The art of describing a vertical exposure of soil, the soil profile, has been standardized over time. The color, texture, structure, and other soil attributes that are included in descriptions can all be interpreted to provide information far beyond their simple tabulation. Even those who are not experts in such interpretation can produce meaningful data much as a patient describes his or her symptoms in lay terms and a skilled physician arrives at a diagnosis. Another point, and one that often confuses the neophyte, is that horizon designators as identified by letters and numbers (e.g., A, Bk2) are interpretations from horizon descriptors. These designators are not absolutely necessary if the description itself is accurate. In fact, an inappropriate designator may more seriously impair communication than would a modest error in a description. The bottom line, therefore, is that although horizon designations are very helpful, their absence does not detract from a good description. A third point is that a description of a soil profile is simply an attempt to separate the vertical section into layers so that there is greater homogeneity within than between layers. Most soils are less isotropic vertically than horizontally, so that a vertical separation reduces variance in measured attributes. The essence of the recognition of a soil layer (= horizon) is therefore a change in one or more properties. Each vertical change has the potential to define a new layer. As in many other activities, those who describe soil profiles can be roughly divided into splitters and lumpers. Some are not satisfied unless they have detected and described every nuance of change, while others require a much higher threshold of change in order to consider it sufficiently different to describe.

Some soil properties that are included in a profile description have not been traditionally emphasized in soil science because of the field's agronomic tradition and bias. For example, the organic surface of many uncultivated soils (e.g., the forest floor) is very different from surface residues from an agronomic crop. Similarly, most land used for agronomy has minimal stones—stony land is simply not suitable for most crops. For ecological studies, a good description of organic horizons and of stone volume and characteristics is very important.

Morphological Description

Standard methods used to describe soil morphology encompass the profile, horizon, and fabric levels of soil organization (Fig. 2.1). The soil profile can be exposed by hand probes or augers, mechanical probes or augers, or soil pits excavated by hand or backhoe. Hand augers or push probes extract disturbed or undisturbed cores of soil whose depth depends on both the length of the auger or probe and the perseverance of the operator. The variety of augers (such as open and closed bucket, Dutch auger, McCauley peat auger, etc.) is designed to accommodate a wide range of soil conditions. Hand methods are the only option in areas where access roads do not exist or where mechanical equipment cannot be used. Because soil cores can be extracted rather quickly, hand augers and probes permit the examination of soil morphological characteristics at many points in the landscape. The disadvantage of their use, however, is that the sample is usually relatively small (a few centimeters in diameter), restricting observation of soil horizonation and structure. Disturbance by the auger or probe may also obscure certain soil morphological characteristics. At sites accessible by vehicle, mechanical augers or probes mounted on trucks or trailers and powered by hydraulic drives allow more rapid coring to deeper soil depths. However, the disturbance of cores remains a problem. A soil profile is best exposed by digging a pit, usually to a depth of approximately 2 meters. Although pits via backhoe are ideal, accessibility and cost can be limiting. The depth to which soils are described depends on the objectives of the study, but a reasonable approach is to describe soils in detail to 100 cm by excavation of a pit, and then to use a bucket auger for the further description of materials to an additional 100 cm depth.

An important consideration when excavating a soil pit is the sun angle. The side of the pit to be described should face the sun to ensure optimal lighting for observation. After the pit is excavated, the pit face must be prepared for description by shaving off the outer few centimeters of soil with a tiling spade or mason's trowel, beginning at the top of the pit and working toward the bottom. This removes artifacts in the pit face created by smearing during the excavation, a common problem with backhoe buckets. After the pit face has been prepared, a portion (usually 0.5–1 m in width) should be "picked," not cut or scraped, using a hand tool such as a blunt knife or trowel. Begin at the top of the pit and work toward the bottom to keep the picked face clean. The goal of this exercise is to break the soil along natural planes of weakness to reveal soil structure. Comparisons can also be made of soil color differences on the ped surfaces (picked surface) and the ped interiors (shaved surface).

After the pit face has been prepared, the best technique is to stand a few meters back from the pit face and observe and delineate the major horizons. Closer examination of changes in soil characteristics can then be used to delineate additional horizons. The specific terminology and methodology for delineating horizons within a soil profile and for describing the morphological characteristics (depth ranges, color, structure, consistence, texture, reaction, and horizon boundary characteristics) are described in the *Soil Survey Manual* (Soil Survey Division Staff 1993) and will not be repeated here. Terminology used to designate master and subhorizons for description and diagnostic horizons for classification is also described in the *Soil Survey Manual* (Soil Survey Division Staff 1993) and in the *Keys to Soils Taxonomy* (Soil Survey Staff 1996). The latter is frequently updated to reflect changes in taxonomy used by the USDA–Natural Resources Conservation Service. Examples of profile descriptions are included in the Appendix of this chapter.

The forest floor in forests and organic litter layers in other ecosystems should be described if present. If appropriate, the type and proportion of rock fragments on the soil surface should also be described. Traditionally in soil science, the measurement point of reference for description was the top of the mineral soil, with organic horizons lying above that zero-point and mineral horizons below. One of the reasons for this standard was that the organic horizon may be ephemeral, either seasonally or with change in flora or fauna. In many cases, for example, by midsummer earthworms have totally consumed all litter material from the previous autumn. In other cases, changes in plant communities may markedly alter the thickness and/or composition of the forest floor. In the most recent recommendation for soil descriptions, however, the soil surface is considered the top of the part of the organic horizon that is at least slightly decomposed (Soil Survey Division Staff 1993). The forest floor is usually described by both color and stage of decomposition, the alteration from the original state of the organic material. The Oi, Oe, and Oa horizons (Soil Survey Division Staff 1993) are approximately equivalent to the L, F, and H layers, respectively (Pritchett and Fisher 1987), and differ in the degree of original plant fiber that they contain. In some cases, separation of the forest floor from the mineral soil material is difficult, with a diffuse gradation between the two. In other cases, changes are abrupt, and the mineral and organic layers can be easily separated. There is no hard rule for this separation. Another important characteristic of the soil surface that should be described in forests is coarse woody debris (see Chapter 11, this volume).

Many soil chemical (e.g., sorption and desorption of metals and organic compounds, exchange processes, weathering reactions, nutrient availability, buffering capacity) and physical (e.g., bulk density, shrink-swell properties, aeration, infiltration, and hydraulic conductivity) processes and properties depend on the nature and relative quantities of the mineral and organic components of the soil (Dixon and Weed 1989). Differences in mineralogy can affect many ecological processes (Sollins et al. 1988). For most medium- to fine-textured soils, the majority of soil chemical properties are determined by mineralogy of clays because of their higher specific surface and charge compared to silts and sands. Relative quantities of carbonates, gypsum, and other more soluble salts are also important because they are much more reactive than the common silicate minerals. Quantitative assessment of the mineralogical components of soils, particularly the clay mineralogy, is a highly technical task, but qualitative to semiquantitative measurements are routinely made in many private and public laboratories, including those of university departments of soil science and geology.

Accurate and precise descriptions of soil profiles can be achieved only with both a basic understanding of pedological concepts and field experience with local soils and landscapes. Because there is an element of subjectivity in describing soil profiles, consistency is vital. Achieving this consistency requires experience, and novices are likely to encounter difficulties without some initial guidance from experienced soil scientists in the local area. If the study site has high value in terms of research, either because of the investment of significant resources or because of the long-term nature of the observations or monitoring, direct field assistance by an expert in soil science is essential. As explained, the separation of soil horizons is part description and part interpretation. Experience with the soils in an area is essential for meaningful interpretation. In addition, some soil properties that are determined in the field, such as texture and consistence, require training and practice with calibration to laboratory data to ensure reproducibility and accuracy. In fact, if the collected data are used to classify the soils, a perspective to be kept in mind is that "no classification is better than a bad classification."

Materials

Materials usually required for soil morphological descriptions include the following:

1. Munsell soil-color book
2. Blunt knife or other implement to pick the soil
3. Tiling spade and geologist hammer
4. Field pH test kit
5. 1 *N* HCl to test for presence of carbonates
6. α,α′-dipridyl to test for presence of ferrous Fe
7. Measuring tape
8. Nails to mark horizon boundaries
9. Field notebook with standardized profile description forms
10. Reference information on terminology for soil profile description
11. Spray-type water bottle
12. Hatchet or clippers for cutting vegetation and roots
13. A hand lens to examine soil fabric
14. Heavy plastic bags to return samples to the laboratory

Field Location

Precise location of field sites is important for site inventory, for repeated visits to the same site, or for spatial analysis using geographic information systems. Global Positioning Systems (GPS) provide a highly accurate and potentially cost-effective means of obtaining locations for a large number of field sites. Accurate positioning is now possible worldwide by nearly continuous coverage of GPS satellites. A variety of hand-held GPS receivers are available. GPS uses a constellation of 24 navigational satellites that have been placed in precise earth orbits by the U.S. military. Low-energy signals (pseudorandom code) are broadcast from synchronized atomic clocks in each satellite, and the distance to the satellites is calculated by comparing time differences between transmission and reception of the signals by the ground receiver, which also contains a synchronized clock. Precise positions are calculated by considering the distance to multiple satellites. Positional errors can be minimized by using two ground-based receivers, one at a known location (base station) and the second at the target. The adjustment of the target position by the error detected at the base station is known as *differential correction* or *differential GPS* (DGPS). Base stations and targets must usually be within 450 km of one another. Differential correction can be made in real time by receiving telemetry from base stations or after data collection using postprocessing techniques. Many hand-held units provide 1–5 m accuracy for a specified portion (usually 65% or 95%) of the observations. For submeter accuracy, DGPS receivers with special features are required. Use of inexpensive, hand-held receivers without differential correction can result in significant positional errors.

Laboratory Analyses

Although site characterization is a field exercise, independent of laboratory operations, some soil properties that are routinely described require laboratory measurement. These measurements are essential for characterizing intensive sites and usually are predicted or inferred for less-intensive sites. Routinely determined properties for soil characterization include particle-size distribution (texture), pH, organic C, exchangeable bases, cation exchange capacity, and bulk density. In many cases, water retention characteristics are also measured, but since they covary with organic C and particle size they can usually be predicted from those properties.

The objective in measuring a soil property is most often to precisely estimate its mean. Because most soil properties are not normally distributed but are more nearly lognormally distributed, this objective requires careful scrutiny. If the properties are lognormally distributed, many fewer samples are usually required to achieve similar precision of their estimated mean than if normality is assumed (Grigal et al. 1991). A second point is that as analytical and statistical procedures have become more refined, our ability to measure precisely in the laboratory and to differentiate statistically among similar observations or treatments has increased. As a result, the perception of meaningful variation and differences in ecosystems has become unrealistic. Because cost must be considered in any assessment of laboratory measurement, less precise but inexpensive laboratory procedures may result in greater overall precision than highly precise but expensive procedures. Greater relative costs of laboratory procedures, compared with field sampling, may lead to increasing the number of field composites and performing fewer laboratory analyses (Mroz and Reed 1991).

Both soil sampling and laboratory procedures are described in detail in other chapters in this volume and will not be further considered here.

Pedotransfer Functions

Extrapolation of ecological information to regional scales requires integration of large data sets collected by different investigators. To standardize data sets, it is useful to fill in data gaps where analyses were not conducted or sample collection was not possible. Although direct measurements are preferred, some studies that require large data sets and hence sample collection and analyses are not practical. Data collected by the National Cooperative Soil Survey and by individual investigators have led to the development of functional relationships among soil properties; these can provide some insight into the interrelationships among soil biological, physical, and mineralogical components. These functional relationships that relate different soil characteristics to one another have been termed pedotransfer functions (Bouma 1989).

Water Retention

A variety of pedotransfer functions have been developed that relate other soil properties to water retention (Rawls et al. 1991). Particle size distribution, organic matter, and bulk density are soil properties that commonly have been used to describe water retention (Rawls et al. 1991).

A synthesis of relationships for surface soils in the literature (Shaykewich and Zwarich 1968; Gupta and Larson 1979; Rawls et al. 1982; De Jong et al. 1983; Rawls et al. 1983) yields

$$P_{33}(\mathrm{cm^3/cm^3}) = 10^{-3} \times [4.12 \times \text{clay (\%)} + 22.09 \times \text{organic matter (\%)} - 1.22 \times \text{sand (\%)} + 174.8 \times \text{bulk density (g/cm}^3)],\ S_{y.x} = 0.004\ \mathrm{cm^3/cm^3},$$

and

$$P_{1.5}(\mathrm{cm^3/cm^3}) = 10^{-3} \times [4.06 \times \text{clay (\%)} + 10.37 \times \text{organic matter (\%)} - 0.33 \times \text{sand (\%)} + 41.3 \times \text{bulk density (g/cm}^3)],\ S_{y.x} = 0.002\ \mathrm{cm^3/cm^3},$$

where

P_{33} = water retention at 33 kPa (1/3 bar),
P_{15} = retention at 1.5 MPa (15 bar)

Available water is often defined as the difference between these values.

Bulk Density

Soil bulk density is an indirect measure of the relative volume of solids and voids in a soil; hence it provides an indication of the soil's ability to store and transport water and gases, as well as an estimate of soil strength. Bulk density (*BD*) is also critically important for determining the mass balance of elements and water within ecosystems. The bulk density of most surface soils is closely related to soil organic matter, and this relationship has been explored and verified many times by pedotransfer functions. A synthesis of relationships in the literature (Curtis and Post 1964; Jeffrey 1970; Adams 1973; Alexander 1980; Grigal et al. 1989) yields

$$BD\ (\mathrm{g/cm^3}) = \mathrm{EXP}[0.23 - 0.037 \times \text{organic matter (\%)}],\ S_{y.x} = 0.05\ \mathrm{g/cm^3}.$$

Other pedotransfer functions have been developed to estimate operationally defined clay content, where standard laboratory procedures developed by midlatitude soil scientists are deficient, and to estimate a suite of soil chemical properties such as cation exchange capacity, base saturation, and pH.

Extrapolation

Classic statistical procedures assume that variation in measured properties is randomly distributed among sample units. In contrast, variation in measured properties in a field setting often is related to the distance between sample locations. In this chapter we have emphasized the predictable variation in soil properties with differences in landform position, soil parent material, soil age, and many other factors. In addition to the expected or predictable variation in soil properties, another part of the variation cannot be attributed to known causes and is therefore termed *random* or *chance variation.* The essence of geostatistical methods is the exploration of the spatial component of variation, and its quantification and subsequent use in estimating properties at unsampled locations. Geostatistical methods were first developed by D. G. Krige for determining the spatial extent of mineral deposits, but since then the techniques have been applied in a wide range of field studies (Warrick et al. 1986; Ver Hoef and Cressie 1993; Robertson et al. 1997).

Geostatistical procedures basically quantify changes with distance in either correlation or covariation of measurements of the same property. In an ideal case, variation increases with distance from a small constant (the nugget) to an asymptotic maximum (the sill). In other words, the nugget is a random component of variation that is unrelated to distance, while the sill is the variation at a distance beyond which measurements are independent of one another. The results of the analysis are used to make unbiased optimal interpolated estimates of properties at unsampled locations (i.e., *kriging;* Trangmar et al. 1985). The results of such interpolations are often presented as maps of properties such as soil organic matter (e.g., Crawford and Hergert 1997). A two-stage analysis can also be conducted, where preliminary data are collected via transect or other scheme, and geostatistics are used to help optimize both intensity and location of a refined sampling scheme.

One of the key concepts in geostatistics is isotropic versus anisotropic variation. Isotropic variation occurs where properties vary in the same way in all directions, so that variation among samples is simply a function of distance. In contrast, variation in most soil properties also has a directional component (e.g., downslope). Where variation among samples has components of both direction and distance, it is termed *anisotropic*. Although analyses are somewhat more complicated in the latter case, recognition of anisotropy is important in using geostatistics in soil science (Crawford and Hergert 1997). As with any sophisticated statistical technique, a rich literature has developed regarding the uses of geostatistics in field studies (see Trangmar et al. 1985; Warrick et al. 1986; ver Hoef and Cressie 1993).

Level of Intensity

The level of detail required for a site characterization varies with the objectives of a specific study, and we therefore recognize three levels of intensity. These range from Level 1, with primary interest in the surface soil and description and sampling by personnel who are relatively inexperienced in soil science, to Level 3, with detailed descriptions by those trained in soil pedogenesis.

At Level 1 intensity, and at the spatial scale of the physiographic region and landform, a detailed site description would not be performed. A general description of a site will be sufficient, and if a soil survey of the area is available it should be used to identify the soil map unit. Each sampling spot should be characterized with respect to position on slope (summit to depression) and slope gradient, slope curvature, and aspect direction. Position on the microrelief should also be noted. If a GPS is used to determine sampling locations, differential correction is probably not necessary. Soil sampling will usually be by sample tube, auger, or tiling spade, but no pit would be dug. Soil morphology would be described only by noting the presence of sharp and obvious changes in soil characteristics within the sampling depth. Although soil mineralogy would not be determined in detail, the carbonate content, using the descriptions of effervescence from the *Soil Survey Manual* (Soil Survey Division Staff 1993) can be estimated.

Both the second and third levels of intensity require similar information that may differ only in the detail and the expertise with which it is collected. A complete description of the physiographic region and landform should be made. The location of each sampling point should be determined using a GPS with differential correction. A complete field description of the shape and orientation of the topographic surface can be achieved by descriptions of local terrain attributes within a local window, coupled with descriptions of hillslope position in both the cross-sectional (downslope profile) and contour (plan) directions. The only field equipment required is a compass, an instrument for measuring slope gradient (Abney level, clinometer, etc.), and a careful eye to discern landscape positions. Descriptions of hillslope position should involve walking across and viewing the hillslope from several different vantage points to minimize bias. If more quantitative information is needed and if a digital terrain model is available, a complete quantitative description of the topographic surface can be achieved by using digital terrain analysis to calculate the fundamental attributes of slope gradient, aspect direction, plan and profile curvature, and catchment area.

In both Levels 2 and 3, detailed descriptions of soil morphology should be made. Soil pits are usually excavated, and complete descriptions are carried out following standard procedures (Soil Survey Division Staff 1993). In the case of Level 2, the recommended description should extend to 100 cm; for Level 3 it should extend to 200 cm. In the case of Level 2, horizons may not be formally designated, and the taxonomic placement of the soil would not be determined. For Level 3, both of these details would be included.

For medium- to fine-textured mineral soils, a qualitative or semiquantitative assessment of clay mineralogy is the most important mineralogical parameter. For soils potentially affected by volcanic tephra deposition, clay mineralogical analyses are considerably more difficult than the standard X-ray diffraction analyses due to the low degree of crystallinity of secondary mineral species. In many cases, chemical and physical measurements can be used to infer clay mineralogy. For coarse-textured mineral soils, the relative amount of weatherable minerals (e.g., feldspars, pyroxenes, amphiboles, micas) compared with quartz can be determined by grain counts based on optical microscopy in thin section.

Conclusions

Descriptions of sites, encompassing descriptions of the landscape and associated soils, must consider the spatial and temporal dimensions of ecological interest. Several distinct scales describe hierarchical levels of organization for the components of the soil-landscape system (Fig. 2.1). Each scale requires a different set of descriptors and elucidates a different set of processes. A good site characterization should provide at least a qualitative description of the environment of a site with respect to light, nutrients, heat, and water, including its movement, providing a unifying theme and rationale for characterization. In addition, such a description should provide the basis for understanding a phenomenon or set of linked phenomena, either attributes or processes. Finally, a good site description should provide a firm basis to move research results from study sites to other areas via interpolation and/ or extrapolation for both science and land management objectives.

Acknowledgments Partial funding for this work was provided by the Minnesota Agricultural Experiment Station.

Appendix: Examples of Soil Profile Descriptions

Sartell Pedon—Mixed, Frigid, Typic Udipsamment

Location: Cedar Creek Natural History Area, Anoka and Isanti Counties, Minnesota. The following is a description of a representative pedon of the Sartell series on a 1.5% nearly plane slope in an old field at an elevation of 280 m, located 502 m west and 597 m north of the southeast corner of Sec. 22, T. 34 N., R. 23 W. (Colors are for moist soils unless otherwise noted.)

- Ap—0–18 cm; dark brown (10YR 3/3) sand; weak, fine and medium, subangular blocky structure; loose; common very fine and fine roots; abrupt, smooth boundary.
- B—18–71 cm; dark yellowish brown (10YR 4/4) fine sand; massive breaking to single grain; loose; few, dark brown (7.5YR 5/4) fillings; few fine roots; gradual, smooth boundary.
- C1—71–107 cm; very pale brown (10YR 7/4), fine sand; single grain; loose; gradual, smooth boundary.
- C2—107–152 cm; light yellowish brown (10YR 6/4) sand; single grain; loose; one reddish brown (5YR 4/4), 1–2 mm thick, irregular, weakly cemented band occurs at about 127 cm; gradual, smooth boundary.
- C3—152–178 cm; very pale brown (10YR 7/4) sand; single grain; loose; one reddish brown (5YR 4/4), 2–3 mm thick, irregular, weakly cemented band occurs at about 152 cm.
- C4—178–254 cm; very pale brown (10YR 7/4) fine sand; single grain; loose.

Parnell Pedon—Fine, Smectitic, Frigid, Typic Argiaquoll

Location: Near Dalton, in Otter Tail County, Minnesota. The following is a description of a representative pedon of the poorly drained Parnell series at the toeslope of a south- to west-facing hillslope. Location is NE 1/4 of SE 1/4, Sec. 10, T. 131 N., R. 42 W. (Colors are for moist soil unless otherwise noted.)

- Ap—0–14 cm; black (2.5Y 2.5/1) loam; weak, fine subangular blocky structure; friable; many coarse and medium roots; no effervescent reaction; clear, smooth boundary.
- A1—14–44 cm; black (2.5Y 2.5/1) silt loam; moderate fine subangular blocky structure; friable; many fine and very fine roots; no effervescent reaction; gradual smooth boundary.
- A2—44–67 cm; black (10YR 2/1) loam with few (<2%) fine prominent dark yellowish brown (10YR 3/4) mottles; moderate coarse and medium subangular blocky structure; friable few fine and very fine roots; no effervescent reaction; gradual smooth boundary.
- Btg1—67–88 cm; very dark gray (10YR 3/1) silty clay loam; moderate medium and fine subangular blocky structure; friable; discontinuous prominent dark gray (10YR 4/1) clay films on faces of peds and in pores; many (>20%) fine prominent reddish brown (5YR 5/4) oxidized rhizospheres; few fine and very fine roots; no effervescent reaction; gradual wavy boundary.
- Btg2—88–102 cm; 55% very dark gray (10YR 3/1) and 40% dark gray (10YR 4/1) clay loam with many (>20%) fine faint dark grayish brown (10YR 4/2) mottles; moderate medium and fine subangular blocky structure; friable; many (>20%) fine prominent reddish brown (5YR 5/4) oxidized rhizospheres; few very fine roots; strong effervescent reaction; clear, abrupt boundary.[1]
- Bkg1—102–115 cm; grayish brown (2.5Y 5/2) clay loam with many (>20%)

fine faint light brownish gray (10YR 6/2) and common (2–20%) medium prominent olive brown (2.5Y 4/4) mottles; strong medium subangular blocky structure; friable; discontinuous black (10YR 2/1) coats in root channels and pores; few very fine roots; strong effervescent reaction; gradual wavy boundary.

- Bkg2—115–145 cm; grayish brown (2.5Y 5/2) loam with many (>20%) fine prominent strong brown (7.5YR 4/6) mottles; strong medium subangular blocky structure; friable; many (>20%) fine prominent gray (10YR 6/1) carbonate threads; few very fine roots; strong effervescent reaction; gradual smooth boundary.
- Bkg3—145–180 cm; grayish brown (2.5Y 5/2) loam with many (>20%) fine prominent strong brown (7.5YR 4/6) and many (>20%) fine prominent gray (10YR 6/1) mottles; strong medium subangular blocky structure; friable; few (<2%) fine prominent black (10YR 2/1) iron-manganese concentrations; no roots; strong effervescent reaction.[2]

Appendix Notes

1. stone line present between Btg2- and Bkg1-horizon in bottom of Btg2-horizon; siliceous and carbonate pebble-sized stones.
2. Bkg3-horizon extends down to 220 cm, where there is a color change to light olive brown (2.5Y 5/3) with grayish brown (2.5Y 5/2) mottles; this may be the C-horizon (texture is loam or clay loam).

References

Adams, W. A. 1973. The effect of organic matter on the bulk and true densities of some uncultivated podzolic soils. *Journal of Soil Science* 24:10–17.

Alexander, E. B. 1980. Bulk densities of California soils in relation to other soil properties. *Soil Science Society of America Journal* 44:689–692.

Austin, M. E. 1972. *Land Resource Regions and Major Land Resource Areas of the United States.* USDA Agricultural Handbook No. 296. U.S. Government Printing Office, Washington, DC, USA.

Bailey, R. G., and C. T. Cushwa. 1981. *Ecoregions of North America.* U.S. Geological Survey, Washington, DC, USA.

Bell, J. C., R. L. Cunningham, and M. W. Havens. 1992. Calibration and validation of a soil-landscape model for predicting soil drainage class. *Soil Science Society of America Journal* 56:1860–1866.

Bell, J. C., R. L. Cunningham, and M. W. Havens. 1994. Soil drainage class probability mapping using a soil-landscape model. *Soil Science Society of America Journal* 58:464–470.

Bouma, J. 1989. Using soil survey data for quantitative land evaluation. *Advances in Soil Science* 9:177–213.

Calus, J. K. 1996. *Soil Survey of Oceana County, Michigan.* USDA Natural Resources Conservation Service and Forest Service. U.S. Government Printing Office, Washington, DC, USA.

Carmean, W. H. 1975. Forest site quality evaluation in the United States. *Advances in Agronomy* 27:209–269.

Crawford, C. A. Gotway, and G. W. Hergert. 1997. Incorporating spatial trends and anisotropy in geostatistical mapping of soil properties. *Soil Science Society of America Journal* 61:298–309.

Curtis, R. O., and B. W. Post. 1964. Estimating bulk density from organicmatter content in some Vermont forest soils. *Soil Science Society of America Proceedings* 28:285–286.

De Jong, R., C. A. Campbell, and W. Nicholaichuk. 1983. Water retention equations and their relationship to soil organic matter and particle size distribution for disturbed samples. *Canadian Journal of Soil Science* 63:291–302.

Dixon, J. B., and S. B. Weed, editors. 1989. *Minerals in Soil Environments. 2d edition.* Soil Science Society of America. Madison, Wisconsin, USA.

Fenneman, N. M. 1928. Physiographic divisions of the United States. *Annals of the Association of American Geographers* 18:261–353.

Giblin, A. E., K. J. Nadelhoffer, G. R. Shaver, J. A. Laundre, and A. J. McKerrow. 1991. Biogeochemical diversity along a riverside toposequence in arctic Alaska. *Ecological Monographs* 61:415–435.

Grigal, D. F., S. L. Brovold, W. S. Nord, and L. F. Ohmann. 1989. Bulk density of surface soils and peat in the North Central United States. *Canadian Journal of Soil Science* 69:895–900.

Grigal, D. F., R. E. McRoberts, and L. F. Ohmann. 1991. Spatial variation in chemical properties of forest floor and surface mineral soil in the North Central United States. *Soil Science* 151:282–290.

Gupta, S. C., and W. E. Larson. 1979. Estimating soil water retention characteristics from particle size distribution, organic matter percent, and bulk density. *Water Resources Research* 15:1633–1635.

Hall, G. F., and C. G. Olson. 1991. Predicting variability of soils from landscape models. Pages 9–24 *in* M. J. Mausbach and L. P. Wilding, editors, *Spatial Variabilities of Soils and Landforms.* Soil Science Society of America, Madison, Wisconsin, USA.

Holmgren, G. G. S. 1988. The point representation of soil. *Soil Science Society of America Journal* 52:712–716.

Jeffrey, D. W. 1970. A note on the use of ignition loss as a means for the approximate estimation of soil bulk density. *Journal of Ecology* 58:297–299.

Jenny, H. 1941. *Factors of Soil Formation: A System of Quantitative Pedology.* McGraw-Hill, New York, New York, USA.

McAuliffe, J. R. 1994. Landscape evolution, soil formation, and ecological patterns and processes in Sonoran Desert bajadas. *Ecological Monographs* 64:111–148.

Milne, G. 1935. Some suggested units of classification and mapping, particularly for East African soils. *Soil Research* 4:183–198.

Moore, I. D. 1992. Terrain analysis programs for the environmental sciences: TAPES. *Agricultural Systems and Information Technology* 2:37–39.

Moore, I. D., P. E. Gessler, G. A. Nielsen, and G. A. Peterson. 1993. Soil attribute prediction using terrain analysis. *Soil Science Society of America Journal* 57:443–452.

Moore, I. D., R. B. Grayson, and A. R. Ladson. 1991. Digital terrain modeling: a review of hydrological, geomorphological and biological applications. *Hydrology Proceedings* 5:3–30.

Mroz, G. D., and D. D. Reed. 1991. Forest soil sampling efficiency: matching laboratory analyses and field sampling procedures. *Soil Science Society of America Journal* 55:1413–1416.

Pritchett, W. L., and R. F. Fisher. 1987. *Properties and Management of Forest Soils. 2d edition.* Wiley, New York, New York, USA.

Rawls, W. J., D. L. Brakensiek, and K. E. Saxton. 1982. Estimation of soil water properties. *Transactions of the American Society of Agricultural Engineers* 25:1316–1320, 1328.

Rawls, W. J., D. L. Brakensiek, and B. Soni. 1983. Agricultural management effects on soil water processes. I. Soil water retention and Green and Ampt infiltration parameters. *Transactions of the American Society of Agricultural Engineers* 26:1747–1752.

Rawls, W. J., T. J. Gish, and D. L. Brakensiek. 1991. Estimating soil water retention from soil physical properties and characteristics. *Advances in Soil Science* 16:213–234.

Robertson, G. P., K. M. Klingensmith, M. J. Klug, E. A. Paul, J. C. Crum, and B. G. Ellis. 1997. Soil resources, microbial activity, and primary production across an agricultural ecosystem. *Ecological Applications* 7:158–170.

Ruhe, R. V., and P. H. Walker. 1968. Hillslope models and soil formation. I: Open systems. Pages 551–560 *in Transactions of the Ninth International Congress of Soil Science.* Volume 4. Adelaide, Australia. International Society Soil Science and Angus Robertson, Sydney, Australia.

Shaykewich, C. F., and M. A. Zwarich. 1968. Relationships between soil physical constants and soil physical components of some Manitoba soils. *Canadian Journal of Soil Science* 48:199–204.

Soil Survey Division Staff. 1993. *Soil Survey Manual.* USDA Agricultural Handbook No. 18. U.S. Government Printing Office, Washington, DC, USA.

Soil Survey Staff. 1996. *Keys to Soil Taxonomy. 7th edition.* USDA–Soil Conservation Service, Washington, DC, USA.

Sollins, P., G. P. Robertson, and G. Uehara. 1988. Nutrient mobility in variable- and permanent-charge soils. *Biogeochemistry* 6:181–199.

Sollins, P., G. Spycher, and C. Topik. 1983. Processes of soil organic-matter accretion at a mudflow chronosequence, Mt. Shasta, California. *Ecology* 64:1273–1282.

Trangmar, B. B., R. S. Yost, and G. Uehara. 1985. Application of geostatistics to spatial studies of soil properties. *Advances in Agronomy* 38:45–94.

Ver Hoef, J. M., and N. Cressie. 1993. Spatial statistics: analysis of field experiments. Pages 319–341. *In* S. M. Scheiner and J. Gurevitch, editors, *Design and Analysis of Ecological Experiments.* Chapman and Hall, New York, New York, USA.

Warrick, A. W., D. E. Myers, and D. R. Nielsen. 1986. Geostatistical methods applied to soil science. Pages 53–82 *in* A. Klute, editor, *Methods of Soil Analysis. Part 1, Physical and Mineralogical Methods.* Agronomy Monograph No. 9. 2d edition. American Society of Agronomy, Madison, Wisconsin, USA.

Whittaker, R. H. 1956. Vegetation of the Great Smoky Mountains. *Ecological Monographs* 26:1–80.

Part I

Soil Physical Properties

3

Soil Water and Temperature Status

Wesley M. Jarrell
David E. Armstrong
David F. Grigal
Eugene F. Kelly
H. Curtis Monger
David A. Wedin

Soil water status and temperature are critical factors that affect the activity of organisms in soils. Soil water content also has important effects on the quantity of water that eventually becomes available to surface water, both through its influence on infiltration/runoff relationships (wet soils have lower infiltration rates) and because of its potential contribution to the groundwater. Soil temperature influences soil water availability, the growth rates of plants and microorganisms, and rates of chemical/biochemical reactions.

A number of useful soil physics texts are available that cover the general theory and application of soil water and temperature (e.g., Baver et al. 1972; Hillel 1980a, 1980b; Marshall and Holmes 1988). In addition, several books contain detailed summaries of current measurement techniques (Klute 1986; Smith and Mullins 1991; Topp et al. 1992). These references should be used to supplement material in this chapter.

Soil water content has been measured for many years. Earliest measurements were semiquantitative. To the experienced fieldworker, the appearance and feel of the soil can provide valuable information on the amount of water it contains, especially the total water content. Wet soils are generally darker than dry soils and feel "smoother" and heavier. However, appearance and feel can be deceiving. Furthermore, they do not provide quantitative measures of soil water status. Earliest measures of soil water content involved collecting a moist field sample, weighing it, drying it in air or in an oven, and determining the weight loss. Weight loss on drying at 100–110 °C is usually attributable to water evaporation. The ratio of weight loss to dry weight has been termed *gravimetric soil water content.*

For soils with comparable textures, gravimetric water content corresponds well to the ability of plants to extract water from the soils. However, gravimetric water

Table 3.1. General Lower Limit of Soil Water Potential for Organism Activity

Organism	Soil Ψ matric, -MPa
Sunflower	1.5
Creosote bush	5.0
Bacteria	2.0
Fungi	3.0
Nematodes	0.15
Protozoa	0.10

Note: For plants, this represents the lower limit for water uptake; for other organisms, processes other than strict survival are limited.

is not a reliable indicator of water availability across soils of differing textures, for example, when comparing clayey and sandy soils. Clays hold relatively high amounts of water even when plants are experiencing drought, while plants can take up water from sandy soils at a very low gravimetric water content.

To deal with water availability to plants and water movement in soils, the concept of *soil water energy potential*—often shortened to *soil water potential*—has been developed (see later discussion). The presence of solutes and the attraction of water molecules to the surfaces of soil particles lower the energy of soil water compared with that of pure water. The plant develops an energy gradient to extract water from the soil. Thus, in order to extract water from a dry, salty soil, the plant itself must become "drier" than the soil, lowering the energy of water in its leaves below that of the soil water. The activity of other soil organisms is also sensitive to soil water potential (Tab. 3.1).

In practical terms, these soil water measurements may be separated into two categories. For ecosystem studies, where knowledge of hydrologic balance is important, measurement of the *total soil water content* is appropriate. For ecophysiological studies, the appropriate measure is the *soil water potential* because it provides information on the instantaneous ability of the plant and microorganisms to extract water from the soil.

When considering the effect of soil water on organism activity at the landscape scale, the availability of water to organisms can be assessed by considering three parameters:

- the total amount of soil water available in the root zone during the growing season, which relates to the total quantity of transpiration that can occur, and therefore total potential net primary production;
- the duration of availability, when temperature and light are adequate for growth, which determines how the phenology of the plant interacts with soil moisture; and
- the depth and spatial location of available water, which determines how plant rooting patterns affect access to water.

Obviously these factors all interact strongly. However, if these attributes of the hydrologic cycle are known, the potential effect of soil water on plant growth can be predicted.

Available Protocols

Water in soils is generally described in terms of either (1) the total amount or (2) the energy status of the soil water, frequently called *water potential.* Decisions about which to measure, and how, are based on the research question of interest.

Water balance measurements depend on accurate assessment of the quantities of water in soil. These quantities are expressed in terms of

- mass water/mass soil (gravimetric water content, usually kg water/kg dry soil), and
- volume of water/volume of soil (volumetric water content, L water/L soil).

To convert from gravimetric to volumetric water content, multiply by the bulk density of the soil (Mg dry soil/m^3 dry soil or g dry soil/cm^3 dry soil). For example, a soil that contains 0.3 kg water/kg soil, with a dry soil bulk density of 1.4 Mg/m^3, has a volumetric water content of 0.42 mL water/cm^3 soil. The water content of soil is usually expressed using the symbol θ, with θ_g referring to gravimetric water content and θ_v referring to volumetric water content.

Soil Water Content

A variety of principles have been applied to estimate volumetric water content in field soils. Two of the most common rely on neutron analysis using a neutron probe and analysis of a soil's dielectric constant using capacitance or time domain reflectometry probes.

The neutron probe technique relies on the ability of hydrogen in water to slow fast neutrons to thermal neutrons. The fraction of neutrons slowed is proportional to water content. The probe is lowered into an access tube permanently installed in the soil, and readings are taken at specific depths, for example, 15, 30, 45, and 60 cm. The probe is held at a given depth for a period of time to count the number of returning thermal electrons. This number is proportional to the water content of the soil. Depending on the degree of textural differences in the field, different calibration curves may be required for different sites. Although it has proven to be rugged and safe in the field, use of the neutron probe requires a radiation license and operator certification with respect to the handling and transport of devices with radioisotopes.

The dielectric constant of soils changes as a function of soil water content (e.g., Whalley et al. 1992), and this constant can be measured by a variety of techniques to provide soil water content estimates. Soil capacitance probes, for example (Tab. 3.2), can be lowered into an access tube similar to that for the neutron probe to measure soil capacitance (a function of the dielectric constant of the soil) at different depths.

Other methods have also been developed for rapid, effective measurements of the soil dielectric constant. The most popular of these relatively new methods is termed *time domain reflectometry* (TDR; Dalton 1992). In TDR, two stainless steel rods are inserted parallel to one another to a given soil depth (10–100 cm); an electrical pulse sent through these wave guides generates an electrical response (read

Table 3.2. Suppliers of Time-Domain Reflectometers, Tensiometers, and Resistance Blocks

Method	Supplier	Model	Datalogger	Instrument Price Range	Sensors (each)
Time domain reflectometer	Soilmoisture	Trase	Possible	$9,200	$15–23
	Campbell Scientific	Water Content Reflectometer	Possible	220	NA
Tensiometer	Soilmoisture	Soilmoisture tensiometer (3′)	No	50	NA
	Irrometer	Tensiometer Model RS (3′)	No	65	NA
Resistance blocks	Soilmoisture	Soilmoisture meter	No	285	11
	Irrometer	Digital readout resistance meter	No	240	24
Thermistor	Onset Computer	Hobo, Stowaway	Built-in	63–329	NA
Laboratory psychrometer	Decagon Devices, Inc.	Decalogger	Built-in	2,500	NA

Notes: List is current as of 1998. Prices listed are for field-capable units without dataloggers, unless otherwise indicated.
NA = not applicable because sensors are integral to instrument.
Soilmoisture Equipment, Inc., Goleta, CA (www.soilmoisture.com)
Campbell Scientific, Inc., Logan, UT (www.campbellsci.com)
Irrometer, Inc., Riverside, CA (www.irrometer.com)
Decagon Devices, Inc., Pullman, WA (www.decagon.com)
Onset Computer Corporation, Pocasset, MA (www.isa.org)

with an oscilloscope) that is characteristic of the dielectric constant for a soil at a particular water content. Probes can also be inserted laterally to provide water contents for a specific soil depth or horizon. The major drawback of TDR at this time is the high initial cost of equipment. However, its convenience makes the method extremely attractive, and further refinements are likely to lower the price.

The dielectric constant of the soil is generally proportional to the water content. Under some conditions, soil salinity will affect the signal, but instruments that are currently available minimize this effect. In fact, some instruments have turned this "interference" into a benefit and have been specifically designed to measure salinity (Dalton 1992).

Soil Water Potential

The energy status of water in soils is expressed as the energy embodied in water molecules per unit volume of water, compared with pure water at standard temperature and pressure. This is described in terms of the following equation (Nobel 1991):

$$\mu_w = \mu_w{}^* + V_w \times \Pi + V_w \times P + m_w \times g \times h$$

where

μ_w = energy potential of water (joules/mole)
$\mu_w{}^*$ = standard energy potential of water (reference is pure water)
V_w = partial molar volume of water (0.018 L/mole)
Π = total osmotic pressure (Pa)
P = hydrostatic pressure (Pa)
m_w = molar mass of water (1000 kg/m^3)
g = acceleration of gravity (kg m/sec^2)
h = height relative to reference (m)

This equation expresses the energy embodied in water, quantifying the effects of solutes, solid surfaces, and gravity.

The net energy potential of the water is usually expressed in terms of the difference between μ_w and μw^*, namely

$$\mu_w - \mu_w{}^* = V_w \times \Pi + V_w \times P + m_w \times g \times h$$

Further, when both sides of the equation are divided by the molar volume of water, we have

$$\mu_w - \mu_w{}^*/V_w = \Pi + P + (m_w \times g \times h)/V_w$$

The term P can represent the pressure relative to the location of the point in question with respect to the gravitational head. The combined term $m_w gh$ refers to the height of capillary rise in a tube of radius r, where r represents the radius of the small-

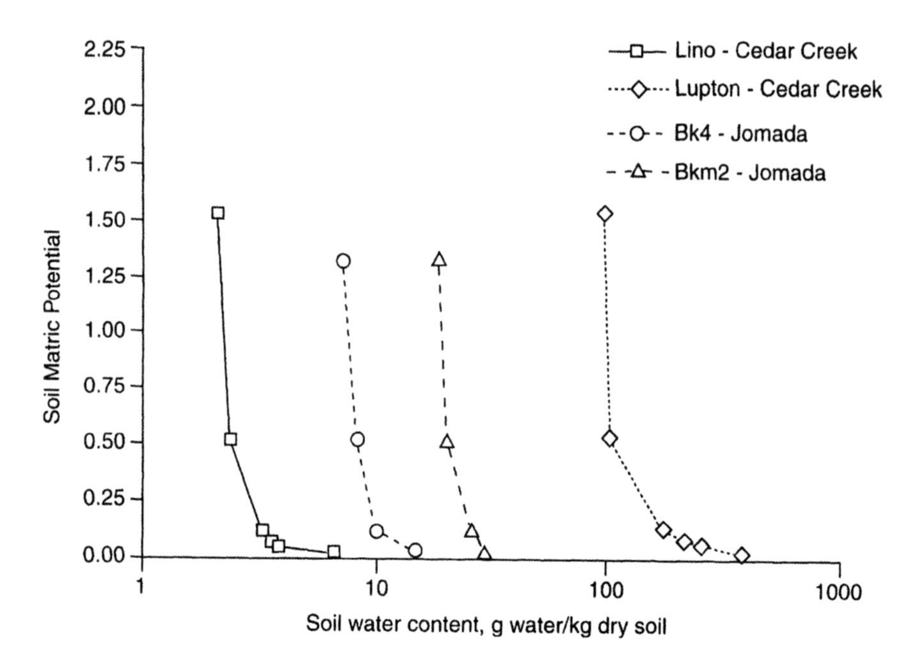

Figure 3.1. A soil water characteristic curve describing the relationship between soil water potential and soil water content.

est water-filled pore. This term expresses the contribution of the matric potential to lowering soil water potential.

The sign of the water potential under these conditions is always negative in unsaturated soils because μ_w is always less than μ_w*.

The units of water potential are energy per unit of volume. In most cases, these units are converted to force/unit area, or pressure. The standard unit for soil water potential is now the pascal (Pa), or newtons/m^2. Most soils, particularly where biological processes are active, have soil water potentials between 0 and −2.0 MPa.

Soil water potential is typically measured using one of several techniques:

- *Tensiometer.* When soil water potential is between 0 and −0.08 MPa, a tensiometer can be used. Tensiometers consist of a ceramic cup attached to a sealed plastic tube, with a vacuum gauge attached to the tube. The tube is filled with water, avoiding entrainment of air bubbles, and sealed tightly. The tensiometer in placed in the soil in a hole slightly smaller in diameter than the tensiometer tube. As the soil dries, water moves out of the tube into the soil, through small pores in the ceramic cup. This creates a tension, or partial vacuum, within the tube, which is reflected by a change in the vacuum gauge reading.
- *Thermocouple psychrometer.* The thermocouple psychrometer is a thermocouple junction surrounded by a ceramic cup. The cup takes up water from the surrounding soil. The water evaporates inside the cup and creates a characteristic

humidity, which is related to total soil water potential by the psychrometric equation

$$\Psi = (R \times T/V_w) \times \ln(rh)$$

where

Ψ = total soil water potential
R = ideal gas constant, 8.31 $J \cdot {}^\circ K^{-1} \cdot mol^{-1}$
T = temperature (°K)
rh = relative humidity of the air

Since *rh* is affected by both the matric potential and the solute potential, the psychrometer measurement senses both. Because the *rh* in most soils is 99% or higher, the measurement of humidity must be very precise, and temperature measurement or control is critical. Psychrometers work best in soils drier than −0.15 MPa.

- *Resistance block.* A solid block of material—frequently gypsum, or $CaSO_4 \cdot 2H_2O$—is cast around two electrodes. Water is absorbed into the pores in the block and dissolves some of the low-solubility gypsum, lowering the resistance to an electrical current passed between the electrodes. The more water that enters the block, the lower the resistance. A standard curve is created that relates the soil matric potential to the resistance in the block.

  Resistance blocks are relatively inexpensive and work in many situations. Their limitations include the fact that the gypsum eventually dissolves and the standard curve shifts; in addition, in saline soils, readings will be artificially high because of high conductance of the soil water; the soil will appear to have a higher water content than it actually has.

The Soil Water Characteristic Curve

The relationship between the soil matric potential and total soil water content is described in a functional relationship called the *soil water characteristic curve* (SWCC). This curve varies as a function of soil texture (Fig. 3.1). This curve does not provide any information on the effects of salts on soil water availability but does provide information on water availability at different moisture contents.

The SWCC determination is especially sensitive to sample handling. Wherever possible, the SWCC is determined in the laboratory using intact soil cores collected in the field. Large pores are critically important in holding water near saturation, making maintenance of soil structure a fundamental concern (the aggregation of single grains of soil particles; see Chapter 4, this volume). If aggregates are broken down and macropores destroyed, the field water content at high water potentials will be underestimated.

The relationship between soil water potential and soil water content depends on whether the soil is in the process of wetting or drying. This asymmetry is termed *hysteresis.* During the wetting phase, there is resistance to water moving from smaller to larger pores, so at a given water potential the soil water content is lower than at true equilibrium. During the drying phase, large pores retain water even when pores around them are drying.

The variability of the SWCC leads to problems in selecting a single method of soil water characterization, particularly if both quantity and energy status are important aspects of the study. If only water potential is measured, a poor SWCC may result in either overprediction or underprediction of total soil water content. If the soil water contains significant salt concentrations, converting total water potentials to total water contents will underestimate water contents, particularly under drier conditions.

Soil Temperature

Soil surface radiation balance can be represented as follows (Buchan 1991):

$$R_n = [(1 - \alpha) \times R_s] + (\epsilon \times L_d) - L_u$$

where

R_n = net radiation
α = short-wave reflection coefficient (albedo)
R_s = incident solar radiation
ϵ = long-wave emissivity of the soil surface
L_d = long-wave radiation input
L_u = long-wave radiation emitted

The net radiation may be partitioned into three terms as

$$R_n = H + (L_v \times E) + G_0$$

where

R_n = net radiation
H = sensible heat flux between soil and air
L_v = latent heat of vaporization
E = mass flux of water evaporated
G_0 = heat flux into soil

We are typically concerned with temperatures throughout the soil profile, since temperature directly affects rates of biological and chemical reactions.

Historically, temperature measurements use the thermal expansion properties of liquids or metals to indicate temperature change. Alcohol, mercury, and bimetallic thermometers are still commonly used. Small, rugged bimetallic thermometers are favored by soil scientists because they can be inserted directly into soils without breaking.

A second approach to temperature measurement is the use of thermocouples, based on the Seebeck effect. If two welded junctions are created with specific metal alloys of wire, current will flow through the wires at a rate proportional to the difference in temperature of the two junctions. If the temperature of one junction is controlled or known, the temperature of the other junction can be determined by measuring the current flowing in the wire. The two junctions together are termed the *thermocouple*.

A third method for measuring temperature involves thermistors, based on another electrical property, resistivity. Thermisters are materials in which resistance to current flow is a function of temperature. These materials are called *thermistors*. Once the relationship between temperature and resistance is known, the temperature of a given location in the soil can be assessed simply by determining the resistance to electron flow through the thermistor.

Sensor Installation—General Considerations

Installation of soil water sensors in the field requires attention to several factors that can influence moisture and temperature measurements. First is the changing fraction of large pores. This can occur either through compaction and loss of large pores in the vicinity of the sensor or by creating and not refilling large pores (air gaps) in the soil during installation. Ideally, the soil adjacent to the sensor should have the same pore size distribution as the bulk soil. This is particularly critical for conditions near saturation, since within this range the large pores fill and the fraction of total porosity present as large pores becomes even more critical. Minimal disturbance to the soil is always recommended. This includes keeping installation hole size to a minimum and not installing sensors when soils are wet and easily compacted or dry and susceptible to air gaps.

Second is the danger of modifying water movement through the soil. This may occur because of the creation of channels of large, continuous pores, e.g., adjacent to tubes or wires installed in the soil. These channels allow preferential flow, potentially accelerating wetting and accentuating drying compared with the bulk soil.

Third, be aware of ways that sensors can affect soil thermal conductivity. To avoid undesirable effects and to protect sensors from weather and animals in general, shield exposed portions of sensors appropriately. At the very least, sensors should be shielded from direct solar radiation.

Gravimetric Water Content

Gravimetric water content (Gardner 1986) is the most direct, laborious, and disruptive method for determining the instantaneous water content in soils. The soil sample is removed from the field, oven-dried, and reweighed. This method is most useful when one must determine water content of a sample collected for some other chemical, physical, or biological analysis.

Materials

1. Field coring device or shovel
2. Ziplock bags or other watertight containers

3. Weighing cans (preferably aluminum with tight-fitting lid, 250 mL capacity)
4. Balance with resolution to 0.01 g

Procedure

1. Collect samples from known depths. The depths selected may be based on soil horizons, in which case the horizon thicknesses and bulk densities must be known to allow conversion to volumetric and spatial units. Alternatively, samples may be collected at fixed depth increments (see Chapter 1, this volume).
2. Transport the samples to the laboratory in sealed containers that prevent evaporation. In the lab, weigh a portion of the field-moist soil (approximately 100 g) into a preweighed drying container with lid (frequently an aluminum can with a tight-fitting lid, although any drying container that maintains constant weight when dried at 105 °C can be used).
3. Place the open container in a drying oven set at 105 °C. When soil is dried to constant weight, remove the container from the oven, seal it with the lid (to avoid rehydration), allow to cool to room temperature, and then weigh the entire can plus lid.

Calculations

Gravimetric Water Content

After the weight of the container plus lid is subtracted from moist and dry samples, the gravimetric water content (always referenced to the oven-dry soil mass) is calculated as

$$\theta_g = [(\text{g moist soil}) - (\text{g dry soil})]/(\text{g dry soil})$$

where θ_g = gravimetric water content as g H_2O/g dry soil.

Volumetric Water Content

Use soil bulk density information (see Chapter 4, this volume) to convert θ_g to volumetric water content θ_v:

$$\theta_v = \theta_g \times BD$$

where

θ_v = volumetric water content as mL H_2O/cm^3 soil or $m^3 H_2O/m^3$ soil
θ_g = gravimetric water content as g H_2O/g dry soil
BD = soil bulk density as g dry soil/cm^3 soil or Mg dry soil/m^3 soil.

If high precision is not required, one can assume clay soils have bulk densities around 1.1 g/cm^3 and soils high in sand have bulk densities nearer 1.7 g/cm^3. However, compaction can significantly increase bulk density, while well-aggregated soils high in organic matter can have much lower bulk densities (well below 1 g/

cm^3). Under these conditions, bulk density should be determined directly to make the conversion accurate.

To convert from volumetric to areal values, θ_v is multiplied by the depth of each horizon for which θ_v has been calculated. For example, if θ_v is 0.42 m^3 water / m^3 soil and the depth increment is 0.2 m, this depth increment contains the equivalent of $(0.42 \times 0.2) = 0.084$ m of water, or 8.4 cm. By adding together the values for each depth increment, the total water storage in the profile can be calculated in terms of centimeters of water. To convert to a volume of water, simply multiply this value by the land area of concern, e.g., a square meter or hectare.

Special Considerations

While they are direct and require little equipment, gravimetric water content measurements are time-consuming and labor-intensive, create substantial disturbance, and can drastically alter the hydrologic cycle. In addition, gravimetric measures do not allow the user to obtain frequent readings from exactly the same site, cannot be automated, and require bulk density measurements for conversion to volumetric terms.

See Chapter 4 (this volume) to calculate water-filled pore space (WFP) from gravimetric water content values.

Time Domain Reflectometry (TDR)

The TDR approach (Dalton 1992) allows continuous or intermittent measurement of total soil water content with minimal disturbance. The technology is advancing quickly, and equipment prices are dropping. TDR represents the best available method for regular monitoring of soil water content. In addition, with appropriate adjustment (e.g., instrument frequency changes), one can also measure soil salinity.

Materials

1. TDR electronics (e.g., Tektronix line tester, Campbell Scientific)
2. Wave guides (probes). Guides can be cut from stainless steel rods or can be purchased commercially.
3. Installation guides

Procedure

1. Probes are inserted in parallel pairs a precise distance apart from one another; they typically are inserted vertically from the soil surface. Commercial probes vary in length from 15 to 70 cm. Differences in measured water contents among probes of different length can be used to estimate water contents over differences in depth increments. For example, a 15–30 cm depth increment can be estimated as the difference between readings from a 0–30 cm depth pair and a 0–15 cm depth pair.

For more precise measurements of soil water at particular depths, probes may be installed horizontally through the sidewalls of a soil pit or trench. Depth increments can be regular or based on major discontinuities in soil properties that affect hydrology, such as texture or structure. In most cases, probes 15–25 cm in length work well. Note that the presence of a pit or trench can severely disrupt local hydrology; the pit should be backfilled to conditions as close as possible to those of the undisturbed site.

If the objective is to understand quantities and patterns of water uptake in the plant root zone, sensors should be distributed through the entire depth of the effective zone of water use. In most cases, a minimum of three sensors equally distributed down a profile is necessary. In deeper or more heterogeneous profiles, other spacings may be necessary to capture the average pattern of water content as a function of depth. Run the sensor wire leads up to the soil surface; protect the lead ends with a waterproof cover.

2. Read water content at a frequency appropriate to the study and site. Sensor systems that are set up for continuous readings are usually interfaced through a multiplexer to a field datalogger unit. In most cases, hourly readings are adequate to describe rates of water loss from the soil through leaching or evapotranspiration. However, if detailed tracking of wetting patterns is a major concern, more frequent sampling intervals may be required, at least for short periods of time.

   While there is a natural tendency to limit readings to the time of most active biological activity (frequently the moist part of the summer), there are good reasons for year-round monitoring. This is especially important where leaching patterns, including water loss, are evaluated.

3. The TDR unit should be periodically checked and calibrated against soil samples of known water content. This should be done initially using both *in situ* and laboratory techniques.

   In situ check. When sensors are installed, a soil core can be taken near the site and sectioned into segments that represent equal distances between sensors or horizon depths. The gravimetric water content of these sections is then determined and compared with the initial TDR readings. Values should be within ±2% for both techniques.

   Laboratory calibration. Insert probes into a variety of soil samples for which water content is known; preferably, soil samples should be intact cores or otherwise minimally disturbed samples from the site of interest. Compare readout of the device with the known water content and make calibration adjustments as needed. Individual cores can be moistened and then allowed to air-dry to provide a wide range of moisture conditions (moisture loss can be monitored gravimetrically).

Calculations

Modern TDRs typically provide direct readouts of volumetric soil water content. Extrapolating to depth of water in the effective root zone uses the same procedures as described earlier for gravimetric water content.

Special Considerations

Calibration can be a problem, especially for organic soils. While uncalibrated values can be obtained that may be useful, additional in-field or in-laboratory calibration is required for accurate measurements of volumetric water content at any particular site.

When choosing sensor sites, select representative areas of the landscape for moisture determinations; avoid anomalously low or high places unless this variation is an important element of the study.

Soil Water Potential—Tensiometers (0 to −0.08 MPa)

No devices are available that measure soil water potential over the entire range of values of interest in biology. As a result, we recommend tensiometers for measuring water potentials under wet conditions (0 to −0.08 MPa) and resistance blocks for measuring potentials in the dry range (−0.10 to −3.0 MPa).

Materials

1. Tensiometer with an appropriate tube length
2. Pump for priming the ceramic cup
3. Deaerated water (boiled)
4. Auger of diameter slightly smaller than tensiometer

Procedure

1. Prepare the tensiometer by placing its ceramic cup end in a large bucket of deaerated water. Unscrew the end cap from the tube, place the vacuum pump mouth over the tube, and pull a vacuum that will bring water through the ceramic cup, displacing air and filling all pores with water. Leave the vacuum on (at about −0.09 MPa) until no bubbles emerge from the ceramic cup (an advantage of using clear tubing material).

   Finish filling the tube with deaerated water and screw the cap on tightly. Make sure no air bubbles are trapped in the water column. If air bubbles are present, the tensiometer will not respond to drying soil, and will always read wetter than actual soil conditions.
2. Bore a hole to the desired depth with the auger. Depths, as discussed previously for the TDR method, are selected as appropriate for characterizing the dynamics of water input and removal from the root zone of the plant. However, in practical terms, few tensiometers are longer than 1.5 m. Insert the tensiometer tube so that it fits snugly in the hole. Inserting the tube at an angle is recommended because soil directly above the ceramic cup is less disturbed. However, under no conditions should there be a gap between the tensiometer tube and the soil, because preferential flow will occur during precipitation events and the area around the ceramic cup will wet faster than the

bulk soil. Poor tube-soil contact can accelerate drying rates as well, since water vapor can rapidly move up and out of the soil.

3. Read the tensiometer at a frequency appropriate to the site and study. This will depend on the rate of soil drying and the specific research questions. If the soil is not near saturation initially, the manometer needle should depart from zero shortly after installation. If the soil is very moist, the needle may stay on zero until drainage or evapotranspiration decreases water potential below −0.002 MPa. If the soil becomes drier than approximately −0.08 MPa around the ceramic cup, air will enter the pores of the cup and the reading will rapidly drop back to zero. At this point the tensiometer has become useless, and it must be removed and refilled before again installing in the soil. The problem can be partly overcome by using ceramic cups with smaller maximum pore sizes, but their resistance to water movement results in slow attainment of equilibrium in soil.

Calculations

The tensiometer reads water potential directly (sometimes potential is described as soil suction), so no additional calculations are necessary.

Special Considerations

Do not install a tensiometer in the landscape or at depths where the soil will be drier than −0.08 MPa most of the time unless you are prepared to frequently remove, recondition, and reinstall the device after rewetting has occurred. In wetland soils, in soils above a shallow groundwater table, or in irrigated sites there may be extended periods when the soil will remain wet. In other cases, use information from the TDR measurements or, for generally dry sites, use gypsum blocks (see next section).

Soil Water Potential—Resistance Blocks (0 to −3.0 Mpa)

The use of resistance blocks (Campbell and Gee 1986) is straightforward. As indicated earlier, resistance blocks reach maximum conductance at approximately −0.1 MPa (−0.2MPa in some cases) and therefore are unresponsive in wet soils. Use tensiometers (as described previously) for wet (0 to −0.48 MPa) soils.

Materials

1. Resistance meter
2. Gypsum or fiberglass resistance blocks with leads of required length
3. Soil auger (3 cm diameter)

Procedure

1. Bore auger holes to desired depths (see the TDR protocol, described earlier). Place the gypsum block in the bottom of the hole, ensuring that if fits snugly

with the soil. Backfill the hole with soil, making sure it is tamped firmly so it does not become a channel for preferential flow. Bring the leads to the surface and tie them together. Connect them directly to the resistance meter/datalogger device and place them in a protective container.

2. Make readings at a frequency appropriate to the study and site. To capture events that involve rapid infiltration, time steps may be on the order of 15 minutes, or at even shorter intervals in very sandy, high-porosity soils. For overall monitoring of soil water status, storing hourly data will provide adequate temporal resolution for most applications.

Calculations

The output may need to be converted from resistance readings to soil water potential using calibration curves provided with the blocks or developed as described below. The potential readings can then be converted to volumetric soil water content estimates using the soil water characteristic curve (see next section). However, depending on the degree of hysteresis exhibited by the soil and the spatial variability in the field, these conversions may result in serious errors in estimating soil water content. Therefore, if soil water content is the desired parameter, it is preferable to measure it directly.

Special Considerations

Over time, the calibration of the blocks may drift due to changes in the chemical composition of the block or to dissolution and loss of gypsum. In most cases the blocks should be removed and recalibrated on a yearly basis. The blocks should be buried in a sample of field soil at a known water potential (e.g., −0.3 MPa based on psychrometric measurements in the laboratory). After the reading has stabilized, if the resistance readings vary more than 5% from the standard curve accompanying the block, a new calibration curve should be developed (at least four points between −0.1 and −2.0 MPa), or the block should be discarded and replaced with a new unit.

The resistance block method is sensitive to soil salinity; any salts entering the block will give artificially wet (low resistance) readings. They are not recommended for soils with saturation extract specific conductance greater than 2 dS/m. Under these conditions, if soil water potential must be measured directly, soil psychrometers are recommended.

Soil Water Characteristic Curve—Psychrometry

In the laboratory, pressure and vacuum plate apparatuses have been used to measure water potential in soil samples (see Klute 1986). One important difference between classical pressure plate methods and the psychrometer is that the latter measures both the matric potential and the solute or osmotic potential—effectively the total potential of the system. The pressure plate measures only the matric potential and is thus less preferred.

Materials

1. Psychrometer
2. Sealable bags

Procedure

1. Add a known amount of water to a variety of dry soils to create a range of samples with known soil moisture contents. Alternatively, determine the gravimetric water content of a range of moist soil samples (see previous section "Time Domain Reflectometry"). Added water must be uniformly distributed throughout the soil. In some cases, freezing the dry soil and mixing with an appropriate mass of crushed ice can lead to a more uniform and rapid equilibration; this is especially appropriate under dry soil conditions.
2. Keep the soil in a sealed bag to allow equilibration for a period of 1–2 days.
3. Place duplicate samples in a psychrometer calibrated following the manufacturer's recommendations. The psychrometer measures the relative humidity of the atmosphere above the samples.

Calculations

The water potential is calculated from the following equation:

$$\mu_w = (R \times T/V_w) \times \ln (p/p_0)$$

where

μ_w = energy potential of water (joules/mole)
R = ideal gas constant, 8.31 J·°K^{-1}·mol^{-1}
T = absolute temperature, °K
V_w = partial molar volume of water, 0.018 L/mole
p = water vapor pressure in equilibrium with the liquid phase
p_0 = saturated water vapor pressure of the liquid phase

This equation highlights the extreme temperature dependence of this relationship, both directly through T and indirectly through p_0. Temperature must thus be known very precisely.

The relationship between soil water potential and soil water content is then graphed as the SWCC. Frequently the relationship is expressed as a semi-log graph, with the soil water potential graphed on a common log scale (see Fig. 3.1).

Soil Temperature

For long-term, continuous measurements in the soil, thermistor sensor arrays are generally preferred. The thermistor is part of an electrical circuit in which the resistance to current flow is proportional to its temperature (Buchan 1991; Taylor and Jackson 1986). Thermistors are rugged, reasonably precise, stable, and inexpensive.

With the advent of inexpensive self-contained datalogging thermistors, continuous monitoring of soil temperature has become easy and inexpensive.

Materials

1. Soil auger, smallest possible diameter
2. Resistance meter appropriately calibrated for specific thermistors to provide direct temperature readout
3. Thermistors with appropriate adequate cable lengths. Alternatively, integral temperature-dataloggers that can be installed in situ (e.g., the Hobo Unit from Onset Computer Corp.; see Table 3.2) can be substituted for items 1 and 2

Procedure

1. Auger a hole to the maximum depth to which sensors will be installed, retaining the soil. The depths selected will depend on the site and study. For near-surface processes, installation at 5, 15, 25, and 50 cm depths will usually suffice. At most sites, soil temperatures do not change diurnally below 1.5 m. At this depth, annual mean temperature is generally equivalent to 2° above the mean annual air temperature at the location.

   Place a thermistor in the bottom of the hole and backfill with soil to the depth of the next thermistor. Place the second sensor in the hole and continue in the same manner until all sensors are in place. Replace soil in approximately the order in which it was removed from the hole.

   Surface shading and soil water status will have dramatic effects on soil thermal properties. Make sure the site selected for installation is typical of the area to be characterized in terms of soil water relations and shading cover. In deserts or savannas, installations under and between perennial woody species may be required.
2. Bring all leads to the surface and connect to the temperature meter/datalogger. Protect the dataloggers from direct sunlight and precipitation.
3. For datalogging, storing 1 hour temperature averages is more than adequate to describe the essentials of diurnal and seasonal heating and cooling patterns. In some cases, less frequent measurement is satisfactory. Data collection should be made sufficiently frequent to provide means over time at various depths, temperature ranges, and initiation and duration of freezing if it occurs.

Calculations

Most thermistor units currently available are designed for meters that read out directly in degrees Celsius, so no calculations are necessary.

Special Considerations

Newer probes have extremely reproducible output, so calibration in the laboratory is less important than in the past. Nevertheless, calibration should be checked annually at two temperatures that bracket the temperature range of interest. For calibration, place probes in a constant-temperature bath and compare the readings to those on a reference thermometer. If probes are not identical (within 0.2 °C of bath temperature), they should be tested at several temperatures to generate individual calibration curves relating bath temperature to output. In this case they must be individually marked in the field so the appropriate calibration or correction can be applied.

Be aware of the potential for installation effects on temperature profiles. Exposure of wire leads to direct sunlight, for example, can affect temperature readings, and precautions should be taken to avoid direct solar radiation. Also, take care to backfill augered soil carefully so as not to change the soil's thermal conductivity with depth.

References

Baver, L. D., W. H. Gardner, and W. R. Gardner. 1972. *Soil Physics.* Wiley, New York, New York, USA.

Buchan, G. D. 1991. Soil temperature regime. Pages 551–612 *in* K. A. Smith and C. E. Mullins, editors, *Soil Analysis: Physical Methods.* Marcel Dekker, New York, New York, USA.

Campbell, G. S., and G. W. Gee. 1986. Water potential: miscellaneous methods. Pages 619–633 *in* A. Klute, editor, *Methods of Soil Analysis. Part 1, Physical and Mineralogical Methods.* 2d edition. Agronomy 9. American Society of Agronomy, Madison, Wisconsin, USA.

Dalton, F. N. 1992. Development of time-domain reflectometry for measuring soil water content and bulk soil electrical conductivity. Pages 143–167 *in* G. Topp, W. Clarke, D. Reynolds, and R. E. Green, editors, *Advances in Measurement of Soil Physical Properties: Bringing Theory into Practice.* Special Publication No. 30. Soil Science Society of America, Madison, Wisconsin, USA.

Gardner, W. H. 1986. Water content. Pages 493–544 *in* A. Klute, editor, *Methods of Soil Analysis. Part 1, Physical and Mineralogical Methods.* 2d edition. Agronomy 9. American Society of Agronomy, Madison, Wisconsin, USA.

Hillel, Daniel. 1980a. *Applications of Soil Physics.* Academic Press, New York, New York, USA.

Hillel, Daniel. 1980b. *Fundamentals of Soil Physics.* Academic Press, New York, New York, USA.

Holtan, H. N., C. B. England, G. P. Lawless, and G. A. Schumaker. 1968. Moisture-tension data for selected soils on experimental watersheds. *Agricultural Research Service Bulletin* 41–144, USDA.

Jury, W. A., W. R. Gardner, and W. H. Gardner. 1991. *Soil Physics.* 5th edition. Wiley, New York, New York, USA.

Klute, Arnold, editor. 1986. *Methods of Soil Analysis. Part 1, Physical and Mineralogical methods.* 2d edition. Agronomy 9. American Society of Agronomy, Madison, Wisconsin, USA.

Marshall, T. J., and J. W. Holmes. 1988. *Soil Physics.* 2d edition. Cambridge University Press, Cambridge, Massachusetts, USA.

Nobel, P. S. 1991. *Physiochemical and Environmental Plant Physiology.* Academic Press, New York, New York, USA.

Rawlins, S. L., and G. S. Campbell. 1986. Water potential: thermocouple psychrometry.

Pages 597–618 *in* A. Klute, editor, *Methods of Soil Analysis. Part 1, Physical and Mineralogical Methods.* 2d edition. Agronomy 9. American Society of Agronomy, Madison, Wisconsin, USA.

Smith, K. A., and C. E. Mullins, editors. 1991. *Soil Analysis: Physical Methods.* Marcel Dekker, New York, New York, USA.

Taylor, S. A., and R. D. Jackson. 1986. Temperature. Pages 927–940 *in* A. Klute, editor, *Methods of Soil Analysis. Part 1, Physical and Mineralogical Methods.* 2d edition. Agronomy 9. American Society of Agronomy, Madison, Wisconsin, USA.

Topp, G., W. Clarke, D. Reynolds, and R. E. Green, editors. 1992. *Advances in Measurement of Soil Physical Properties: Bringing Theory into Practice.* Special Publication No. 30. Soil Science Society of America, Madison, Wisconsin, USA.

Whalley, W. R., T. J. Dean, and P. Izzard. 1992. Evaluation of the capacitance technique as a method for dynamically measuring soil water content. *Journal of Agricultural Engineering Research* 52:147–155.

4

Soil Structural and Other Physical Properties

Edward T. Elliott
Justin W. Heil
Eugene F. Kelly
H. Curtis Monger

Soil structure influences ecosystem processes by controlling the storage and flow of water, gases, and heat, by controlling the availability of nutrients, and by providing the spatial habitat for organisms. Soil structure can be characterized in terms of particles and their arrangements, their strength of aggregation, and pore space characteristics.

Many soil processes are highly correlated with bulk density, such as water infiltration and movement, heat transfer, aeration, and root penetrability. Particle size distributions affect water storage, heat transfer, and structural stability. Analyses of soil aggregation provide information on microbial activity, since microbial activity contributes to building and maintaining the structural stability of soils. Measurements of water-filled pore space are used to estimate limitations to microbial activity.

Available Methods

Methods for bulk density measurements and particle size analysis are common and widely used. The various methods are described and compared clearly in Klute (1986). In most cases the intact-core method for measuring bulk density is appropriate and therefore recommended here. Depending on soil structure and texture, however, other methods may be necessary. For example, it may not be possible to obtain an intact, uncompacted core because of rocky, sandy, or heavy clay soils. The excavation method is more appropriate for rocky soils and sandy soils, and the clod method may be more appropriate for heavy clay soils. By also measuring field water content, it is possible to use bulk density measurements to calculate the percent water-filled pore space (WFP), which provides valuable information on limits to microbial activity.

The hydrometer method is widely used for soil textural analysis, but the pipette method is often preferred because it is a more direct measurement. However, unless particle density is also measured, the pipette method is not significantly more accurate than the hydrometer method, which we recommend because it provides equivalent results with less effort. A modification of the hydrometer method also enables measurement of particulate organic matter with slightly more effort.

Aggregate stability and size distribution measurements are also reviewed by Klute (1986) and more recently by Beare and Bruce (1993). It is possible to characterize aggregation in many ways, and standard methods are not as widely accepted as those for bulk density and texture. Measurements can be made on dry or wet aggregates, and different types of forces can be applied to measure aggregate strength. Particular methods are used to provide different types of information, depending on the purpose of measuring aggregate strength. For example, measuring the destruction of aggregates caused by rapid wetting of dry aggregates (slaking) is related to erosion potential in many soils. In another procedure (Kemper and Rosenau 1986), a selected size class of aggregates is carefully wetted, and then wet sieved through a single sieve to determine the fraction of aggregates that is destroyed through sieving. Another common procedure is to slowly wet the whole soil via capillary or vapor action and then wet sieve in a series of sieves to measure the size distribution of aggregates. We recommend this latter procedure (described in the Aggregate Size Distribution section, below) because it also provides an aggregate size distribution, which can provide a basis for making inferences about the influence of soil organic matter and other factors on aggregation. We have also included calculations for pore size distributions, which can be calculated based on water retention data obtained in Chapter 3, this volume, without any additional measurements.

Soil Bulk Density, Total Porosity, and Water-Filled Pore Space

The core method of bulk density is applicable to a wide variety of soils. This method is appropriate for any soil into which a core can be driven without considerable compaction and then removed intact. If it is not possible to take a field core, first consider using the excavation method, which is also described here. If the soil is a heavy clay, and neither of these two methods will work, refer to Blake and Hartge (1986a) for the clod procedure. Sampling frequency will depend on the type and use of the soil. Bulk density may change during the year, especially in tilled or trafficked sites. Bulk density may also change with moisture content in shrinking or swelling soils.

Materials

Hammer-type core sampler with thin-walled metal sleeves that slide into the coring tube; this type of sampler is available commercially. It should be at least 5 cm in diameter, but diameters up to 15 cm are acceptable. The length should be appropriate for the depth to be sampled. Alternately, thin-walled metal cylinders such as aluminum irrigation pipe sharpened at one end can be used together with an appropriate-sized mallet.

Procedure

The following procedure assumes that a hammer-type core sampler is being used. The coring tube of this sampler should contain thin-walled sleeves that slide out of the coring tube with the contained soil. The sleeves may be sectioned so that the soil core can be more conveniently cut into sections of appropriate length. To avoid compressing the soil while sampling, it is best to sample at moderate water contents. At higher water contents, the soil tends to stick to the cylinder, causing compaction, and at lower water contents it is difficult to drive the cylinder into the ground.

1. Clean litter from the soil surface so that the cylinder can pass freely into the mineral soil.
2. Drive the cylinder into the soil to the desired depth. While dropping the hammer, it is helpful to keep a firm downward pressure on the core assembly to minimize soil disturbance.
3. Either excavate around the cylinder (especially when a cylinder without a sleeve and hammer is used) or lift out the cylinder. It is helpful to insert a shovel or trowel under the cylinder to ensure that the soil does not fall out the bottom of the core. Check to make sure that the soil level inside the core is the same as outside. If it is noticeably lower, then compaction has occurred and one should consider discarding the core. If compaction is minor and limited to within the core, bulk density and derivative values will not be compromised; if, on the other hand, compaction has pushed soil below the core tip, then bulk density will be underestimated. If it is not possible to obtain uncompacted cores, then use the excavation method described in the section "Special Considerations," below.
4. Remove the cylinder from the sleeve and cut the soil flush with the bottom (for surface samples) or both ends (for sectioned or subsurface samples) using a knife or spatula.
5. If it is necessary to keep the core intact (for other analyses), it is helpful to place plastic or rubber disks on the ends of the core and secure them with rubber bands. If water-filled pore space is to be measured, cores should be sealed with polyethylene film or another barrier to water loss.
6. Oven-dry the soil at 105 °C overnight until there is no more mass loss, and weigh.

Calculations

Bulk Density

$$\text{Bulk density (g/cm}^3) = W \,/\, V$$

where

W = oven-dry soil weight in grams
V = volume of core in cm^3

Bulk density typically ranges between 0.6 and 1.8 g/cm^3 and more typically ranges between 1.0 and 1.4 g/cm^3.

Total Porosity

Total porosity (S_t) can be calculated from bulk density measurements assuming a particle density of 2.65 g/cm^3 for most mineral soils. This density value is appropriate for most soils except those that are volcanically derived; when in doubt, check directly (Blake and Hartge 1986b).

$$\text{Total porosity (\%)} = [1 - (\text{bulk density / particle density})] \times 100$$

where

particle density = 2.65 g/cm^3

Water-Filled Pore Space (WFP)

When field water content is also measured, the percent WFP can be calculated. Linn and Doran (1984) show that microbial activity is at a maximum at around 60% WFP. Above this WFP value, microbial activity may be limited by reduced aeration; below this value, microbial activity may be limited by water availability.

$$\%\ \text{WFP} = [P_W \times (D_B / S_t)] \times 100$$

where

% WFP = percent water-filled pore space
P_W = water content ([g water / g dry soil] × 100)
D_B = bulk density (g/cm^3)
S_t = total porosity (%)

Special Considerations

When an uncompacted core cannot be obtained with the coring method, an alternative is the excavation method. It is accurate and can be used on almost any soil:

1. Excavate a volume of soil, approximately 10 cm × 10 cm × 10 cm. Save the soil for mass measurement.
2. Fill the excavated hole with either sand or water to determine the volume of soil excavated. First determine the volume of an excess amount of sand or water that will be used to fill the hole. Sand may be poured directly into the ex-

cavation. Alternatively, water can be poured in after the hole is lined with thin plastic film. Fill the excavation until the sand or water is flush with the surface of the soil.

3. Determine the volume of sand or water that remains after filling the excavation. Calculate the volume of material used (V) by subtracting this amount from the original volume.
4. Oven-dry the soil at 105 °C overnight until there is no more mass loss and weigh it. The bulk density calculation is the same as for the core method described earlier.

Particle Size (Soil Texture) Analysis

Soil particle size analysis (also called texture or sand-silt-clay analysis) by the hydrometer or Buoycous method is based on the principle that soil particles settle out of a solution at a rate proportional to their size. Sand particles settle more quickly than silt-sized particles, which settle more quickly than clay.

In order to simplify this procedure we make two assumptions that will be appropriate in most circumstances. First we assume that the analysis temperature is 20 °C. Next we assume a clay content of 5–50%, which allows us to use a sedimentation parameter calculated using a clay content of about 25%. These assumptions allow us to take just one hydrometer reading at 12 hours and simplify calculations. However, if the clay content is outside of the range of 5–50% special considerations apply (see the section "Special Considerations," below).

Sample pretreatment may be necessary if the soil contains an appreciable amount of carbonates or iron oxides, or has a very high (>5%) organic matter content. If it is necessary to remove carbonates, soluble salts, iron oxides, or organic matter, refer to Gee and Bauder (1986) for procedures.

Since the coarse particulate organic matter (CPOM) method (Cambardella and Elliott 1992) can be very easily combined with particle size analysis, we include a description of the method here (see also Chapter 5, this volume). If one is not interested in CPOM measurements, these steps can be ignored.

Materials

1. 250 mL plastic bottles
2. 5% sodium hexametaphosphate. It is not necessary to use reagent-grade; some dishwasher additives (e.g., Calgon) are 5% sodium hexametaphosphate.
3. Small-diameter 53 μm sieve
4. Sedimentation cylinders (1 L). These can be purchased from major scientific supply houses. Uncalibrated versions are typically less expensive and can be calibrated with 1000 g distilled water.
5. Preweighed aluminum pans
6. Sedimentation plunger. This can be purchased or constructed from a metal rod threaded on one end, several washers, a nut, and a piece of semistiff black rub-

ber (e.g., a truck inner tube) or plastic disk of a diameter about 4 mm smaller than the sedimentation cylinder.
7. Mortar and pestle or ball mill (for CPOM analysis only)
8. Reciprocal shaker

Procedure

1. Weigh 40 g of soil into 250 mL plastic bottles. Also set aside a sample for moisture analysis (see Chapter 3, this volume).
2. Add 100 mL of 5% sodium hexametaphosphate (a dispersal agent) to the bottles and shake overnight on a reciprocal shaker.
3. Wash the sample through a 53 μm sieve using <1 L distilled water. Collect the sample and water that passes the sieve and place in a 1 L sedimentation cylinder. Place the captured particles (consisting of sand and coarse particulate organic matter) in preweighed aluminum pans and dry at 60 °C.
4. Add distilled water to bring the suspension in the sedimentation cylinder to 1000 mL. Stir the suspension for 30 seconds with the sedimentation plunger, using long strokes. A drop of amyl alcohol can be added to reduce foam if present.
5. Add 100 mL of 5% sodium hexametaphosphate to a second sedimentation cylinder, add distilled water to 1000 mL, and mix. This cylinder will serve as a blank control.
6. Twelve hours after stirring, insert a hydrometer and record the level of the hydrometer both in the soil suspension (R_{12}) and in the blank (R_L) cylinders.
7. If coarse particulate organic matter (CPOM) is to be determined, pour the silt/clay slurry in the sedimentation cylinders into a large pan. Evaporate the water from the slurry at 60 °C in a forced-air oven. If CPOM is not to be determined, a faster method for drying the slurry is to add several milliliters of 0.5 mol/L $MgCl_2$ to the sedimentation cylinder to flocculate clays, wait 24 hours for the clay to settle, and then siphon off and discard the clear supernatant, wash the slurry into a pan, and dry at 60 °C.
8. For CPOM determination, determine total C and N in the sample (NP_C) using the dry combustion method described in Chapter 5, this volume. At the same time, determine organic C and N in a whole soil sample. See Chapter 5, this volume, for calculations.

Calculations

Particle Size Density

$$\text{Sand percentage} = (\text{sand mass} / C_o) \times 100$$
$$\text{Silt percentage} = 100 - (\text{sand percentage} + \text{clay percentage})$$
$$\text{Clay percentage} = (C/C_o) \times 100$$

where

Sand mass = dry mass of material captured on the 53 μm seive
$C = R_{12} - R_L$
R_{12} = uncorrected 120 minute hydrometer reading
R_L = hydrometer reading of the blank solution
C_o = dry mass of the soil sample

Special Considerations

For a more accurate clay determination, take hydrometer readings at 1.5 ($R_{1.5}$) and 24 (R_{24}) hours. The percentage of 2 μm particles ($P_{2\mu m}$) is calculated using the following equations that assume a temperature of 20 °C (refer to Gee and Bauder [1986] for calculations using other temperatures).

$$P_{2\mu m} = [m \times \ln(2/X_{24})] + P_{24}$$

where

$m = (P_{1.5} - P_{24}) / \ln(X_{1.5}/X_{24})$
$X_{24} = 1000 \times [0.00019 \times (-0.164\, R_{24} + 16.3)]^2 \times t_{24}^{\ 2}$
$X_{1.5} = 1000 \times [0.00019 \times (-0.164\, R_{1.5} + 16.3)]^2 \times t_{1.5}^{\ 2}$
$P_{24} = [(R_{24} - R_L) / C_o] \times 100$
$P_{1.5} = [(R_{1.5} - R_L) / C_o] \times 100$
t_{24} = 1440 minutes
$t_{1.5}$ = 90 minutes

Aggregate Size Distribution

Methods for measuring waterstable aggregates have been reviewed by Beare and Bruce (1993). The following procedure from Elliott (1986) is for wet sieving of the whole soil. The stability of aggregates is very sensitive to the wetting procedure, so it is important to follow the procedure carefully to get consistent results. This procedure is written for a hand sieving method, although it has been automated (Beare and Bruce 1993).

Materials

1. Ceramic pressure plate
2. 8 mm, 2 mm, 250 μm, and 53 μm sieves (other sizes may be used as required)
3. Pans in which to sieve
4. Aluminum pans in which to dry sieved aggregates

Procedure

1. Collect an intact soil sample, preferably of 8 cm diameter or greater to avoid edge effects on aggregate destruction. After air drying, the sample should be

carefully broken apart and gently passed through an 8 mm sieve. See the discussion of air drying in the section "Special Considerations," below.

2. Weigh 100 g of soil. Take a moisture sample at the same time (see Chapter 3, this volume) to calculate the oven-dry weight of the air-dried soil (W).
3. Air-dried soil may be either wet sieved directly, with aggregate disruption occurring due to soil slaking, or prewetted to near field capacity, in which case a higher proportion of aggregates remain intact. In the former case, slaking occurs due to the buildup of air pressure in the center of aggregates as water enters capillary pores and aggregates are disrupted along planes of weakness. In the latter case, the air is slowly displaced with water and the slaking is avoided. A comparison of both methods provides a measure of aggregate stability (see the section "Special Considerations," below).

   For capillary prewetting, place the soil on a ceramic pressure plate with a hanging water column about 2 cm below the plate surface. Use deionized water to slowly wet the soil on the ceramic plate. There should never be any standing water on the ceramic plate. This wetting can be very slow and may take as long as a day to wet the aggregates; do not hurry this procedure. Where a pressure plate is not available, an alternative method for wetting soils is given in the section "Special Considerations," below.
4. Place the soil in a 2 mm sieve (other sieve sizes may be used as required). Fill a pan of slightly larger diameter than the sieve with water to a depth sufficient to submerse the sieve screen to 4 cm. Submerse the soil on the screen in the pan for 5 minutes before beginning to sieve. Then gently move the sieve 3 cm vertically 50 times over a period of 2 minutes. The sieve screen should just break the surface of the water so that aggregates are being raised slightly out of the water on each stroke but the screen is barely (1–2 mm) moving above the surface of the water.
5. Transfer the material remaining on the sieve to another container, dry at 60 °C overnight and weigh this fraction ($W_{>2.0}$).
6. Place the 250 μm sieve in a pan filled with water as for the 2 mm sieve; fill with the slurry that passed through the 2 mm sieve. Collect the captured 250 μm fraction, dry, and weigh ($W_{2.0mm}$).
7. Repeat step 6 for the 53 μm sieve, drying and weighing the captured material ($W_{250\mu m}$).
8. Dry and weigh the slurry that passed through the 53 μm sieve to obtain the <53 μm fraction ($W_{<53\mu m}$)

Calculations

$$Agg_{size\ class} = [W_{size\ class}/W] \times 100$$

where

$Agg_{size\ class}$ = percent of water-stable aggregates in a particular size class
$W_{size\ class}$ = dry weight of water-stable aggregates in a particular size class
W = oven-dry weight of whole soil

Special Considerations

The initial moisture content of the soil sample will strongly affect the aggregate size distribution obtained, since soil moisture influences the amount of slaking (aggregate bursting) that occurs when wet sieving. Elliott (1986) found it very useful to compare slaked versus capillary-wetted size distributions for whole soils to gauge the relative stability of specific aggregate size classes. To ensure standard comparisons, we suggest air drying all soil samples before analysis, since the slaked sample requires air drying and the use of field-moist samples introduces considerable variability into measurements of slaked aggregate size distributions. However, Beare and Bruce (1993) have obtained useful results by capillary wetting field-moist samples and also found that these distributions vary somewhat from those obtained by either slaking or capillary wetting of comparable air-dried samples. For some soils, particularly those that rarely come to a near-air-dried state in the field, it may be inappropriate to air-dry samples for aggregate analysis. In this case, one should capillary-wet field-moist samples, although one will not then be able to compare slaked versus air-dried aggregate distributions. However, other means of disruption may be employed so that comparison of different aggregate distributions can be made.

If one does not have access to a ceramic pressure plate for capillary wetting, filter paper can be used instead. Some additional error may be introduced, but the results are generally comparable. Place 50 g soil on an 11 cm filter paper in a large petri plate. Calculate the amount of water required to bring the sample to field capacity and add an additional 5%. With the soil in the center of the paper, slowly add the water around the edge of the paper, a few drops at a time. After all the water is added, place the cover on the petri plate and store in a refrigerator overnight until ready for wet sieving. Transfer the soil to the first sieve and proceed as described earlier.

Pore Size Distributions

Pore size distributions can be directly calculated from water desorption curves (Danielson and Sutherland 1986). Although other methods are available (Darbyshire et al. 1993), we recommend the water desorption method because the data available from the soil water characteristic (or desorption) curve in Chapter 3, this volume, can be used to calculate pore size distributions without any additional measurements.

Materials and Procedure

Produce a soil water characteristic curve as per instuctions in Chapter 3, this volume.

Calculations

The largest size of pore that still holds water at given water tensions can be calculated at each desorption point by

$$r_p = 2000 \times d \,/\, DP$$

where

r_p = largest neck radius of pore (μm) filled with water at DP
d = surface tension of water (72.75×10^3 J μm^2)
DP = hydraulic pressure in kPa

The percent pore space (S) in a given size class expressed on a whole-soil basis is calculated by

$$S_{r1,\, r2} = [(W_2 - W_1)/V_t] \times 100$$

where

$S_{r1,\, r2}$ = pores with neck radii between r_1 and r_2 as a percent of soil volume
W_1 = mass of core at pressure corresponding to r_1 in grams
W_2 = mass of core at pressure corresponding to r_2 in grams
V_t = total volume of the soil core in cm^3

Soil Pore Architecture

Thin and polished section analysis is a method for directly viewing the undisturbed architecture of soil pores and particles. This is accomplished by soaking a soil with low-viscosity resin that then is allowed to harden. Once the resin hardens, the impregnated soil is cut with a rock saw, polished, and viewed macroscopically (polished section), or mounted on a microscope slide, ground to about 30 μm, and viewed microscopically (thin section).

Both types of sectioning provide a means to measure pore areas, shapes, and tortuosities. Pores are made more visible by adding dyes or fluorescent chemicals to the epoxies (FitzPatrick 1993). Image analysis software can be used to characterize soil porosity numerically. Three-dimensional images can be created using software that stacks several two-dimensional cross sections together. In addition to soil porosity, this method provides a means of viewing the spatial distribution of roots, fauna, microorganisms, soil organic matter, and minerals.

Materials

1. Kubiena boxes
2. Acetone
3. Spurr's Low-Viscosity Embedding Media
4. Rock saw
5. Rock polisher
6. Microscope slide
7. Epoxy

Procedure

Several methods for preparing soil thin sections have been developed by geologists and soil micromorphologists (Bulloch et al. 1985; FitzPatrick 1993; Humphries 1992; Wilding and Drees 1990). Preparing good polished blocks and thin sections takes some skill, so refer to these references for more details. The method described here is used in the pedology lab at New Mexico State University.

1. Carefully collect soil peds or soil samples in Kubiena boxes and transport to the lab without crushing. Aluminum foil, tissue, or egg cartons may be helpful for transporting fragile soil peds.
2. Place soil sample in a disposable container. If soil sample is moist, add acetone first to displace water.
3. Add Spurr's Low-Viscosity Embedding Media. Blue dyes or fluorescent chemicals can be added to enhance observation of porosity (FitzPatrick 1993). If possible, deaerate the embedding resin under vacuum before adding it to the sample, and add the resin while the sample is under vacuum. This will improve impregnation and reduce disturbance of structure. Alternatively, place the soil sample with resin in a vacuum chamber and apply vacuum until frothing stops.
4. Transfer sample to oven to cure (apply heat according to Spurr's directions).
5. After hardening, cut sample with rock saw to create a smooth, flat surface. If the sample is to be mounted for thin sectioning, cut it in the shape of a block, with dimensions on one side that are just smaller than the size of the glass slide.
6. Polish the surface to be mounted with a commercial rock polisher or on a glass plate with wettable sandpaper or grinding compound. Begin with coarse grit. Progressively polish with finer grit until surface has a mirror finish. The surface must be flat for mounting, so use care not to round the edges.
7. Thoroughly clean the surface to be mounted. Using a stiff paper clip or a weight, clamp the section on a microscope slide with epoxy and let it cure. Use care not to entrap air bubbles.
8. Cut the bulk of the sample off the slide. Grind and polish the remaining sample to the desired thickness. Apply desired stains to enhance observation of mineralogy or biological components (Bulloch et al. 1985).
9. Coverslips can be attached permanently with epoxy or temporarily with a drop of oil.
10. The polished block or thin section can be photographed, and the image can then be digitized using a scanner to characterize the pore space using image analysis software. Images can be captured using petrographic microscopes, cameras with magnifying lenses under UV light (using a UV dye in the resin), or using electron microscopes at low magnification.

References

Beare, H. B., and R. R. Bruce. 1993. A comparison of methods for measuring water-stable aggregates: implications for determining environmental effects on soil structure. *Geoderma* 56:87–104.

Blake, G. R., and K. H. Hartge. 1986a. Bulk density. Pages 363–376 *in* A. Klute, editor, *Methods of Soil Analysis. Part 1, Physical and Minerological Methods.* 2d edition. American Society of Agronomy, Madison, Wisconsin, USA.

Blake, G. R. and K. H. Hartge. 1986b. Particle Density. Pages 377–382 *in* A. Klute, editor, *Methods of Soil Analysis. Part 1, Physical and Minerological Methods.* 2d edition. American Society of Agronomy, Madison, Wisconsin, USA.

Bulloch, P., N. Fedoroff, A. Jongerius, G. Stoops, and T. Tursina. 1985. *Handbook for Soil Thin Section Description.* Waine Research Publications, Wolverhampton, England, UK.

Cambardella, C. A., and E. T. Elliott. 1992. Particulate soil organic-matter changes across a grassland cultivation sequence. *Soil Science Society of America Journal* 56:777–783.

Cambardella, C. A., and E. T. Elliott. 1993. Methods for physical separation and characterization of soil organic matter fractions. *Geoderma* 56:449–457.

Danielson, R. E., and P. L. Sutherland. 1986. Porosity. Pages 443–462 *in* A. Klute, editor, *Methods of Soil Analysis. Part 1, Physical and Minerological Methods.* 2d edition. American Society of Agronomy, Madison, Wisconsin, USA.

Darbyshire, J. F., S. J. Chapman, M. V. Cheshire, J. H. Gauld, W. J. McHardy, E. Paterson, and D. Vaughan. 1993. Methods for the study of interrelationships between microorganisms and soil structure. *Geoderma* 56:3–23.

Elliott, E. T. 1986. Aggregate structure and carbon, nitrogen, and phosphorus in native and cultivated soils. *Soil Science Society of America Journal* 50:627–633.

FitzPatrick, E. A. 1993. *Soil Microscopy and Micromorphology.* Wiley, Chichester, UK.

Gee, G. W., and J. W. Bauder. 1986. Particle-size analysis. Pages 383–412 *in* A. Klute, editor, *Methods of Soil Analysis. Part 1, Physical and Minerological Methods.* 2d edition. American Society of Agronomy, Madison, Wisconsin, USA.

Humphries, D. W. 1992. *The Preparation of Thin Sections of Rocks, Minerals, and Ceramics.* Microscopic Handbook 24. Oxford University Press, Oxford, UK.

Kemper, W. D., and R. C. Rosenau. 1986. Aggregate stability and size distribution. Pages 425–442 *in* A. Klute, editor, *Methods of Soil Analysis. Part 1, Physical and Minerological Methods.* 2d edition. American Society of Agronomy, Madison, Wisconsin, USA.

Klute, A., editor 1986. *Methods of Soil Analysis. Part 1, Physical and Minerological Methods. 2d edition.* American Society of Agronomy, Madison, Wisconsin, USA.

Linn, D. M., and J. W. Doran. 1984. Effect of water-filled pore space on carbon dioxide and nitrous oxide production in tilled and nontilled soils. *Soil Science Society of America Journal* 48:1267–1272.

Wilding, L. P., and L. R. Drees. 1990. Removal of carbonates from thin sections for microfabric interpretations. *Developments in Soil Science* 19:613–620.

Part II

Soil Chemical Properties

5

Soil Carbon and Nitrogen

Pools and Fractions

Phillip Sollins
Carol Glassman
Eldor A. Paul
Christopher Swanston
Kate Lajtha
Justin W. Heil
Edward T. Elliott

Soil organic matter (SOM) is important as a major source of most nutrients, especially nitrogen (the one major plant nutrient not supplied by weathering), and as a source of cation exchange capacity, especially in highly weathered soils. SOM also contributes to good soil structure, thus promoting drainage, water-holding capacity, aeration, and root penetration, and provides strong control of soil pH, especially in highly weathered soils. In addition, soil organic matter is a major global C source and sink. Jenkinson et al. (1991) calculated that an increase in global temperature of 0.03 °C/yr could increase CO_2 release from soil by about 1 Gt C/yr over the next 60 years; by comparison, fossil fuel burning currently releases about 5 Gt C/yr.

Soil nitrogen availability limits plant growth in many terrestrial ecosystems, yet increased N deposition due to atmospheric pollution has been implicated in the decline of forest productivity in some heavily industrialized parts of the northern temperate zone (Schulze 1989). The relation between total soil N and nitrogen availability to plants is still unclear, in large part because the nature of soil organic N is only now beginning to be understood.

In this chapter we present standard methods for determination of total soil C and N and measurement of total organic and inorganic soil C. Methods are also presented for separating SOM into biologically meaningful fractions. Field sampling methods and soil standards are discussed in Chapter 2, this volume.

Available Methods

Total Carbon and Nitrogen Analysis

Dry (Dumas) combustion is the most suitable method for routine analysis of total C and N in soil, litter, and plant samples. The Walkley–Black technique, a wet combustion method, is no longer recommended because it can underestimate soil C by 20–30% (Nelson and Sommers 1982) and can give spurious results in highly reduced soils unless precautions are taken (Snyder and Trofymow 1984). The Walkley–Black method is also laborious and produces toxic wastes. Measuring carbon content by measuring mass loss following high-temperature combustion (the loss-on-ignition, [LOI] method) is easily performed but has serious shortcomings. LOI measures volatilization of all material, including the water retained in clay structures at standard ovendrying temperature (105 °C). Allophane, for example, was found to lose 26% of its mass upon heating from 100 to 800 °C; gibbsite lost 54% (Gardner 1986). Nonetheless, because of its low cost, LOI may have a place in the soils lab when large numbers of similar samples must be processed. LOI should always, however, be checked against a dry-combustion method.

Soil Organic Matter Fractionation

Because SOM is so heterogeneous, much effort has focused on methods for isolating biologically meaningful pools ("separates"), which can then be analyzed for composition or biological availability. There are two general approaches: chemical extraction and physical separation. The latter can be based on particle size, particle density, degree of physical protection, or a combination of the three. It is also possible to estimate the size of fast turnover, slow turnover, and inactive SOM pools based on biological assays (see Chapter 13, this volume).

Chemical fractionation relies traditionally on solubility in acid and base, producing fractions termed *humic acid* (base extractable, acid insoluble), *fulvic acid* (base extractable, acid soluble), and *humin* (insoluble in base). Unfortunately, these procedures do not produce separates that differ consistently in their biological properties. Despite many studies, none have documented clear or consistent variation in age or amount of these chemical fractions by soil type or in response to management (Duxbury et al. 1989). Although the procedure continues to give useful results for fractionating dissolved organic C (DOC) from soil solution and streams, as a procedure for soils, it is not recommended here. Two chemical fractionation methods that we do recommend are extraction in cold water and acid hydrolysis (boiling 6N HCl). The former appears to provide a useful measure of microbially degradable C, while acid hydrolysis leaves a residue that gives a consistently old ^{14}C date and can thus be regarded as a slow turnover pool (Paul et al. 1997).

Physical fractionation methods separate the soil into component particles, many of which are aggregates consisting of smaller particles bound together by various agents of different binding strengths (Tisdall 1996). Aggregates can be dispersed with chemicals that remove binding agents or by mechanical action such as shaking or ultrasonic vibration. After dispersion the particles are sorted by density and/

or size. The effectiveness of dispersion techniques increases with the amount of energy applied. It may be convenient to think there is some ultimate dispersion point beyond which no decrease in particle size is possible, but in practice this is unachievable: (1) with enough energy even clay micelles are disrupted, and (2) many of the smallest particles can be highly reactive and reflocculate after dispersal (see Christensen 1992). The problem of achieving and maintaining dispersion is especially critical in soils rich in amorphous constituents such as allophane, ferrihydrite, and Al-OM complexes. Since complete, permanent dispersion is never possible, the effectiveness of a given method for dispersion can only be judged by the usefulness of the resulting data.

Sample pretreatment greatly affects dispersibility and thus the results of any physical fractionation procedure. Air drying soils rich in amorphous constituents often causes aggregation that cannot be reversed even with vigorous sonication (e.g., Sollins 1989). For other soils (e.g., quartz sand soils), fractionation results may be essentially unaffected even by oven drying at 105 °C. For both chemical and physical fractionation methods, the resulting pools are operationally defined (defined only by the method). There is thus no single right answer; the usefulness of the results can only be judged by how biologically meaningful they prove. Biologically meaningful could mean, for example, that the separates (i.e., the fractions after separation) play distinctly different roles in any of the functions listed earlier for soil organic C, differ consistently across landscapes and ecosystems, or change consistently with management or disturbance.

Characterization of the Molecular Structure of SOM

Recent technological advances permit considerable characterization of the molecular structure of SOM. Broad categories are often defined, such as carbohydrates, aromatics, and alkyls, rather than specific compounds. To some extent, lignin-related compounds can be distinguished from other aromatics. These techniques can be used effectively on whole soil samples or on physically or chemically isolated soil fractions. Often they have been used to document changes in composition of the SOM with depth in profile, generally by assuming that the OM in deeper positions is older, as has been shown by ^{14}C dating (Scharpenseel and Schiffmann 1977). Some of the most intriguing results have been obtained by comparing soils of differing age, degree of weathering, or mineralogy. All SOM characterization techniques require expensive equipment and extensive training and thus are not described in detail here.

The oldest of the techniques is high-temperature pyrolysis, which splits molecules into fragments that can then be identified by mass spectrometry or other techniques (e.g., Kögel-Knabner et al. 1992). Because pyrolysis creates quite small fragments, only limited inferences are possible about the original compounds. Nonetheless, this technique has provided some of the best early evidence for the large amount of alkyl compounds in SOM (Theng et al. 1989).

^{13}C nuclear magnetic resonance (NMR) spectrometry, with cross-polarization and magic-angle spinning, quantifies the amount of C in each of several typical molecular positions in solid-phase soil samples and in aqueous SOM extracts (Kinchesh et al. 1995; Preston 1996). Most spectra can be divided into alkyl (10–45 ppm), O-alkyl (45–110 ppm), aromatic (110–160 ppm), and carbonyl (160–200 ppm) regions (Baldock and Preston 1995). Alkyl C consists of long polymethylene (CH_2) chains (e.g., fatty acids, waxes, and resins), along with varying amounts of shorter polymethylene branches and methyl groups. The O-alkyl region includes C in carbohydrates such as cellulose and hemicelluloses, as well as the oxygenated and methoxyl C in phenylpropane lignin units and some amine C. Aromatics comprise both C, H, and O substituted rings and most lignin C. The carbonyl region includes C in carboxylic, amide, ester, ketone, and aldehyde groups.

^{13}C-NMR has several limitations. First, some C atoms produce similar resonance signals despite quite different positions within organic molecules; thus some peaks are ambiguous. Second, paramagnetic elements such as iron cause interference, in effect raising detection limits for carbon. To some extent this last problem can be offset by increasing counting time, but a better solution has been to find soils that are low in iron yet still suitable for the intended study. Typical of the recent ^{13}C-NMR work on soils are indications that there is sequential conversion of carbohydrate to aromatics and then to alkyls (see Baldock and Preston 1995), a process that may be arrested or at least slowed in soils with abundant free Al (Sollins et al. 1996).

C and N Analysis by Dry Combustion

Most dry-combustion C and N (CN) analyzers oxidize samples at high temperature (approx. 1000 °C), then measure the CO_2 and N gases evolved by infrared gas absorption (IRGA) analysis or gas chromatography (GC). Depending on the individual instrument, the maximum allowable sample size may be as small as 20 μg. The maximum sample size depends on the C concentration, which may require some initial data before a strategy can be chosen. No hard-and-fast rules can be offered for sample size because the precision and accuracy needed for any individual sample depend on the overall sampling and data analysis scheme. Use of small samples, however, always requires careful attention to subsampling and especially to grinding.

Materials

1. CN analyzer
2. Tin sample capsules
3. Microbalance
4. Soil or rock mill (e.g., Spex mixer-mill)
5. Gas pressure gauge

Procedures

High-temperature multiple-sample dry-combustion analyzers are manufactured by several companies including LECO and Carlo-Erba. The Carlo-Erba NA 1500 elemental analyzer is discussed here. The detection limit is 10 ppm, and measurements are reproducible to better than ±0.1% absolute value. Sample mass needed for analysis may range from 0.5 to 30 μg depending on the nature of the material. Because such a small sample is needed, material must be homogenized thoroughly by grinding several hundred grams of soil to pass a 40- to 60-mesh screen. A typical sample run comprises one or two "bypass" samples of high concentration to condition the columns, two "blanks" consisting of empty tin sample cups, three standards of known C and N composition to calibrate the instrument (EDTA is used commonly), and three to five check standards scattered throughout the sample run. Typically, 39 unknowns can be included in one run of 50 samples. Extra sample trays may be purchased and set up to make consecutive runs more convenient.

Samples are weighed into tin capsules, which are loaded into an autosampler that drops the capsule plus sample into a combustion column maintained at 1020 °C. The sample and container are flash combusted in a temporarily enriched atmosphere of O_2. The combustion products are carried by a carrier gas (helium) past an oxidation catalyst of chromium trioxide kept at 1020 °C inside the combustion column. To ensure complete oxidation, a layer of silver-coated cobalt oxide is placed at the bottom of the column. This catalyst also retains interfering substances produced during the combustion of halogenated compounds. The combustion products (CO_2, CO, N, NO, and water) pass through a reduction reactor in which hot metallic copper (650 °C) removes excess O_2 and reduces N oxides to N_2. These gases, together with CO_2 and water, are next passed through magnesium perchlorate to remove water, then through a chromatographic column to a thermal conductivity detector. The detector generates an electrical signal proportional to the concentration of N or C present. This signal is graphed on a built-in recorder and ported to a computer, which integrates the area under each curve and converts it to concentrations after each sample is run.

Before the start of each run, pressure should be checked to ensure against gas leaks. Gas flow rates (helium, oxygen, and air) are checked with a stopwatch and set to the correct values. Routine maintenance involves removing the slag (residue from combustion of the tin sample capsules) from the top of the combustion column after 150 samples, then refreshing the top 10 cm of the column with CrO_3. The combustion column and its chemicals can be used for 350–425 samples. The reduction column can be used for up to 900 samples, or until its copper is three-fourths spent as indicated by change to a black color. The moisture trap must be changed every 300–350 samples.

Calculations

Most CN analyzers read out directly in concentration units.

Special Considerations

The need to measure both bulk density and stone content cannot be overstated (see Chapters 1 and 4, this volume). Typically, only the <2 mm soil is analyzed for C and N. We recommend, however, that at least one subsample of the 2–10 mm material be ground and analyzed. If this coarse soil fraction contains significant C and

N, then the fraction should be analyzed in all field samples. The 2–10 mm fraction will usually contain coarse roots or other woody debris. Since these pools may be the focus of work by other investigators at a site (see Chapters 11 and 19, this volume), care must be taken to ensure that sites do not double-count C and N when calculating ecosystem totals on an areal basis.

Volcanic rocks tend to sorb C and N strongly. Thus at volcanic sites we recommend that at least a subsample of the >10 mm fraction be analyzed also, focusing on coarse aggregates (e.g., shot) and the most weathered rocks. Material will need to be weighed wet and a subsample dried (see Chapter 3, this volume).

Carbonates

Soils in arid and semiarid environments often contain large amounts of carbonates, as do soils derived from carbonate parent materials regardless of climate. Carbonates decompose during dry combustion, releasing CO_2 that is counted as soil C during dry-combustion analysis. Carbonates, however, do not play any of the roles listed previously for soil organic C; thus the amounts of carbonate (inorganic) and organic C need to be measured separately.

Most carbonate methods involve treatment with strong mineral acid to convert carbonates to CO_2. The residue is then analyzed to provide an estimate of the organic C, and the inorganic C is estimated by difference. The acid, however, can solubilize organic as well as carbonate C, so the solution must be evaporated and the soil completely dried to fully measure the organic C. H_2SO_4 is not recommended because it is too strong an oxidant, cannot easily be removed by evaporation, and may damage CN analyzers. HCl can be removed by evaporation, but the residual Cl^- likewise damages CN analyzers. Some researchers remove the Cl^- by washing, but this also removes water-soluble organic C.

At present the simplest solution is to use phosphoric acid and measure the CO_2 released with a gas chromatograph or infrared gas analyzer. Alternatively, small samples of soil can be weighed into metal boats (CN-analyzer sample holders), the acid added directly to the boat, and the boat and sample run through a CN analyzer once the released CO_2 has bubbled off and the sample has redried.

Recent work with ^{13}C abundance indicates that grinding is necessary in some soils rich in carbonates to ensure complete carbonate removal (C. Van Kessel, personal communication). The grinding step may be omitted if an initial trial shows that it does not increase the amount of C removed by the acid hydrolysis.

Dissolved Organic Nitrogen (DON)

Dissolved organic nitrogen (DON) is an important component of the soil solution, and in many undisturbed forests it is the main vector for N loss from the soil via leaching (Sollins and McCorison 1981; Lajtha et al. 1995). Precipitation may also contain a significant amount of DON, an input that has largely been ignored by ecologists.

Several methods exist for the analysis of DON and total N, including Kjeldahl digestion. We prefer a persulfate digestion technique because it is simpler, requires less specialized equipment, does not produce a toxic waste that is difficult to dispose, and is more sensitive than the standard Kjeldahl technique. In addition, persulfate N includes DON, nitrate, and ammonium, whereas Kjeldahl N includes only DON and ammonium.

A standard method for total N analysis of seawater is given by D'Elia et al. (1976). This procedure was modified by Ameel et al. (1993) to increase sensitivity in freshwater samples; we recommend the use of the Ameel et al. (1993) procedure, which is outlined here for the analysis of soil solution. Dissolved organic phosphorus (DOP) can be analyzed with the same extract, as described later in this chapter.

This method presumes that soil solution has already been collected, via either extraction (see Chapter 6, this volume) or lysimetry (see Chapter 9, this volume).

Materials

1. 40 mL borosilicate glass screw-top vials with rubber-lined caps.
2. Teflon cap liners (optional)
3. 0.148 M $K_2S_2O_8$; dissolve 20 g $K_2S_2O_8$ in 500 mL deionized water
4. 3 M NaOH; dissolve 12 g NaOH in 100 mL deionized water
5. Standard organic solutions; at least 4 concentrations including 0 μg N/L. For DON, use an organic N standard such as urea or EDTA; for DOP, use an organic P standard such as ATP.
6. Standard NO_3^-–N solutions at the same concentrations as for standard organic solutions, described earlier, including 0 μg N/L
7. Spike solution; a 200 μg N/L^1 standard organic solution
8. A means for measuring NH_4^+–N, NO_3^-–N, and (as needed) orthophosphate in solution samples (see Chapter 7, this volume, for phosphorus)

Procedure

1. Set aside a separate predigestion aliquot for inorganic N analysis (NH_4^+ and NO_3^-)
2. Add 15 mL of a nitrate standard, organic standard, spiked sample, or sample to a 40 mL digestion vial. Spiked samples and organic standards are used to determine digestion efficiency (see the section "Special Considerations," below); add 15 μL of the organic spike solution to the 15 mL of sample in a digestion vial.
3. Add 5 mL of the 0.148 mol/L persulfate solution and 0.25 mL of the 3 mol/L NaOH solution to each digestion vial, and seal the vials.
4. Autoclave at 121 °C and 17 psi for 55 minutes.
5. Cool, then add 0.25 mL of the 3 mol/L NaOH solution.
6. Analyze samples for NO_3^-–N using the nitrate standards also brought through the digestion procedure. If measuring DOP, also analyze for orthophosphate.
7. Analyze the predigestion samples for NO_3^-–N and NH_4^+–N.

Calculations

$$\text{DON } (\mu\text{g N/L}) = (\text{Persulfate N / Digestion Efficiency}) - \text{Inorganic N}$$

where

Persulfate N = NO_3^-–N in solution after persulfate digestion (μg NO_3–N/L); based on nitrate standards also brought through the digestion procedure.

Digestion efficiency (*DE*) can be calculated as either

(1) *DE* = (NO_3^-–N in spiked sample − NO_3^-–N in unspiked sample) / (0.200 μg/L organic-N spike); or

(2) *DE* = (NO_3^-–N in organic standard) / (NO_3^-–N in equivalent nitrate standard). Use an equivalent formula if analyzing for phosphorus. For any given sample, use the average digestion efficiency for all spiked samples within that run.

Inorganic N = NO_3^-–N + NH_4^+–N in solution prior to digestion (μg N/L); based on separate inorganic N analysis of the predigestion aliquot set aside in step 1.

Special Considerations

1. To determine digestion efficiency, 10% of all samples from each run should be analyzed with the spike solution. Alternately, compare the organic standards with the inorganic (NO_3^-–N) standards to determine efficiency (see the section "Calculations," above). Digestion efficiencies will normally be >90% but will vary among sample runs. If efficiency drops below 80%, suspect a bad batch of persulfate and discard results. Accuracy may be measured using a quality control check standard from a commercial supplier or with the organic standards.
2. All standards for colorimetric analysis are brought through the digestion procedure. Solutions are analyzed for nitrate and orthophosphate colorimetrically.
3. Seawater samples or samples high in Mg should be run according to the D'Elia et al. (1976) procedure to remove the $Mg(OH)_2$ precipitate that can form.
4. The Ameel et al. (1993) procedure assumes that there is no evaporation of liquid from the tubes; this should be verified, and if evaporation is detected, then vials must be brought to a standard volume with deionized water before analysis.

Water-Extractable Carbon

Materials

1. 200–300 mL plastic bottle
2. Centrifuge

3. 0.2 μm acid-washed polycarbonate filters
4. DOC analyzer

Procedure

1. Shake 50 g field-moist soil overnight in 150 mL distilled water (dH_2O) (200–300 mL plastic bottle).
2. Centrifuge.
3. Filter the supernatant through a 0.2 μm acid-washed polycarbonate membrane (chosen to remove virtually all organisms except viruses).
4. Analyze the solution for DOC with any of several DOC analyzers (e.g., Dohrmann, Shimadzu).

Calculations

Most CN analyzers read out directly in concentration units.

HCl—Insoluble Organic Carbon

Refluxing with 6N HCl yields a chemically resistant residue that gives much older ^{14}C dates than the bulk soil C (Paul et al. 1997; Follett et al. 1997). The acid hydrolysis solubilizes amino compounds, pectins, and most cellulose, and also releases C as CO_2 by decarboxylation. Acid hydrolysis does not degrade aromatics present in lignin or soil humic structures. Soils often contain lignin derived from recent plant residues. Acid hydrolysis, therefore, does not provide complete separation between old and new C, so it is important to remove identifiable plant residues before analysis. The amount of soil C remaining as non-acid-hydrolyzable C is related to the clay content and type of parent materials (Paul et al. 1998), and can be used to define a resistant SOM pool (see "Long-Term Respiration Potential (SOM Pool Sizes)" in Chapter 13, this volume).

Materials

1. 250 mL round-bottomed flasks with heating mantles or Kjeldahl digestion tubes with block digester
2. Reflux condensers for flasks or digestion tubes
3. Vacuum filtration unit with Whatman no. 50 or other hardened fine-pore filters or centrifuge
4. Soil or rock mill
5. NaCl solution (1.2 g/mL)

Procedure

1. Remove identifiable plant materials by flotation in NaCl (1.2 g/mL), then by hand-picking under a 20× microscope.
2. Grind soil to 100-mesh.

3. Remove any inorganic carbonates as described earlier.
4. Place soil and HCl in 250 mL round-bottomed flasks fitted with heating mantles or in Kjeldahl digestion tubes heated in a digestion block. Reflux condensers fitted to the digestion flasks or tubes with ground-glass joints provide the best reflux. The acid:soil ratio should be 10:1. Up to 100 mL acid can be accommodated in flasks. Kjeldahl tubes should not be filled above the level of the block (10–20 mL).
5. Heat the mixture overnight in the temperature-controlled digestion block, allowing the acid to reflux. Swirl the tubes occasionally to wash down soil that collects on the tube walls. After hydrolysis, rinse the reflux condensers with distilled water.
6. Isolate the residue by vacuum filtration through preweighed Whatman no. 50 or other hardened fine-pore filters, then wash with water and dry. Alternatively, isolate the residue by repeated centrifugation and washing.
7. Analyze for total C or specific C isotopes.

Special Considerations

Note that use of hardened filter papers permits the residue to be scraped off nearly quantitatively. If the residue is to be ^{14}C-dated, however, we recommend glass fiber filters even though recovery will not be as complete.

Density Fractionation

Density fractionation takes advantage of the fact that the density of soil particles reflects differences in the ratio of organic materials, which are light, to mineral materials, most of which are heavy (see Gregorich and Janzen 1996). The lightest particles comprise mainly organic debris—fragments of dead roots and leaves. In many soils, however, amorphous mineral materials adsorb on the surfaces of the debris, increasing its density. Secondary minerals (e.g., clays) have reactive surfaces that accumulate thick coatings of organic materials. Because these organic coatings are light, the secondary mineral particles tend to be of intermediate density. In addition, secondary mineral particles, along with their organic coatings, often form the components of aggregates. Such aggregates are also of intermediate density. Primary mineral particles are generally the heaviest, both because the minerals themselves may be heavy and because the surfaces of such particles tend to be unreactive and thus tend not to accumulate thick organic layers.

Soil particles span a continuum of densities, with the majority falling in the range of 1–2 g/cm^3. Many soils have bimodal density distributions with a trough around 1.5–1.8 g/cm^3. Thus 1.6 to 1.7 is often a convenient cutoff for separating a light and heavy fraction (Ladd and Amato 1980; Spycher et al. 1983). This cutoff is not applicable to all soils: some may not show a bimodal distribution or may include light rocks such as pumice in the light fraction (Sollins et al. 1983). Each group of soils should be checked initially by sequential separation at a series of densities from about 1.2–1.9 g/cm^3. The resulting light and heavy fractions should ideally be analyzed for ash content (muffle furnace), C content, and C:N ratio. The density above which ash content of the light fraction increases markedly or C content decreases, or at which the difference in C:N ratio of light and heavy fraction peaks, will be optimal for separating a biologically meaningful light fraction.

The simplest method for density fractionation is to disperse soil in a heavy liquid, then aspirate the floating material from the surface of the liquid (Strickland and Sollins 1987). Several heavy liquids are popular, including Si suspension (Meijboom et al. 1995), sucrose, NaI, and Na polytungstate (NaPT). NaPT is the generally preferred flotation medium (Baldock et al. 1990). It is much less reactive than NaI and thus solubilizes much less C. It is also much less viscous than sucrose solutions of the same density and does not introduce readily metabolizable C into the system. The Si suspension is not recommended because it does not allow separation at densities much greater than 1.37 g/cm^3.

Sample preparation and dispersion procedures are critical and must be described carefully if work is to be reproducible. Dispersive energy that is applied should be expressed as joules/mL, calculated based on wattage of the instrument, volume of the suspension, and time sonicated. Applied dispersive energy has been measured directly as pressure exerted downward in the liquid (as measured with an electronic balance) and heat buildup in the liquid (Christensen 1992). Note that C-containing surfactants should not be used because they sorb onto mineral surfaces and inflate C values.

Materials

1. 10 mL volumetric pipet
2. Sodium polytungstate [$Na_6(H_2W_{12}O_{40})\cdot H_2O$]
3. 400 mL tall-form (Berzelius) beakers
4. Ultrasonic probe
5. Mixer (e.g., Hamilton Beach bench mixer, Scovil soil mixer, or Sorvall Omni Mixer)
6. Buchner flasks and funnels (1 L)
7. Whatman no. 50 or other hardened fine-pore filters
8. Vacuum source

Procedure

1. Soil samples may be stored field moist at 4 °C until needed. Subsample to determine moisture content. Fractionation should be done in triplicate.
2. Prepare the NaPT solution by adding about 1050 g/L dH_2O. Remove a 10 mL aliquot with a volumetric pipet and weigh (a 1.7 g/cm^3 solution should weigh 17.0 g). Add NaPT or dH_2O as necessary to adjust the density to the desired value.
3. Suspend 40 g field-moist soil in 200 mL of the NaPT solution. Disperse soil with mixer (1800 rpm for 0.5 minute). Rinse any soil adhering to the post and blades with NaPT solution.
4. Allow samples to settle for 48 hours at room temperature. The suspended light

fraction is then aspirated through a Tygon hose (1.0 cm i.d.) attached to a 1 L Buchner flask and a strong vacuum source. Care must be taken to avoid disturbing the heavy fraction (sediment) because several centimeters of NaPT will be aspirated with the light fraction into the Buchner flask, decreasing the headspace of the density solution. A second Buchner flask may be connected to the first to avoid loss of light fraction and to protect the vacuum source from moisture and organic particles. Wash the captured light fraction into a separate container and store at 4 °C.

5. Add NaPT solution to the heavy fraction to bring the suspension back to the original volume, and disperse the heavy fraction again (as described earlier). Allow the suspension to sit for 48 hours. Aspirate the new light fraction and combine with the previously collected light fraction.
6. Pour the combined light fraction onto Whatman no. 50 filters (2.7 μm retention), vacuum filter, and rinse the light fraction three times with 100 mL dH_2O. Set aside the rinse solution if desired for DOC or DON analysis.
7. Remove NaPT from the heavy fraction by adding at least 200 mL dH_2O and mixing thoroughly. In many soils the suspended particles may be allowed to settle and the clear supernatant aspirated into a Buchner flask, taking care not to disturb the heavy fraction. In sandy soils it may be possible to wash the heavy fraction onto filter paper and rinse. In clay-rich soils, the suspension should be rinsed into a centrifuge tube, spun down, and the supernatant aspirated. For all soils, the rinse solutions should be saved and analyzed appropriately if C or N recovery is to be gauged. The rinse procedure is repeated at least three times, more often for soils with high clay content.
8. Wash the heavy and light fractions into preweighed tins with dH_2O, dry at 105 °C, and weigh. Material may be ground and analyzed for total C and N as described earlier.

Calculations

Most CN analyzers read out directly in concentration units.

Special Considerations

The use of tall, narrow beakers is strongly recommended: it allows the soil:solution ratio to be maximized, saving considerably on the cost of the NaPT, prevents splashing during mixing, and maximizes the vertical separation between the floating light fraction and settled heavy fraction. The use of a mechanical mixer, rather than an ultrasonicator, is based on the finding that, in many soils, sonication yields essentially no light-fraction material beyond that released by mechanical mixing. This assertion may not hold for all soils and should be checked for each group of soils before a method is finalized.

The procedure can be modified to allow sequential separation at a series of densities. An initial separation may be made at 1.0 g/cm^3 using dH_2O. The resulting heavy fraction can then be resuspended in NaPT at a higher density, and the process repeated.

Although NaPT is expensive, a procedure to recycle used NaPT has recently been developed (J. Six and J. Jastrow, personal communication, 1997). The used NaPT, adjusted to about 1.4 g/cm^3, is passed through a column (6 cm i.d.) packed, from top to bottom, with 3 cm glass wool, 25 cm activated charcoal, 1 cm glass wool, 7.5 cm cation exchange resin (sodium-form), 1 cm glass wool, 7 cm activated charcoal, and 3 cm glass wool. The column should be kept moist and rinsed with 1 L dH_2O for every 2 L of NaPT.

Golchin et al. (1994) distinguished between a free and protected or occluded light fraction. The free light fraction is that obtained after soil is dispersed by gentle inversion only, without mechanical mixing or sonication. The resulting heavy fraction is then resuspended in the same density medium and dispersed by sonication. The additional light fraction recoverable after sonication is termed protected or occluded. ^{13}C-NMR revealed compositional differences between the free and occluded light fractions.

Coarse Particulate Organic Matter

Size fractionation assumes that smaller particles, because of their larger specific surface area, are more reactive chemically and biologically. As discussed in detail in Chapter 4, this volume, particle size separation relies on the decrease in sedimentation rate with increasing particle size. Particle shape and density also affect sedimentation rate; thus the clay size fraction will inevitably contain light fraction material that is larger than the upper size cutoff for clay. Procedures for solubilizing organic matter and amorphous mineral coatings are inappropriate as pretreatments when separating organic matter pools because they will alter the distribution of the organic matter among the pools.

Many methods have been proposed that use particle size separation to isolate soil organic matter pools. The usefulness of these procedures has been much debated, and the one recommended here is separation of sand-size organic matter (coarse particulate organic matter [CPOM]). In grassland and cultivated soils, this pool responds strongly to changes in management and vegetation (Cambardella and Elliott 1992, 1993).

The sieving procedure described here is equivalent to the sedimentation procedure described in Chapter 4, this volume, for separating coarse particulate inorganic matter (CPIM) from bulk soil. Use the sedimentation procedure if also performing particle size (texture) analysis.

Materials

1. 5 g/L Na hexametaphosphate
2. Reciprocating shaker
3. 53 μm (250-mesh) sieve
4. 50 mL plastic bottles
5. Al drying pans
6. Soil or rock mill

Procedure

1. Hand pick and discard identifiable plant material.
2. Weigh 10 g air-dry soil (previously sieved to pass a 2 mm screen) into a 50 mL bottle.
3. Add 30 mL Na hexametaphosphate solution.
4. Shake on a reciprocating shaker for 18 hours.
5. Pass the suspension through a 53 μm sieve, rinse the retained material several times with DW, transfer it quantitatively to a preweighed aluminum pan, and oven-dry.
6. The CPOM is finely ground and analyzed for C and N by dry combustion as described above.

Calculations

Most researchers will want to express CPOM-C on a whole soil basis. For CPOM-N, substitute N for C in the following calculations.

For the Seiving Procedure

$$\text{CPOM-C} = C_s \times W_s \times 10$$

where

CPOM-C = coarse particulate organic matter C (g C/kg soil)
C_s = %C of sand fraction
W_S = dry mass of sand fraction (g fraction/g soil); CPOM mass is assumed to be insignificant relative to the mass of the sand fraction

For the Sedimentation Procedure

$$\text{CPOM-C} = [C_{ts} - (C_f \times W_f)] \times 10$$

where

C_{ts} = total soil C (%)
C_f = silt + clay fraction C (%)
W_f = mass of silt + clay fraction (g fraction/g soil)

Special Considerations

The CPOM fraction may contain large amounts of mineral grains. In many soils such grains will contain relatively little C and N. Lacking C and N, they will dilute C and N concentrations but will not affect C:N ratios. In some soils, however, especially those of volcanic origin, the sand grains will contain large amounts of adsorbed SOM and fine aggregates containing both adsorbed and occluded SOM. In such soils a density fractionation step should be added to separate true CPOM (wide

C:N ratio) from the OM adsorbed on mineral particles and occluded within aggregates (narrow C:N ratio).

The CPOM method includes drying at 50 °C, which can cause irreversible aggregation in soils that do not normally dry, especially those rich in amorphous weathering products or in Fe and Al (hydr)oxides (e.g., Sollins 1989). For such soils the procedure will need to be modified to use field-moist soil.

The extent to which light fraction and CPOM are equivalent remains unresolved. Some CPOM particles are heavy (>1.8 g/cm^3) and would thus not be included in a light fraction (Gregorich and Janzen 1996). The other issue is whether a significant amount of the light fraction consists of particles finer than sand size (<50 μm). Barrios et al. (1996) compared density and size fractions in a Kenyan Alfisol but performed density fractionation only on the whole soil and on the sandsize fraction. Thus there is no way to determine what proportion of the light fraction was in silt- and clay-sized particles. Turchenek and Oades (1978) report that, for one soil after sonication, coarse clay (0.4–2 μm) that was light (<1.8 g/cm^3) accounted for 12.1% of total soil C.

Young and Spycher (1979) report the only data of which we are aware on density fractionation after particle size fractionation of unsonicated soil. They worked with seven soils of greatly differing mineralogy. Generalizations are difficult because they used different density cutoffs in each soil, but for the three in which their density cutoff was ≤1.8 g/cm^3, 11–22% of total SOM was in clay-sized light fraction material. Although the proportion this represents of total light fraction material in all size fractions was not reported, it seems clear that, in these three soils, CPOM excludes a major portion of the light fraction material, and thus CPOM and light fraction are not equivalent.

Acknowledgements We thank Harold Collins, Knute Nadelhoffer, Robert Ahrens, and others in our working group for their many helpful comments.

References

Ameel, J. J., R. P. Axler, and C. J. Owen. 1993. Persulfate digestion for determination of total nitrogen and phosphorus in low-nutrient waters. *American Environmental Laboratory* 10:1–11.

Baldock, J. M., J. M. Oades, A. M. Vasallo, and M. A. Wilson. 1990. Solid-state CP/MAS ^{13}C NMR analysis of particle size and density fractions of a soil incubated with uniformly labeled ^{13}C-glucose. *Australian Journal of Soil Research* 28:193–212.

Baldock, J. A., and C. M. Preston. 1995. Chemistry of C decomposition processes in forests as revealed by solid-state carbon-13 nuclear magnetic resonance. Pages 89–117 in W. W. McFee and J. M. Kelly, editors, *Carbon Forms and Functions in Forest Soils*. Soil Science Society of America, Madison, Wisconsin, USA.

Barrios, E., R. J. Buresh, and J. I. Sprent. 1996. Organic matter in particle size and density fractions from maize and legume cropping systems. *Soil Biology and Biochemistry* 28:185–193.

Cambardella, C. A., and E. T. Elliott. 1992. Particulate soil organic-matter changes across a grassland cultivation sequence. *Soil Science Society of America Journal* 56:777–783.

Cambardella, C. A., and E. T. Elliott. 1993. Carbon and nitrogen distribution in aggregates from cultivated and native grassland soils. *Soil Science Society of America Journal* 57:1071–1076.

Christensen, B. T. 1992. Physical fractionation of soil and organic matter in primary particle size and density separates. *Advances in Soil Science* 20:1–90.

Coleman, D. C., and B. Fry, editors. 1991. *C Isotope Techniques.* Academic Press, San Diego, California, USA.

D'Elia, C. F., P. A. Steudler, and N. Corwin. 1976. Determination of total nitrogen in aqueous samples using persulfate digestion. *Limnology and Oceanography* 22:760–764.

Duxbury, J. M., M. S. Smith, and J. W. Doran. 1989. Soil organic matter as a source and a sink for plant nutrients. Pages 33–67 *in* D. C. Coleman, J. M. Oades, and G. Uehara, editors, *Dynamics of Soil Organic Matter in Tropical Ecosystems.* University of Hawaii Press, Honolulu, Hawaii, USA.

Follett, R. F., E. A. Paul, S. W. Leavitt, A. D. Halvorson, D. Lyon, and G. A. Peterson. 1997. Carbon isotope ratios of Great Plains soils and in wheat-fallow systems. *Soil Science Society of America Journal.* 61:1068–1077.

Gardner, W. H. 1986. Water content. Pages 493–544 in A. Klute editor, *Methods of Soil Analysis. Part 1, Physical and Mineralogical Methods.* 2d edition. American Society of Agronomy, Madison, Wisconsin, USA.

Golchin, A., J. M. Oades, J. O. Skjemstad, and P. Clarke. 1994. Study of free and occluded particulate organic matter in soils by solid state ^{13}C CP/MAS NMR spectroscopy and scanning electron microscopy. *Australian Journal of Soil Research* 32:285–309.

Gregorich, E. G., and H. H. Janzen. 1996. Storage of soil carbon in the light fraction and macroorganic matter. Pages 167–190 *in* M. R. Carter and B. A. Stewart, editors, *Structure and Organic Matter Storage in Agricultural Soils.* Lewis Publishers, Boca Raton, Florida, USA.

Jenkinson, D. S., D. E. Adams, and A. Wild. 1991. Model estimates of CO_2 emissions from soil in response to global warming. *Nature* 351:304–306.

Kinchesh, P., D. S. Powlson, and E. W. Randall. 1995. ^{13}CNMR studies of organic matter in whole soils. 1. Quantitation possibilities. *European Journal of Soil Science* 46:125–138.

Kögel-Knabner, I., J. W. de Leeuw, and P. G. Hatcher. 1992. Nature and distribution of alkyl carbon in forest soil profiles: implications for the origin and humification of aliphatic biomacromolecules. *Science of the Total Environment* 117/118:175–185.

Ladd, J. N., and M. Amato. 1980. Studies of nitrogen immobilization and mineralization in calcareous soils. I. Changes in the organic nitrogen of light and heavy subfractions of silt and fine clay size particles during nitrogen turnover. *Soil Biology and Biochemistry* 12:185–189.

Lajtha, K., B. Seely, and I. Valiela. 1995. Retention and leaching losses of atmospherically-derived nitrogen in the aggrading coastal watershed of Waquoit Bay, Massachusetts. *Biogeochemistry* 28:33–54.

Meijboom, F. W., J. Hassink, and M. van Noordwijk. 1995. Density fractionation of soil macroorganic matter with silica suspensions. *Soil Biology and Biochemistry* 27:1109–1111.

Nelson, D. W., and L. E. Sommers. 1982. Total carbon, organic carbon, and organic matter. Pages 539–579 *in* A. L. Page, R. H. Miller, and D. R. Kenney, editors, *Methods of Soil Analysis: Chemical and Microbiological Properties.* ASA Monograph 9. American Society of Agronomy, Madison, Wisconsin, USA.

Paul, E. A., H. P. Collins, D. Harris, U. Schulthess, and G. P. Robertson. 1998. The influence of biological management inputs on carbon mineralization in ecosystems. *Applied Soil Ecology* 3:1–13.

Paul, E. A., R. F. Follett, S. W. Leavitt, A. Halvorson, G. A. Peterson, and D. J. Lyon. 1997.

Radiocarbon dating for determination of soil organic matter pool sizes and dynamics. *Soil Science Society of America Journal.* 61:1058–1067.
Preston, C. M. 1996. Applications of NMR to soil organic matter analysis: history and prospects. *Soil Science* 161:144–166.
Scharpenseel, H. W., and H. Schiffmann. 1977. Radiocarbon dating of soils, a review. *Z. Pflanzenernahr. Bodenkd* 40:159–174.
Schulze, E. D. 1989. Air pollution and forest decline in a spruce (Picea abies) forest. *Science* 244:776–782.
Snyder, J. D., and J. A. Trofymow. 1984. A rapid accurate wet oxidation diffusion procedure for determining organic and inorganic carbon in plant and soil samples. *Communications in Soil Science and Plant Analysis* 15:587–597.
Sollins, P. 1989. Factors affecting nutrient cycling in tropical soils. Pages 85–95 *in* J. Proctor, editor, *Mineral Nutrients in Tropical and Savanna Ecosystems. British Ecological Society Special Publication No. 9.* Blackwell Scientific, Oxford, UK.
Sollins, P., P. Homann, and B. Caldwell. 1996. Stabilization and destabilization of soil organic matter: mechanisms and controls. *Geoderma* 74:65–105.
Sollins, P., and F. M. McCorison. 1981. Nitrogen and carbon solution chemistry of an old growth coniferous forest watershed before and after cutting. *Water Resources Research* 17:1409–1418.
Sollins, P., G. Spycher, and C. Topik. 1983. Processes of soil organic-matter accretion at a mudflow chronosequence, Mt. Shasta, California. *Ecology* 64:1273–1282.
Spycher, G., P. Sollins, and S. Rose. 1983. Carbon and nitrogen in the light fraction of a forest soil: vertical distribution and seasonal patterns. *Soil Science* 135:79–87.
Strickland, T. C., and P. Sollins. 1987. An improved method for separating light- and heavy-fraction organic material from soils. *Soil Science Society of America Journal* 51:1390–1393.
Theng, B. K. G., K. R. Tate, and P. Sollins. 1989. Constituents of soil organic matter in temperate and tropical soils. Pages 5–31 *in* D. C. Coleman, J. M. Oades, and G. Uehara, editors, *Dynamics of Soil Organic Matter in Tropical Ecosystems.* University of Hawaii Press, Honolulu, Hawaii, USA.
Tisdall, J. M. 1996. Formation of soil aggregates and accumulation of soil organic matter. Pages 57–96 *in* M. R. Carter and B. A. Stewart, editors, *Structure and Organic Matter Storage in Agricultural Soils.* Lewis Publishers, Boca Raton, Florida, USA.
Turchenek, L. W., and J. M. Oades. 1978. Organo-mineral particles in soils. Pages 137–144 *in* W. W. Emerson, R. D. Bond, and A. R. Dexter, editors, *Modification of Soil Structure.* Wiley, Chichester, UK.
Young, J. L., and Spycher, G. 1979. Water dispersible soil organic-mineral particles. I. Carbon and nitrogen distribution. *Soil Science Society of America Journal* 43:324–328.

6

Exchangeable Ions, pH, and Cation Exchange Capacity

G. Philip Robertson
Phillip Sollins
Boyd G. Ellis
Kate Lajtha

Almost all nutrients taken up by plants and microbes are taken up in their ionic form from the soil solution, and knowledge of the size and composition of the soil solution, together with knowledge of nutrient turnover rates, can provide valuable insight into soil nutrient availability and into other biogeochemical processes such as rates of weathering and leaching losses. While nutrient pool size per se can be a poor indicator of availability, changes in pool sizes can indicate changes in ecosystem processes that bear closer examination. Moreover, in all soils the potential for significant hydrologic loss of mobile nutrients is directly related to the size of the exchangeable nutrient pool and the proportion in solution at any given time.

In this chapter we present standard methods for the extraction of the major cations and anions from soil, the measurement of exchangeable acidity and soil pH, and cation exchange capacity. Methods for the extraction of dissolved-only nutrients (those ions only in the solution phase as opposed to total exchangeable nutrients that have both solution and surface adsorbed phases) are provided elsewhere (see Chapter 9, this volume), as are methods for available phosphorus (see Chapter 7, this volume) and dissolved organic nitrogen and carbon (see Chapter 5, this volume).

Exchangeable Ions

The determination of exchangeable ions in soil requires that ions on soil exchange sites be forced into a solution in which they can be effectively measured. Generally this involves flooding the exchange sites on clay and organic matter surfaces of a soil with ions from an extractant, usually a strong salt solution. The extractant, now containing exchangeable soil ions in addition to ions from the added salt, is separated from soil by filtering or centrifugation and is then analyzed for the ions of interest.

Choice of salt for the extractant solution will depend on the target ions. Extractant ions must effectively displace soil ions from exchange sites and must not interfere with subsequent chemical analysis of the extracted solution. For most soils 1.0 mol/L KCl is a reasonable extractant if K^+ is not the target ion. KCl is the most common extractant for inorganic nitrogen, for example. For total cation analysis, NH_4OAc is useful because both NH_4^+ and Ac will volatilize and thus not accumulate on spectrometer burners. $BaCl_2$ provides a reasonable but expensive alternative for simultaneous extraction of both K^+ and NH_4^+; NaCl will also work for most ions. Special extractants and precautions are needed for trace metal species (e.g., Pb, Cu), however, and will not be covered here (see various chapters in Sparks et al. 1996). For most major cations (e.g., NH_4^+, K^+, Mg^{+2}, Ca^{+2}, Al^{+3}) and anions (e.g., NO_3^-, SO_4^{-2}), any of these extractants will work well.

In general, soils should be extracted when fresh from the field, prior to drying. This is especially important for ions that can undergo rapid microbial transformations during storage (e.g., NH_4^+, NO_3^-, PO_4^{-3}) or undergo volatilization or some other chemical transformation at elevated temperatures (e.g., NH_4^+, PO_4^{-3}). Where immediate extraction is not possible, or when the ion of interest is unlikely to undergo significant microbial transformation or precipitation reaction, samples can be dried at low temperature (<45 °C). Prior to extraction, soil should be sieved and mixed as described in Chapter 1, this volume.

Materials

1. 120 mL screw-cap polyethylene extraction cups and lids (e.g., urinalysis cups) or equivalent
2. Extractant: For ions other than K^+ and Cl^-, use 1 mol/L KCl (74.6 g/L). For K^+ and ions other than NH_4^+, use 1 mol/L NH_4OAc at pH 7.0 (add 77.1 g NH_4OAc to 950 mL deionized water, adjust pH to 7.0 with acetic acid or aqueous ammonia, and bring volume to 1.0 L with deionized water).
3. Sample vials sufficient to hold extracted sample (e.g., 20 mL screw-cap polyethylene scintillation vials)
4. 10 cc Luerlok polyethylene syringes, reusable Gelman syringe filter holders (25 mm diameter), and Gelman Type A/E glass-fiber filter paper (25 mm diameter) or other filter known to be free of target ions. Alternately, use standard glass or plastic funnels, a funnel holding rack, and glass-fiber filter paper to fit the funnel.

Procedure

1. Weigh triplicate 10.0 g subsamples of fresh sieved soil into each of three extraction cups. Soil should be passed through at least a 4 mm sieve; 2 mm (undried) is preferred if soil is sufficiently dry (see Chapter 1, this volume).

2. Weigh separate soil samples for gravimetric moisture analysis (see Chapter 3, this volume).
3. Add 100 mL extractant to each cup, cap, and shake vigorously for 1 minute. Also add 100 mL extractant to three cups without soil to serve as blanks. Allow to equilibrate overnight (12–24 hours).
4. Reshake extraction cups and allow to settle for 45 minutes.
5. Remove 10 mL solution from the extraction cup into a syringe. Place a preloaded filter holder on the syringe and filter the solution directly into a sample vial. Use a separate syringe and filter holder for each replicate sample to avoid cross-contamination. Alternatively, filter each extract through a clean glass fiber filter paper held in a funnel.
6. Label vials and store in refrigerator or freezer until analysis.
7. Analyze appropriately as per specific instrumentation (see the section "Special Considerations," below).

Calculations

Gravimetric Basis—Element Mass

$$\mu\text{g element/g soil} = (C \times V) / W$$

where

C = concentration of ion in extract in mg/L, as provided by standards used for elemental analysis
V = volume of extract (mL), e.g., 100 mL, plus water in soil sample.
W = dry mass of soil, e.g., 10.0 g (as in step 1) less its water content as determined in step 2, above.

Gravimetric Basis—Element Moles of Charge

$$\text{cmol}_c \text{ element/kg} = (C_g \times n) / (10 \times A)$$

where

C_g = element mass on gravimetric basis, as μg element/g soil
n = valence of ion
A = atomic mass of ion

Areal Basis—Element Mass

$$\text{g element/m}^2 = C_g \times BD \times SD$$

where

C_g = element mass on gravimetric basis as μg element/g soil
BD = bulk density as g soil/cm^3 soil or Mg soil/m^3 soil (see Chapter 4, this volume)

SD = sample depth in m, e.g., 0.2 m for standard Level 1 depth of 20 cm (see Chapter 1, this volume)

Special Considerations

1. If a shaker table is available, 1 hour on a shaker table can substitute for the shaking and overnight equilibration described here.
2. Filter holders should be loaded using care to avoid touching filters or other surfaces that will contact the extractant; use forceps and wear latex gloves as needed (but beware of glove talc as a potential contaminant). For large numbers of samples it is convenient to have on hand the same number of syringes and reusable preloaded filter holders as extraction cups.
3. Centrifugation (e.g., 10 minutes at 1000 g) can substitute for filtration, but take care to avoid floating pieces of organic debris when sampling from the centrifuge tubes.
4. Procedures for the analysis of specific ions will depend on available instrumentation and manufacturer-specific protocols. In general, colorimetric techniques are recommended for NH_4^+, NO_3^-, and Si^{+4}, ion chromatography for SO_4^{-2}, and atomic absorption (AA) spectrometry or inductively coupled plasma atomic emission spectrometry (ICP-AES) for other cations. Ion-specific electrodes are rarely appropriate due to their general insensitivity and to differential interference from other ions in the soil solution matrix; where used, they should always be calibrated against colorimetric values for each soil examined. Ion chromatography (IC) and high-performance liquid chromatography (HPLC), ideally suited for leachate and rainwater samples, are generally not preferred for soil extracts because of complications introduced by the high concentration of salt in the extractant solution. It is essential to include blanks and check standards as part of the analysis and to make all analytical standards in the extractant solution (see Chapter 1, this volume).
5. When preparing NH_4OAc, add aqueous ammonia under a hood in a room or building not used for the preparation of samples for NH_4^+ analysis; volatilized NH_3 will readily contaminate samples and surfaces exposed to the atmosphere.
6. As noted in Chapter 1, values should also be expressed on an areal basis for making ecosystem-level comparisons.

Soil pH

Soil pH is a measure of hydrogen ion activity in the soil solution. pH is generally measured electrometrically, although chemical means can provide a coarse indication (e.g., ± 0.5 pH units for pH strips). The electrometric pH reading is a product of complex interactions between the electrode and the soil suspension; differences in the soil:water extraction ratio, the electrolyte concentration of the soil suspension, and the spatial placement of the electrode can all affect measured pH.

Materials

1. Extraction cups or beakers (50–100 mL)
2. Deionized water
3. pH meter with standard combination electrode
4. Standard pH buffers, available commercially (pH 4.0, 7.0, and 10.0)
5. Squirt bottle and beaker for rinsing the electrode between samples

Procedure

1. Calibrate the electrode as per the manufacturer's instructions. Always use two buffers that span the pH range of the sampled soils.
2. Weigh duplicate 15.0 g subsamples of fresh (undried), sieved soil into each of two extraction cups. Use a 2 mm sieve if the soil is sufficiently dry; otherwise 4 mm will suffice (see Chapter 1, this volume).
3. Add 30 mL deionized water and stir the mixture well.
4. Allow the mixture to stand for 30 minutes to equilibrate with atmospheric CO_2, stir again, then read pH to nearest 0.1 pH unit. The soil slurry should be gently swirled while taking the measurement.
5. Check electrode stability every 10–12 samples with a standard buffer solution and recalibrate electrode as needed. Rinse the electrode well between samples, and especially well following immersion in the buffer solution.

Calculations

No further calculations are necessary for reporting pH. To convert pH to $[H^+]$, use the equation

$$[H^+] = 10^{-pH}$$

Special Considerations

1. Depending on the buffer capacity of an individual soil, pH readings may be extremely sensitive to electrode condition. Always store electrodes in recommended salt solution between uses. The electrode needs to be open to air during measurement (there is often a small rubber vent cap near the top of the electrode tube) and closed during storage. Combination electrodes should be kept filled with electrolyte solution as per the manufacturer's instructions. If the pH reading of soil does not stabilize to 0.1 units within a few minutes, consider reconditioning or replacing the electrode.
2. Depending on the level of biological activity in a soil, measurement of pH may need to be performed within hours (e.g., tropical sites) or can be postponed indefinitely (e.g., desert sites).
3. Use as a check sample a dried soil analyzed previously.
4. In some situations it may also be useful to measure pH in a weak $CaCl_2$ solution, in which case the procedure can be performed as described earlier but substituting 0.01 mol/L $CaCl_2$ for deionized water. This procedure is useful

chiefly for historical comparisons where earlier measurements may have used a salt extractant to correct for variable salts in soil. It is not the preferred method today even in agricultural soils (Thomas 1996).

Exchangeable Acidity and Aluminum

Aluminum is a predominant cation in many soils and can be a critical variable in establishing effective cation exchange capacity (ECEC) values. In exchange reactions its characterization is complicated by the coexistence of complex multiphase Al components that make its measurement specific to a particular extractant and potentially difficult to interpret (Bertsch and Bloom 1996). In practice, the most common extractant for calculations of ECEC and Al saturation is 1 mol/L KCl. For ECEC determinations it is not necessary to differentiate between exchangeable Al^{+3} and H^+, and analysis can stop after step 4 of the following procedure. Al mobility in soil depends on its speciation in addition to total concentration; see Driscoll et al. (1985, 1989) for appropriate speciation methods.

Materials

1. Extraction cup or beaker (100 mL)
2. Buchner funnel with prewashed Whatman no. 42 filter paper (see the section "Special Considerations," below)
3. Burrette for titrations
4. 1 mol/L potassium chloride (74.6 g KCl/L) stored in plastic carboy (150 mL per soil sample)
5. Phenolpthalein solution, 1 g phenolphthalein dissolved in 100 mL ethanol
6. 0.1 N sodium hydroxide (4.0 g NaOH/L)
7. 1 mol/L potassium fluoride (58.1 g KF/L) (for determination of exchangeable Al)
8. 0.1 N hydrochloric acid (HCl) (for determination of exchangeable Al)

Procedure

1. Weigh 10.0 g fresh (undried), sieved soil into a 100 mL extraction cup, add 25 mL KCl, stir or shake well, and let sit for 30 minutes. Use a 2 mm sieve if the soil is sufficiently dry; otherwise 4 mm will do (see Chapter 1, this volume).
2. Determine moisture content of a separate subsample as per Chapter 3, this volume.
3. Filter through a Buchner funnel and then wash with five successive 25 mL aliquots of KCl, for a total of 150 mL KCl per soil sample.
4. To the filtrate add 5 drops of phenolphthalein and titrate with 0.1 N NaOH to the first permanent pink end point. Record the volume of NaOH solution used.
5. Repeat step 4 for a blank solution of 150 mL KCl, also washed through a Buchner funnel. Record the volume of NaOH used.

6. For exchangeable aluminum, note the volume of NaOH added in step 4, add to the titrated solution 10 mL of 1 mol/L KF, and titrate with 0.1 N HCl until the pink color disappears. If the solution turns pink again after 30 minutes, add additional HCl and re-evaluate after 30 minutes. Record the volume of HCl used.

Calculations

$$\text{Exchangeable acidity (cmol/kg)} = (NaOH_{dif}/W) \times (0.1 \text{ mmol H}^+/\text{mL NaOH}) \times (0.1 \text{ cmol H}^+/\text{mmol H}^+) \times (10^3 \text{ g soil/kg soil})$$

where

$NaOH_{dif}$ = mL of NaOH added to sample filtrate less mL of NaOH added to blank solution
W = g dry soil (e.g., 10.0 g less its water content as determined in step 2, above)

$$\text{Exchangeable Al (cmol/kg)} = (HCl_{sample}/W) \times (0.1 \text{ mmol Al}^{+3}/\text{mL HCl}) \times (0.1 \text{ cmol Al}^{+3}/\text{mmol Al}^{+3}) \times (10^3 \text{g soil/kg soil})$$

where

HCl_{sample} = mL of HCl added to sample
W = g dry soil based on moisture determination

$$\text{Exchangeable H}^+ = \text{Exchangeable acidity} - \text{Exchangeable Al}$$

Special Considerations

Where exchangeable Al values are low and filter materials could introduce error, use a 0.45 μm polycarbonate filter.

Cation Exchange Capacity

The measurement of cation exchange capacity (CEC) is complicated by the fact that CEC is affected by both pH and the ionic strength of the soil solution, especially in highly weathered soils and other soils rich in Al and Fe oxides, hydroxides, and amorphous clays. Traditional methods for measuring CEC include adjustment of soil pH to 7.0, which will misrepresent CEC in soils with variable charge minerals or substantial organic matter. The compulsive exchange method (Gillman 1979) avoids both pH and ionic strength problems but is very laborious. For routine work and comparisons among soils dominated by 2:1 clay minerals, we recommend summing exchangeable cations to provide a measure of ECEC.

Exchangeable cations include both base cations (K^+, Ca^{+2}, Mg^{+2}, and Na^+) and the acidic cations (Al^{+3} and H^+). Other cations such as NH_4^+ and trace metals are usually present in relatively minor amounts and can be effectively ignored. In calcareous and saline soils ECEC will overestimate CEC because measured cations will also include those solubilized from mineral deposits. In these soils a more specialized technique is appropriate (e.g., Amrhein and Suarez 1990, described in Sumner and Miller 1996).

Materials and Procedure

1. Measure base cations (K^+, Ca^{+2}, Mg^{+2}, Na^+) as described in the section "Exchangeable Ions," above; ammonium acetate is the preferred extractant for AA and ICP analysis.
2. Measure exchangeable acidity (Al^{+3} and H^+) as described in the section "Exchangeable Acidity and Aluminum," above. It is not necessary to measure either ion separately.

Calculation

$$\text{ECEC (cmol}_c\text{/kg)} = exch\ K^+ + exch\ Ca^{+2} + exch\ Mg^{+2} + exch\ Na^+ + exch\ acidity$$

where

exch K^+ etc. = concentrations of individual ions expressed as cmol_c/kg dry soil

Special Considerations

Na^+ is likely to be important mainly in arid and coastal soils and can often be ignored in other environments.

References

Amrhein, C., and D. L. Suarez. 1990. Procedure for determining sodium-calcium selectivity in calcareous and gypsiferous soils. *Soil Science Society of America Journal* 54:999–1007.

Anderson, J. M., and J. S. I. Ingram, editors. 1993. *Tropical Soil Biology and Fertility: A Handbook of Methods.* CAB International, Oxford, UK.

Bertsch, P. M., and P. R. Bloom. 1996. Aluminum. Pages 517–550 *in* D. L. Sparks, A. L. Page, and P. A. Helmke, editors, *Methods of Soil Analysis. Part 3, Chemical Methods.* Soil Science Society of America, Madison, Wisconsin, USA.

Driscoll, C. T., N. van Breemen, and J. Mulder. 1985. Aluminum chemistry in a forested spodosol. *Soil Science Society of America Journal* 49:437–444.

Driscoll, C. T., B. J. Wyskowski, P. DeSteffan, and R. M. Newton. 1989. Chemistry and transfer of aluminum in a forested watershed in the Adirondack region of New York, USA. Pages 83–105 *in* T. E. Lewis, editor, *Environmental Chemistry and Toxicology of Aluminum.* Lewis Publishers, Chelsea, Michigan, USA.

Gillman, G. P. 1979. A proposed method for the measurement of exchange properties of highly weathered soils. *Australian Journal of Soil Research* 17:129–139.

Sparks, D. L., A. L. Page, and P. A. Helmke, editors. 1996. *Methods of Soil Analysis: Chemical Methods. Part 3.* American Society of Agronomy, Madison, Wisconsin, USA.

Sumner, M. E., and W. P. Miller. 1996. Cation exchange capacity and exchange coefficients. Pages 1201–1229 *in* D. L. Sparks, A. L. Page, and P. A. Helmke, editors, *Methods of Soil Analysis. Part 3, Chemical Methods.* Soil Science Society of America, Madison, Wisconsin, USA.

Thomas, G. W. 1996. Soil pH and soil acidity. Pages 475–490 *in* D. L. Sparks, A. L. Page, and P. A. Helmke, editors, *Methods of Soil Analysis. Part 3, Chemical Methods.* Soil Science Society of America, Madison, Wisconsin, USA.

7

Soil Phosphorus

Characterization and Total Element Analysis

Kate Lajtha
Charles T. Driscoll
Wesley M. Jarrell
Edward T. Elliott

Phosphorus is a major element in soil organic matter, and in natural terrestrial ecosystems is derived from the weathering of minerals in parent rock material. It is usually the second most limiting nutrient for terrestrial primary production (after nitrogen) and is often the primary limiting nutrient in freshwater ecosystems. Due to its critical role in controlling aquatic primary production, there is an increased concern over runoff of P from terrestrial to aquatic ecosystems and the role of P in affecting water quality.

Studies of P cycling and availability pose a challenge to agronomists and ecologists, in part because P occurs in soils in many different physicochemical forms that are operationally difficult to define, and in part because P is involved in a myriad of biological and chemical processes. The cycling of P can be controlled by inorganic chemical reactions, but in many systems the turnover of organic P controls the availability of P to plants through organic matter decomposition or the release of P from microbial biomass. Phosphate sorption by various soil constituents is a dominant process in maintaining very low concentrations of P in soil solutions, particularly in mineral soils. Tight control of P concentrations by adsorption creates difficulty for the measurement of the amount of P available for plant uptake or P release upon organic matter mineralization. Whereas important pools and fluxes of N are readily measured in soils, the different physicochemical forms of soil P are more difficult to categorize or define operationally. As a result, it is challenging to define or measure bioavailable P.

Understanding the cycling of P through soils has also proven challenging to ecologists. In contrast to measurements of N mineralization, the determination of changes in extractable P pools over time yields little information, in part due to the high background of sorbed mineral P and in part due to the lack of a single extract that can measure the newly released P. Although researchers have used various techniques, including the use of ^{32}P, to measure the mineralization of organic P in soils, these approaches either have not been widely tested or have met with some skepticism by other researchers, and they are not yet standard enough to be recommended here (for examples and comments see Walbridge and Vitousek 1987; Zou et al. 1992, 1995). Sorption/desorption also plagues measurements of solution or available P as well as measurements of organic P.

Due to these difficulties, it is important to realize that many pools of P such as "available" P are functional concepts rather than measurable quantities. Many measurements of P are simple to conduct (e.g., extractable P, sorbed P) but are more difficult to interpret. The measurement of total P in soils is one of the few measures of this element that is both operationally and theoretically well defined.

In this chapter we will discuss (1) the measurement of labile, or readily available, P in soils; (2) a current recommended method to estimate soil organic P; (3) total P and total element analysis by fusion; (4) an index of P sorption; and (5) a discussion of a current method to estimate operationally defined fractions of P in soils for use in specific research questions. Finally, procedures for measuring inorganic and organic P in solution are given.

General procedures used for sample collection, processing, and determining soil pools on an areal basis are provided in detail in Chapter 1, this volume. However, the sampling and handling protocols associated with the measurement of pools of P deserve special consideration. Due to the strong geochemical reactions of P in soils, certain measures of P, such as the estimation of total P pools, might best be analyzed by soil profile horizon rather than by depth, since distinct horizons will differ strongly in strength of geochemical reactions. Additionally, the preservation of soils prior to analysis, if samples cannot be analyzed or handled immediately while still field-moist, is an issue that is still debated by soil scientists. The drying of soils may present problems for the analysis of specific P fractions such as available or microbial P, as well as sorbed P. In many cases, particularly in heavy clays, soils may aggregate irreversibly upon even air drying, making them difficult to handle. Although grinding soils solves the aggregation problem, grinding exposes previously unexposed surfaces and may lyse microbial P, resulting in an apparent increase in measured labile P (Potter et al. 1991). In such cases soils should be analyzed for available P while still field-moist if at all possible; researchers have kept soils in plastic bags, refrigerated or cooled, without measurable changes in properties such as P sorption potential (Parfitt et al. 1989). However, other studies have shown that even short-term moist storage at 4 °C decreased labile P pools (Chapman et al. 1997), just as sieving soil has been shown to increase labile P.

Drying may also result in the conversion of P into physicochemical fractions that are fairly resistant to redissolution. Changes in field moisture and air drying have been shown to have pronounced effects on microbial biomass (Powlson and Jenkinson 1976; Potter et al. 1991), although adjustment of moisture content to approximately 60% of field capacity, followed by incubation, reduces this drying effect on microbial P (Potter et al. 1991). Of course, drying is less of a problem for soils that dry naturally under field situations. If analysis of field-moist soils is not possible, soils should be air-dried rather than oven-dried. However, since even air drying of specific soils, such as many tropical soils, can result in unalterable changes in soil chemistry, individual researchers should be aware of the properties of their soils before assuming that air drying will not cause problems. In all cases, final values of available P concentrations should be converted to an oven-dry (105 °C) basis.

Available P by Laboratory Resin Extraction

Problems with the definition of a "bioavailable" pool of P go beyond the measurement problems caused by the complexity of P chemistry in soils. Research has shown that different plants can extract different quantities of P from the same soils, and can derive this P from different physicochemical pools (Lajtha and Harrison 1995). Thus pools of available P differ among plant species. However, researchers have found useful indices of P availability and thus fertilizer requirements for many crop species by developing different chemical extractants for use in specific soil types. Unfortunately, these different extractants do not necessarily yield specific soil P fractions that are useful in understanding P cycling in an ecosystem or allow for comparisons across ecosystems and soil types. Because the purpose of this volume is to suggest methods that can be used across ecosystems and soil types, these extracts, commonly used in agronomic applications, will not be discussed here.

Anion-exchange resins, enclosed in permeable nylon or polyethylene mesh bags, have been used in both field and laboratory situations, and across a wide variety of soil types, as an index of labile P. The resins are assumed to simulate the ion uptake action of plant roots and, like active roots, provide a strong sink for P released into solution. In the bicarbonate form, anion-exchange resins will most closely mimic the chemical conditions of the rhizosphere, where bicarbonate accumulates and buffers soil pH (Sibbesen 1978). The relative affinity of anion-exchange resins for bicarbonate is low, thus favoring exchange with ortho-P, and resins in the bicarbonate form extract more ortho-P than resins in either the chloride form or the hydroxyl form (Lajtha 1988).

Researchers have used both Teflon-based anion-exchange membranes (Saggar et al. 1990; Schoenau and Huang 1991; Abrams and Jarrell 1992) and iron oxide strips (Lin et al. 1991; Menon et al. 1989; Sharpley 1991) in laboratory assessments of P availability. The anion-exchange membranes can certainly be used in place of the resin bags, but it is essential that a sufficient number of membranes of a sufficient size be used to ensure an excess of exchange sites. Resin strips and iron oxide strips give nearly the same result for labile P in poorly weathered soils; however, the iron oxide strip removes 5 to 10 times more P than do resins in highly weathered soils. Clearly the iron oxide strips are better competitors for P in these highly weathered soils, but they do not necessarily better mimic the action of plant roots. Myers et al. (1995) discuss some of the problems with the preparation and use of the iron oxide strips. These problems include abrasion and removal of the FeO coating during extractions, problems with different pore sizes of the papers, and adherence of soil particles to the strips. Thus here we recommend the use of resin bags and/or resin membrane filters over the use of iron oxide filters.

Materials

1. Dowex 1-X8 anion-exchange resin beads (or any type I resin), 20–50 mesh (or the largest available bead size); OR anionic resin strips, 2 × 6 cm (available commercially, e.g., BDH, marketed through VWR or through Soil, Plant, Water Quality Inc., 12505 NW Cornell Rd., Portland OR 97229)
2. Nylon or polyethylene small-mesh fabric (if using resin beads)
3. Salad spinner or open-basket hand centrifuge (if using resin beads)
4. 0.5 mol/L $NaHCO_3$
5. 0.5 mol/L HCl
6. 250 mL polyethylene bottles
7. Mechanical shaker
8. 50 μg P/mL stock solution: 0.2195 g oven-dried primary standard-grade KH_2PO_4 dissolved in 1000 mL deionized water. Stored in a polyethylene bottle, refrigerated, with a few drops of chloroform.
9. P solution standards. 0–4 μg P/mL : 0.01 μg P/mL, 0.1 μg P/mL, etc. Up to two times the expected maximum concentration of samples is diluted from the primary stock daily.

Procedures

Resin Bag Preparation

1. Sew resin bags (with a surface area approximately 50 cm^2, or 4–5 × 8–10 cm) from undyed stocking material or small-mesh rigid polyester screen. Any porous nylon or polyethylene material with a mesh size smaller than the resin bead may be used, but it is critical that dyed stocking material not be used because dye leached from the stockings may interfere with P analysis.
2. Weigh 4 g of dry-weight equivalent (wet/dry conversions are usually given on the bottle) resin, or 10 meq total anion exchange capacity (if total anion exchange capacity equivalents are given) into each bag.
3. Convert the anion resin to the bicarbonate form by shaking bags for 10 minutes in three successive 100 mL 0.5 mol/L $NaHCO_3$ solutions, rinsing with deionized water between each $NaHCO_3$ equilibration. Alternatively, resins may be converted to the bicarbonate form in large batches before placing in bags, but bicarbonate-form resins do not store as long as the hydroxyl-form resins, even when refrigerated.
4. Rinse bags thoroughly in deionized water and spin dry in a hand centrifuge or a salad spinner (do not air- or oven-dry). If a salad spinner is used, mesh should be placed around the open edges of the spinner so that the bags do not protrude through the slots and tear.

Resin Membrane Preparation

1. Cut membranes into strips of a standard dimension, usually 2 × 6 cm, and convert to the bicarbonate form as for resin beads. Membranes are often eas-

ier to use because they can be rinsed directly under deionized water and may be shaken rather than spun dry.

2. Membranes may be stored after rinsing in weak HCl, and then refrigerated in Ziplock bags. Because they are not completely abrasion resistant they need to be checked before each reuse.

Use of Bags and Membranes

1. Place 100 mL deionized water and 10 g of soil in a polyethylene bottle along with two resin strips or a resin bag, cap bottle lightly, and shake gently for 18 hours.
2. Remove bags, rinse thoroughly in deionized water, and spin (bags) or shake (membranes) dry.
3. Place bags or membranes in clean 250 mL bottles with 100 mL of 0.5 mol/L HCl and extract for 1 hour. Because bicarbonate still associated with the resin will be protonated by HCl, CO_2 will outgas, and extracting bottles will need to be uncapped periodically to release trapped CO_2, especially for the first 15 minutes.
4. Prepare standards by shaking bags or resin strips in 100 mL of the solution standards and extracting using the same procedure as described for soil solutions. Extracts and standards are then analyzed for ortho-P using methods described in the "Inorganic P in Solution" section, below.

Calculations

$$\mu g \text{ P/g soil} = C \times F \times 1/f_{dry\ soil} \times 1/f_P$$

where

C = concentration of P in the extract solution as μg P/mL extract
F = mL extract/g soil at field moisture content (e.g., 100 mL/10g soil)
$f_{dry\ soil}$ = the fraction of field-moist soil that is dry soil, or (oven-dry mass / fresh mass);
f_p = the fractional recovery efficiency of the ion exchange material as determined from the extracted standards, or (recovered P/standard P).

Special Considerations

If resin bags are to be recharged and reused, the extraction of ortho-P needs to take place with intact bags. The extraction of intact bags is time-consuming, and CO_2 is not readily released from nylon bags. If resins are to be recycled but the nylon or polyethylene bags disposed, bags can be cut and the resins placed into bottles before the addition of acid. After acid extraction, resins can be reconverted to the bicarbonate form and reused. Resins will degrade if completely dried; they should be refrigerated in weak (0.5 mol/L) HCl for storage. The efficiency of recycled resins should be monitored periodically using standard solutions as part of the standard protocol, described earlier.

Many researchers have used resin bags, prepared as described earlier, to monitor P availability in the field (Lajtha 1988; Giblin et al. 1994). For these applications, care should be taken to have all resin bags of uniform size because the extent of sorption depends more on total surface area than on total volume or weight of resin. Bags are flattened and placed horizontally in soil. It is generally assumed that water percolates vertically through the soil and the bags, yet it has been shown that if the bag material is of a hydrophobic fabric or if the mesh size is very small the movement of water will be impeded. Bags have been left in the field for periods of 2 weeks to 4 months before retrieval, although Giblin et al. (1994) found that phosphate was desorbed under field conditions if bags were left for long periods, and thus deployment times on the order of a few weeks are recommended. It is not recommended that field-placed resins be reused because some decomposition and fouling of the organic resin material occurs.

Anion-exchange resin–impregnated membranes have been used under field situations as well, and they appear to be simpler and easier to use and maintain than resin bags (Abrams and Jarrell 1992; Cooperband and Logan 1994). Ion-exchange membranes are flat, and thus they enhance the interaction of the exchange surface with soil, although they may inhibit percolation through exchange sites. This technique is described in detail in Chapter 9, this volume.

Soil Organic P

Even though plants assimilate inorganic P, organic P is an important source of inorganic P in most soils. This organic pool is affected by weathering and soil age, parent geochemistry, management practices and cultivation, and organic matter dynamics (Walker and Syers 1976; Tiessen et al. 1983; Stewart and Tiessen 1987). Organic P is potentially available to plants or microbes, or soil sorption sites, after mineralization. The turnover of organic P is, in part, dependent on the mineralization of organic carbon pools, although soluble pools of organic P can be mineralized by soil enzymes, thus disconnecting soil C and P cycles (McGill and Cole 1981). Due to the difficulties of measuring organic P mineralization, an estimate of organic P pools serves as an index of potentially available P over a longer time scale than that obtained from a resin extract, although the lability of this pool is highly variable (Tiessen and Stewart 1985). Because of this variability, "potentially mineralizable" organic P in soils is not analogous to the mineralizable N pools that are easily measured in field or laboratory incubations (see Chapter 13, this volume). Phosphorus fractionation techniques, such as the Hedley fractionation described below, may serve to better differentiate specific pools of organic P with different turnover times.

There are many extraction methods for the determination of organic P. These methods vary in complexity and time required. Perhaps the most common current method for measuring organic P is the dry combustion method of Saunders and Williams (1955) as modified by Walker and Adams (1958). This method is simple, requires few steps, and requires no special apparatus. The basic technique uses a strong acid extraction of ignited and nonignited soil samples. However, recoveries using this method are erratic and vary from Ca-dominated to Fe/Al-dominated soils and with weathering intensity. In particular, the ignition method may be highly inaccurate for weathered soils rich in Fe and Al oxides. Several researchers have observed negative values in tropical soils, perhaps because ignition changes the chemistry of oxides by driving off water of hydration, changing water and hydroxide balance at mineral surfaces, and thus affecting sorption. A more detailed discussion of potential errors with this method can be found in Olsen and Sommers (1982).

Mehta et al. (1954) described a sequential extraction procedure using HCl and NaOH. Bowman (1989) described a much simpler and faster procedure involving sequential extractions of H_2SO_4 and NaOH.

The alkaline EDTA extraction procedure of Bowman and Moir (1993) is simple, should work equally well in soils across a broad range of pH and clay content, and compares well with the Bowman (1989) procedure. It is recommended that this procedure be used to calibrate any other method; it is quite likely, in calcareous or low P sorbing soils, that the ignition method and the EDTA method will yield similar results. The principle behind the Bowman and Moir (1993) method is that an alkaline solution solubilizes soil organic matter (SOM) and associated organic P. EDTA chelates metal cations (and thus reduces cationic bridges with SOM) to increase the efficiency of SOM extraction. Persulfate is used to oxidize the extracted solution from dissolved organic P into ortho-P, which can then be measured colorimetrically. Organic P is calculated as the difference between persulfate (total) P and extracted inorganic P.

Materials

1. Electric ball mill (e.g., Spex mixer-mill)
2. Mechanical mixer
3. Incubator, heating water bath, or oven
4. 50 mL polypropylene screw-cap centrifuge tubes
5. Centrifuge plus adapters
6. Extracting solution: 0.25 mol/L NaOH in 0.05 mol/L Na_2EDTA

Procedure

1. Dry soils at 105 °C until no additional weight is lost, and store in a warm oven or dessicator.
2. Dried soils should be finely ground (<1 mm mesh) with an electric ball mill.
3. Weigh 0.5 g soil into a 50 mL polypropylene screw-top centrifuge tube (less mass should be used for highly organic or calcareous soils), add 25 mL of extracting solution, cap the vessel, and mix about 15 seconds with a touch mixer.
4. Loosen caps (to minimize gas buildup inside tubes) and place for 10 minutes in an oven, incubator, or water bath that has been preheated to 85 °C.
5. After 10 minutes, tighten caps and continue heating for a total heating time of 2 hours.

6. Remove tubes, cool until comfortable to handle, remix, and centrifuge at 10,000 rpm for 10 minutes. (Tabletop centrifuges often operate at lesser speeds; any speed is acceptable, but at lower speeds the time required for filtration increases.)
7. Filter the supernatant through no. 40 filter paper. Reserve one aliquot for inorganic P (ortho-P) determination (see "Inorganic P in Solution" section, below) and one for total P (persulfate) digestion (see "Analysis of Dissolved Organic P . . . " section, below).

Calculations

Convert both inorganic ortho-P and total P extract concentrations to soil quantities as:

$$\mu\text{g P/g soil} = C \times F$$
$$\text{organic P} = \text{total P} - \text{inorganic P}$$

where

C = concentration of P in extract solution as μg P/mL extract
F = mL extract/g dry soil (e.g., 25 mL extract/0.5 g dry soil)

Microbial P

In many soils, the cycling of P is highly dependent on microbial population dynamics, with labile P released only during episodic population crashes (Chapin et al. 1978). Microbial P may constitute a significant fraction of total P in highly organic soils (Walbridge 1991), yet might be insignificant in less organic soils (Lajtha and Schlesinger 1988). Problems with the measurement of microbial P include the calculation of a K_p, or extraction efficiency coefficient, which accounts for P not extracted after fumigation due to incomplete lysis and to soil sorption of released P. K_p values will vary with different soils and will be most problematic in high P fixing soils.

The method proposed here was developed by Hedley and Stewart (1982), following the chloroform fumigation technique of Anderson and Domsch (1978) first adapted by Cole et al. (1978) to estimate microbial P. In the Hedley and Stewart (1982) procedure, removal of labile P by resin extraction before the $NaHCO_3$ extraction step was found to increase the accuracy of the measurement of microbial P.

In the Hedley and Stewart (1982) procedure, dried and ground soil was used. However, microbial populations can change significantly with either storage or drying, and thus immediate processing of field-moist soils is strongly preferred. Fresh samples of field soils that are not ground require sample sizes of 2–5 g to allow for soil heterogeneity and for correction to oven-dry weight. Since soil:solution ratios need to be between 1:30 and 1:60, larger centrifuge tubes or bottles may be required. If immediate processing is not possible, then refrigeration is still preferable to air drying soils; air-dried soils should be rewetted and incubated before use (Powlson and Jenkinson 1976).

Materials

1. Resin strips (prepared as described in the section "Available P by Laboratory Resin Extraction," above).
2. 50 mL screw-cap polypropylene centrifuge tubes (250 mL centrifuge bottles if field-moist soils are used)
3. Chloroform ($CHCl_3$)
4. 0.5 mol/L $NaHCO_3$ adjusted to pH 8.5
5. 0.5 mol/L HCl
6. Centrifuge
7. Mechanical shaker with basket for holding tubes horizontally
8. 0.45 μm filters with syringe filter holders (e.g., Swinnex or Gelman filter holders) and 50 mL syringes
9. Acid-washed vials with polyethylene or polypropylene screw caps, 20–50 mL capacity

Procedure

1. Filters must be cleaned before use. This is accomplished by either filtering approximately 150 mL deionized water through the filter just prior to use or, preferably, soaking all filters in 0.5 mol/L HCl and then rinsing in deionized water several times before use. In the latter case, deionized water should still be passed through the filter before use to remove any remaining acid. If filters are not dried before use, about 5 mL of the extract that is being filtered should be passed through the filter and discarded before the filtered extract is saved for analysis.
2. Place duplicate 0.5 g finely ground, oven-dried soil subsamples in separate centrifuge tubes and add 30 mL deionized water and two resin strips in the bicarbonate form to each. If field-moist soils are used, 2–3 g soil and 150 mL deionized water should be used.
3. Cap the tubes and shake gently for 16 hours.
4. Remove resin strips, gently rinsing soil particles back into the tube with 1–2 mL deionized water. Swirl tubes gently while vertical before placing in a centrifuge in order to dislodge soil particles lodged in the cap or the tops of tubes. Centrifuge tubes for 10 minutes at 10,000 rpm, or at lower speeds for longer times. Decant the supernatant carefully so as not to lose soil and discard. If there is interest in measuring the resin P pool, the resin strips should be rinsed thoroughly in deionized water, placed in clean 50 mL tubes with 30 mL of 0.5 mol/L HCl, and shaken gently for 2 hours. The resulting solution is saved for analysis.
5. Add 1 mL $CHCl_3$, in a hood, to one tube (the "fumigated" tube), cap, and shake for 1 hour. Remove the cap and allow the chloroform to evaporate overnight under the hood.

6. Allow the other, nonfumigated tube to sit overnight, uncapped; refrigeration is suggested to avoid microbial P release, although this was not mentioned in the original procedure.
7. Add 30 mL of 0.5 mol/L $NaHCO_3$ (or 150 mL to bottles containing field-moist samples) to the fumigated and unfumigated tubes, cap, and shake for 16 hours. After 16 hours centrifuge the tubes and filter and save the supernatants.
8. Analyze both supernatants ("fumigated" and "unfumigated") for total (persulfate) P (described later in this chapter).

Calculations

$$\mu g\ P/g \text{ in microbial biomass} = [(\textit{Total P} \text{ in fumigated soil}) - (\textit{Total P} \text{ in unfumigated soil})]/k_p$$

where

Total P = μg P/g dry soil as determined by acid persulfate digestion (see later)
k_p = extraction efficiency = 0.4 (see the section "Special Considerations," below)

Special Considerations

This extraction removes only part of the microbial P that is lysed by the chloroform fumigation, and thus the k_p, or extraction efficiency factor, differs from 1.0. Ideally, a k_p factor is measured for each soil type. However, because this procedure involves culturing appropriate fungi and bacteria that are radiolabeled (Hedley and Stewart 1982), it cannot be casually undertaken for each site or study. Over a wide variety of soils, k_p factors have been found to range from 0.32 to 0.47, and thus an approximate k_p factor of 0.4 may be used with some degree of accuracy. While k_p factors are generally lower in highly weathered soils, the relationship between k_p and sorption capacity is not quantitative and thus cannot be modeled. However, some authors have used sorption capacity in place of a measured k_p factor.

P Sorption

Sorption is considered to be the most important process controlling P availability in soils. The dominant sorption reactions are those with Fe-Al oxides and hydroxides and Ca-carbonates, but the strength of sorption can vary considerably among soils. Since sorption is to some degree reversible, sorbed P may be a source of plant-available P either immediately or over a longer term. However, the degree of reversibility is not well understood, and there is no single easy measurement of P sorption.

The approach taken by most researchers to measure sorption has been to run a series of sorption studies at constant temperatures, or sorption isotherms, for a set period of time and then fit a model to the data in order to get a single number, or index, that can be compared across sites. Models include the simple Langmuir equation (Olsen and Watanabe 1957), from which a sorption maximum can be calculated, but this model does not fit empirical curves of sorption very well. Several authors have modified this simple equation to include distinct sorption surfaces or reactions (e.g., Holford and Mattingly 1975). The Freundlich equation, which describes a model of adsorption in which the affinity for adsorption decreases exponentially as adsorbed P increases (Barrow 1978), seems to fit many highly P-deficient soils. However, no one model is currently the accepted norm in all soils. Sorption maxima are often calculated, but soils rarely show a distinct sorption maximum; an initial, rapid period of sorption is generally followed by slower rates of P uptake. Studies have shown a long-term migration of P into aggregate or particle surface sorption sites that are of decreasing accessibility to solution (Willett et al. 1988; Fardeau and Jappe 1980; Tiessen et al. 1991), and this makes modeling the dynamics of P sorption complex. For the simple indices described here, however, we use a standard reaction time of 24 hours.

All of these models require that P sorption be measured over a range of P concentrations in solution so that curves can be generated. Bache and Williams (1971) suggested that valid comparisons could be made among different soils with an index that measured sorption at only one single high P addition, making the measurements faster and easier. Here we recommend that simple P sorption curves be generated and that the Bache and Williams (1971) index be calculated so that soils may be compared. Once the curves are made, models may be fitted if desired, depending on the questions being addressed.

Most of the original procedures recommend that sieved, oven-dried soils be used. However, oven drying soils with high clay contents could cause irreversible aggregation, making their handling nearly impossible. Drying may result in the conversion of fairly lightly sorbed P into physicochemical fractions that are fairly resistant to redissolution, thus causing major errors in estimates of already sorbed P. In such cases soils should be analyzed for available P while still field moist. Grinding soils is not recommended, since new mineral surfaces are thus exposed that would not be exposed under field conditions. Authors working in relatively young or unweathered soils have not reported significant problems in the use of air-dried or oven-dried soils, although comparisons to results with field-moist soils are rarely reported.

Materials

1. 50 mL screw-cap polypropylene centrifuge tubes
2. Centrifuge with adapters
3. Chloroform ($CHCl_3$)
4. Acid-washed vials with polyethylene or polypropylene screw caps, 20–50 mL capacity
5. 0.01 mol/L KCl
6. 1000 μg P/mL stock solution: 4.39 g oven-dried primary standard-grade KH_2PO_4 dissolved in 1000 mL deionized water. Stored in a polyethylene bottle, refrigerated, with a few drops of chloroform.
7. P working solutions: 0–200 μg P/mL solutions should be diluted daily from the stock solution and made up in 0.01 mol/L KCl.

Procedure

1. Add 30 mL of working solution to 3.0 g of oven-dry equivalent, unground soil in a 50 mL centrifuge tube with one drop of chloroform or toluene. Each soil at each working solution level should be run in duplicate.
2. Shake tubes for 24 hours at a constant temperature (approx 25 °C), centrifuge at 10,000 rpm for 30 minutes, and decant the supernatant, filtering if necessary.
3. Analyze solutions for ortho-P (described later in this chapter).

Each soil at each working solution level should be run in duplicate, and at least three working solutions, plus blank solution, should be used to generate the sorption isotherm curve. The range of working solution concentrations used does not affect intersite comparisons even if the range differs among studies, since curves of soil sorption *vs.* resulting equilibrium solution are plotted.

Calculations

Sorbed P

$$X_s = (s - c) \times F$$

where

X_s = sorbed P at working solution concentrations (μg P/g soil)
s = μg P/mL of original working solution
c = μg P/mL in equilibrium solution
F = mL working solution/g dry soil (e.g., 30 mL/3 g dry soil)

Note that if isotopically exchangeable P can be measured (see Bache and Williams 1971 and "Special Considerations," below) or if resin-extractable P is measured, then the correct final equation is

$$X_s = [(s - c) \times F] - E$$

where

E = isotopically exchangeable or resin-extractable P, as μg P/g soil

Sorption Curves

Sorption curves are plotted as X versus c. At this point, curves may be fitted by eye or by using the models discussed earlier. A somewhat artificial absorption maximum (X_m) may be calculated using the Langmuir equation, which can be expressed as

$$c/X = c/X_m = 1/kX_m$$

A plot of c/X against c should give a straight line of slope $1/X_m$; k is a constant that relates to the bonding energy. The Freundlich equation may be written as

$$X = a \times c^{1/n} \text{ or } \log(X) = \log(a) + (1/n) \times \log(c)$$

where a and n are constants, and a plot of log (X) versus log (c) would yield a straight line, and the slope can be calculated as an index for comparison.

Phosphorus Adsorption Index

The phosphorus adsorption index (*PAI*) of Bache and Williams (1971) is calculated from sorption with a working solution of 150 μg P/mL, and is calculated as

$$PAI = X/\log(c)$$

This index is used rather than a simple measure of X at 150 μg P/mL to correct for the different P equilibrium concentrations (c) that will result from differential sorption capacities of soils that are being compared.

Special Considerations

Many researchers have used background electrolyte solutions such as $CaCl_2$ for pH stabilization and to maintain electrochemical neutrality, although Carreira and Lajtha (1997) have shown that this leads to erroneous results in neutral to high pH soils and recommend KCl as a background solution. Strength of the background electrolyte has been shown to affect sorption, and although researchers have used KCl at concentrations that can range from 0.02 mol/L (Bache and Williams 1971) to 0.1 mol/L (Barron et al. 1988) in sorption studies, we here recommend a standard concentration of 0.01 mol/L KCl.

Tiessen et al. (1991) added an interesting twist to this method by determining not only P sorbed from a water solution but also the amount of P that is sorbed strongly enough so as not to be extractable by anion-exchange resin. After pouring off the supernatant following the centrifugation step, resin-extractable P is determined using 30 mL deionized water and two resin strips in the bicarbonate form (*as above*). This resin-extractable P (in μg P/g soil) is subtracted from the calculation of sorbed P as described in "Available P by Laboratory Resin Extraction," above. This is not a standard method, but curves of sorbed P (X) and sorbed P that is not resin-extractable vs. c on the same graphs can be informative.

A problem may arise for soils with high amounts of already sorbed P, such as highly fertilized agricultural soils. Because P that is adsorbed in a soil sample includes both the P sorbed from the equilibration solution and this initially sorbed P, the measurement of initially sorbed P is important if accurate adsorption capacities are to be calculated. However, because the measurement of the initially sorbed P is both difficult and controversial (i.e., there is no general agreement on whether this term should be measured with a resin extract or with stronger extractants such as a weak acid extract), many authors have ignored the initially sorbed P because it is a small percentage of total sorption capacity, or have produced nonlinear sorption curves when initially sorbed P is significant. Bache and Williams (1971) found that a 24 hour measure of isotopically exchangeable P (referred to as E) could accurately correct for this initially sorbed P in fertilized soils. However, the use of ^{32}P in lab sorption studies will never be routine and thus cannot be recommended as a standard measurement here. This problem with initially sorbed P will be most serious for plotted sorption curves (see the calculation for sorption curves, above), but will not be as significant for the Bache and Williams (1971) single-addition method (see the calculation for phorphorus adsorption index, above), which was created in part to swamp out the effect of initially sorbed P. Thus the Bache and Williams index may prove to be the most useful index for comparing a large number of soils. However, for plotted sorption curves we recommend the analysis of resin P and using this value as E in the calculation for μg P sorbed g^{-1} soil, above.

Total P and Total Element Analysis by Fusion

Various methods for the analysis of total P have been proposed, including both wet and dry digestion techniques. Unfortunately, many of these methods are plagued with poor recoveries for certain types of soils, and thus they are not useful for cross-system comparisons. One fairly common and accurate procedure for total P analysis is fusion with Na_2CO_3 (Brenner et al. 1980; Olsen and Sommers 1982). The lithium metaborate fusion procedure has the advantage of being simpler, and it allows for the simultaneous total analysis of most major and minor ions (Thompson and Walsh 1983). In addition, less expensive graphite crucibles may be used instead of the platinum or nickel crucibles that are used for the Na_2CO_3 fusion, and fusions can be conducted on multiple samples in a muffle furnace rather than by hand, individually, over a flame. Although the Na_2CO_3 fusion technique certainly yields accurate results, we recommend the lithium metaborate fusion procedure here.

Materials

1. Ultrapure graphite crucibles (e.g., from Ultra Carbon, Bay City, MI; 517-894-2911)
2. Soil or rock mill (e.g., Spex mixer-mill) for powdering rock
3. Reagent-grade $LiBO_2$
4. Muffle furnace (small is better than large due to the high heat used in this procedure)
5. Long furnace tongs, furnace gloves, protective eyewear, etc.
6. 10% HNO_3 solution by volume (200 mL conc. HNO_3 in approx. 1 L deionized water, bring to 2 L final volume). Note: if trace elements are to be determined simultaneously, Ultrapure or trace metal grade HNO_3 must be used.
7. 150 mL acid-washed beakers with watch glass lids
8. Stir plates and Teflon-coated stir bars
9. 100 mL or 250 mL acid-washed volumetrics

Procedure

1. All soils must be very finely powdered prior to fusion. Using 10 g soil, 1–2 minutes are required for grinding in most rock mills. Consistency should be that of very fine talcum powder, without contamination by larger particles.

2. Place powdered soil (0.25 g) and 0.75 g lithium metaborate in a prefired graphite crucible. Mix the powders thoroughly with a thin spatula until the color is uniform and independent bands of soil or lithium metaborate cannot be observed.
3. Using extreme caution and heat-resistant gloves, eye protection, and long tongs, place crucibles in a muffle furnace that has been preheated to 1000 °C. Fusion is usually completed in 10–15 minutes and is observed as a glowing white (not red) bead. Fusion time is furnace-specific, and trial runs should be conducted to determine the minimum time required to achieve complete fusion in order to extend crucible life.
4. While fusion is proceeding, place beakers with 50 mL of 10% HNO_3 on individual stir plates with stir bars, and start stirring just before opening the furnace. As each crucible is removed from the furnace, the molten bead is immediately poured into the acid and a watch glass placed over the beaker to reduce evaporation and contamination. The molten bead dissolves within 1–2 hours.
5. Transfer solutions to volumetric flasks and make up to either 100 mL or 250 mL with deionized water, depending on the estimated total P in the soil or the concentrations of the other elements that are being analyzed.
6. Analyze solutions for ortho-P. Standards are made in the same normality of HNO_3 as final sample solutions. If 250 mL volumetrics are used, standards should be made over the range of 0–1000 μg P/mL. Note that these levels are very high, and dilution of solutions will be necessary before ortho-P analysis in most cases.

Calculations

$$\mu\text{g P/g soil} = C \times F$$

where

C = concentration of P in the extract solution as μg P/mL extract
F = mL extract/g dry soil (e.g., 250 mL/0.25 g dry soil)

Special Considerations

For best results, graphite crucibles should be briefly ashed two or three times before use and wiped clean with lint-free Kimwipes. New crucibles with unconditioned walls and older crucibles (those with a reduction in wall thickness of 50% after repeat use) may have a tendency to retain small beads of flux. In most cases these beads can be scraped and added to the 10% HNO_3 solution, but they may not dissolve. For soils rich in Fe, pouring may be difficult and the amount of $LiBO_2$ flux may need to be increased to 1 g.

Platinum crucibles may be used as well, although they are significantly more expensive. In this case, crucibles with sample and flux are heated as described earlier, allowed to cool, and placed whole in the beaker with 10% HNO_3.

Soil P Fractionation

Soil P fractionation is too complicated to recommend as a routine soil procedure; instead, it is a focused research tool that provides insight on the biogeochemistry of P in terrestrial ecosystems. Researchers may wish to examine organic versus inorganic fractions of P in soils, or available versus recalcitrant P fractions, and how soil P fractions change with cultivation (e.g., Hedley et al. 1982; Tiessen et al. 1983) or in different landscape positions (e.g., Lajtha and Schlesinger 1988). Cross-site comparisons might also be useful; extensive soil survey comparisons have already been made (e.g., Tiessen et al. 1984; Cross and Schlesinger 1995).

There is a long history of procedures that have been developed to fractionate phosphorus pools in soils and in sediments. The Chang and Jackson (1957) procedure was used widely for many years and was subsequently modified by Williams et al. (1967). However, there are problems associated with several extractants used in these procedures (for a discussion see Tiessen and Moir 1993), and they are not recommended.

Perhaps the most widely used method in recent years for soil P fractionation is the Hedley fractionation (Hedley et al. 1982), outlined in modified form in Figure 7.1. The Hedley procedure was designed to examine pools of P that have fairly clearly defined chemical properties. Note, however, that this fractionation procedure is evolving, and many modifications have been proposed. For example, many researchers do not include the sonification step, although this pool ranges from 10% to 40% of P extracted by the previous step. The determination of microbial P is often deleted from this procedure in part due to erratic results in acid, highly weathered soils (Potter et al. 1991). Tiessen and Moir (1993) recently suggested a hot concentrated HCl step to follow the dilute HCl step to extract some of the organic P that remains and that is otherwise counted as part of the residual P pool in the original procedure. Perrott (1992) found that where exchangeable Ca was present in soils, precipitation of calcium phosphate and calcium-organic matter complexes occurred during the NaOH extraction; dissolution of this precipitated calcium phosphate during the subsequent acid extraction resulted in an overestimation of Ca-bound P. Perrott (1992) suggested prewashing soils with a buffered NaCl-EDTA solution. Certainly this step should be considered for soils with high concentrations of exchangeable Ca. It is likely that more modifications and caveats to this procedure will appear in the future. Unfortunately, as the fractionation procedure is modified, it becomes difficult to compare results among studies. Moreover, it is virtually impossible to compare results obtained using different fractionation procedures that have used different extractants altogether.

The principles behind the extraction steps in the Hedley et al. (1982) procedure are straightforward. The pool measured by the resin strips is fairly labile inorganic P that is directly exchangeable with the soil solution (see section "Available P by Laboratory Resin Extraction," above). The alkaline bicarbonate extract (Olsen et al. 1954) provides a measure of relatively labile and plant-available P sorbed onto soil surfaces, as the bicarbonate mimics the respiration activity of plant roots and will depress the activity of Ca^{2+} in high Ca soils. This step is more effective in relatively unweathered soils than in tropical soils, however. NaOH extracts amorphous and

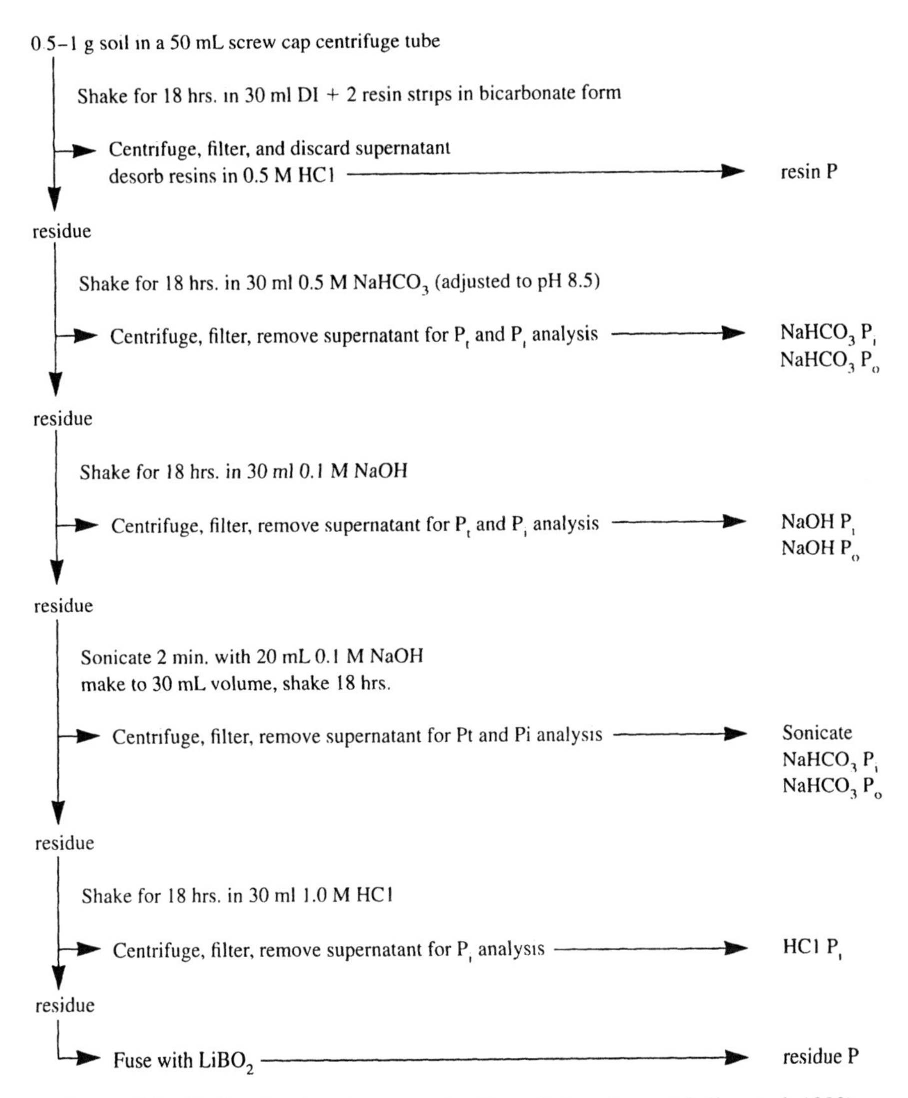

Figure 7.1. Hedley fractionation procedure for soil phosphorus (Hedley et al. 1982).

some crystalline Fe and Al phosphates, as well as P strongly bound by chemisorption to Fe and Al compounds. There is also a large organic pool of P in this fraction in some soils. Ultrasonification with fresh NaOH enables extraction of P held at internal surfaces of soil aggregates. Relatively insoluble Ca-P minerals, including apatite, are extracted with HCl. Residual P is a combination of organic and inorganic forms, and is determined by fusion (see earlier discussion) or as the difference between total P and the sum of the fractions determined.

The question of drying, and grinding, of soils emerges in this procedure as well (see earlier discussion). The original procedure suggested using dried and ground soils, although Potter et al. (1991) noted problems with dried versus field-moist soils. Lajtha and Schlesinger (1988) found few significant differences between using ground and unground soils for fairly unweathered arid soils. However, most researchers recommend using dried and ground soils to reduce sample heterogeneity, realizing that such pretreatment will influence results. High-clay soils either must be analyzed while still field-moist or, if dried, must be ground. Certainly grinding will release new mineral surfaces, which will be important for poorly weathered soils, and drying will affect microbial pools and estimates of sorbed P in many cases. Because using field-moist soils may be impossible in many cases, we recommend using dried and ground soils for ecosystem intercomparisons—with full knowledge of the associated problems and assumptions.

Materials

1. 50 mL polypropylene centrifuge tubes with screw-cap lids
2. Centrifuge with adapters for the tubes
3. Sonicator, probe type rather than bath
4. Mechanical shaker with basket for holding tubes horizontally
5. 0.45 μ filters with syringe filter holders (e.g., Swinnex or Gelman filter holders) and 50 mL syringes
6. Acid-washed vials with polyethylene or polypropylene screw caps, 20–50 mL capacity
7. Resin strips (prepared as described the section "Available P by Laboratory Resin Extraction," above)
8. 0.5 mol/L HCl
9. 0.5 mol/L $NaHCO_3$ adjusted to pH 8.5
10. 0.1 mol/L NaOH
11. 1.0 mol/L HCl
12. Concentrated HCl (optional)

Procedure

1. Filters must be cleaned before use. This is accomplished either by filtering approximately 150 mL deionized water through the filter just prior to use or, better yet, by soaking all filters in 0.5 mol/L HCl and then rinsing in deionized water several times before use. In the latter case, deionized water should still be passed through the filter prior to use to remove any acid residue. If filters are not dried before use, then about 5 mL of the extract that is being filtered should be passed through the filter and discarded before filtered extract is saved for analysis.
2. Each step outlined in Figure 7.1 is conducted on a separate day. Place 1 g finely ground, oven-dried soil in a centrifuge tube and add 30 mL deionized water and two resin strips converted to the bicarbonate form. Cap tubes and shake gently for 18 hours.
3. Remove resin strips, making sure that soil particles are gently rinsed back into the tube with 1–2 mL deionized water. Swirl tubes gently while vertical to dislodge soil particles lodged in the cap or the tops of tubes, then centrifuge for 10 minutes at 10,000 rpm, or at lower speeds for longer times. Discard the

supernatant, pouring out as much solution as possible by hand without losing soil. Carefully aspirate the remaining supernatant into the syringe, which is then fitted with the filter holder containing a cleaned filter. Soil remaining on the filter is scraped back into the tube or rinsed off with some of the 30 mL of the next extractant, and the tubes are ready for the next extractant. Resins are rinsed thoroughly in deionized water, placed in clean 50 mL tubes with 30 mL 0.5 mol/L HCl, and shaken gently for 2 hours. The resulting solution is saved for analysis.

4. For the rest of the extracts, the supernatant is saved after centrifugation for both dissolved inorganic and dissolved total P analysis (see later discussion). As much suspension-free supernatant as possible is decanted into acid-washed vials, but all cloudy supernatant should be filtered. Remaining supernatant is filtered as described earlier, with soil remaining on the filter returned to the tube.

Special Considerations

The fractions will consist of solutions with different pH values and levels of extracted organic matter. Therefore, solutions should be neutralized before P analysis (see below) or else analysis reagents should be made up to account for the various levels of acidity. In many soils the $NaHCO_3$ and the NaOH fractions will be highly colored. These solutions should be acidified to <pH 2 with a few milliliters of 1 mol/L H_2SO_4 before analysis of inorganic P or the dissolved organic matter will precipitate in the acid conditions of the P reagents. The exact number of milliliters needed should be determined either stoichiometrically or else separately using blank extracting solution, since pH electrodes should never be used in solutions that are to be analyzed. Solutions are refrigerated for 30 minutes and then brought back to room temperature, and aliquots are removed for analysis once organic matter has precipitated and solutions are clear. Alternatively, solutions may be centrifuged.

Analysis of Dissolved Organic P and Total Dissolved P by Acid Persulfate Digestion

Several persulfate digestion procedures can be used for the analysis of total dissolved phosphorus in soil solutions or other aquatic samples, and by difference, dissolved organic P. The principle behind persulfate digestion is to use potassium (or ammonium) persulfate to oxidize dissolved organic P to ortho-P, and total P is measured as soluble reactive phosphorus (see earlier discussion). Organic P is then calculated as total P less inorganic P in the solution. The procedure that will be described here is an acid persulfate digest (Bowman 1989) that should be used only for the oxidation of P (not N) in solutions or extracts. A common alternative when both dissolved organic N and dissolved organic P are to be determined in soil solutions or extracts is the alkaline persulfate digest (Chapter 5, this volume; Ameel et al. 1993; D'Elia et al. 1976).

Johnes and Heathwaite (1992) describe a very fast and simple procedure for the simultaneous digestion of total N and total P in aqueous samples using persulfate microwave digestion. Because this procedure requires an automated microwave that is adjustable in 1% power increments and specialized Teflon vessels, it will not be described here. However, if the user has access to an automated microwave system (e.g., CEM Digestion Systems), the microwave technique is the easiest and fastest procedure for total dissolved N and P analysis.

Materials

1. 25 mL volumetric flasks (or double all volumes and use 50 mL volumetrics), or block-digestion tubes
2. Potassium persulfate, low-N ($K_2S_2O_8$)
3. 5.5 mol/L H_2SO_4
4. Hot plate or block digestor
5. Organic standards (may include glycerophosphate, ATP, p-nitrophenyl phosphate, sodium inositol hexaphosphate, etc.)

Procedure

1. Pipette an appropriate aliquot or standard (e.g., 1 mL) into a 25 mL volumetric flask or block-digestion tube, then add 0.5 g $K_2S_2O_8$ and 1 mL of 5.5 mol/L H_2SO_4 to each flask. Start with an aliquot of 1 mL sample; increase volume if concentrations are too low, and adjust the normality of the acid reagent accordingly.
2. Digest samples on a hot plate or in a block digestor (approx. 150 °C) for 20–30 minutes. Digestion is complete once vigorous boiling subsides. After solution cools, deionized water is added to volume.
3. Analyze total P as ortho-P after neutralization (see "Inorganic P in Solution," below). The malachite green method is recommended.
4. Two organic P standards, two replicate analyses, two spikes, and two blanks (deionized water) should be brought through the entire procedure for each batch of 25 sample unknowns to control for contamination of reagents and contamination during the digestion process, as well as to check for digestion completion. Spikes are 0.05 μg P/mL (15 μL of the 50 μg P/mL stock, for low-P solutions) or 0.1 μg P/mL (30 μL of the 50 μg P/mL stock, higher P solutions) added to samples to check for recoveries and matrix interferences. When analyzing for ortho-P, regular standards made up in blank solution are run to determine oxidant contamination and volume loss.

Calculations

$$\mu g\ P/mL \text{ in solution} = C \times F$$

where

C = concentration of P in the digestion solution as μg P/mL digest
F = mL digestion solution/mL sample (e.g., 25 mL/1 mL).

Special Considerations

An alternative approach for large numbers of samples or low-P solutions is to use an autoclave as for the alkaline persulfate digestion. In this case, 5 mL sample are added to a 50 mL volumetric or a 40 mL glass screw-top vial. 10 mL of 0.9 mol/L H_2SO_4 and approximately 0.8 g $K_2S_2O_8$ are added. Vials are capped tightly, or volumetrics are covered with tin foil, and are autoclaved at 121 °C and 17 psi for 50 minutes. After cooling, vials are brought to volume with deionized water, and soluble reactive phosphorus (see later) is measured after neutralization. For very low levels of P in solution, more sample solution can be added. The normality of the acid reagent should be adjusted accordingly.

Inorganic P in Solution—Murphy and Riley Procedure

The level of inorganic P in solutions measured by the colorimetric techniques described later is often referred to as soluble reactive phosphorus (SRP) to distinguish it from orthophosphate, or "free" phosphate, in solution. Phosphate sorbed onto Al- or Fe-colloidal complexes may not be directly available to bacteria or to phytoplankton, and perhaps not to roots in soils except over the long term, but this form of P is partially measured by both the Murphy and Riley (1962) procedure and the malachite green procedure due to the high acidity of the reagents. Acid hydrolyzable organic P is also measured with these colorimetric procedures. While this consideration does not represent a problem for ecosystem-level studies of element budgets or fluxes, it has been a major impediment to studies of P cycling using ^{32}P in soils and aquatic ecosystems. Unfortunately, measures of SRP do not provide a good estimate of the size of the unlabeled P compartment because SRP can be 2–10 times the true ortho-P pool.

Several techniques have been developed to try to estimate true "free" ortho-P in solutions, often using plant bioassays or algal cultures (Rigler 1966). Such measures are beyond the scope of this chapter. SRP here is used as an estimate for free ortho-P, with an understanding of the limitations of the results.

Two methods are commonly used to measure SRP in solution, although several others exist. The Murphy and Riley (1962) procedure, as modified by Watanabe and Olsen (1965), is commonly adapted for automated analysis. The manual procedure is given here; most autoanalyzers come with instructions for automated procedures. The malachite green procedure (described next) is significantly more sensitive than the Murphy and Riley (1962) procedure but has the disadvantage that malachite green stains plastic ware, countertops, and skin.

Materials

1. Ammonium molybdate
2. Antimony potassium tartrate
3. Ascorbic acid
4. 50 mL volumetrics for samples

5. 500 mL (2), 1000 mL (1), and 2000 mL (1) volumetrics for reagents
6. Spectrophotometer
7. 50 μg P/mL stock solution: 0.2195 g oven-dried primary standard-grade KH_2PO_4 dissolved in 1000 mL deionized water. Stored in a polyethylene bottle, refrigerated, with a few drops of chloroform.
8. P solution standards. Up to two times the expected maximum concentration of samples is diluted from the primary stock daily.

Procedure

1. The following solutions should be made:
   A. H_2SO_4 (2.5 mol/L). Add 278 mL concentrated H_2SO_4 to approximately 1000 mL deionized water in a volumetric that is partly submerged in cold water. When cool, bring to 2000 mL. This solution is extremely stable.
   B. Ammonium molybdate (40.0 g in 1000 mL deionized water). If kept refrigerated it will precipitate, but the mixture can be shaken and used without problem. This solution is stable for several months.
   C. Antimony potassium tartrate (1.454 g in 500 mL deionized water). This solution is stable.
2. In a 500 mL volumetric, add 250 mL solution A, 75 mL solution B, 2.64 g ascorbic acid, and 25 mL solution C. Solution C reduces interferences from silica and thus is critical for total P analyses, but adding solution C is less critical for freshwater samples. Bring to 500 mL with deionized water. This mixed reagent should be kept in a dark bottle and needs to be made fresh daily; sensitivity is noticeably lower after 8 hours.
3. Many solutions will not be neutral (e.g., resin extracts, total P fusions, most solutions from the fractionation procedure). Neutralization can be accomplished in two ways. In the first method, after 40 mL of the solution is added to the 50 mL volumetric, a few drops of paranitrophenol indicator are added. If the solution is acid, then add 4 mol/L NaOH dropwise until the solution turns yellowish, then add 0.25 mol/L H_2SO_4 dropwise until the solution just turns clear, and make to volume. If the solution is alkaline, just add acid until the solution plus indicator turns clear, and then make to volume. The second method involves adjusting the acidity of the mixed reagent to compensate for the acidity or alkalinity of solutions to be analyzed. Because the final pH of the solution plus reagent is critical for complete color development, the amount of acid added to the final mixture from acidic solutions can be subtracted from the amount of acid added to the final solution by the mixed reagent.
4. To 40 mL of a neutral (pH 5–7.5) solution (see later) in a 50 mL volumetric, add 8 mL of mixed reagent, bring to volume with deionized water, wait exactly 10 minutes, and read at 880 nm (most sensitive) or 660 nm (less sensitive) with a spectrophotometer. For high-P samples, less sample solution and more diluent can be used as long as the final volume remains constant across samples. In many spectrophotometers, path lengths can be increased for greater sensitivity. Note that all volumes can be reduced to save solution and

reagents, but proportions should be maintained. The standard curve should span the range of measured samples.

Special Considerations

It is very easy to contaminate collecting bottles or laboratory equipment with phosphates. Thus any material that will come in contact with solutions that will be analyzed for P, such as collection bottles, filters, and bottles containing reagents, should not be washed with laboratory detergents. In addition, some materials such as Parafilm have been found to contain high concentrations of P; even Parafilm placed over acidic solutions has been shown to contaminate solutions. Bottles and other equipment should be acid-washed before use and rinsed thoroughly with deionized water, and plastic gloves should be worn whenever handling solutions or glassware that will be used for P analyses.

If solutions are highly stained with dissolved organic matter (e.g., alkaline soil extracts), this organic matter needs to be precipitated before analysis or else it will precipitate in the acidic reagents used for analysis. These solutions should be acidified to $<$pH 2 and refrigerated for 30 minutes; after the cleared sample returns to room temperature, an aliquot is removed for analysis with a pipette from the surface. The exact number of milliliters needed should be determined on a separate blank solution using a pH meter; never use a pH electrode in solution that is to be analyzed. If solutions are only lightly colored, reading at 880 nm rather than 660 will reduce interferences. A nonreagent blank may be run to account for any remaining color in the solution.

Inorganic P in Solution—The Malachite Green Procedure

The malachite green procedure for inorganic P is more sensitive than the Murphy and Riley (1962) procedure and should be used where warranted because of low concentrations. The manual procedure of Ohno and Zibilske (1991) is offered here; an automated procedure can be found in Fernandez et al. (1985) that is fast and simple. Note, however, that there is a misprint in the Fernandez et al. (1985) paper: reagent B should be in 8 mol/L acid, not 4 mol/L as printed. Thus the volume of acid used should be doubled from their directions.

Materials

1. Concentrated H_2SO_4
2. 1000 mL volumetrics
3. 20 mL vials
4. Ammonium para-molybdate
5. Polyvinyl alcohol
6. Malachite green
7. Hotplate
8. 50 μg P/mL stock solution: 0.2195 g oven-dried primary standard-grade

KH_2PO_4 dissolved in 1000 mL deionized water. Store in a polyethylene bottle, refrigerated, with a few drops of chloroform.

9. P solution standards. Up to two times the expected maximum concentration of samples is diluted from the primary stock daily.

Procedure

1. Prepare reagent 1: Add 106 mL concentrated H_2SO_4 to 500 mL deionized water—never the reverse—in a 1000 mL volumetric flask. Dissolve 17.55 g ammonium para-molybdate in the acid and bring to volume.
2. Prepare reagent 2: Heat approximately 800 mL deionized water to 80°C, and add 3.5 g polyvinyl alcohol and stir until dissolved. Add 0.35 g malachite green and stir until dissolved. Cool to room temperature and dilute to 1000 mL with deionized water.
3. Pipette 10 mL of sample or standard into a 20 mL vial. Add 2 mL of reagent 1 and mix. After 10 minutes add 2 mL reagent 2 and mix. After 30 minutes read absorbance at 630 nm in a 20 mm cuvette and compare with standard curve.

Special Considerations

See the description of the Murphy and Riley (1962) procedure in the section "Inorganic P in Solution" (above) for procedures to use for samples that are highly colored with organic matter.

Summary

The study of P cycling and availability in soils has been hampered by the difficulty in measuring theoretically defined pools. Most pools are defined operationally or by specific extracts and indices that we hope are correlated with our theoretical fractions. Problems with the definition and measurement of "bioavailable" P are perhaps the greatest challenge to studies of P cycling, although resin P may prove to be the best index developed to date. Similarly, so far it has proven difficult to reliably measure organic P mineralization. Detailed analyses of specific P fractions by sequential leaching techniques may prove to be the best tool for studying P biogeochemical cycles across ecosystems; these techniques are still evolving.

References

Abrams, M. M., and W. M. Jarrell. 1992. Bioavailability index for phosphorus using ion exchange resin impregnated membranes. *Soil Science Society of America Journal* 56:1532–1537.

Ameel, J. J., R. P. Axler, and C. J. Owen. 1993. Persulfate digestion for determination of total nitrogen and phosphorus in low-nutrient waters. *American Environmental Laboratory* 10/93:1–11.

Anderson, J. P. E., and K. H. Domsch. 1978. Mineralization of bacteria and fungi in chloroform-fumigated soils. *Soil Biology and Biochemistry* 10:207–213.

Bache, B. W., and E. G. Williams. 1971. A phosphate sorption index for soils. *Journal of Soil Science* 22:289–301.

Barron, V., M. Herruzo, and J. Torrent. 1988. Phosphate adsorption by aluminous hematites of different shapes. *Soil Science Society of America Journal* 52:647–651.

Barrow, N. J. 1978. The description of phosphate adsorption curves. *Journal of Soil Science* 29:447–462.

Bowman, R. A. 1989. A sequential extraction procedure with concentrated sulfuric acid and dilute base for soil organic phosphorus. *Soil Science Society of America Journal* 53:362–366.

Bowman, R. A., and J. O. Moir. 1993. Basic EDTA as an extractant for soil organic phosphorus. *Soil Science Society of America Journal* 57:1516–1518.

Brenner, I. B., A. E. Watson, G. M. Russell, and M. Concalves. 1980. A new approach to the determination of the major and minor constituents in silicate and phosphate rocks. *Chemical Geology* 28:321–330.

Carreira, J. A., and K. Lajtha. 1997. Factors affecting phosphate sorption along a Mediterranean, dolomitic soil and vegetation sequence. *European Journal of Soil Science* 48:139–149.

Chang, S. C., and M. L. Jackson. 1957. Fractionation of soil phosphorus. *Soil Science* 84:133–144.

Chapin, F. S., III, R. J. Barsdate, and D. Barel. 1978. Phosphorus cycling in Alaskan coastal tundra: a hypothesis for the regulation of nutrient cycling. *Oikos* 31:189–199.

Chapman, P. J., C. A. Shand, A. C. Edwards, and S. Smith. 1997. Effect of storage and sieving on the phosphorus composition of soil solution. *Soil Science Society of America Journal* 61:315–321.

Cole, C. V., E. T. Elliott, H. W. Hunt, and D. C. Coleman. 1978. Trophic interactions in soils as they affect energy and nutrient dynamics. V. Phosphorus transformations. *Microbial Ecology* 4:381–387.

Cooperband, L. R., and T. J. Logan. 1994. Measuring *in situ* changes in labile soil phosphorus with anion-exchange membranes. *Soil Science Society of America Journal* 58:105–114.

Cross, A. F., and W. H. Schlesinger. 1995. A literature review and evaluation of the Hedley fractionation: applications to the biogeochemical cycle of soil phosphorus in natural ecosystems. *Geoderma* 64:197–214.

D'Elia, C. F., P. A. Steudler, and N. Corwin. 1976. Determination of total nitrogen in a queous samples using persulfate digestion. *Limnology and Oceanography* 22:760–764.

Fardeau, J. C., and J. Jappe. 1980. Choix de la fertilisation phosphorique des sols tropicaux: emploi du phosphore 32. *Agronomie Tropicale* 35:225–231.

Fernandez, J. A., F. X. Niell, and J. Lucena. 1985. A rapid and sensitive automated determination of phosphate in natural waters. *Limnology and Oceanography* 30:227–230.

Giblin, A. E., J. A. Laundre, K. J. Nadelhoffer, and G. R. Shaver. 1994. Measuring nutrient availability in arctic soils using ion exchange resins: a field test. *Soil Science Society of America Journal* 58:1154–1162.

Hedley, M. J., and J. W. B. Stewart. 1982. Method to measure microbial phosphate in soils. *Soil Biology and Biochemistry* 14:377–385.

Hedley, M. J., J. W. B. Stewart, and B. S. Chauhan. 1982. Changes in inorganic and organic soil phosphorus fractions by cultivation practices and by laboratory incubations. *Soil Science Society of America Journal* 46:970–976.

Holford, I. C. R., and G. E. G. Mattingly. 1975. The high- and low-energy phosphate adsorbing surfaces in calcareous soils. *Journal of Soil Science* 26:407–417.

Johnes, P. J., and A. L. Heathwaite. 1992. A procedure for the simultaneous determination of total nitrogen and total phosphorus in freshwater samples using persulfate microwave digestion. *Water Research* 26:1281–1287.

Lajtha, K. 1988. The use of ion-exchange resin bags for measuring nutrient availability in an arid ecosystem. *Plant and Soil* 105:105–111.

Lajtha, K., and A. F. Harrison. 1995. Strategies of phosphorus acquisition and conservation by plant species and communities. Pages 139–147 *in* H. Tiessen, editor, *Phosphorus in the Global Environment: Transfers, Cycles, and Management.* SCOPE 54. Wiley, Chichester, UK.

Lajtha, K., and W. H. Schlesinger. 1988. The biogeochemistry of phosphorus cycling and phosphorus availability along a desert soil chronosequence. *Ecology* 69:24–39.

Lin, T. H., S. B. Ho, and K. H. Houng. 1991. The use of iron-oxide impregnated filter paper for the extraction of available phosphorus from Taiwan soils. *Plant and Soil* 133:219–226.

McGill, W. B., and C. V. Cole. 1981. Comparative aspects of cycling of organic C, N, S and P through soil organic matter. *Geoderma* 26:267–286.

Mehta, N. C., J. O. Legg, C. A. I. Goring, and C. A. Black. 1954. Determination of organic phosphorus in soils: 1. Extraction methods. *Soil Science Society of America Proceedings* 18:443–449.

Menon, R. G., L. L. Hammond, and H. A. Sissingh. 1989. Determination of plant-available phosphorus by the iron hydroxide–impregnated filter paper (P_i) soil test. *Soil Science Society of America Journal* 53:110–115.

Murphy, J., and J. P. Riley. 1962. A modified single solution method for determination of phosphate in natural waters. *Analytica Chimica Acta* 27:31–36.

Myers, R. G., G. M. Pierzynski, and S. J. Thien. 1995. Improving the iron oxide sink method for extracting soil phosphorus. *Soil Science Society of America Journal* 59:853–857.

Ohno, T., and L. M. Zibilske. 1991. Determinations of low concentrations of phosphorus in soil extracts using malachite green. *Soil Science Society of America Journal* 55:892–895.

Olsen, S. R., C. V. Cole, F. S. Watanabe, and L. A. Dean. 1954. *Estimation of Available Phosphorus in Soils by Extraction with Sodium Bicarbonate.* USDA Circular No. 939. USDA, Washington, DC, USA.

Olsen, S., and L. Sommers. 1982. Phosphorus. Pages 403–430 *in* A. Page, R. Miller, and D. Keeney, editors, *Methods of Soil Analysis: Chemical and Microbiological Properties.* 2d edition. American Society of Agronomy, Madison, Wisconsin, USA.

Olsen, S. R., and F. S. Watanabe. 1957. A method to determine a phosphate adsorption maximum of soil as measured by the Langmuir isotherm. *Proceedings of the Soil Science Society of America* 21:144–149.

Parfitt, R. L., L. J. Hume, and G. P. Sparling. 1989. Loss of availability of phosphate in New Zealand soils. *Journal of Soil Science* 40:371–382.

Perrott, K. W. 1992. Effect of exchangeable calcium on fractionation of inorganic and organic soil phosphorus. *Communications in Soil Science and Plant Analysis* 23:827–840.

Potter, R. L., C. F. Jordan, R. M. Guedes, G. J. Batmanian, and X. G. Han. 1991. Assessment of a phosphorus fractionation method for soils: problems for further investigation. *Agriculture, Ecosystems and Environment* 34:453–463.

Powlson, D. S., and D. S. Jenkinson. 1976. The effects of biocidal treatments on metabolism in soil. II. Gamma irradiation, autoclaving, air-drying and fumigation. *Soil Biology and Biochemistry* 8:179–188.

Rigler, F. 1966. Radiobiological analysis of inorganic phosphorus in lakewater. *Verhandlungen. Internationale Vereinigung fur Theoretische und Angewandte Limnologie* 16:465–470.

Saggar, S., M. J. Hedley, and R. E. White. 1990. A simplified resin membrane technique for extracting phosphorus from soils. *Fertilizer Research* 24:173–180.

Saunders, W. M. H., and E. G. Williams. 1955. Observations on the determination of total organic phosphorus in soil. *Journal of Soil Science* 6:254–267.

Schoenau, J. J., and W. Z. Huang. 1991. Anion-exchange membrane, water, and sodium bicarbonate extractions as soil tests for phosphorus. *Communications in Soil Science and Plant Analysis* 22:465–492.

Sharpley, A. N. 1991. Soil phosphorus extracted by iron-aluminum-oxide–impregnated filter paper. *Soil Science Society of America Journal* 55:890–892.

Sibbesen, E. 1978. An investigation of the anion-exchange resin method for soil phosphate extraction. *Plant and Soil* 50:305–321.

Stewart, J. W. B., and H. Tiessen. 1987. Dynamics of soil organic phosphorus. *Biogeochemistry* 4:41–60.

Thompson, M., and J. N. Walsh. 1983. *A Handbook of Inductively Coupled Plasma Spectrometry.* Blackie and Son, Glasgow, Scotland.

Tiessen, H., E. Frossard, A. R. Mermut, and A. L. Nyamekye. 1991. Phosphorus sorption and properties of ferruginous nodules from semiarid soils from Ghana and Brazil. *Geoderma* 48:373–389.

Tiessen, H., and J. O. Moir. 1993. Characterization of available P by sequential extraction. Pages 75–86 *in* M. R. Carter, editor, *Soil Sampling and Methods of Analysis.* Lewis Publishers, Boca Raton, Florida, USA.

Tiessen, H., and J. W. B. Stewart. 1985. The biogeochemistry of soil phosphorus. Pages 463–472 *in* D. E. Caldwell, J. A. Brierley, and C. L. Brierley, editors, *Planetary Ecology.* Van Nostrand Reinhold, New York, New York, USA.

Tiessen, H., J. W. B. Stewart, and C. V. Cole. 1984. Pathways of phosphorus transformations in soils of differing pedogenesis. *Soil Science Society of America Journal* 48:853–858.

Tiessen, H., J. W. B. Stewart, and J. O. Moir. 1983. Changes in organic and inorganic phosphorus composition of two grassland soils and their particle size fractions during 60–90 years of cultivation. *Journal of Soil Science* 34:815–823.

Walbridge, M. R. 1991. Phosphorus availability in acid organic soils of the lower North Carolina coastal plain. *Ecology* 72:2083–2100.

Walbridge, M. R., and P. M. Vitousek. 1987. Phosphorus mineralization potentials in acid organic soils: processes affecting ^{32}P isotope dilution measurements. *Soil Biology and Biochemistry* 19:709–717.

Walker, T. W., and A. F. R. Adams. 1958. Studies on soil organic matter. I. Influence of phosphorus content of parent materials on accumulation of carbon, nitrogen, sulfur, and organic phosphorus in grassland soils. *Soil Science* 85:307–318.

Walker, T. W., and J. K. Syers. 1976. The fate of phosphorus during pedogenesis. *Geoderma* 15:1–19.

Watanabe, F. S., and S. R. Olsen. 1965. Test of an ascorbic acid method for determining phosphorus in water and $NaHCO_3$ extracts from the soil. *Proceedings of the Soil Science Society of America* 29:677–678.

Willett, I. R., C. J. Chartres, and T. T. Nguyen. 1988. Migration of phosphate into aggregated particles of ferrihydrite. *Journal of Soil Science* 39:275–282.

Williams, J. D. H., J. K. Syers, and T. W. Walker. 1967. Fractionation of soil inorganic phosphate by a modification of Chang and Jackson's procedure. *Proceedings of the Soil Science Society of America* 31:736–739.

Zou, X., D. Binkley, and B. A. Caldwell. 1995. Effects of dinitrogen-fixing trees on phosphorus biogeochemical cycling in contrasting forests. *Soil Science Society of America Journal* 59:1452–1458.

Zou, X., D. Binkley, and K. G. Doxtader. 1992. A new method for estimating gross phosphorus mineralization and immobilization rates in soils. *Plant and Soil* 147:243–250.

8

Analysis of Detritus and Organic Horizons for Mineral and Organic Constituents

Mark E. Harmon
Kate Lajtha

The analysis of the chemical composition of plant matter (leaves, stems, roots, and detritus) is critical for studies of nutrient turnover in ecosystems and of the biotic pools of important biogeochemical elements, as well as for understanding nutrient and chemical limits to plant growth. Understanding the organic constituents of plant litter and their transformation into those forming soil organic matter is also a critical need given their link to the biogeochemistry of other elements and the storage of carbon in ecosystems.

The chemical analysis of plant materials for mineral constituents is fairly straightforward, although different materials may need either different pretreatment or slightly different digestion methods. In contrast, the analysis of organic constituents is not straightforward because of both the underlying complexity of the compounds themselves and the variety of methods that can be used, none of which are perfect. Our intent in this chapter is to review these methods and recommend standard protocols. In the case of organic constituents, we recognize that these recommendations may become dated rapidly if the most modern instrumentation becomes generally available.

Sample Pretreatment

Aboveground plant materials, including plant litter, organic layers, and woody detritus, should be dried at 65 °C in an oven before homogenizing. For plant litter this may take 48 hours, whereas woody debris may take up to a week. Complete drying should be monitored by measuring sample weight loss; when weights have stabilized for 24 hours, drying is complete. Once dry, materials should be homogenized by grinding in a Wiley mill or similar grinder for macroelement analysis. Hand grinding with a mortar and pestle may be warranted for trace element analysis. Materials should be ground to pass a 60-mesh (0.246 mm) screen for carbon-hydrogen-nitrogen (CHN) or carbon-nitrogen-sulfur (CNS) combustion analyses, or a 20-mesh (0.833 mm) to 40-mesh (0.417 mm) screen for wet digestion. Samples should be stored either in a warm oven or in a desiccator to prevent rehydration prior to elemental analysis.

Analysis of root materials is slightly more complex. Roots removed from soils need to be cleaned of soil without unduly fragmenting the roots or leaching water-soluble compounds, and thus should not be dried before processing. In addition, Fe and various other metals may form insoluble coatings on root surfaces, particularly in soils with fluctuating water tables. Such coatings are virtually impossible to remove, and thus interpretation of the chemical analysis of these elements from roots must include this pool of surface-bound, nonorganic material. Field- or greenhouse-moist roots can be placed in a beaker or other acid-washed, cleaned glass or polyethylene container and gently swirled in a phosphate-free, dilute detergent solution. Only a small amount of detergent is needed, and plant roots do not need to be mechanically rubbed, since the purpose of this step is to break the surface tension. After rinsing, roots are swirled in a 0.01 mol/L NaEDTA solution for 5 minutes, which complexes cations bound at the surfaces, including metals. After rinsing in deionized water, roots should be placed immediately in paper bags or small envelopes and dried for 24 hours at 65 °C. Once dry, roots can be ground and stored as discussed previously.

Available Methods

Organic Matter Digestion for Mineral Analysis

There are several published methods for organic matter digestion and chemical analysis, including both dry ashing and various wet digestion/ashing techniques. We will not discuss dry ashing techniques because although these are quite fast and simple to perform, they may cause loss of elements due to volatilization, sorption on crucible surfaces, or particulate loss. Dry ashing can never be used for the analysis of volatile elements such as nitrogen or sulfur. Nitrogen and sulfur are most easily analyzed on a CHN or CNS analyzer, although wet ashing and analysis are still possible if access to such analyzers is limited. Wet digestion techniques vary in the oxidants and acids used to oxidize organics and dissolve chemical constituents. In this chapter we avoid procedures that use perchloric acid because it is extremely dangerous to use, it requires specialized hoods, and much safer alternatives exist. We recommend two fairly straightforward wet digestion methods, one for major element analysis and one for trace element analysis.

Organic Constituents

Numerous methods exist for the analysis of organic constituents in detritus and organic horizons. Although only two are recommended as standard procedures, the organic chemistry of detritus and organic horizons is extremely complex. Depending on the question being asked, additional analysis methods may be needed. Three classes of methods are used to determine organic constituents: (1) mass loss methods that determine composition by sequential extraction, hydrolysis, and oxidation; (2) end-product methods that analyze the chemical constituents resulting from oxidation or thermal decomposition; and (3) nondestructive methods based on the spectral properties of samples.

Mass Loss Methods

Gravimetric determinations of organic fractions after various extraction, hydrolysis, and oxidation steps are collectively called proximate analysis. Given the complex nature of organic compounds to be found in undecomposed and decomposed plant litter, these methods define constituents operationally based on their resistance to various chemical treatments. This classic method has routinely been applied to determine organic constituents of forest products (Ryan et al. 1990). In this method the first step is to extract polar and nonpolar compounds, followed by hydrolysis of the remaining material in heated 72% H_2SO_4 (approx. 12 M). The remaining mass, minus ash, is the so-called Klason, or acid-resistant, lignin.

Proximate analysis may not produce well-defined chemical constituents. That is, lignin as measured by proximate analysis may contain other chemical constituents such as condensed tannins. This mismatch is less problematic for undecomposed gymnosperm wood, the substrate for which proximate analysis was originally developed, but the mismatch increases as one applies the analysis to angiosperm or decomposing wood because the lignin in these substrates can be partially hydrolyzed by acid (Effland 1977). The mismatch between chemical composition and operational definitions becomes an especially significant issue for leaves and fine roots, which contain aliphatic compounds such as cutin and suberin, as well as condensed tannins that may be acid-resistant but not extracted by the solvents usually used in proximate analysis (Trofymow et al. 1995).

An alternative proximate analysis method was developed by Van Soest (1963a, 1963b, 1965) to quantify the fiber and lignin content of livestock feed and forage. This method uses boiling acid-detergent solutions to extract simple sugars and hemicelluloses, leaving the acid-detergent fiber (ADF), which is then hydrolyzed in 72% H_2SO_4 at room temperature to remove the cellulose. The remaining material, adjusted for ash content, is termed the acid-detergent lignin fraction (ADL). In samples rich in waxes and cutins, the mass remaining after the H_2SO_4 hydrolysis may be oxidized with potassium permanganate ($KMnO_4$) to remove the lignin (Van Soest and Robertson 1981). The remaining material is ashed, and the ash-corrected mass is considered to be cutin. The Van Soest method was commonly used by ecologists (Fogel and Cromack 1977; Melillo et al. 1982) until the mid-1980s when the forest products–based proximate analysis increased in popularity (McClaugherty et al. 1985; Aber et al. 1990; Ryan et al 1990). Although the Van Soest method is easier than traditional forest products analysis, the fact that polar and nonpolar extractives are not considered does limit the interpretations of constituents that are possible. These extractions can, however, be used in conjunction with the Van Soest method to give a more complete analysis of constituents (Ryan et al. 1990; Gallardo and Merino 1993).

End-Product Methods

The end products of classic proximate analysis can be further analyzed, but this is rarely done in ecological studies. An alternative method for decomposing organic samples into constituent fractions is alkaline cupric-oxide oxidation (Hedges and Parker 1976). The oxidation products are then analyzed by capillary gas chromatography (Hedges and Parker 1976; Hedges and Mann 1979), gas chromatography and mass spectroscopy in series (Goñi and Hedges 1990a), or high-performance liquid chromatography (Kögel-Knabner and Ziegler 1993). The technique has been applied mainly to lignin and less often to cutin (Goñi and Hedges 1990a,b).

The main goal of the studies that employ end-product methods has been to identify the source of the organic portion of marine sediments. This is possible because the ratios of oxidation products (e.g., syringyl:vanillyl phenols in the case of lignin, or C_{14}:C_{16} fatty acids in the case of cutin) can indicate derivation from angiosperm versus gymnosperm taxa and/or woody versus nonwoody tissues. Similar applications are possible for plant detritus (Hedges et al. 1988) and organic horizons (Kögel-Knabner 1986; Kögel-Knabner and Ziegler 1993). The main drawback of the technique is that the oxidation itself is time-consuming and the efficiency of the oxidation process in producing detectable end products is difficult to determine (Goni and Hedges 1990a). A faster thermochemolysis method uses tetramethylammonium hydroxide (TMAH) (Hatcher et al. 1995). This method appears more suited to large numbers of samples and may become standard with time.

In addition to chemical treatments, heat can be used to decompose the organic fractions in a sample. One such method that has recently shown great promise is analytical pyrolysis (Kögel-Knabner et al. 1992; Preston et al. 1994), in which samples are heated gradually to an upper limit of 320–750 °C. The pyrolysis products are then analyzed with a mass spectrometer or a gas chromatograph–mass spectrometer linked in series. Of all the methods we reviewed, this system gives the most detail on the chemical structure of constituents such as carbohydrates, lignin, fatty acids, and aromatic and aliphatic esters and how they are transformed by decomposition. A second method involving thermal decomposition of organic fractions is differential scanning calorimetry combined with differential thermogravimetry (Reh et al. 1990). In this system, extremely small samples (3–5 mg) are heated at a constant rate from 100 to 800 °C to burn the organic compounds at their characteristic combustion temperatures. The amount of heat released at each temperature (calorimetry) and rate of mass loss (thermogravity) are used to identify the amount of each compound present in the sample. This method holds great promise, because even the form of constituents (e.g., amorphous versus crystalline cellulose) can be determined.

Nondestructive Methods

Near-infrared reflectance (NIR) spectroscopy is a nondestructive technique that uses the reflectance of dried, ground organic material to indirectly determine concentrations of ash, nitrogen, and organic constituent contents of dried, ground samples. Although this method has gained widespread use for qualitative analysis of oilseed, grains, and forages, its acceptance as an ecological analysis tool has been slow (Wessman et al. 1988a). Numerous tests, however, have shown its applicability to fresh as well as decomposed plant litter (Wessman et al. 1988a; McClellan et al. 1991a,b; Joffre et al. 1992; Gillon et al. 1993). The technique is based on the fact that individual plant constituents have characteristic absorbance properties in the near-infrared spectral region (i.e., 1100–2500 nm). While these peaks are clearly defined in pure compounds, they are not as well defined in plant material. Reflectance spectra of plant litter and organic horizons exhibit overlapping absorption peaks that correspond to overtones and combinations of C-H, O-H, or N-H chemical bonds. Material-specific calibration of absorption spectra against wet chemistry data enables one to determine the abundance of a wide range of constituents. Once this calibration is completed, however, an extremely large number of samples can be assayed in a short time.

Nuclear magnetic resonance (NMR) spectroscopy is another nondestructive means to determine organic constituents of plant litter, soil organic horizons, and mineral soil (Preston 1993; Baldock and Preston 1995; Preston 1996). The technique, which was originally developed for use in organic chemistry and biochemistry, is based on the fact that atomic nuclei have a characteristic magnetic moment or spin that can be altered by radio frequency waves in a strong magnetic field (Jardetzky and Roberts 1981). The frequency of radio waves required to alter the spin is affected by the chemical bonding associated with the nuclei, resulting in the so-called chemical shift. Phenolic bonds, for example, modify the energy required to alter the spin differently than alkyl bonds.

With the advent of Fourier transform techniques and application of the cross-polarization magic angle spinning (CPMAS) method, NMR has become a powerful tool for the analysis of the complex nature of organic matter in soil and litter as it can be applied to dilute solutions, complex solids, extracts, and gels. For organic compounds the nuclei examined are ^{13}C, but other nuclei such as ^{15}N and ^{31}P can also be examined. Although NMR is an excellent analysis tool, and should be applied more widely, the instrumentation and training needed to use it are generally not available to ecologists. This is indeed unfortunate because the method has already been used to examine many long-held hypotheses concerning the chemical nature of soil organic matter (Kögel-Knabner et al. 1992; deMontigny et al. 1993; Preston et al. 1994; Preston 1996).

Suggested Standard Methods

Recommending a single method for the analysis of organic constituents is difficult given that none of the currently available methods is without problems of either chemical precision (e.g., gravimetric determinations) or availability (e.g., NMR). Indeed, recommending a single method is counter to the recent, healthy trend to examine organic constituents from several approaches (Kögel-Knabner et al. 1992; de Montigny et al. 1993; Preston et al. 1994). Nonetheless, we have selected two complementary methods for general application based on their availability and training requirements. These are the forest products–based proximate analysis (Ryan et al.

1990) and NIR spectroscopy (Wessman et al. 1988a). Proximate analysis has been widely used in the past, and despite problems of interpretation (e.g., is the acid-resistant fraction really lignin?) the method can be applied to a wide range of materials with a minimum of expensive equipment and training. Results of proximate analysis for litter have also been the basis for many existing litter decomposition models (Aber et al. 1990; Parton et al. 1994), a trend that is likely to continue for some time. NIR spectroscopy has been selected because it is extremely fast and can be applied to more samples than would be possible with other methods. Given its calibration to proximate analysis, it is subject to the same limitations of interpretation; however, this also makes it completely consistent with current litter decomposition models. Finally, NIR spectroscopy has the advantage that it can be linked to remote sensing (Wessman et al. 1988b), allowing one to potentially examine large-scale patterns of aboveground litter quality.

H_2SO_4–H_2O_2 Digestion for Plant Major Element Analysis

Perhaps the most common technique for plant material analysis is wet digestion using sulfuric acid and hydrogen peroxide in a block digester. As with any sulfuric acid digest, $CaSO_4$ may precipitate, and thus this digest should not be used for calcium analysis. In addition, $CaSO_4$ may precipitate in the glass tubes after repeated use, and thus tubes used for H_2SO_4 digests may never be used for calcium analysis.

Materials

1. 20- or 40-position block digester with tube rack
2. 50, 70, or 75 mL calibrated block-digestion tubes
3. Concentrated reagent-grade H_2SO_4
4. H_2O_2 (30%), reagent-grade, low P

Procedure

1. Weigh 200–300 mg of dried, ground material into acid-washed block-digestion tubes. This weight may need to be doubled in the case of woody material. Care should be taken to ensure that powders are placed near the bottom of the tube and do not adhere near the top.
2. Add 5 mL concentrated H_2SO_4 to each tube, swirling to wet the material and to wash down any powder from the sides.
3. Add 2 mL H_2O_2 very slowly and carefully to each tube, swirling constantly to reduce the vigorous boiling that will ensue.
4. When all tubes have finished boiling (1–5 minutes), place the rack in the block that has been preheated to no more than 170 °C, manually turn the heater on, and digest for 1 hour. The temperature must reach 230 °C before the heater is turned off, but if this temperature is reached before the hour, turn off the heater and allow tubes to sit until the end of the hour.
5. Remove the rack from the block, turn off the heater, and allow the tubes to cool. When tubes are cool to the touch, add another 2 mL H_2O_2 to each tube.

6. When the block has cooled to 175 °C or below, turn on the heater, place the rack with tubes in the block, and digest for an additional 2 hours, taking care that the final temperature does not exceed 350 °C. At least 1 hour of this final digest should be at 330–350 °C to ensure complete removal of the H_2O_2 because H_2O_2 interferes with Murphy and Riley (1962) phosphorus analysis.
7. After tubes are cool, solutions are saved in acid-washed polyethylene or glass vials. Analysis of nitrogen, phosphorus, potassium, magnesium, sulfur, and various trace elements can be performed from this digest.

Calculations

The concentration in mg element/g tissue (C_{tissue}) of an element (C) is calculated as

$$C_{tissue} = C_{digest} \times V_{digest} / M_{dry}$$

where

C_{digest} = the concentration of the digest (mg/L)
M_{dry} = the dry mass of the sample digested (g)
V_{digest} = the volume of the calibrated digestion tube, typically 0.05, 0.07, or 0.075 L

For roots or organic materials that might have high ash contents, results are often expressed on an ash-free dry-mass basis. Ash is determined using a muffle furnace (see later), and fractional ash-free dry mass ($F_{ash\text{-}free}$) is calculated as:

$$F_{ash\text{-}free} = (M_{dry} - M_{ash}) / M_{dry}$$

where

M_{dry} = the dry mass of sample
M_{ash} = the ash mass of sample

Ash-free concentration of elements of samples ($C_{ash\text{-}free}$) in units of mg element/g ash-free tissue is calculated as

$$C_{ash\text{-}free} = (C_{digest} \times V_{digest}) / (M_{dry} \times F_{ash\text{-}free})$$

where C_{digest}, V_{digest}, M_{dry}, and $F_{ash\text{-}free}$ are defined above.

Special Considerations

A standard plant sample (e.g., from National Institute of Standards and Technology [NIST]) and at least two sample replicates should be brought through the digestion procedure with every batch of 40 samples. Wood standards are not available from NIST; however, three wood standards ranging in amount of decay are available from Phillip Sollins or Mark Harmon at Oregon State University (see Contributor List, this volume, for addresses). Finally, blank matrix material for the automated analysis should be made by following the preceding procedure without adding tissue material.

Digestion of woody material may be incomplete with the procedure described earlier, leading to an underestimate of mineral element concentrations. Addition of either K_2SO_4 or Na_2SO_4 can raise the boiling temperature sufficiently to increase the recovery of nitrogen, phosphorus, and other elements from woody material (Dan Binkley, personal communication).

Digestion for Trace Element Analysis

Microwave or hot plate digestion of plant materials is the most commonly used procedure for trace element or micronutrient analysis. Nitric acid is a more powerful oxidant than sulfuric acid, and thus peroxide is rarely needed. Obviously, nitrogen cannot be determined in these digests. This technique is particularly useful if automated microwave digestion equipment is available. If not, commercially available microwave ovens may be used with microwave Parr bombs that are available from Cole-Parmer. The microwave procedure is the easiest digestion procedure, although the resulting digestate is strongly acid, which may cause problems for analysis if dilution is not possible. Hot-plate digestion is the most low-tech procedure, but it requires more operator time. In all cases, samples can be evaporated to dryness and redissolved in weak HNO_3 to avoid the excess acid problem.

Here we describe digestion using a hot plate. If there is access to an automated microwave digestion system, follow the instructions included with the machine, since models vary. If microwave digestion is done manually, the technique is equally simple, but microwave digestion should only be attempted in ovens made for this purpose, because the power increments on standard microwave ovens are generally not fine enough and they lack safety features. Specific methods for Parr bomb digestion depend on the size of bomb purchased; directions are included with the specific bombs. Although microwave digestion is easy and convenient, more samples may be processed at the same time using a hot plate.

Materials

1. 20 mL closed Teflon vials (available from Cole-Parmer)
2. Large, adjustable hot plate
3. Concentrated HNO_3, ultrapure or trace metal grade
4. 25 mL calibrated volumetric flasks (polypropylene or Teflon for trace metal analysis)

Procedure

1. With large hot plates, up to 20 samples may be digested at one time. Place 200 mg of dried and ground plant material in the vials and add 2 mL concentrated HNO_3.
2. Close the lids and digest the mixture on a hot plate for 1 hour at ~120 °C. This

is generally the lowest heat setting on commercial hot plates and is the point at which a light reflux, or condensation, is first observed on lids.

3. If colorimetric analysis is to be employed, the digestion can stop here. Samples are then quantitatively transferred to 25 mL volumetric flasks and brought to volume with deionized (DI) water.
4. For analysis in which acid concentrations are a problem (e.g., graphite-furnace atomic-absorption spectrometry, inductively coupled plasma (ICP) spectrometry, or ICP-Mass Spectrometry (ICP-MS)), evaporate digestates to dryness at moderate heat (75–105 °C) taking care not to char the residue. Bring residues up to volume (usually 20 mL, although this will vary with the sensitivity needed for analysis) with 3% HNO_3. Note that if samples are evaporated to dryness, then the residue may be redissolved by adding a known mass of HNO_3 rather than bringing samples to volume in a small volumetric flask. Adding acid by mass is significantly more accurate than using a small volumetric flask. Because variable amounts of HNO_3 are reduced and thus lost during the digestion step, digest solutions cannot be brought to volume using known weights of acid unless samples are first brought to dryness.

Calculations

To determine the concentration of a trace element in a material, use the same calculations as for macroelement analysis, described earlier. The only difference is that the digest volume for trace element analysis is typically 0.02 L. For roots or materials that might have high ash contents because of adhering soil, results should be expressed on an ash-free dry-weight basis as described under the preceding procedure.

Special Considerations

Although some researchers have added several drops of H_2O_2 to the NO_3 digestion, this is generally more useful for animal tissues with high lipid contents than for plant materials.

Teflon vials used for trace metal analysis should be cleaned by boiling in 25% aqua regia (1:3 HNO_3:HCl). Solutions should be stored in polyethylene bottles that have been heated for 48 hours in 5–10% HCl.

A plant material standard (e.g., from NIST) should be processed through the entire digestion procedure with every batch of 20 samples, along with one blind replicate and at least one blank. The analysis of standards is crucial to identify problems with the procedure such as incomplete digestion or contaminated reagents.

Forest Products–Based Proximate Analysis

The forest products–based proximate analysis recommended is a series of extraction, hydrolysis, and oxidation steps (Fig. 8.1). These are used to determine gravimetrically the proportions of general classes of organic constituents.

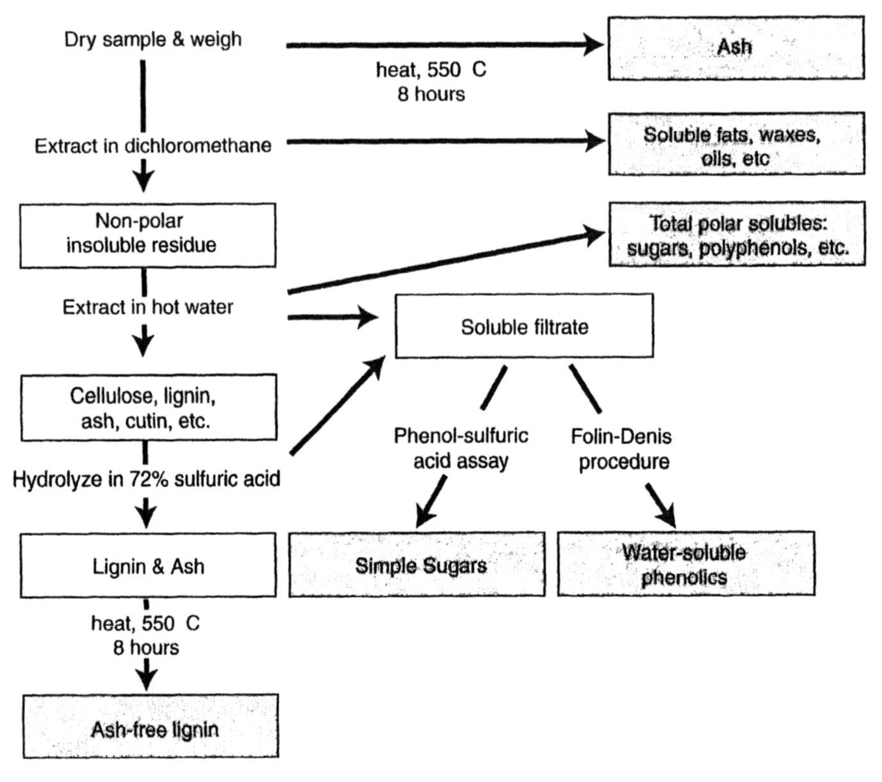

Figure 8.1. Flow diagram of steps used in proximate analysis. The open boxes indicate intermediate components separated by chemical and/or heat treatments. The shaded boxes indicate components resulting from final calculations. Modified from Ryan et al. (1990).

Materials

The materials for the various extractions and hydrolysis steps overlap; therefore, materials specific to each are indicated by the abbreviations to the right of the materials list (NPE = nonpolar extractives; PE = polar extractives; SA = sugar analysis; TA = tannin analysis; AH = acid hydrolysis; ASH = ash).

1. Pre-ashed fritted glass filtering crucibles (TA, AH, ASH)
2. Pre-ashed Gooch filtering crucible (NPE, PE, AH)
3. Wire hangers to suspend extraction thimbles (NPE)
4. BD-20 or BD-40 block-digestion tubes; straight tubes without volume markings are acceptable (NPE, PE)
5. Rubber stoppers with cold finger setup (see Figure 8.2) (PE, SA, TA, AH, ASH)
6. Walter crucible holders (NPE)
7. Buchner flask (NPE)
8. 100 mL beakers (NPE)
9. 15 mL round-bottomed Pyrex test tubes (AH)
10. 125 mL Erlenmeyer flask (AH)

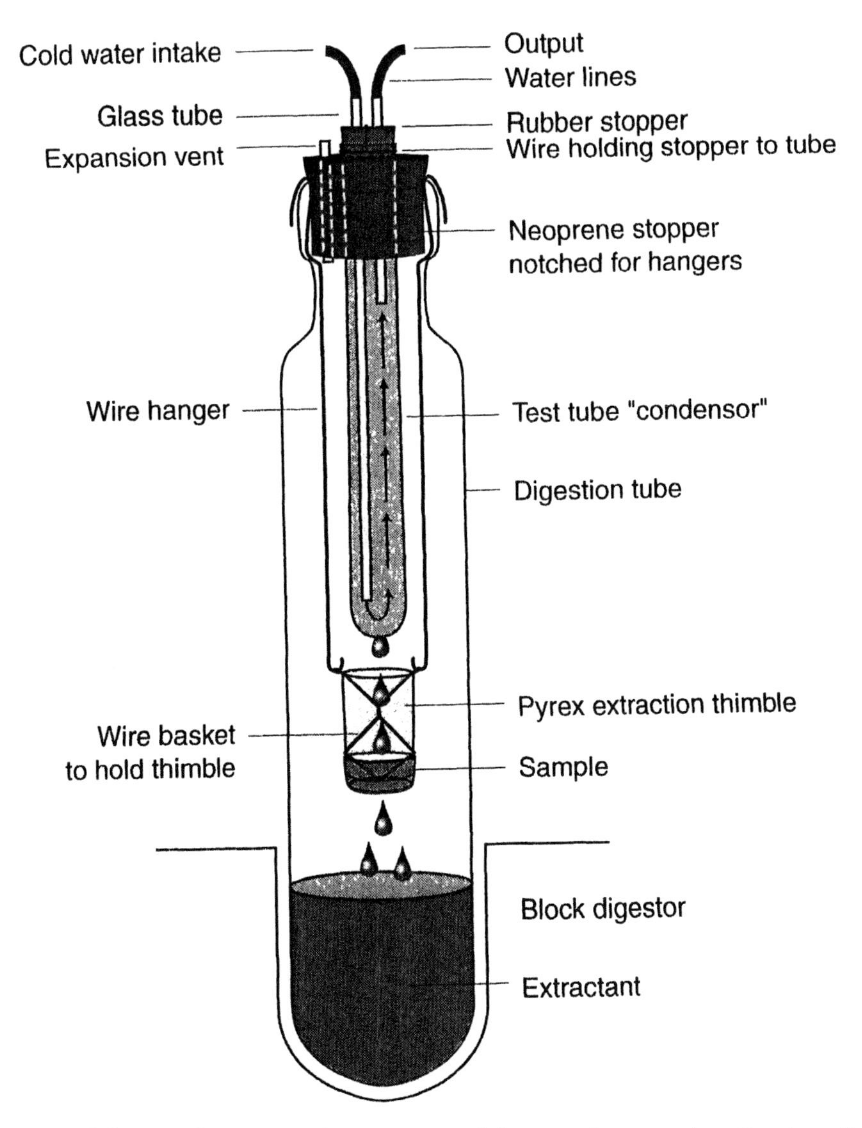

Figure 8.2. Cold finger extraction system used for nonpolar extraction. Based on the system developed by C. A. McClaugherty (personal communication).

11. 250 mL Erlenmeyer flask (PE, AH, SA, SA)
12. Glass funnels, small (AH)
13. 50 mL volumetric flasks (TA)
14. 500 mL volumetric flask (TA)
15. 1000 mL volumetric flasks (SA, TA)
16. Eppendorf pipettes, calibrated to known volume (SA, TA)
17. Block digester or sonicating bath (NPE, PE)
18. Autoclave or hot plate (AH)

19. Heating water bath (TA)
20. Desiccator (NPE, PE, AH, ASH)
21. Drying oven (NPE, PE, AH, ASH)
22. Muffle furnace (ASH)
23. Analytical balance (NPE, PE, AH)
24. Spectrophotometer (SA, TA)
25. Dichloromethane (CH_2Cl_2) (NPE)
26. Deionized water (PE, SA, TA, AH)
27. H_2SO_4 concentrated and 72% (AH)
28. Dextrose (SA). Dextrose standards used in the sugar analysis procedure are based on a primary standard of 200 μg of dextrose/mL. To prepare this, take 200 mg of dextrose that has been oven-dried at 105 °C and cooled in a desiccator for 30 minutes and add it to 1000 mL of deionized water in a 1 L volumetric flask. This stock solution is then diluted to form seven standards that systematically increase from 0 to 50 μg/mL.
29. Phenol, redistilled reagent grade or phenol solution of known density (SA). If phenol solution of known density (mg/mL) is not available, then phenol solution can be made by adding 20 g of deionized water to 80 g of redistilled reagent-grade phenol that has been placed in a beaker. Phenol is very toxic and must be handled with rubber gloves under a hood.
30. 1 M Na_2CO_3 (TA)
31. Folin-Denis reagent or Folin-Ciocalteau's reagent. Folin-Denis reagent is made by adding, in the following order, 50 g of sodium tungstate, 10 g of phosphomolybdic acid, and 25 mL of orthophosphoric acid to 375 mL of deionized water in a 1 L flat-bottomed flask fitted with a reflux condenser. The flask with several glass beads inside is refluxed on a hot plate for 2 hours, cooled, and emptied into a 500 mL volumetric flask, which is brought to volume with deionized water.
32. Phenol reagent, available through Sigma (TA)
33. Sodium tungstate for the Folin-Denis reagent (TA)
34. Phosphomolybdic acid for the Folin-Denis reagent (TA)
35. Orthophosphoric acid for the Folin-Denis reagent (TA)
36. Tannic acid standards (TA). Prepare standards based on a stock solution of 100 mg of tannic acid in 1000 mL of deionized water. This results in a concentration of 0.1 mg tannic acid/mL. Standards are made in the range of 0–0.6 mg tannic acid /mL by pipetting 0–6 mL of the stock solution into a 50 mL volumetric flask and adding deionized water to up to the final volume. These tannin standards are not stable and must be made fresh each day an analysis is conducted.

Procedure

Nonpolar Extractives

The nonpolar extraction, an adaptation of the method described by TAPPI (1976), removes oils, waxes, and fats from samples. Cutin and suberin are not effectively removed by this extraction. Note that CH_2Cl_2 used in the extraction is volatile, moderately flammable, and toxic. It should only be used under a fume hood and handled with gloves. Any CH_2Cl_2 remaining should be saved and disposed of properly (i.e., do not pour down the sink!).

The TAPPI method employs Soxhlet extractors, an expensive system that is difficult to use with many samples. To increase the number of samples that can be processed, one can use either of the following two methods:

1. Place approximately 1 g of sample in a pre-ashed, preweighed Gooch filtering crucible and record the weight. Use a number 2 pencil to mark the number of the sample (other markings will be removed by the solvent). Place 75 mL of CH_2Cl_2 in a block-digester tube that contains the extraction thimble held up by wire hangers. Under a fume hood place the block-digester tube in a cold block-digester and seal it with a rubber stopper containing a cold finger connected to a cold water circulating system (Fig. 8.2). The block-digester temperature is set to 56 °C, and once the CH_2Cl_2 begins to boil (at 40 °C), condense on the cold finger, and drip onto the sample, the extraction is continued for 5 hours. Allow the block digester to cool before removing the extraction thimbles.
2. Place approximately 1 g of sample in a pre-ashed, preweighed Gooch filtering crucible and record the weight. Use a number 2 pencil to mark the number of the sample (other markings will be removed by the solvent). Set the Gooch filtering crucible inside a 100 mL beaker. Under a fume hood add 30–40 mL of CH_2Cl_2 into the Gooch crucible, pouring slowly so that the solvent moves through the crucible without overflowing. Set the 100 mL beakers with crucibles into a sonication bath, making sure they are packed tightly to avoid tipping during the sonication. One may need to use water-filled beakers to properly pack the beakers with samples. Sonicate for 30 minutes at 60–70 watts of power. Remove the Gooch crucible and suction off the CH_2Cl_2 using a Buchner flask fitted with a Walter crucible holder. Pour off the CH_2Cl_2 remaining in the 100 mL beakers into a waste container and repeat the extraction twice more, using fresh CH_2Cl_2 for each extraction. Once the extraction is completed, place the extraction thimbles in a drying oven placed in a fume hood and dry at 50 °C for 12 hours. Cool in a desiccator and weigh the extraction thimble and sample to determine the mass lost during extraction.

Polar Extractives

The polar extractives procedure, an adaptation of TAPPI (1981), removes water-soluble polyphenols and simple sugars but not condensed tannins. The residue remaining after the nonpolar extraction is removed from the extraction thimble with a spatula.

1. Carefully weigh approximately 500 mg of the residue and place in a clean block-digestion tube. Depending on the size of the block-digester tube, add 25 or 75 mL of deionized water to the tube, being careful to wash all adhering fibers from the sides of the tube.

2. Place the tube in a block digester preheated to 104 °C, adjusting the temperature to allow gentle boiling for 3 hours.
3. Once the extraction is completed, filter the water and residue into pre-ashed, preweighed Gooch filtering crucibles.
4. Wash the fiber residues with deionized water several times. The filtrate should be captured in a 50–250 mL volumetric flask if sugar and tannin contents are to be determined, brought up to final volume with deionized water, and saved.
5. Dry the fiber residues and filtering crucible at 50 °C for 12 hours.
6. Cool in a desiccator and determine the total weight remaining minus the weight of the crucible. The mass lost is the polar extractives.

Acid Hydrolysis

The H_2SO_4 hydrolysis process follows that described by Effland (1977) and removes cellulose and hemicellulose.

1. Weigh out 200 mg of the oven-dried, extracted fiber and place it in a 15 mL round-bottomed Pyrex test tube.
2. Add 2 mL of 72% H_2SO_4 to the tube and heat in a water bath set at 30 °C for 1 hour, mixing occasionally to assure complete dissolution.
3. Add 6 mL of deionized water to the test tube and transfer the solution and remaining fiber to a 125 mL Erlenmeyer flask.
4. Use another 50 mL of deionized water to rinse the test tube thoroughly, transferring all the water and fiber to the Erlenmeyer flask. The flask is covered by a small glass funnel to reduce water loss.
5. A secondary hydrolysis is then carried out in an autoclave set at 120 °C for 1 hour. If an autoclave is not available, the secondary hydrolysis can be carried out on a hot plate, boiling the solution for 4 hours. If the hot plate system is used, add deionized water periodically to maintain volume.
6. The resulting solution is filtered through a pre-ashed, preweighed Gooch filtering crucible, saving the filtrate if sugar content is to be determined.
7. Wash the fibers caught in the crucible with deionized water to remove the acid. As with polar extractions, capture the filtrate in 100 or 250 mL volumetric flasks and bring up to final volume before storing.
8. Dry the crucible and fibers at 50 °C for 12 hours.
9. Cool in a desiccator and weigh, subtracting the weight of the crucible.

Permanganate Oxidation

The mass removed by acid hydrolysis is considered to be hemicellulose and cellulose, while the remaining fraction is acid-resistant lignin, cutin, suberin, and remaining ash. To determine the fraction of acid-resistant material that is either cutin or suberin, the acid-resistant fraction is oxidized with potassium permanganate ($KMnO_4$) because lignin is soluble in this substance. The remaining material should be cutin (for leaves) or suberin (for roots) and ash that has not dissolved in the preceding extraction, hydrolysis, and oxidation steps.

Total Sugars

To determine the sugar content of either the polar extractives or the H_2SO_4 hydrolyzed fiber, we recommend the phenol–sulfuric acid colorimetric method described by DuBois et al. (1956). This method is based on the fact that sugars and their derivatives produce a yellow color with an absorbance peak at 490 nm in the presence of phenol and strong sulfuric acid. The filtrate resulting from either the polar extraction or acid hydrolysis is diluted to fall within the range of standards. A 20 to 1 dilution is a good starting point when filtrates were prepared in a 250 mL volumetric flask, and a 50 to 1 dilution is a good starting point when a 100 mL volumetric flask is used.

1. 2 mL of the diluted filtrate is pipetted into a Pyrex test tube, with duplicates of each filtrate and a standard prepared simultaneously.
2. 80 mg of phenol is added into each tube using a precalibrated Eppendorf pipette and the solution mixed with a touch mixer (e.g., Vortex). The volume of phenol to be added will depend on the exact density of the stock solution used. Phenol is very toxic; it must be handled under a hood and by individuals wearing gloves.
3. 5 mL of concentrated H_2SO_4 is rapidly added to each tube with a large-bore pipette to ensure good mixing.
4. Place the tubes under a hood for 10 minutes, then mix them with a Vortex and place in a 25–30 °C water bath for 20 minutes. Absorbance is measured at 490 nm using a spectrophotometer.

Total Polyphenols

The Folin-Denis method (Allen et al. 1974) is used to determine the quantity of water-soluble polyphenols in the polar extracts.

1. Add 0.5–3 mL of the polar extract with an Eppendorf pipette into a 50 mL volumetric flask. The amount used will range from 0.25 to 3 mL depending on the type of material being analyzed and the size of the volumetric flask used to prepare polar extracts; the key point is to have concentrations within the range of the standards.
2. Record the amount added to determine the dilution factor. For standards add 1 mL to the 50 mL volumetric flasks.
3. Add deionized water to each flask containing extracts of standards so that it is two-thirds full, and add 2.5 mL of Folin-Denis or Folin-Ciocalteau's phenol reagent to the flask, allowing it to sit 3 minutes.
4. Add 10 mL of 1 mol/L Na_2CO_3 solution to the volumetric flask and bring up to full volume with deionized water.
5. Shake 10 times and place in a 25 °C water bath for 25 minutes.
6. Examine the flasks carefully for precipitate. Those with precipitate will have to be either remade or spun down with a centrifuge.
7. Read at 760 nm with a spectrophotometer.

Ash

The amount of ash in samples is determined by the standard muffle-furnace method. One gram of sample is oven dried, cooled in a desiccator, and then added to a ceramic crucible that has been preweighed. The sample is heated in a muffle furnace at 450–550 °C for 4 hours, allowed to cool, then placed in a desiccator. The fraction of ash is determined as the ratio of mass remaining after ashing to the initial sample. At a minimum, the fraction of ash should be determined for the entire sample. For some types of material, such as wood, ash content is minimal ($<1\%$) and might be ignored. Leaf and root material, however, can have very high ash contents, and adjustments must be made to correctly calculate the organic fractions. It is also highly recommended that the ash content of the acid-resistant fraction (ARF) be determined because some of the ash may dissolve during the extraction and hydrolysis processes.

Calculations

The following calculations are based on the assumption that the ash fraction is not removed by the extraction or hydrolysis steps. We feel this assumption is preferred over the assumption that ash is removed equally by each treatment. More precise estimates of the effect of ash content can be made by comparing the initial ash content to that remaining after the extraction and hydrolysis steps are completed.

The fraction of the sample consisting of ash (A) is

$$A = M_{final}/M_{initial}$$

where M_{final} and $M_{initial}$ are the sample mass corrected for the mass of the crucible after and before heating in the muffle furnace, respectively.

The fraction in NPEs is calculated as

$$\text{NPE} = M_{anpe}/[M_{initial} \times (1 - A)]$$

where

M_{anpe} = the mass after nonpolar extraction
$M_{initial}$ = the initial mass of the sample
A = the proportion of ash of the initial sample

The fraction in PEs is calculated as

$$\text{PE} = M_{ape}/[M_{anpe} \times (1 - A)]$$

where

M_{ape} = the mass after polar extraction
M_{anpe} = the mass after nonpolar extraction
A = the ash content of the initial sample

M_{ape} is calculated as

$$M_{ape} = M_{anpe} \times M_{act\text{-}ape}/M_{act\text{-}anpe}$$

Where $M_{act\text{-}ape}$ and $M_{act\text{-}anpe}$ are the actual mass remaining after polar extraction and the actual mass of polar extract free fiber used in the analysis (approximately 500 mg), respectively. This corrects for the fact that less than the full amount of non-polar extracted fiber was used in the analysis.

The proportion in ARF is calculated as

$$\text{ARF} = M_{aah}/[M_{ape} \times (1 - A)]$$

where

M_{aah} is the mass after acid hydrolysis
M_{ape} is the mass after polar extraction
A is the initial ash content

M_{aah} is calculated as

$$M_{aah} = M_{ape} \times M_{act\text{-}aah}/M_{act\text{-}ape}$$

Where $M_{act\text{-}aah}$ and $M_{act\text{-}ape}$ are the actual mass of fiber remaining after acid hydrolysis and the actual mass of polar extract free fiber used in the analysis (approximately 200 mg), respectively. This corrects for the fact that less than the full amount of polar extracted fiber was used in the analysis.

The proportion in polar extract sugar is calculated after developing a linear regression between the absorbance of the dextrose standards and their concentration. This regression is used to determine the concentration of dextrose equivalents (CDE) in the samples in units of μg/mL based on the absorbance value of the sample. The total mass in grams of the sugar (M_{sugar}) in the sample is calculated as

$$M_{sugar} = CDE \times V_{extract} \times DF \times 10^6$$

where

CDE = concentration of dextrose equivalents as mg/mL
$V_{extract}$ = the total volume of extract or filtrate resulting from the polar extraction or acid hydrolysis (mL)
DF = the dilution factor used to prepare the samples

The proportion of the total sample in polar extractive sugars (PES) or acid hydrolyzed sugars (AHS) is computed as

$$\text{PES} = M_{sugar}/[M_{initial} \times (1 - A)]$$

$$\text{AHS} = M_{sugar}/[M_{initial} \times (1 - A)]$$

where

M_{sugar} = the mass in sugar of either the polar extracts or acid hydrolysates
$M_{initial}$ = the initial mass of the sample
A = the initial ash content of the sample

The fraction in water-soluble polyphenols (WSP) is calculated by first developing a regression between the absorbance of the tannic acid standards (mg/mL). This means that polyphenol concentration is given in tannic acid equivalents and not the concentration of the polyphenol compounds actually present. The concentration of "tannin" (*CT*) of each sample is then calculated from the absorbance value. The total mass (in grams) of tannin (M_{tannin}) in the polar extracts is calculated as

$$M_{tannin} = CT \times V_{extract} \times DF \times 10^3$$

where

CT = tannin concentration in mg/mL
$V_{extract}$ = the total volume of extract or filtrate resulting from the polar extraction or acid hydrolysis (mL)
DF = the dilution factor used to prepare the samples

The proportion of polar extractive tannin (PET) is computed as

$$\text{PET} = M_{tannin}/[M_{initial} \times (1 - A)]$$

where

M_{tannin} = the mass in tannin of the polar extracts
$M_{initial}$ = the initial mass of the sample
A = the initial ash content of the sample

Special Considerations

We strongly recommend duplicate analysis of all samples. Given the number of samples being analyzed, it is extremely easy for a small amount of fiber to be overlooked and thus influence results. Duplicate samples that differ markedly should be rerun to determine which value is correct.

As with any proximate analysis of organic constituents, one must be aware that the operational definition (e.g., acid resistance) may not exactly match the chemical definition (lignin). Therefore, one has to be extremely careful about the interpretation of data generated by proximate analysis. This may not be a major concern if the general aspects of decomposition are to be modeled, where emphasis is on general classes of organic constituents of litter (labile versus resistant). For models considering the biochemical nature of the organic constituents and how they are altered by decomposition, the level of resolution offered by proximate analysis is probably inadequate. Therefore, CuO oxidation, NMR, or other methods will be required.

NIR Analysis of Organic Fractions

NIR spectroscopy is a rapid, nondestructive method for determining the major organic constituents of fresh and decomposing plant litter. It may also be used to determine nitrogen concentrations. In the most recent instruments, reflectance is determined from 400 to 2500 nm at intervals of 2 nm. Although this range includes some of the visible spectrum (400–700 nm), it has provided some useful information in terms of correlations (Gillon et al. 1993).

Prior to analysis, samples need to be dried and ground to pass a 20- or 40-mesh sieve. The determination of NIR spectral properties takes 2–3 minutes per sample. Before the proportion of organic constituents can be determined by NIR spectroscopy, a calibration to wet chemistry methods (e.g., proximate analysis) must be made. This is the most time-consuming step in the NIR method. To be most useful, the calibration samples must be representative of the overall population of interest, these samples must be accurately analyzed in terms of wet chemistry, and the correct mathematical processing of the spectral data must be determined.

Materials

1. NIR spectrometer with spinning sample module
2. Personal computer
3. Software for spectral analysis and calibration
4. 10–20 sample cups with quartz glass windows
5. Sample cup covers (paperboard)
6. Black iodized aluminum washers to reduce effective size of cell for small samples
7. Marking pens
8. Tweezers

Procedure

In most cases those interested in the NIR method will have to consult laboratories that possess the required equipment and expertise. While ecological laboratories generally do not possess this equipment, NIR spectrometers have been used extensively in crop and food sciences. Therefore, the equipment may be available locally. The procedure can be divided into two steps: development of routine regression equations from calibration samples, and subsequent scanning of samples for routine analysis.

1. To scan samples with large quantities, take approximately 2 g of dried, powdered, and well-mixed sample and place it in the sample cup with quartz window face down. To ensure the quartz window is not scratched, fill the sample cup on a soft surface. Place a paperboard cover on top of the filled sample cup and push firmly into place. Mark the sample number on the paperboard and turn the sample cup over and inspect to see that the sample material fills the entire window and the paperboard is not visible. Although it is possible to fill

a single sample cup, determine its reflectance, empty the cup, and then repeat the process sample by sample, it is most efficient to fill 10–20 cups, then determine their reflectance.

2. In the case of samples that are less than 2 g in weight, one can use a simple system to reduce effective size of the quartz window. This involves placing an aluminum washer that has a black finish similar in color to the walls in the sample cups. This "micro-cup" is then filled with the smaller sample and the procedure continues as usual.

   Follow the instructions that come with the spectrometer and software to determine the NIR reflectance spectra. The usual procedure consists of opening the door of the spinning sample module, placing the sample cup in two clamps, and closing the door to analyze the sample. A ceramic surface, which acts as a standard, is presented to the detectors when the door is opened. When the door is closed, the sample is presented to the detector. Once the sample is analyzed, open the door, take the sample out, and remove the paperboard cover with the tweezers to empty the sample cup. Because this is a nondestructive method, it is recommended that the sample be saved for reanalysis. One should be able to scan 80–100 samples per day.
3. Samples to be used for developing the calibration equations can be analyzed using the proximate analysis described earlier or for nitrogen using the method described in Chapter 7, this volume. A minimum of 50 samples are recommended to develop correlations between NIR reflectance and the wet chemical analysis. If the entire population of samples is scanned before the wet chemistry is performed, the newer analysis software that is available can be used to select representative samples (Infrasoft International 1993).

Calculations

To report actual spectral data reflectance (R) is usually converted to absorbance (A) using the following equation:

$$A = \log (1/R)$$

Because NIR spectrometers usually come with analysis software (e.g., Infrasoft International 1993), we will not describe all the combinations of mathematical procedures or treatments that can be used to create calibration equations. In most cases those wishing to develop their own prediction equations will initially need to seek outside expertise or become trained in the procedures. Here we will review the basic options involved in creating prediction equations so that novices have some foundation on which to base their initial decisions.

The overall approach is to try a number of mathematical procedures to determine which has the best fit to the wet chemistry calibration data. The first consideration is whether to use the original spectral data or to use the first or second derivative. Using the derivatives of the spectra is usually preferred because it eliminates effects caused by particle arrangement and moisture (Wessman et al. 1988a). In determining the derivatives, one must decide the segment length, expressed as the number of data points, over which the derivative is to be determined. The second consideration is choosing the algorithm to be used to determine the calibration equation: either stepwise regression (SR) or partial least squares (PLS). In SR, the wavelengths most highly correlated to the chemical constituent are added to the calibration equation, and this process continues until the addition of a wavelength does not increase the variation explained by the equation. The PLS algorithm is a combination of principal components analysis and multiple linear regression. This is advantageous because all the spectral data are included in the principal components analysis, whereas in the SR algorithm only a few wavelengths may be used (Bolster et al. 1996). When presenting the calibration equations, it is important to specify the derivative used, its segment length, and the calibration algorithm used.

Special Considerations

The quality of the calibration data is a major limitation in using NIR reflectance spectra to determine organic constituents of fresh plant and decomposing litter. The interpretation of results from NIR analysis has all the limitations of the original wet chemical methods (e.g., acid-resistant material may not be entirely lignin). The NIR spectrometer is extremely delicate and should be kept in a vibration- and dust-free environment. We therefore recommend that sample cups be filled and emptied under a hood or in a room separate from the instrument.

Acknowledgments We wish to thank Brad Dewey, Carol Glassman, Chuck McClaugherty, John Pastor, and Andrea Ricca for sharing their knowledge of organic proximate analysis. Dan Binkley, David Coleman, Phillip Sollins, and two anonymous reviewers provided helpful comments on the manuscript. Funding was provided by grants from the National Science Foundation.

References

Aber, J. D., J. M. Melillo, and C. A. McClaugherty. 1990. Predicting long-term patterns of mass loss, nitrogen dynamics, and soil organic matter formation from initial fine litter chemistry in temperate forest ecosystems. *Canadian Journal of Botany* 68:2201–2208.

Allen, S. E., H. M. Grimshaw, J. Parkenson, and C. Quarmby. 1974. *Chemical Analysis of Ecological Materials.* Blackwell Scientific, Oxford, UK.

Baldock, J. A., and C. M. Preston. 1995. Chemistry of carbon decomposition processes in forests as revealed by solid-state carbon-13 nuclear magnetic resonance. Pages 89–117 *in* W. W. McFee and J. M. Kelly, editors, *Carbon Forms and Functions in Forest Soils.* Soil Science Society of America, Madison, Wisconsin, USA.

Bolster, K. L., M. E. Martin, and J. D. Aber. 1996. Determination of carbon fraction and nitrogen concentration in tree foliage by near infrared reflectance: a comparison of statistical methods. *Canadian Journal of Forest Research* 26:590–600.

deMontigny, L. E., C. M. Preston, P. G. Hatcher, and I. Kögel-Knabner. 1993. Comparison of humus horizons from two ecosystem phases on northern Vancouver Island using ^{13}C CPMAS NMR spectroscopy and CuO oxidation. *Canadian Journal of Soil Science* 73:9–25.

DuBois, M., K. A. Gilles, J. K. Hamilton, P. A. Rebers, and F. Smith. 1956. A colorimetric method for determination of sugars and related substances. *Analytical Chemistry* 28:350–356.

Effland, M. J. 1977. Modified procedure to determine acid-insoluble lignin in wood and pulp. *TAPPI* 60(10):143.

Fogel, R., and K. Cromack Jr. 1977. Effect of habitat and substrate quality on Douglas-fir litter decomposition in western Oregon. *Canadian Journal of Botany.* 55:1632–1640.

Gallardo, A., and J. Merino. 1993. Leaf decomposition in two mediterranean ecosystems of southwest Spain: influence of substrate quality. *Ecology* 74:152–161.

Gillon, D., R. Joffre, and P. Dardenne. 1993. Predicting the stage of decomposing leaves by near infrared reflectance spectroscopy. *Canadian Journal of Forest Research* 23:2552–2559.

Goñi, M. A., and J. I. Hedges. 1990a. Cutin-derived CuO products from purified cuticles and tree leaves. *Geochimica et Cosmochimica Acta* 54:3065–3072.

Goñi, M. A., and J. I. Hedges. 1990b. Potential application of cutin-derived CuO reaction products for discriminating vascular plant sources in natural environments. *Geochimica et Cosmochimica Acta* 54:3073–3081.

Hatcher, P. G., M. A. Nanny, R. D. Minard, S. D. Dible, and D. M. Carson. 1995. Comparison of two thermochemolytic methods for the analysis in decomposing gynosperm wood: the CuO oxidation method and the method of thermochemolysis with tetramethylammonium hydroxide (TMAH). *Organic Geochemistry* 23:881–888.

Hedges, J. I., R. A. Blanchette, K. Weliky and A. H. Devol. 1988. Effects of fungal degradation on the CuO oxidation products of lignin: a controlled laboratory study. *Geochimica et Cosmochimica Acta* 52:2717–2376.

Hedges, J. I., and D. C. Mann. 1979. The characterization of plant tissues by their lignin oxidation products. *Geochimica et Cosmochimica Acta* 43:1803–1807.

Hedges, J. I., and P. L. Parker. 1976. Land-derived organic matter in surface sediments from the Gulf of Mexico. *Geochimica et Cosmochimica Acta* 40:1019–1029.

Infrasoft International. 1993. NIRS 2 Version 3.0: Routine operation and calibration software for near infrared instruments.

Jardetzky, O., and G. C. K., Roberts. 1981. *NMR in Molecular Biology.* Academic Press, New York, USA.

Joffre, R., D. Gillon, P. Dardenne, R. Agneessens, and R. Biston. 1992. The use of near-infrared reflectance spectroscopy in litter decomposition studies. *Annales des Sciences Foresti'eres* 49:481–488.

Kögel-Knabner, I. 1986. Estimation and decomposition pattern of the lignin component in forest humus layers. *Soil Biology and Biochemistry* 18:589–594.

Kögel-Knabner, I., P. Hatcher, E. W. Tegelaar, and J. W. de Leeuw. 1992. Aliphatic components of forest soil organic matter as determined by solid state ^{13}C NMR and analytical pyrolysis. *The Science of the Total Environment* 113:89–106.

Kögel-Knabner, I., and F. Ziegler. 1993. Carbon distribution in different compartments of forest soils. *Geoderma* 56:515–525.

McClaugherty, C. A., J. Pastor, J. D. Aber, and J. M. Melillo. 1985. Forest litter decomposition in relation to soil nitrogen dynamics and litter quality. *Ecology* 66:266–275.

McClellan, T. M., J. D. Aber, M. E. Martin, J. M. Melillo, and K. J. Nadelhoffer. 1991a. Determination of nitrogen, lignin, and cellulose content of decomposing leaf material by near infrared reflectance spectroscopy. *Canadian Journal of Forest Research* 21:1684–1689.

McClellan, T. M., M. E. Martin, J. D. Aber, J. M. Melillo, K. J. Nadelhoffer, and B. Dewy. 1991b. Comparison of wet chemistry and near infrared reflectance measurements of carbon fraction chemistry and nitrogen concentration in foliage. *Canadian Journal of Forest Research* 21:1689–1694.

Melillo, J. M., J. D. Aber, and J. F. Murtore. 1982. Nitrogen and lignin control of hardwood leaf litter decomposition dynamics. *Ecology* 63:621–626.

Murphy, J., and J. P. Riley. 1962. A modified single solution method for determination of phosphate in natural waters. *Analytica Chimica Acta* 27:31–36.

Parton, W. J., D. S. Ojima, C. V. Cole, and D. S. Schimel. 1994. A general model for soil organic matter dynamics: sensitivity to litter chemistry, texture and management. Pages 147–167 *in* R. B. Bryant, editor, *Quantitative Modeling of Soil Forming Processes.* Publication No. 39. Soil Science Society of America, Madison, Wisconsin, USA.

Preston, C. M. 1993. The NMR user's guide to the forest. *Canadian Journal of Applied Spectroscopy* 38:61–69.

Preston, C. M. 1996. Applications of NMR to soil organic matter analysis: history and prospects. *Soil Science* 161:144–166.

Preston, C. M., R. Hempfling, H. R. Schulten, M. Schnitzer, J. A. Trofymow, and D. E. Axelson. 1994. Characterization of organic matter in a forest soil of coastal British Columbia by NMR and pyrolysis-field ionization mass spectrometry. *Plant and Soil* 158:69–82.

Reh, U, W. Kratz, G. Kraepelin, and C. Angehrn-Bettinazzi. 1990. Analysis of leaf and needle litter decomposition by differential scanning calorimetry and differential thermogravimetry. *Biology and Fertility of Soils* 9:188–191.

Ryan, G. M., J. M. Melillo, and A. Ricca. 1990. A comparison of methods for determining proximate carbon fractions of forest litter. *Canadian Journal of Forest Research* 20:166–171.

TAPPI. 1976. Alcohol-benzene and dichloromethane solubles in wood and pulp. *TAPPI Official Standard* T204 OS-76.

TAPPI. 1981. Water solubility of wood and pulp. *TAPPI Official Test Method* T207 OM-81.

Trofymow, J. A., C. M. Preston, and C. E. Prescott. 1995. Litter quality and its potential effect on decay rates of materials from Canadian forests. *Water, Air and Soil Pollution* 82:215–226.

Van Soest, P. J. 1963a. Use of detergents in analysis of fibrous feeds. I. Preparation of fiber residues of low nitrogen content. *Association of Official Analytical Chemists* 46:825–829.

Van Soest, P. J. 1963b. Use of detergents in analysis of fibrous feeds. II. A rapid method for the determination of fiber and lignin. *Association of Official Analytical Chemists* 46:829–835.

Van Soest, P. J. 1965. Use of detergents in analysis of fibrous feeds: III. Study of effects of heating and drying on yield of fiber and lignin in forages. *Association of Official Analytical Chemists* 48:785–790.

Van Soest, P. J., and J. B. Robertson 1985. *Analysis of Forages and Fibrous Foods: A Laboratory Manual for Animal Science.* Cornell University Publications, Ithaca, New York, USA.

Wessman, C. A., J. D. Aber, D. L. Peterson, and J. M. Melillo. 1988a. Foliar analysis using near infrared reflectance spectroscopy. *Canadian Journal of Forest Research* 18:6–11.

Wessman, C. A., J. D. Aber, D. L. Peterson, and J. M. Melillo. 1988b. Remote sensing of canopy chemistry and nitrogen cycling in temperate forest ecosystems. *Nature* 335:154–156.

9

Collection of Soil Solution

Kate Lajtha
Wesley M. Jarrell
Dale W. Johnson
Phillip Sollins

The collection and analysis of the in situ soil solution is important for studies of pedological processes, environmental quality monitoring, and nutrient cycling (Zabowski and Ugolini 1990). Soil solution measurements are relevant for plant uptake and nutrient availability concerns, and estimates of solution fluxes from ecosystems are needed to balance ecosystem nutrient budgets and for research questions addressing losses of elements via leaching. Although laboratory soil extracts using chemical extractants or resin bags may measure an index of time-integrated nutrient availability, such extracts cannot measure fluxes within or between ecosystems.

Soil solution can be measured either by collecting field-moist soils and extracting solutions in the laboratory or by collecting solution in the field, usually with lysimeters. We will discuss three soil solution collection and measurement techniques here: field exchange resin membranes, soil lysimeters, and laboratory extraction of soil solution. Each technique measures something slightly different, and the technique needs to match the specific research question. Zero-tension lysimeters may be most appropriate for measuring fluxes through the soil profile and absolute losses from the system, while tension lysimeters can measure soil solution chemistry by depth and are indicated for questions relating to solution-solid phase equilibria or plant nutrition (Lajtha et al. 1995; Marques et al. 1996). Field resin measurements reflect both diffusion coefficients and mobile soil concentrations, and are often used to measure in situ nutrient availability. Laboratory extraction of soil water is generally less invasive and time-consuming than lysimeter installation and maintenance, and is useful for measurements of the intensity of soil solution. Detailed discussion of methods, as well as comparisons among methods, is offered in each section.

Field Resin Membranes

Many authors have used field-placed resin bags, prepared as for laboratory resin bags (described in Chapter 7, this volume) to monitor in situ nutrient availability. An alternative technique is the use of anion and cation-exchange resin impregnated membranes to measure nutrient bioavailability in the field (Abrams and Jarrell 1992; Cooperband and Logan 1994). Since ion-exchange resins have the potential to mimic nutrient uptake by plant roots, resins placed in the field can provide a measure of nutrient supply in soils (Huang and Shoenau 1996).

Ion-exchange membranes (IEMs) are well-defined planar, ion-sink surfaces. As such, they offer a specific geometry, unlike mesh bags filled with resin beads. In addition, the exchanger surface is in direct contact with soil particles, eliminating potential interferences from mesh bag material. The membranes are less likely to disrupt the flow of water through soils than are resin bags, and there is less soil disturbance with membrane placement in soils. However, as for resin bags, the amount of a nutrient sorbed by the membranes is not a quantitative measure of a pool size or of nutrient mineralization but rather is a correlate of the amount of labile nutrients in soils.

Membrane-bound NO_3^- has been shown to be highly correlated with soil NO_3^- concentrations and net soil nitrification (Subler et al. 1995). However, Giblin et al. (1994) found that nutrient accumulations on ion-exchange resins did not correlate well with other measures of NH_4^+, NO_3^-, or phosphate availability in an arctic ecosystem, although landscape differences in N versus P accumulation corresponded well with N:P ratios in soils and soil solutions. Lajtha (1988) found that total mineral N sorbed by resins was correlated with laboratory N mineralization rates over a desert landscape, but this relationship was weak, and P accumulation was not related to other measures of P availability. Accumulation onto resins is significantly affected by water flux, whereas field and laboratory measures of mineralization do not allow for changes in water flux. Thus resins might well pick up subtle differences in ion supply in the field when water flux changes rapidly. Because resins cannot measure pool sizes or fluxes, this technique is recommended only when an index of ion supply in the soil solution is needed.

Advantages to the IEM technique include

- soil is not removed and is only minimally disturbed using this technique, allowing repeated measurements on a fairly small area;
- chemical processing is simple;
- analysis includes a dynamic component that cannot be reproduced in the laboratory;
- little waste is generated in the laboratory; and
- the analytical equation can be specifically solved to produce a "universal" quantity.

However, there are also several concerns with this technique related to IEM sensitivity to the field environment:

- IEMs are slightly sensitive to temperature (10% change per 10 °C temperature change);

- IEMs are sensitive to soil water content (Nye and Tinker 1977); and
- IEMs are potentially sensitive to the total concentration of anions; in most cases, this means Cl^-, SO_4^{-2}, and occasionally HCO_3^-.

Materials

1. Ion-exchange resin membranes (available from a variety of sources including Soil-Plant-Water Quality, 125054 NW Cornell Rd., Portland, OR 97229). Both anion and cation IEMs are available; use anion-exchange resins to determine phosphate and nitrate; use cation-exchange resins to determine potassium, calcium, sodium, magnesium, and other divalent metals. Ion sinks are available in a range of sizes, from 0.45 m^2 sheets to 5 cm × 5 cm squares, or even smaller. The size used depends on the field sampling desired and ease of handling in the laboratory. The actual parameter determined, M_t, is independent of size, although a large ratio of perimeter to area can result in unacceptable edge effects. The 5 cm × 5 cm size is recommended.
2. 10 cm diameter petri plates, plastic or glass
3. Heavy-duty putty knife
4. 0.5 mol/L $NaHCO_3$
5. 0.5 mol/L HCl

Procedure

1. Before use, chloride-saturated anion-exchange membranes must be converted to the bicarbonate form. Shake resin sheets for 10 minutes in three successive solutions of 0.5 mol/L $NaHCO_3$, rinsing with deionized water between each solution. Although the bicarbonate form is less stable, it is preferred for determining P availability, since P affinity for the resin is low relative to Cl^- and OH^-.

   Cation-exchange membranes are usually supplied in the H^+-saturated form. Before use, the membranes should be rinsed thoroughly with fresh 0.5 mol/L HCl solution.
2. Prior to installation, ion sinks can be labeled by placing nylon monofilament line through a hole in one corner, tying it off with a knot, and connecting a label to the other corner. The label may be made of any material that maintains its integrity in the field. Membranes are inserted into slits opened in the soil with a broad-bladed tool like a putty knife. They can be placed at any depth; in most cases, the primary root zone is the region of greatest interest. Ideally they should be placed at a slight angle from the vertical, e.g., 15–30°, since this creates better soil-membrane contact. If desired, one cation sheet and one anion sheet can be placed back-to-back in each slit. They can then be treated as a unit through the desorption and analysis phases.
3. The membranes should not be left in soil longer than 100 hours in most cases. Beyond this time the membrane may no longer maintain a near-zero concentration of phosphate or nitrate at its surface. If the concentration near the surface becomes significant relative to the soil concentration, the simple model

described later no longer applies and interpretation is complicated. Even with resin bags, Giblin et al. (1994) noted that long deployment times gave lower estimates of nutrient availability than did a series of shorter deployments; they also noted that nitrate and phosphate could be desorbed from resins in the field.

The ion sink should be removed from the soil gently, although mild scraping of the soil from the surface causes little change in the amount extracted. Clinging soil particles should also be removed gently, and the membrane rinsed with deionized water to remove any additional soil. A small amount of soil on the membrane will not cause problems in the extraction except for trace metal determinations.

4. Keep the IEMs moist, e.g., in a Ziplock bag with a few drops of deionized water, prior to desorption. However, in most cases drying does not appear to adversely affect sorption properties. Dab the ion sink dry with a clean cloth, placed in 25 mL of 0.5 M HCl in a petri plate, and gently shake for 20 to 30 hours. The desorption sample solution can be stored in polyethylene bottles for analysis using appropriate laboratory techniques.

Calculations

The defined planar geometry allows simple mathematical analysis of results. Vaidyanathan and Nye (1966) attempted to determine the effective P diffusion coefficient, D_{eff}, in soil by applying the following relationship to uptake by an exchanger sheet:

$$D_{eff} = \frac{1}{4} M_t^2/(4 \times t \times c^2)$$

where

D_{eff} = effective diffusion coefficient, expressed as cm^2/sec

M_t = mass sorbed on planar surface after time t, expressed as $\mu mole/cm^2$; divide the total amount of ion extracted from the resin, μmoles, by the surface area of the resin. For a 5 cm × 5 cm sheet, with one side in contact with soil, the area is 25 cm^2. If both sides of the ion sink are exposed to soil, then both sides are counted in the area term (50 cm^2 for the preceding example).

t = time after placement in soil in seconds

c = effective diffusible P concentration in soil, as $\mu mole\ p/cm^3$ soil

Since M_t and t are known from analysis, rearranging the preceding equation allow us to calculate a term c^2D_{eff}, designated the *ion sink bioavailability factor* (Abrams and Jarrell 1992):

$$c^2D_{eff} = \frac{1}{4} M_t^2/(4 \times t)$$

Special Considerations

Huang and Shoenau (1996) describe the construction of an IEM probe that makes it easy to insert the membranes into the soil and ensures minimum disturbance of the surrounding soil, as well as permitting easy retrieval. Membranes are attached to long stakes with the bottoms of the stake pointed for easy insertion, and a single probe spans the entire soil profile. After probes are retrieved (Huang and Schoenau [1996] used field placement times of 2 hours), sections of membranes corresponding to specific depths or horizons can be cut out and analyzed separately. To best compare basic fertility among several similar sites, ion sinks should be inserted after a soaking rain or irrigation, to make water content more comparable among treatments.

Lysimeters

Many types of lysimeters have been used in both agricultural and natural settings, and they collect water either with or without applied tension to extract water. Tension lysimeters are generally smaller and relatively easy to install, and they collect water from the soil matrix. They have been made of ceramic, glass, Teflon, and other materials, and partially filter water that enters the lysimeter. Zero-tension lysimeters collect gravitational water, generally have significantly larger collection areas than tension lysimeters, and have been constructed from pans or PVC pipe, among other materials.

Tension lysimeters and zero-tension lysimeters collect different pools of water, and thus deciding which lysimeter type to use in specific studies is not necessarily a simple matter. Any differences in the chemical composition of water collected by tension and zero-tension lysimeters could lead to biases in estimates of nutrient fluxes if one or the other types are used.

Zero-tension lysimeters can collect only saturated flow or macropore flow, not gravitational or matric flow (which occurs at 0.01–0.03 MPa tension). Water will not enter the collection vessel of a zero-tension lysimeter unless the water is saturated at some stage along the collection pathway, but this is probably true of the majority of water moving through the soil profile. Tension lysimeters, on the other hand, could in theory collect both matric and saturated flow components; however, in practice, the hydraulic conductivity of tension lysimeters is probably too low to proportionally sample saturated flow in many cases. This all remains a matter of conjecture, however, insofar as there have been no systematic studies of the degree to which tension lysimeters bias against saturated flow via macropores. Haines et al. (1982) compared volumes and chemistry of soil solutions collected by tension and zero-tension lysimeters at Coweeta, North Carolina. They found that the zero-tension lysimeters collected seven times more solution than the tension lysimeters in the litter, probably because the zero-tension lysimeters are more efficient at collecting macropore flow. In the deeper horizons, however, the zero-tension lysimeters collected 50% less water because they miss unsaturated flow.

Tension lysimeters could also, in theory, collect soil water at the appropriate volumes if their tension is set to exactly that of the soil. Commercially available Prenart systems include the option of having lysimeter tension set to that measured with tensiometers. With lysimeters set at a constant tension, however, there is almost always either an underestimate or an overestimate of soil water flux because the tension at the lysimeter usually differs from that of the soil, which can vary. This is sometimes referred to as *coning*. If tension is too low, water will move around the lysimeter and flow though the soil as unsaturated flow in the matrix, which is always the case in the zero-tension lysimeter.

Several authors have compared the chemical composition of soil solutions collected by zero-tension versus tension lysimeters (Haines et al. 1982; Nyberg and Fahey 1988; Swistock et al. 1990; Hendershot and Courchesne 1991; others summarized in Marques et al. 1996). Although several found that soil solutions collected with the tension lysimeters had higher concentrations, as one would expect since they should collect a more tightly bound fraction of soil water, this varied a great deal depending on the ion examined and the site. In general, there have not been clear patterns of differences between lysimeter types across the many studies, although nitrate often appears to be elevated in tension versus zero-tension lysimeters.

Zero-Tension Lysimeters

Zero-tension lysimeters can be used to assess both the quality and the quantity of water leaching through the soil profile, and thus are critical for complete ecosystem elemental budgets. However, collection efficiencies (volume of water collected divided by percolating volume, which is separately calculated from a water balance model) are often low, and thus zero-tension lysimeters may not be appropriate for use in fairly dry systems or when large volumes of water are needed. They have the advantage of continuously sampling moving water rather than sampling water only when tension is applied. However, this also means that water may collect in storage bottles between collection events, and even with preservatives there is the possibility of nutrient or elemental transformation or loss via denitrification, volatilization, or flocculation.

Published collection efficiencies of zero-tension lysimeters are generally less than 10% (Radulovich and Sollins 1987). Radulovich and Sollins (1987) found that by increasing catchment area to 2500 cm^2 and by pushing the lysimeter rim upward into the soil, collection efficiency was increased to 36% under grass and 17% under forest, and the failure rate of lysimeters was also substantially decreased. Jemison and Fox (1992) found a mean collection efficiency of about 50% for pan lysimeters even larger than those used by Radulovich and Sollins (1987) that were placed at a depth of 1.2 m in an agricultural soil in Pennsylvania; they also noted a large variation among lysimeters. The greater efficiencies of large pan area lysimeters is likely due to both the greater chances of collecting preferential flow water and a lower proportion of flow around the edges of the pan. Thus, it would appear that matric potential–driven water flow in soils can be a significant proportion of total soil water flux. Even in a highly sandy soil with high infiltration rates, Seely et al. (1997) found that large catchment area lysimeters at 15 cm depth captured only 25–75% of flow. Efficiency at 50 cm depth was reduced to 15–25% of calculated flow, and efficiencies at 100 cm were under 10%.

Materials and Procedure

1. Because of low collection efficiencies, we recommend constructing lysimeters to be the largest size possible for the money and labor available. Jemison

and Fox (1992) used 0.5 m^2 catchment areas, while Radulovich and Sollins (1987) used 0.25 m^2 areas; we recommend areas within this range. However, collection efficiency is never 100%, and thus chemical data must be combined with a water balance model (see later discussion) for calculation of element flux.

The material used to construct the lysimeter should be chemically inert. Pan-type lysimeters are often constructed of aluminum. If these are custom made, the height of the side walls can be varied and thus favor water flow to the lowest corner of the pan with the outlet port. Seely et al. (1998) constructed lysimeters from 10 cm diameter PVC pipe that was cut in half lengthwise and had caps at each end. The latter design has the advantage of easier installation into long but narrow tunnels, although edge area:volume is greater than in square pan lysimeters. If asymmetrical pan lysimeters are constructed, outlet ports are placed at the lowest corner. If pipe is used, outlet ports are installed at one end, and lysimeters must be installed at an angle to ensure gravity flow of water to collection bottles. To prevent soil from collapsing into the lysimeter, lysimeters are filled with polypropylene pellets, acid-washed silica sand, or other inert materials to a level a few millimeters below the edges of the lysimeter.

2. Install lysimeters from large pits. To measure element flux from below the rooting zone, lysimeters must be placed at a depth below where at least 90% of roots are found. This should be determined in advance and will vary among ecosystems. Pits should be dug to at least 0.5 m below the lowest lysimeter depth, and should be sufficiently wide for easy manipulation of collection bottles. In many soils wooden support structures inside the pit will be needed if pits are to be maintained for several years. Plywood pit covers also protect the pits from disintegration. Side tunnels from the pit faces are excavated at the appropriate depths for lysimeter installation so that each lysimeter collects solution water from underneath an undisturbed soil profile. At least 50 cm space between the pit face and the edge of the lysimeter is needed to avoid edge effects, and lysimeters placed at different depths should not overlap.
3. Place lysimeters in the tunnel and push them up against the bottom of the soil horizon to maximize contact. Because the fill material does not come up to the top edge of the lysimeter, the top of the lysimeter will cut into the soil profile, and soil will fill the very top of the lysimeter. This last step is critical because matric potential will change between the soil and the lysimeter fill material, and water flow tends to follow matric potential; thus water will tend to flow laterally around the outside edges of the lysimeter unless a physical barrier (i.e., the top edge of the lysimeter wall) is present. Boards or other materials are often used to add pressure to the bottom of the lysimeter to ensure a close contact with the bottom of the soil profile.
4. Connect the outlet port via Tygon tubing to collection bottles that sit at the base of the pit. The tubing should enter the bottle through a tightly fitted hole in a cap. A smaller hole with smaller-diameter tubing must also be placed in the cap as a pressure equilibration port, and the tubing should be wrapped in circles to minimize evaporation losses. The volume capacity of the collection

bottle will depend on the area of the pit and the estimated maximum precipitation per area for each rainfall event. Because bottles will sit for several hours, or perhaps days, between collection, the possibility exists for nutrient loss or transformation. The addition of several drops up to 1 mL of chloroform will prevent microbial transformation, but chloroform evaporates and thus will need to be renewed often. It is important to collect water after each rainfall event.

5. Pump and discard several water collections. Although disturbance to the overlying soil pit is minimal, most researchers have suggested an equilibration period to counteract disturbance effects. Some authors have suggested equilibration periods of up to 2 years. We recommend a period of at least 6 months with indicators of disturbance, such as elevated NO_3^- leaching, used to judge when disturbance effects are past.

Tension Lysimeters

Perhaps due to their relative ease of installation and their premade commercial availability, tension lysimeters have been used more extensively than have zero-tension lysimeters. Large soil pits do not need to be dug because lysimeters can be installed from soil cores. However, the use of tension lysimeters requires disturbing the soil column immediately above the tube lysimeter, and in contrast to zero-tension lysimeters, it is not clear what area of soil is being sampled with a tension lysimeter, although the depth of soil water collection can be regulated.

Lysimeter installation always involves some degree of soil disturbance, even when performed from tunnels as in the plate system or at an angle as in Prenart™ lysimeters. This disturbance can result in anomalously high soil solution concentrations of nitrate and/or silica bicarbonate (the latter in soils with large amounts of weatherable minerals; Liator 1988; Shephard et al. 1990; Lajtha et al. 1995). The best way to account for this effect is to simply wait until several year's data can be collected and regular seasonal patterns can be observed, allowing the anomalous period to be identified. The waiting period for this can range from 2 to 4 months to 2 years. Johnson (1995b) found a very large nitrate pulse (>7,000 μeq L^{-1}), which lasted over a year after lysimeter installation in a beech forest soil in the Great Smoky Mountains.

The most common commercial tension lysimeters consist of a PVC or other tube of inert material that is of a variable length, with a round-bottomed ceramic cup at the bottom that serves as the filtering membrane for soil water. A neoprene access tube, fitted into the PVC tube by a rubber stopper, extends above the surface of the soil and is connected to the vacuum source for water collection. Ceramic cups can be purchased and lysimeters can be customized for specific applications (e.g., Stone and Robl 1996). Alternatively, lysimeters can be made of the "plate" type: a ceramic plate can be installed in the soil sideways from a soil pit, as for zero-tension lysimeters, with direct connection to an access tube.

Other, more inert materials than ceramic have been used for the collection-filtering membrane of both tube and plate lysimeters, including fritted glass, fritted stainless steel, glass-steel mixtures, and Teflon or Teflon-glass mixtures. These materials address concerns raised by some about the chemical inertness of porous ceramic, even with acid leaching or equilibration as pretreatments Zimmerman et al. 1978; McGuire et al. 1992). The material used for the soil interface must be hydrophilic in order to maintain the capillary tension necessary to keep tension between rain events. If the lysimeter material is hydrophobic, lysimeters may be coated with hydrophilic materials such as silica flour. By their very nature, these materials interact to some degree with the solutions passing through them. Liator (1988) provides a comprehensive review of chemical interactions with various types of lysimeters; see also Grossmann and Udluft (1991) and McGuire et al. (1992). Suffice it to say that results vary depending on contact time, the ion in question, and soil characteristics.

We have found that fritted glass or Prenart Teflon-glass lysimeters equilibrate with most ions rather quickly and are slightly superior to ceramic in terms of phosphate retention. Prenant lysimeters are also small and can be installed at an angle, thus minimizing soil disturbance. Glass is clearly less desirable in cases where Si or B (borosilicate glass) is of interest but would be superior to the alundum in ceramic in cases where Al is of interest. Krejsl et al. (1994) found that excessive filtering in ceramic lysimeters made them unsuitable for collecting and quantitatively measuring microbial constituents; they recommended using sand-filled or fritted glass lysimeters. In cases where soil pits can be dug, plate lysimeters might be preferable because the overlying soil column is left relatively intact; when many lysimeters must be employed or when soil disturbance is to be kept to a minimum, tube lysimeters are probably preferable.

The amount of tension that should be applied to draw soil water into the tubes has also come into question; it should be remembered that radically different tensions will draw on different sources of soil water, with potential repercussions for chemical analysis. Because tension is applied, it cannot be assumed that the water collected by tension lysimeters is chemically equivalent to water that leaches through the soil profile, although it is this latter quantity that is to be measured in ecosystem-level budget analyses. Tension lysimeters can collect too little water if there is saturated flow that is flowing in faster than the hydraulic conductivity of the lysimeter material. Coning toward the tension lysimeters (too much water) can occur if tension is set too high.

Finally, as for zero-tension lysimeters, an accurate water balance model must be constructed for each site to translate soil solution concentrations into ecosystem-level fluxes.

Materials

1. Soil water samplers of desired lengths. A wide variety of premade lysimeters of different ceramics and lengths can be purchased from Soilmoisture Equipment Corp. (P.O. Box 30025, Santa Barbara, CA 93105; 805-964-3525); Prenart Equipment ApS (Buen 14, DK-2000 Frederiksberg, Denmark, phone: +45 3874 1664) makes lysimeters of a variety of materials.
2. 2–4 inch soil corer (a larger soil core will be needed for more rocky soils)
3. Bentonite clay (optional, available from Soilmoisture Equipment Corp.)

4. 200-mesh silica sand (available from Soilmoisture Equipment Corp. or hardware stores)
5. Field soil sieve to remove pebbles and rocks from backfill material
6. Ehrlenmeyer flask with two-hole stopper for soil water collection
7. Vacuum hand pump or battery-operated pump with a tension gauge

Procedure

Lysimeters come with fairly detailed installation instructions for different situations, and virtually all materials needed for installation and collection can be purchased from the manufacturer. The following procedure is for the commonly used ceramic cup lysimeters; Prenart systems come with pointed ends for insertion, and directions are provided. Grossman and Udluft (1991) also provide installation guidelines. We recommend using the most complex, but the safest, installation method that isolates the ceramic cup from the soil below the depth to be measured, and that guards against channeling of water down the installation hole.

1. Dig soil cores to the desired depth, and sieve the extracted soil for use as backfill material.
2. Pour a small quantity of wet bentonite clay into the bottom of the core to isolate the sampler from the soil below.
3. Pour a small layer of silica sand into the hole and insert the lysimeter, followed by at least 6 inches of silica sand to completely cover the ceramic cup of the lysimeter.
4. Add a small quantity of bentonite clay to guard against channeling of water down the installation hole, and backfill the hole with the sieved native soil with continuous tamping with a metal rod to ensure that large air pockets do not form. The main concern with installation is that the ceramic cup should be in tight, intimate contact with the soil (or with the silica sand that is in contact with the soil) so that soil moisture can move readily from the pores of the soil through the pores in the ceramic cup and into the soil water sampler. It may be necessary to protect the top of the lysimeter and the access tubing from native fauna with metal screening.
5. After equilibration, soil water may be collected in a number of ways. In general, tension is applied using a hand pump, and the lysimeter is allowed to draw in soil moisture for 12–24 hours. A pinch clamp at the end of the neoprene access tube allows the hand pump to be removed while leaving the lysimeter under tension. To retrieve the collected water from the lysimeter into a collection flask, tension may be applied to tubing that is inserted into one hole of the two-hole stopper in the Ehrlenmeyer flask, while plastic tubing from the other hole of the stopper is inserted into the lysimeter via the neoprene access tube.

Special Considerations

Lysimeters can be adapted for under-snow sampling by adding vent and sample collection lines to the collection bottles or to the lysimeter tube itself (Johnson et al.

1977; Johnson 1995a). The vent and sample collection lines are elevated on poles or on trees and closed while the vacuum is on and samples are being collected. During collection, both lines are opened, and samples are simply withdrawn from the sample line with a vacuum pump and collection vessel from above the snow-pack.

Calculations for Both Zero-Tension and Tension Lysimeters

To obtain nutrient flux data from lysimeters, one must obtain a volume-weighted average annual concentration for the site measured and multiply this by the "true" estimate of water flux obtained either from a model or from the use of Cl^- as an inert tracer. In the latter case, it is assumed that Cl^- flux into the system equals Cl^- flux out, and soil solution water balance is calculated as the one unknown variable:

$$\text{Soil water flux} = [Cl^-(dep)/SS(Cl^-)] \times F$$

where

$Cl^-(dep)$ = chloride deposition
$SS(Cl^-)$ = weighted average soil solution chloride concentration
F = the factor for converting volumes of water to centimeters

Many hydrology models are available that differ substantially in the attention paid to processes such as interception of rain and snow, snowmelt, and saturated/preferential flow. These processes differ in importance by site, and thus a model suitable for one site may not be suitable for another. Because of this, we make no attempt to recommend hydrology models. Instead, we list criteria by which a model might be selected, given site hydrologic characteristics.

Most hydrology models are geared toward predicting streamflow (hydrographs) and thus devote considerable attention to subsoil and channel processes. Such processes are irrelevant to predicting flow past surface soil lysimeters; thus in choosing a model, it is important to pay close attention to the model representation of surface soil and litter moisture processes, and to the processes that control them. Specifically, note the following:

- For soils in which most of the water drains via saturated and/or preferential flow, the model needs to deal with such flow; models that assume that waters drain via unsaturated flow are not recommended for sites at which this does not happen.
- If a snowpack forms at the site, then the model must deal with the timing of snowmelt; moreover, if lysimeters are located under the canopy, the snowmelt model has to deal with the influence of the canopy on the snowpack energy balance.
- If lysimeters are located under different types of plant cover, then the model must consider effects of plant cover on transpiration, interception, and evaporation from the canopy.
- Finally, if lysimeters are located directly beneath the litter layer, then the model needs to predict drainage from the litter layer.

Most hydrology models are tested by comparison with streamflow data. Given our interest here in predicting flow past lysimeters, we recommend that the model be tested instead by comparison with time-series data on soil moisture for the layers above the lysimeters (see Chapter 3, this volume). In addition, comparison of simulated and measured throughfall, snowpack moisture, and litter moisture content is highly recommended. Annual or seasonal streamflow totals can provide a valuable check on model predictions of evapotranspiration, but a model's ability to predict short-term (i.e., daily or hourly) variation in streamflow is generally not a good test of its ability to predict flow past lysimeters.

Laboratory Collection of Soil Water from Soil Samples

Soil solution has been extracted from soils in a number of ways in the laboratory. In many cases, a simple extraction using a low ratio of water to soil provides an adequate estimate of soil solution composition. A relatively large soil sample can be used, which helps to minimize effects of heterogeneity in the soil. This procedure can be performed, which works best for field-moist samples.

Saturation Paste Extract

The saturation extract method is commonly used to extract soluble salts from soils (Richards 1954; Janzen 1993). The composition of this extract is generally closely related to that at field-moist water contents, with the advantage that the solution can readily be extracted using a Buchner funnel vacuum system. The technique can be used to assess the quantities of nitrate and other nonsorbing ions in the soil. For elements that are highly buffered in soils, such as P, cations, and trace metals, the effects of soil water content on concentration in solution are small.

The method also allows simultaneous determination of two physical parameters that can be informative: saturation percentage and saturated bulk density. Saturation percentage is a reasonable indicator of the texture of the soil and micropore space (sands are low, clays are high in saturation percentage). The saturated bulk density can indicate the average particle density of the soil solids; this is especially useful where coarse organic matter constitutes a significant (>5–10%) volumetric fraction of the soil.

We recommend that the analysis be performed on field-moist soils. Frozen samples can be thawed and analyzed, although it is preferable to analyze the sample within 48 hours of collection. The final saturated water content is corrected for the sample's field-moist water content. Air drying the field soil can precipitate salts that redissolve only slowly upon rewetting, if at all.

Materials

1. Balance weighing to 1 kg
2. 500 mL graduated plastic, glass, or aluminum container
3. 500 mL graduated cylinder

4. Spatula
5. Buchner (vacuum) funnel apparatus

Procedure

1. Determine the tare weight of the beaker and add 400 g of field-moist soil.
2. Determine the gravimetric water content of a separate subsample as per Chapter 3, this volume.
3. Add deionized water gradually with regular mixing with a stainless steel spatula. Allow samples to stand without mixing at several points in the procedure. The sample will be saturated when free water at the soil surface causes the surface to change from a dull sheen to a brighter glistening. In most cases, the soil will flow slightly when the beaker is tilted, but it will not drip out. After the soil appears to be saturated, it should be checked after an hour to determine if it needs more water. Once it is saturated, the beaker is weighed and the volume of saturated soil in the containers estimated.
4. After the sample has been fully saturated, allow it to equilibrate for 4 ± 1 hours. The saturation paste is then filtered through Whatman no. 42 filter paper in a 10 cm Buchner funnel. Smaller-diameter funnels are more likely to clog and result in slow filtration rates.
5. Refrigerate the filtrate until analysis. After pH and electrical conductivity have been determined, the filtrate can be stabilized further with 2 mL of 1 mol/L HCl.

Calculations

Solution composition: In most instances, results are expressed in terms of "concentration in the saturation extract," in mg/L or mol/L. For nonsorbing species such as nitrate, composition may be related back to dry mass of soil and expressed as mg NO_3^-–N/kg dry soil.

Saturation percentage (estimate of pore volume) = 100 × [(mass of saturation paste + beaker) − mass dry soil − mass beaker]/mass dry soil

Saturated bulk density = (mass dry soil)/volume saturated soil.

Centrifugation

In soils that are not excessively dry during most months of the year, centrifugation of soil water has proved to be an easy and effective way to collect soil water. Most collectors are handmade, thus requiring a large commitment to initial startup time. A common design uses a standard centrifuge tube that has been fitted with an internal screen to isolate particles from the centrifuged soil water, although others have simply decanted the solution successfully (Giesler et al. 1996). Clearly the largest problem with this technique will be limitations to the volume of soil water that may be collected, and thus this procedure is not recommended for chemical analyses that require large amounts of soil water. Soon and Warren (1993) discuss remoistening field-moist soils to 90% of field capacity and reequilibrating the samples for 48 hours before attempting to extract the soil solution. However, it is most likely that mineralization and dilution, and thus sorption-desorption, will occur, and significantly change the chemistry of the extracted solution, so we do not recommend it here for measures of the in situ soil solution.

Zabowski and Ugolini (1990) compared low-tension lysimeter and centrifuged soil solutions in a subalpine Spodosol over the course of a year. Differences in centrifuge speeds, corresponding to soil solutions held with tensions of about 0–30 kPa versus 30–3000 kPa by soils, did not seem to affect solution chemistry, suggesting that micropore water was fairly constant. However, lysimeter water and centrifuge-collected soil solutions did vary, with centrifuge solutions generally yielding higher concentrations at certain times of the year. The authors suggested that lysimeters are more likely to collect preferential (macropore) flow water, and micropore water, collected by centrifugation, would be more affected by biological activity and would have longer residence times in the soil. Giesler et al. (1996) compared the chemistry of zero-tension versus centrifuged collected soil solutions and concluded that centrifugation would avoid the hydrologic anomalies introduced by lysimeters and thus would be a more accurate reflection of the true soil solution.

There is no standard procedure for this technique. In one design, 60 mL centrifuge tubes have been cut in half crosswise, one half fitted with a mesh screen to support a filter, and then the two halves fastened together again. A glass fiber or smaller filter is placed on top of the screen, and soil is added to the top of the tube; after centrifugation, soil water is collected from the base of the tube. For drier soils, large centrifuge bottles and a high-speed refrigerated centrifuge (capable of >10,000 g) is recommended. Giesler et al. (1996) centrifuged soils for 80 minutes in centrifuge bottles without screens or filters, but we suggest that the time required, as well as the necessary g-force, will vary depending on the soil examined. Soon and Warren (1993) used the following materials:

Materials

1. The bottom half of a disposable 60 mL plastic syringe (the top half is cut off)
2. The bottom half of a 50 mL polypropylene centrifuge tube
3. Centrifuge with horizontal 50 mL container rotors. High speed and refrigeration may or may not be needed, depending on the soil and the moisture content.
4. Glass wool
5. Optional: no. 42 filter paper cut into 27 mm diameter disks or glass fiber filters of the largest pore size available
6. Parafilm
7. 1 mol/L HCl
8. 0.45 μm filters (e.g., Gelman or Nucleopore)

Procedure

1. The cut disposable syringe is used as the soil container. Plug the drainage hole with glass wool and line the bottom with a paper or glass fiber disk. Rinse this

entire apparatus by pouring in several milliliters of 1 mol/L HCl, followed by several deionized water rinses. Dry by centrifuging for several minutes or overnight in a warm oven. If glass fiber disks are used, the wool, the tube, and the filter can all be acid-washed separately beforehand, and the glass filters can be ashed.
2. Place approximately 25 g of field-moist soil into the top of this container.
3. Place the bottom half of a 50 mL polypropylene centrifuge tube into the rotor shield to serve as a solution collecting cup, and place the soil container into the shield and into the cup. Be sure to balance opposite sides of the rotor. Centrifuge at the maximum speed of the centrifuge for at least 30 minutes; experimentation will determine the time needed to extract the maximum amount of solution.
4. Immediately filter the solution through acid-washed 0.45 μm filters and store frozen or acidified for further analysis.

Immiscible Displacement with Centrifugation

An immiscible displacement technique (Whelan and Barrow 1980; Soon and Warren 1993) is more elaborate than the centrifugation-only technique, but it may be best suited for soils where centrifugation alone does not yield sufficient solution for chemical analysis. A dense, immiscible liquid is used to displace the soil solution, which, when centrifuged, floats on top of the immiscible liquid. Disadvantages of this technique include the fact that most of the immiscible liquids used are either relatively toxic or require special material centrifuge tubes to avoid tube dissolution, and generally, the use of a high-speed centrifuge. Advantages include a high yield of soil solution and the use of intact centrifuge tubes without filters or glass wool. The various immiscible liquids used are discussed in detail in Whelan and Barrow (1980) and Soon and Warren (1993). We will follow the recommendation of Whelan and Barrow (1980) in using tetrachloroethylene (C_2Cl_4).

Materials

1. Polyallomer 50 mL centrifuge tubes with caps
2. Tetrachloroethylene
3. Transfer pipettes or disposable syringes, 10 mL
4. High-speed centrifuge with 50 mL rotor

Procedure

1. Weigh 20 g field-moist soil into acid-washed polyallomer tubes.
2. Add 20 mL tetrachloroethylene to each tube. Balance the centrifuge by adding drops of tetrachloroethylene to one pair of tubes until the members of the pair are within 0.002 g of one another.
3. Centrifuge the capped tubes for 1 hour at approximately 20,000 × g.
4. After centrifuging, remove the displaced soil solution with either a transfer pipette or else a 10 mL syringe, and filter through a 0.45 (m filter prior to storing the samples as above.

References

Abrams, M. M., and W. M. Jarrell. 1992. Bioavailability index for phosphorus using ion exchange resin impregnated membranes. *Soil Science Society of America Journal* 56:1532–1537.

Cooperband, L. R., and T. J. Logan. 1994. Measuring *in situ* changes in labile soil phosphorus with anion-exchange membranes. *Soil Science Society of America Journal* 58:105–114.

Giblin, A. E., J. A. Laundre, K. J. Nadelhoffer, and G. R. Shaver. 1994. Measuring nutrient availability in arctic soils using ion exchange resins: a field test. *Soil Science Society of America Journal* 58:1154–1162.

Giesler, R., U. S. Lundstrom, and H. Grip. 1996. Comparison of soil solution chemistry assessment using zero-tension lysimeters or centrifugation. *European Journal of Soil Science* 47:395–405.

Grossmann, J., and P. Udluft. 1991. The extraction of soil water by the suction cup method: a review. *Journal of Soil Science* 42:83–93.

Haines, B. L., J. B. Waide, and R. L. Todd. 1982. Soil solution nutrient concentrations sampled with tension and zero-tension lysimeters: report of discrepancies. *Soil Science Society of America Journal* 46:658–661.

Hendershot, W. H., and F. Courchesne. 1991. Comparison of soil solution chemistry in zero tension and ceramic-cup tension lysimeters. *Journal of Soil Science* 42:577–583.

Huang, W. Z., and J. J. Schoenau. 1996. Microsite assessment of forest soil nitrogen, phosphorus, and potassium supply rates in-field using ion exchange membranes. *Communications in Soil Science and Plant Analysis* 27:2895–2908.

Janzen, H. H. 1993. Soluble salts. Pages 161–166 *in* M. R. Carter, editor, *Soil Sampling and Methods of Analysis.* Lewis Publishers, Boca Raton, Florida, USA.

Jemison, J. M., Jr. and R. H. Fox. 1992. Estimation of zero-tension pan lysimeter collection efficiency. *Soil Science* 154:85–94.

Johnson, D. W. 1995a. Soil properties beneath Ceanothus and pine stand in the eastern Sierra Nevada. *Soil Science Society of America Journal* 59:918–924.

Johnson, D. W. 1995b. Temporal patterns in soil solutions from a beech forest: Comparison of field and model results. *Soil Science Society of America Journal* 59:1732–1740.

Johnson, D. W., D. W. Cole, S. P. Gessel, M. J. Singer, and R. V. Minden. 1977. Carbonic acid leaching in a tropical, temperate, subalpine, and northern forest soil. *Arctic Alpine Research* 9:329–343.

Krejsl, J., R. Harrison, C. Henry, N. Turner, and D. Tone. 1994. Comparison of lysimeter types in collecting microbial constituents from sewage effluent. *Soil Science Society of America Journal* 58:131–133.

Lajtha, K. 1988. The use of ion-exchange resin bags to measure nutrient availability in an arid ecosystem. *Plant and Soil* 105:105–111.

Lajtha, K., B Seely, and I. Valiela. 1995. Retention and leaching losses of atmospherically-derived nitrogen in the aggrading coastal watershed of Waquoit Bay, MA. *Biogeochemistry* 28:33–54.

Liator, M. I. 1988. Review of soil solution samplers. *Water Resourses Research* 24:727–733.

Marques, R., J. Ranger, D. Gelhaye, B. Pollier, Q. Ponette, and O. Goedert. 1996. Comparison of chemical composition of soil solutions collected by zero-tension plate lysimeters with those from ceramic-cup lysimeters in a forest soil. *European Journal of Soil Science* 47:407–417.

McGuire, P. E., B. Lowery, and P. A. Helmke. 1992. Potential sampling error: trace metal adsorption on vacuum porous cup samplers. *Soil Science Society of America Journal* 56:74–82

Nyberg, R. C., and T. J. Fahey. 1988. Soil hydrology in lodgepole pine ecosystems in southeastern Wyoming. *Soil Science Society of America Journal* 52:844–849.

Nye, P. H., and P. B. Tinker. 1977. *Solute Movement in the Soil-Root System.* University of California Press, Berkeley, California, USA.

Radulovich, R., and P. Sollins. 1987. Improved performance of zero-tension lysimeters. *Soil Science Society of America Journal* 51:1386–1388.

Richards, L. A., editor. 1954. *Diagnosis and Improvement of Saline and Alkali Soils.* USDA Agricultural Handbook No. 60. U.S. Government Printing Office, Washington, DC, USA.

Seely, B., K. Lajtha, and G. Salvucci. 1998. Transformation and retention of nitrogen in a coastal forest ecosystem. *Biogeochemistry* 42:325–343.

Shephard, J. P., M. J. Mitchell, T. J. Scott, and C. T. Driscoll. 1990. Soil solution chemistry of an Adirondack Spodosol: lysimetry and N dynamics. *Canadian Journal of Forest Research.* 20:818–824.

Soon, Y. K., and C. J. Warren. 1993. Soil solution. Pages 147–159 *in* M. R. Carter, editor, *Soil Sampling and Methods of Analysis.* Lewis Publishers, Boca Raton, Florida, USA.

Stone, D. M., and J. L. Robl. 1996. Construction and performance of rugged ceramic cup soil water samplers. *Soil Science Society of America Journal* 60:417–420.

Subler, S. J., J. M. Blair, and C. A. Edwards. 1995. Using anion-exchange membranes to measure soil nitrate availability and net nitrification. *Soil Biology and Biochemistry* 27:911–917.

Swistock, B. R., J. J. Yamona, D. R. DeWalle, and W. E. Sharpe. 1990. Comparison of soil water chemistry and sample size requirements for pan *vs.* tension lysimeters. *Water, Air, and Soil Pollution* 50:387–396.

Vaidyanathan, L. V., and P. H. Nye. 1966. The measurement and mechanism of ion diffusion in soils. I. An exchange resin paper method for measurement of the diffusive flux and diffusion coefficient of nutrient ions in soils. *Journal of Soil Science* 17:175–183.

Whelan, B. R., and N. J. Barrow. 1980. A study of a method for displacing soil solution by centrifuging with an immiscible liquid. *Journal of Environmental Quality* 9:315–319.

Zabowski, D., and F. C. Ugolini. 1990. Lysimeter and centrifuge soil solutions: seasonal differences between methods. *Soil Science Society of America Journal* 54:1130–1135.

Zimmerman, C. F., M. T. Price, and J. R. Montgomery. 1978. A comparison of ceramic and teflon *in situ* samplers for nutrient pore water determinations. *Estuarine Coastal Marine Science* 7:93–97.

Part III

Soil Biological Processes

10

Soil CO_2, N_2O, and CH_4 Exchange

Elisabeth A. Holland
G. Philip Robertson
James Greenberg
Peter M. Groffman
Richard D. Boone
James R. Gosz

The composition of the earth's atmosphere is largely determined by the exchange of trace gases with the oceanic and terrestrial biosphere. Human-induced perturbations to the earth's system are changing the atmospheric composition at unprecedented rates. The gases of particular interest can be divided into two categories: those species that are radiatively active and thus influence the radiation balance of the earth (the so-called greenhouse gases: nitrous oxide [N_2O], carbon dioxide [CO_2], methane [CH_4], chlorofluorocarbons [CFCs], halogenated chlorofluorocarbons [HCFCs], and ozone [O_3]), and those species that are photochemically active and thus influence the troposphere's ability to cleanse itself (its "oxidizing capacity": hydroxyl radical OH, hydroperoxy radical HO_2, O_3, active nitrogen (NO_x–NO + NO_2), carbon monoxide [CO], CH_4, and volatile organic carbon compounds other than methane [VOCs]). The atmospheric lifetime of the radiatively active species is often longer than 10 years, extending to as long as 120 years for N_2O in the troposphere (the lower 10–16 km of the atmosphere), whereas the lifetime of the photochemically active species is 1–3 days (Tab. 10.1). Because of their very long lifetimes, N_2O, CFCs, HCFCs, and other halogenated compounds persist long enough to be transported to the stratosphere, where they play an important role in stratospheric ozone depletion. Some gases, notably O_3 and CH_4, are both photochemically and radiatively active. Both radiatively active and photochemically active gases are central to global change research; these gases are rapidly altering atmospheric composition and chemistry, in part as a result of changes in biospheric activity.

Trace gas exchange can also provide an important pathway for ecosystem inputs and losses of nitrogen and carbon. Methane emissions constitute 7% of annual aboveground net primary productivity for wetlands (Aselmann and Crutzen 1989).

Table 10.1. Summary of Some Greenhouse Gases Affected by Human Activity Including Concentration, Lifetimes, and Rates of Increase

	CO_2	CH_4	N_2O	CFC-12
Preindustrial concentration	280 ppm_v	700 ppb_v	275 ppb_v	zero
Concentration in 1992	335 ppm_v	1714 ppb_v	311 ppb_v	503 ppt_v*
Recent rate of increase				
1980s	0.4%/yr	0.8%/yr	0.25%/yr	4%/yr
1990–1992	0.1%/yr	0.27%/yr	0.16%/yr	—
Atmospheric lifetime (years)	(50–200)**	(12–17)***	120	102

*ppt_v = part per trillion by volume.
**No single lifetime for CO_2 can be defined because of the different rates of uptake by different sink processes.
***This has been defined as an adjustment time which takes into account the indirect effect of methane on its own lifetime.

Nitrogen gas losses (NH_3, NO_x, N_2O, and N_2) can be as much as 50% of fertilizer input into agricultural ecosystems and may be equally significant in unmanaged ecosystems. Models that do not incorporate these loss pathways can substantially overestimate net primary production (NPP) and long-term storage of both carbon and nitrogen (Schimel et al. 1997; W. Parton and W. Hunt, personal communications). NO_x and ammonia inputs to the atmosphere generate nitrogen deposition elsewhere. Nitrogen deposition is increasing exponentially, with a total of 80 Tg y^{-1} of N deposited on today's earth (Holland et al. 1997). Quantification of both inputs and loss pathways is central to describing the state of an ecosystem as well as possible trajectories for change.

In this chapter we focus on the measurement of the exchange, or the net flux, of radiatively active gases that are produced and consumed in soils of natural and agricultural ecosystems: CO_2, N_2O, and CH_4. The specific processes responsible for the production and consumption of these gases is treated in other chapters of this volume, including Chapters 11, 13, and 14. Without exception, the estimated global budget of each of these trace gases has changed substantially as we have incorporated the results of research performed between the late 1980s and the mid-1990s (IPCC 1995). The rates of atmospheric increase for the three gases have been dynamic (Tab. 10.1). One of the most variable terms in each budget has been the estimated fluxes from natural and agricultural ecosystems. Terrestrial biogenic sources are thought to constitute between 40% and 80% of the global sources of these gases (depending on the gas) and are known to be dynamic, but controls as well as patterns of fluxes are poorly known. Thus, periodic measurements using standard techniques at a variety of sites will provide much needed information.

Available Methods

A wide variety of techniques are available for the measurement of trace gases on spatial scales that range from a single point to measurements that can be integrated over square kilometers. At the finest scale there are measurements of soil atmosphere concentrations using probes (stainless steel or Teflon tubes) placed at various depths in the soil. For flux measurements at spatial scales ranging from 0.1 to 1.0 m^2, enclosures (chambers) are placed on the surface of the soil, allowing gas to accumulate over time and enabling the calculation of accumulation. For flux measurements at larger spatial scales, fluxes can be estimated using micrometeorological measurements on towers. Such measurements characterize the vertical gradient and flux of a gas integrated over areas ranging from 0.5 to 100 ha (spatial scales for this method are dependent on the height of the measurements). Although they are not flux measurements, Fourier transform infrared (FTIR) spectrometer and differential absorption (LIDAR) measure gas concentrations and integrate over distances as long as several kilometers . Aircraft sampling integrates over a larger surface area; the footprint of integration depends on both the flight altitude and meteorological conditions (Matson and Harriss 1995; Desjardins et al. 1993).

When compared, different methods provide similar estimates of fluxes. Aircraft and tower flux measurements of carbon dioxide fluxes over the Konza LTER tallgrass prairie were highly correlated (Desjardins et al. 1993). Comparison of N_2O fluxes using different chamber techniques (closed versus open and chambers of different volumes), different micrometeorological techniques (eddy covariance, flux gradient and conditional sampling using two tunable diode lasers [TDL], an FTIR and a gas chromatograph), and chamber versus micrometeorological techniques show reasonable agreement provided the patchiness of the landscape is taken into account when examining micrometeorological measurements from different wind directions (Christensen et al. 1996; Hargreaves et al. 1996; Ambus et al. 1993).

The two most frequently used techniques for measuring surface-atmosphere gas exchange are micrometeorological and enclosure techniques. For experiments where the goal is to produce an estimate of trace gas exchange over large areas (greater than a few square meters), micrometeorological techniques are preferable because they incorporate much of the meter-to-meter variation (Lenschow 1995; Baldocchi 1991). Micrometeorological techniques have evolved considerably, and there are now methods for avoiding the stringent sampling requirements and sampling frequencies of eddy correlation measurements (Lenschow 1995; Businger and Oncley 1990). Multiyear deployments of eddy flux measurements at the Harvard Forest LTER have been highly successful in providing insights in regional carbon exchange and storage on both an intra-annual and an interannual basis (Goulden et al. 1996; Wofsy et al. 1993). Enclosure (chamber) techniques enable the study of factors driving meter-to-meter variation in fluxes, facilitate manipulative studies, and are often a critical complement to eddy flux measurements.

We recommend enclosure techniques because they are relatively inexpensive, simple to operate, require less data analysis and manipulation than micrometeorological methods, and use equipment that can easily be moved from one location to another, thus allowing sampling of many locations within a landscape. As is pointed out in many other chapters in this volume, the scientific question posed is the primary determinant of the choice of techniques and the optimal experimental design.

Enclosure Technique for Measuring CO_2, CH_4, and N_2O Fluxes

Enclosure designs are of two basic types, static and flow-through. Static designs typically contain a small port to permit sampling and a small vent to permit equilibration of internal and external atmospheric pressures. Flow-through designs may be steady-state (in which the enclosure is swept with air drawn from a source of known concentration resulting in a "steady" concentration gradient across the air-soil interface within the enclosure) or non–steady-state (in which the trace gas concentration gradient diminishes in response to continual concentration changes within the enclosure). Static chambers such as those recommended here are by nature non-steady-state enclosures. A review of the possibilities and the considerations needed for each type of enclosure is provided in Denmead (1979), Kanemasu et al. (1974), Jury et al. (1982), Hutchinson and Livingston (1993), and Livingston and Hutchinson (1995). The most commonly used technique for measurement of N_2O and CH_4 fluxes is to periodically sample a static vented chamber with subsequent gas chromatographic analysis of the gas sample (Fig. 10.1).

Materials

1. Permanent collars made of PVC, stainless steel, or aluminum. See below for design recommendations.
2. Soil knife to circumscribe collar location (optional depending on the type of site)
3. Enclosures with a vent, sampling port, and a mechanism for securing and sealing to permanent collars (see section on "Enclosure and Collar Design and Construction," below and Fig. 10.1)
4. Seven conditioned polypropylene or nylon syringes fitted with one-way stopcock valves to transport gas samples to laboratory (and storage vials if the gas samples are to be stored for more than 24 hours) for each gas and each enclosure to be sampled. One syringe will be used in step 6, one syringe will be used to sample the initial or time-zero concentration, and the remaining five will be used for sampling the gas concentration within the chamber. (See the section "Special Considerations," below, for a more complete description.
5. Instrument for gas analysis. A gas chromatograph (GC) equipped with the appropriate detectors can be used for N_2O, CH_4, and CO_2 analysis; alternatively, an infrared gas absorption (IRGA) analyzer can be used for CO_2 analysis.
6. Four standards for calibration of the GC and a canister of at least one of the standards to transport to the field site for checking the effects of storage and transport on gas concentration.

Procedure

1. Insert the collars into the soil at least 1 week prior to sampling to mitigate placement disturbance. Collars should be inserted 5–10 cm into the soil.
2. Put the enclosure in place and record the time.

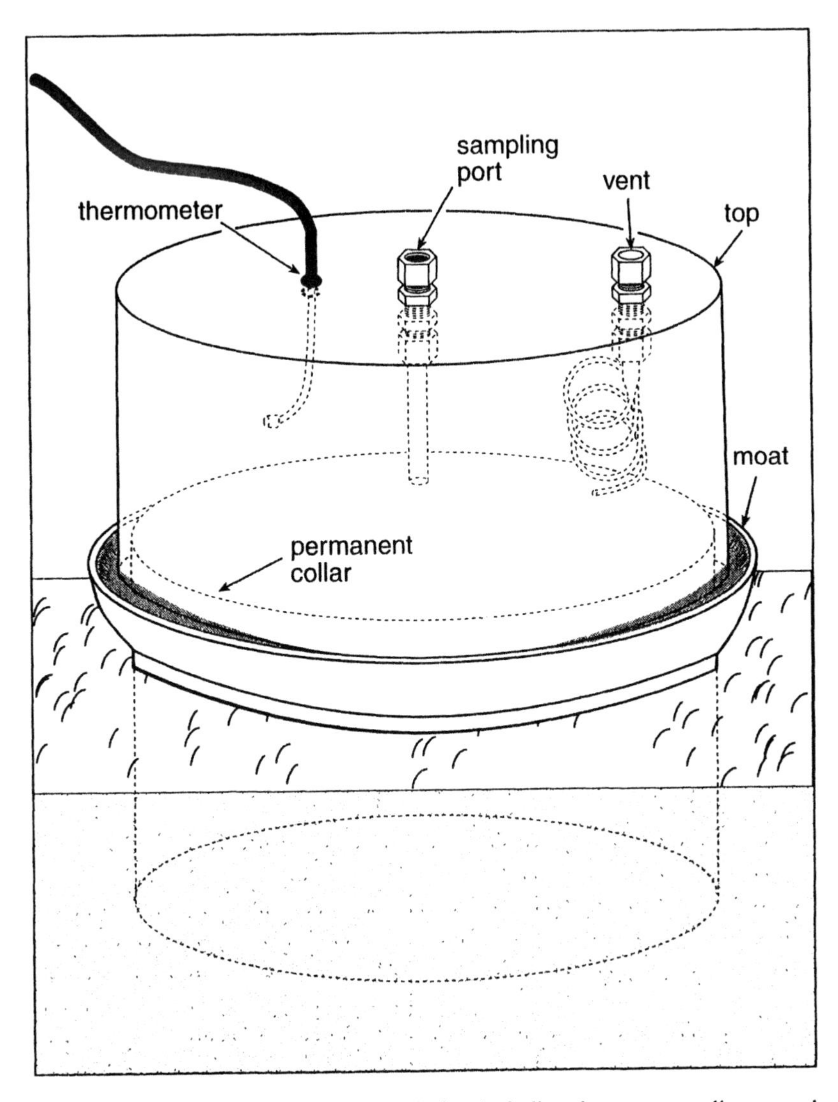

Figure 10.1. The recommended enclosure design including the vent, sampling port, thermometer, and moat for sealing the chamber to its permanent collar. Enclosure, permanent collar, and all fittings should be manufactured with inert materials: PVC, stainless steel, or aluminum. For measurement of reactive trace species, the enclosure can be lined with Teflon, but all materials should be tested to ensure that they do not adsorb, desorb, or otherwise react with the gas of interest. The moat of the permanent collar should be partially (and carefully) filled with water before placement of the enclosure. The water serves as a gas diffusion barrier, thus providing an effective seal. Vent dimensions and sampling port design are detailed in the text. Gastight fittings for the sample port are 1/4 inch Swagelock or Parker fittings. The enclosure can be tested in the laboratory by placing the enclosure in a shallow tray of water and testing for stable gas concentrations over time.

3. Establish the time-zero concentration of the gas by taking 10–20 7–8 mL air samples with a 10 mL syringe equipped with a one-way stopcock.
4. For each gas of interest, sample 7–8 mL of the enclosure volume using a 10 mL syringe equipped with a one-way stopcock valve every 10 minutes for 50 minutes. Record the time of each sampling. This sampling technique has been shown to be sufficient for most situations. In a few cases with very low fluxes, it may be necessary to sample less frequently over a longer period; for cases with very high fluxes, it may be necessary to sample more frequently over a shorter period. The size of the sample taken always exceeds the minimum needed for analysis. For example, a 5 mL sample injection for a 1 mL sample loop is used to ensure that connections and valve are purged with the air sample. We recommend that a single sample be taken for each gas to be analyzed (with the exception of GCs set up for simultaneous injection and analysis).
5. Before removing the enclosure, measure the height of the enclosure over the soil surface in at least four places for each enclosure. Enclosure height must be measured when collars are first used or when disturbed by freeze-thaw or other events.
6. Periodically sample the field standard as for samples.
7. In the laboratory, analyze the sample by gas chromatography. Inject the sample through a valve equipped with a sample loop of the appropriate size (e.g., 1 mL for CH_4 and N_2O and 0.1 mL for CO_2). See section on "Instruments and Other Analytical Considerations," below, for a more detailed discussion.

Analysis and Instruments

Calibration and Standards

Proper instrument calibration and use of standards are common to the analysis of all three gases and require great care. We recommend a four-point calibration with different standards that span the anticipated range of concentrations to be measured including atmospheric concentrations (note that for CH_4 uptake studies, atmospheric concentrations will be the high value). Calibration frequency will depend on the stability of the individual instruments; typically calibration is needed every 15–20 samples. The Environmental Protection Agency (EPA) standard for Quality Assurance/Quality Control (QA/QC) calls for 20% of all analyses to be standards.

All standards must be handled with caution and stored in pressurized containers. Standards should be intercompared against other standards in the laboratory, or other laboratories, and should be traceable to standard gases available from the National Institute of Standards and Technology (NIST, formerly the National Bureau of Standards [NBS]). Adequate standards are available from several commercial suppliers (including Matheson Gas Products, P.O. Box 89, 530 Watson Street E., Whitby Ontario, Canada, L1N 5R9; and Scott Specialty Gases, 1290 Combermere St., Troy, MI 48083). Stainless steel regulators should be used to avoid introducing impurities to the standard, and where possible regulators should be dedicated to a specific standard tank (this is essential for reactive compounds like NO and NO_2).

Instrumentation and Other Analytical Considerations

Introduction of a gas sample into the GC may be performed by direct injection through a septum or by use of a two-position gas chromatographic valve with a fixed volume sample loop (Fig. 10.2). We recommend the use of a sample loop connected to a GC sample valve to ensure introduction and analysis of a constant volume of sample. When all gases are to be measured simultaneously, it is useful to purchase and/or reconfigure a GC for simultaneous injection of all the gases. More information on how to do this is available in Sitaula et al. (1992). The detectors can be set up in series or parallel with different considerations for each. However, if the interest is in measuring CO_2 fluxes alone, then consider using a portable CO_2 analyzer.

Carrier gas purity is essential for gas chromatographic analysis of any trace gas but is particularly important for measurement of nitrous oxide and for methane consumption. Purchase of gases that are at least 99.999% pure (ultrahigh purity [UHP]) is recommended. Even so, the two carrier gases commonly used for the measurement of methane:(N_2 and He) are often contaminated by 0.5 ppm of methane against a background of 1.7 ppm (Tab. 10.1). Where necessary, N_2 and He carrier gases can be cleaned of CH_4 either by using an in-line catalytic converter or by passing the carrier gas through a molecular sieve trap (1.27 cm diameter tubing containing coarse mesh molecular sieve, 1.6–3.2 mm [$\frac{1}{16}$ – $\frac{1}{8}$ inch] pellets) placed in a dewar flask containing liquid nitrogen. Traps should be cleaned nightly by placing them, with carrier flow, in a heating mantle filled with silica sand heated to 200 °C.

Carbon dioxide. CO_2 fluxes can be measured by techniques ranging from soda lime absorption of CO_2 to IRGA. Both soda lime and base trap (NaOH and KOH) absorption tend to underestimate high CO_2 fluxes and overestimate low CO_2 fluxes as a result of varying absorption efficiencies (Coleman 1973; Nadelhoffer and Raich 1992; Nay et al. 1994). As a result, we recommend against the use of base traps for routine in situ flux measurements. They can, however, provide an accurate 24-hour integrated flux estimate in some soils, but should be used with caution and at a minimum must be calibrated against instantaneous flux measurements (Nadelhoffer and Raich 1992).

For enclosure measurement of CO_2 fluxes, IRGA analysis is fastest. GC analysis using either a thermal conductivity detector (TCD) or a flame ionization detector (FID and a Carbosieve column) allows simultaneous analysis of more than one gas (Crill et al. 1995). The FID GC technique is similar to that for CH_4 as CO_2 is converted to CH_4 by a methanizer placed before the FID and then measured as CH_4. Methanizers may be purchased commercially (Alltech Associates, Inc. 2051 Waukegan Rd., Deerfield, IL 60015).

Other CO_2 measurement systems include those manufactured by PP Systems, United Kingdom, and Licor Inc., Canada, or Analytical Development Corporation (ADC; http://www.crowcon.com/adc.html). For studies focusing on CO_2 flux alone, a commercial soil respiration chamber linked to an IRGA system (LI-COR, Inc., P.O. Box 4425, Lincoln, NE 68504, http://www.licor.com/index.htm; Norman et al. 1992) may be adequate, although users should be aware that (1) field calibration of the LI-COR (or others) is needed to correct for temperature dependencies of the instrument, (2) the measured flux may be affected by the instrument flow rate

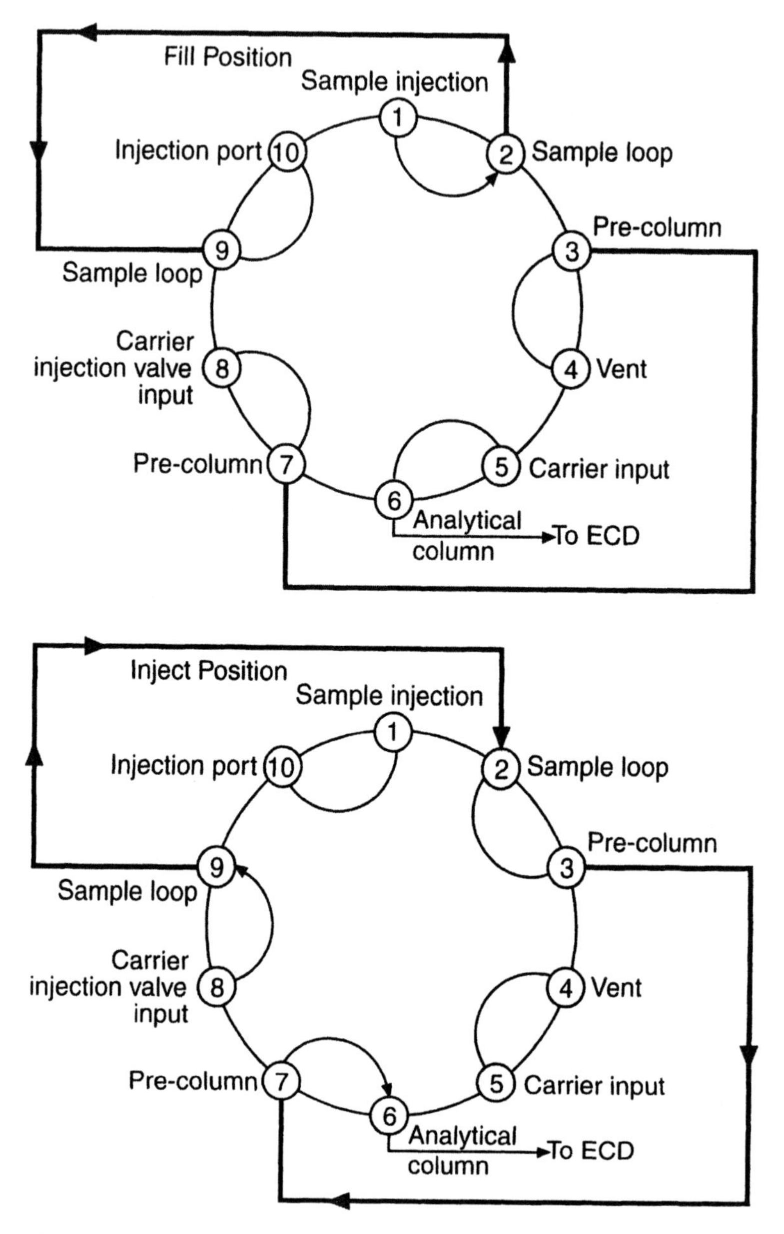

Figure 10.2. Two-position valve layout (with flow diagrams for both positions), including a sample loop, injection and exhaust ports, a pre-column, and an analytical column. Sample is injected into the sample loop with the valve in position 1 (top). Thus, the contents of the sample loop are swept into the pre-column and onto the column when the valve is switched to position 2 (bottom). When the valve is switched back to position 1 (top), the flow in the pre-column is reversed (or back-flushed) and the elements of the gas sample that could prolong the analysis are removed. A simpler six-port valve can be used in cases where a pre-column is not necessary.

because of potential overpressurization of the soil atmosphere, and (3) chamber volumes may be inappropriately small for some soils. Our recommendations assume that the focus of the study is soil or system respiration; to accomplish this, the enclosure must be made of a material that prevents light penetration.

Nitrous oxide. Measurement of N_2O concentrations requires a GC equipped with an electron capture detector (ECD). Gas filter correlation, TDL, IRGA, and photoacoustic IRGA techniques (Ambus and Robertson 1998; Crill et al. 1995) have been applied with varying degrees of success, but the ECD remains the most common and reliable instrument for N_2O analyses. Typically, GC columns used for N_2O analyses are 3–4 m stainless steel tubes (3.175 mm [$\frac{1}{8}$ inch] outer diameter) packed with Porapak QS 60–80 mesh often divided into a 2 m pre-column and a 2 m analytical column. The pre-column allows for faster analyses because air components that are not of interest (water and fluorocarbons) can be flushed off the pre-column, preventing the transfer of later eluting components onto the analytical column. Some ECDs are sensitive to CO_2, allowing simultaneous quantification of CO_2 fluxes. An oxygen/water trap should be placed in the carrier gas line upstream of the GC to prevent damage, because the ECD is susceptible to damage by electrophilic materials including O_2 and freon. Good carrier gases include P5 (95% argon and 5% methane), UHP N_2, and zero air (hydrocarbon-free air). Because copper and brass both react with N_2O, stainless steel fittings, tubing, and valves are required.

Methane. A flame ionization detector in a GC is the most reliable means of measuring methane concentrations, although other techniques, including TDL and gas correlation cells, are available (Crill et al. 1995). One example GC setup includes a 2 m molecular sieve 5A (60–80 mesh), of 2.16 mm inner diameter, with a 0.3 m pre-column to reduce contamination and interference.

Calculations

Calculations of rates of trace gas exchange are based on a difference in the concentration of the gas over time. The calculations required for estimating either net production or net consumption of a trace gas are conceptually straightforward but can be complicated by the fact that the concentration gradient between the soil and the atmosphere begins to diminish immediately upon deployment of the enclosure (Hutchinson and Livingston 1993). Further complications are introduced by the disruption of the atmospheric boundary layer; the importance of these disruptions to the estimated flux increases with the length of time a chamber is in place (Healy et al. 1996). The two factors combined argue for as short a deployment time for the enclosures as logistics will allow. The reader is referred to Livingston and Hutchinson (1995) for a review of trace gas flux calculation considerations.

All measured concentrations should be converted to mass units and corrected to field conditions through application of the Ideal Gas Law:

$$C_m = (C_v \times M \times P)/(R \times T)$$

where

C_m = the mass/volume concentration, e.g., μg CO_2–C/L enclosure, which is equivalent to mg CO_2–C/m^3 enclosure

C_v = the volume/volume concentration (trace gas concentration expressed as a part per million or billion by volume, i.e., ppm_v or ppb_v, respectively, also called mixing ratio), e.g., μL CO_2/L enclosure or ppm_v CO_2

M = the molecular weight of the trace species, e.g., 12 μg CO_2–C/μmol CO_2 or 28 μg N_2O–N/μmol N_2O

P = barometric pressure (P is expressed here in atmospheres), e.g., 1 atm

T = air temperature within the enclosure at the time of sampling, expressed as °K (°K = °C + 273.15)

R = the universal gas constant (0.0820575 L atm·°K·mole)

Thus converted, atmospheric mixing ratios of CH_4 (1.714 ppm_v) and N_2O (311 ppb_v) become 0.855 mg CH_4–C/m^3 and 362 μg N_2O–N/m^3 at standard temperature and pressure (20 °C, 1 atm), and 360 ppm_v CO_2 becomes 179 mg CO_2–C/m^3. The small battery-powered weather stations manufactured by Davis Instruments (Davis Instruments, 3465 Diablo Ave., Hayward, CA 94545; http://www.davisnet.com/) have been particularly helpful for measurements of temperature and barometric pressure, but see also Chapter 3, this volume. If barometric pressure and temperature do not change significantly over the course of the enclosure measurement period (the usual case), it is not necessary to measure either variate precisely to accurately calculate a gas flux.

The converted concentration values are then used to calculate the flux of interest. The most commonly used equation assumes a constant flux (f) and a linear increase in trace gas concentration (C) over time (t):

$$f = V \times C_{rate}/A$$

where

f = gas flux as mass·m^{-2}·h^{-1}, e.g., mg CO_2–C·m^{-2}·h^{-1}

V = the internal volume of the enclosure, including collar volume, expressed as m^3

A = the soil area the enclosure covers, expressed as m^2

C_{rate} = change in concentration of gas (C_m) over the enclosure period, expressed as mass·m^{-3}·h^{-1}, e.g., mg CO_2–C·m^{-3}·h^{-1}

The calculation of gas concentration change should include only data for the time period of linearly increasing trace gas concentrations in the chamber. Thus, C_{rate} is the slope of the best-fit line for the regression of gas concentration (mass/m^3) versus time (h). Each flux series should be graphed and evaluated for linearity; individual point measurements should be carefully checked and discarded if outside confidence bounds. The change in gas concentration within an enclosure will taper off after some period of time, and inclusion of points taken after this period can lead to underestimates of the flux. The recommended units for expression of the flux are μg N_2O–N·m^{-2}·h^{-1}, mg CH_4–C·m^{-2}·h^{-1}, and mg CO_2–C·m^{-2}·h^{-1}.

Special Considerations

Enclosure and Collar Construction and Design

The objectives of enclosure design are to be as noninvasive as possible, and to avoid pressure or temperature changes and excessive trace gas concentration increases. Enclosures are constructed in two parts: (1) a permanent collar, which is put in place prior to sampling and left in place for the duration of the sampling period (in some cases years), and (2) the enclosure itself, which is placed on the collar (over the soil) for the short period over which gas samples are removed (usually an hour or so; Fig. 10.1).

A thorough discussion of chamber geometry is provided in Matthias et al. (1978). The enclosure's surface area to volume ratio determines its sensitivity. A rectangular or cylindrical rigid enclosure (usually 500–900 cm^2) is typical. The surface area should be designed to minimize soil disturbance. Larger surface areas have the advantage of capturing more of the local soil heterogeneity. All construction materials (including sealants) must be made of inert materials that do not react or "bleed" with the gases to be measured. Aluminum, stainless steel, and PVC (which must be opaque to prevent light penetration and temperature rises) have been used successfully for measurement of the nonreactive gases considered here (but may not be suitable for measurement of NO_x, NH_x, and some VOCs and sulfur species). Sealants should always be tested for gas production or interference before use—e.g., different formulations of silicone caulk can differentially bleed N_2O, consume CH_4, or produce interfering peaks upon GC-ECD analysis.

Aluminum and stainless steel collars can be machined, and plastic can be molded to include a channel or "moat" that can be filled with water to provide a gastight seal between the enclosure and the permanent collar. In some cases, a water-filled moat may not be appropriate (e.g., where the landscape is steeply sloped or when the experimental objectives require winter sampling). In these cases, we recommend that the enclosure fit tightly over the ring and be sealed on the outside by elastic material (e.g., a rubber bicycle tube).

Enclosure sampling ports are typically a swagelock fitting with a septum (e.g., Alltech Associates, Inc., 2051 Waukegan Rd., Deerfield, IL 60015; stock no. 15418) attached to a piece of stainless steel tubing that extends sufficiently far into the enclosure to allow sampling of the well-mixed atmosphere in the enclosure's center. A vent tube volume of between 15 and 35 cm^3 is adequate for most barometric conditions and enclosures (Hutchinson and Mosier 1981).

The preferred design of an enclosure includes a sampling port, a properly sized vent, a permanent collar, and a moat to provide a gastight seal between the permanent collar and enclosure (Fig. 10.1). When using the recommended permanent collar and enclosure design it is useful to establish how the gas fluxes change over time following the placement of the collar, particularly for shorter-term studies of a few weeks to months. In cases where the experimental goal is short-term measurement of fluxes from a remote site, it may be desirable to use more portable enclosures that have a skirt secured by an inner tube filled with sand. Again, however, these sorts of measurements are likely to be complicated by difficulties in establishing a gas-tight seal between the soil and atmosphere.

Enclosure Considerations

Leaks, mixing, and temperature control problems are common to all enclosures. The leaks are likely to occur in two locations: where the enclosure top meets the base (prevented by use of the moat sealing system depicted in Figure 10.1) or where base ring meets the soil. Leakage may be a greater problem in very dry soil because air-filled pore space increases dramatically, but leakage under most circumstances will be minimal from a vented static enclosure so long as temperature differences (between the inside and outside of the enclosure) are small (approx. 1 °C) and sampling periods are short (<1 hour). Temperature differences between the ambient air outside the enclosure and the air within the chamber can be minimized by reducing deployment time, by constructing the enclosure out of a light-colored material (e.g., white PVC), or by covering the enclosure with reflective insulation.

Disturbance of the enclosure site can be a problem for all trace gases. Insertion of a permanent collar that allows repeated sampling of the same location may cause substantial root damage in some sites, so collars should be put in place at least 1 week prior to sample collection. Soil disturbance (e.g., compaction) associated with collar placement or sampling activities should be minimized; frequent sampling may require a semipermanent boardwalk or other protective platform. Internal pressures generated by heavy footsteps near enclosures over organic soils (such as those where methane is commonly measured) can force gases out of the soil.

Sampling Considerations

Polypropylene syringes (available from medical suppliers) are inexpensive, allow the generation of the pressure differential required for extraction of a sample, and usually present minimal contamination problems when pretreated (Mosier and Klemedtsson 1994). Nylon syringes have been tested and used successfully but are more expensive. Algal growth in glass syringes has been reported in wet or humid environments.

All butyl rubber products, including the plungers of syringes, can adsorb and desorb gases and therefore need to be conditioned to avoid contamination of CH_4 and N_2O samples in particular; this is easily done by baking separated plungers and barrels at 50–55 °C overnight.

Sample storage in syringes should be limited (not more than 24 hours) because gases diffuse through the polypropylene walls of the syringe (see section on "Sample Storage," below). Nylon one-way stopcock valves (Baxter catalog no. K71, Baxter Diagnostics Inc., Scientific Products Division, 1430 Waukegan Road, McGaw Park, IL 60085-6787) are recommended for sealing syringes during transport; alternatively, silicone stoppers can be placed over the needle. Side-port needles to avoid needle coring are preferred but are not mandatory. Pressure differences during transport can cause leakage where field temperatures are different from laboratory temperatures, where samples are flown from the field to the laboratory, or when samples are transported across different elevations. To protect against such changes, the syringe can be slightly pressurized with a rubber band over the syringe and plunger. Where samples are stored in vials rather than syringes, vials can be overpressurized with excess gas during sampling. As a further precaution, the field standard canister should be sampled in the field and treated as a sample.

Sampling Strategy

Successful enclosure sampling requires consideration of several temporal scales. First is the number of samples and frequency of sampling once the enclosure is put in place. We recommend a deployment time of no shorter than 4–5 minutes and no longer than 1 hour. Continuous sampling instruments (e.g., CO_2 using a flow-through IRGA system connected to a data logger) have shown that fluxes are usually perturbed for the first 30–60 seconds following the placement of the enclosure, probably due to the pressure fluctuations induced by the enclosure placement. The period over which sampling should occur is the period over which there is a linear increase in gas concentration over time. Usually an hour is sufficient to document a significant flux. The potential influence of deployment time on the calculated flux has been evaluated by Healy et al. (1996), who have shown that long deployment times lead to significant underestimates of the flux.

Sample Storage

When samples need to be stored prior to analysis, it is critical to store sample standards in the same way as field samples to identify potential problems. Storage should be approached carefully. Sealed glass or polished stainless steel containers are best for long-term storage but are prohibitively expensive for the large sample numbers required for enclosure flux measurements (Tans et al. 1989). Less expensive alternatives include Venoject and Vacutainer blood sampling vials, but these suffer from differential (cross-batch) problems with sample absorption, leakage, and contamination of the samples during vial processing, and contamination caused by the butyl rubber seals. Boiling the butyl rubber seals in water for 20–30 minutes usually minimizes contamination problems, but not always; we recommend against the use of blood sampling vials altogether. As a word of caution, the manufacture of Vacutainers in particular can include processing in an atmosphere of N_2O. A more extensive discussion of the problems associated with sample storage is provided in Mosier and Klemedtsson (1994). Crimp-top vials designed for automated analysis of headspace volumes and Wheaton serum vials have been used successfully in some laboratories, but each batch of septa material must be carefully evaluated for absorption and contamination. All storage containers should be adequately purged with inert gas and evacuated before use or purged with sample to minimize cross-contamination.

Miscellaneous Considerations

Enclosure-to-enclosure variation in fluxes is considerable, and the resulting measurements are usually not normally distributed. Estimates of the number of enclosures required to characterize the flux of a trace gas from a given area range between 50 and 100 depending on the flux: the larger the flux, the greater the variance (Starr et al. 1995; Robertson 1993; Parkin 1987; Burton and Beauchamp 1985; Foloronuso and Rolston 1984; Robertson and Tiedje 1984). Because of labor and time constraints, only rarely are a sufficient number of enclosures deployed for confident characterization of a site. Flux measurements often follow a Poisson density function with a few high values and increasing variance with increasing flux (Matson et al. 1990). As a result, the measurements should be analyzed with the appropriate statistical techniques that accommodate nonnormally distributed data (see Chapter 1, this volume).

Fluxes of all of these gases can vary diurnally, seasonally, and interannually depending on climate, substrate availability, and the rates of the processes responsible for producing and consuming the gases. Diurnal changes in fluxes can be substantial and should be evaluated for individual sites (Mosier et al. 1991). The overall sampling strategy should build on the soil information available at a given site and be sufficiently specific to address the scientific question posed (see Chapter 1, this volume). In many cases, gas fluxes peak during seasonal transitions or immediately following precipitation or fertilization. If the goal of the study is to develop an annual estimate of flux, the *minimum* sampling requirement is once per month with more frequent sampling during the time of peak flux. For sites where the peak flux is in the spring, this requires increasing the sampling frequency to weekly or biweekly. For sites where peak fluxes follow precipitation or fertilization, hourly sampling may be required to fully characterize the response. In all cases, decisions about sampling frequency should be based on the understanding of the dynamics of the underlying processes and the objective of the study (see Chapter 14, this volume, for discussion of denitrification; Chapter 11, this volume, for discussion of decomposition; and Chapter 13, this volume, for discussion of C and N mineralization and nitrification).

Ancillary Data

Other ancillary measurements helpful to the interpretation of trace gas measurements include soil temperature at the surface and at 2.5 and 5 cm, and water-filled pore space (see Chapter 3, this volume); soil organic carbon and nitrogen content of the surface layers (see Chapter 5, this volume); soil texture (see Chapter 4, this volume); and N mineralization and mineralizable carbon (see Chapter 13, this volume). Soil pH and redox are particularly helpful for interpretation of methane production.

Acknowledgments James Sulzman assisted with the preparation of the final manuscript. Pamela Matson, Elizabeth Sulzman, Peter Harley, and three anonymous reviewers provided thorough and thoughtful reviews that significantly enhanced the quality of this manuscript. E. A. Holland acknowledges the support of NCAR and the Niwot Ridge NSF LTER program during the completion of this manuscript. The National Center for Atmospheric Research is sponsored by the National Science Foundation.

References

Ambus, P., H. Clayton, J. M. R. Arah, K. A. Smith, and S. Christensen. 1993. Similar N_2O flux from soil measured with different chamber techniques. *Atmospheric Environment* 27:121–123.

Ambus, P., and G. P. Robertson. 1998. Automated near-continuous measurement of CO_2 and N_2O fluxes with a photoacoustic infrared spectrometer and flow-through soil cover boxes. *Soil Science Society of America Journal* 62:394–400.

Aselmann, I., and P. J. Crutzen. 1989. Global distribution of natural freshwater wetlands and rice paddies, their net primary productivity, seasonality and possible methane emissions. *Journal of Atmospheric Chemistry* 8:307–359.

Baldocchi, D. D. 1991. Canopy control of trace gas emissions. Pages 293–333 *in* T. D. Sharkey, E. A. Holland, and H. A. Mooney, editors, *Trace Gas Emissions by Plants.* Academic Press, San Diego, California, USA.

Burton, D. L., and E. G. Beauchamp. 1985. Denitrification rate relationships with soil parameters in the field. *Communications Soil Science Plant Analysis* 16:539–549.

Businger, J. A., and S. P. Oncley. 1990. Flux measurement with conditional sampling. *Journal of Atmospheric and Oceanic Technology* 7:349–352.

Christensen, S., P. Ambus, and F. G. Wienhold. 1996. Nitrous oxide emission from an agricultural field: comparison between measurements by flux chamber and micrometeorological techniques. *Atmospheric Environment* 30:4183–4189.

Coleman, D. C. 1973. Compartmental analysis of total soil respiration; an exploratory study. *Oikos* 24:195–199.

Crill, P., J. H. Butler, D. J. Cooper, and P. C. Novelli. 1995. Standard analytical methods for measuring trace gases in the environment. Pages 164–205 *in* P. A. Matson, editor, *Biogenic Trace Gases: Measuring Emissions from Soil and Water.* Blackwell Science, Cambridge, UK.

Denmead, O. T. 1979. Chamber systems for measuring nitrous oxide emission from soils in the field. *Soil Science Society of America Journal* 43:89–95.

Desjardins, R., R. Hart, J. Macpherson, P. Schuepp, and S. B. Verma. 1992. Aircraft- and tower-based fluxes of carbon dioxide, latent and sensible heat. *Journal of Geophysical Research* 97:18,477–18,485.

Desjardins, R., P. Rochette, and E. Puttey. 1993. Measurement of greenhouse gas fluxes using aircraft and tower based techniques. *In* L. A. Harper, A. R. Mosier, J. M. Duxbury, and D. E. Rolston, editors, *Agricultural Ecosystem Effects on Trace Gases and Global Climate Change.* American Society of Agronomy, Crop Science Society of America, Soil Science Society of America, Madison, Wisconsin, USA.

Foloronuso, O. A., and D. E. Rolston. 1984. Spatial variability of field-measured denitrification gas fluxes. *Soil Science Society of America Journal* 48:1214–1219.

Goulden, M. L., J. W. Munger, and S. C. Wofsy. 1996. Exchange of carbon dioxide by a deciduous forest: response to interannual climate variability. *Science* 271:1576–1579.

Hargreaves, K. J., F. G. Wienhold, L. Klemedtsson, J. R. M. Arah, I. J. Beverland, D. Fowler, B. Galle, D. W. T. Griffith, U. Skiba, K. A. Smith, M. Welling, and G. W. Harris. 1996. Measurement of Nitrous Oxide emission from agricultural land using micrometeorological methods. *Atmospheric Environment* 30:1563–1571.

Healy, R. W., R. G. Striegl, T. F. Russell, G. L. Hutchinson, and G. P. Livingston. 1996. Numerical evaluation of static-chamber measurements of soil-atmosphere gas exchange: identification of physical processes. *Soil Science Society of America Journal* 60:740–749.

Holland, E. A., B. H. Braswell, J.-F. Lamarque, A. Townsend, J. M. Sulzman, J.-F. Müller, F.

Dentener, G. Brasseur, H. I. Levy, J. E. Penner, and G. Roelofs. 1997. Variations in the predicted spatial distribution of atmospheric nitrogen deposition and their impact on carbon uptake by terrestrial ecosystems. *Journal of Geophysical Research* 102:15,849–15,866.

Hutchinson, G. L., and G. P. Livingston. 1993. Use of chamber systems to measure trace gas fluxes. *In* L. A. Harper, A. R. Mosier, J. M. Duxbury, and D. E. Rolston, editors, *Agricultural Ecosystem Effects on Trace Gases and Global Climate Change.* American Society of Agronomy, Crop Science Society of America, Soil Science Society of America, Madison, Wisconsin, USA.

Hutchinson, G. L., and A. R. Mosier. 1981. Improved soil cover method for field measurement of nitrous oxide fluxes. *Soil Science Society of America Journal* 45:311–316.

IPCC. 1995. *Climate Change 1994 Radiative Forcing of Climate Change and an Evaluation of the IPCC IS92 Emission Scenarios.* Cambridge University Press, Cambridge, UK.

Jury, W. A., J. Letey, and T. Collins. 1982. Analysis of chamber methods used for measuring nitrous oxide production in the field. *Soil Science Society of America Journal* 46:250–256.

Kanemasu, E. T., W. L. Powers, and J. W. Sij. 1974. Field chamber measurements of CO_2 flux from soil surface. *Soil Science* 118:233–237.

Kolb, C. E., J. C. Wormhoudt, and M. S. Zahniser. 1995. Recent advances in spectroscopic instrumentation for measuring stable gases in the natural environment. Pages 259–290 *in* P. A. Matson and R. C. Harriss, editors, *Biogenic Trace Gases: Measuring Emissions from Soil and Water.* Blackwell Science, Cambridge, Massachusetts, USA.

Lenschow, D. H. 1995. Micrometeorlogical techniques for measuring biosphere-atmosphere trace gas exchange. Pages 126–163 *in* P. A. Matson and R. C. Harriss, editors, *Biogenic Trace Gases: Measuring Emissions from Soil and Water.* Blackwell Science, Cambridge, Massachusetts, USA.

Livingston, G. P., and G. L. Hutchinson. 1995. Enclosure-based measurements of trace gas exchange: applications and sources of error. Pages 14–51 *in* P. A. Matson and R. C. Harriss, editors, *Biogenic Trace Gases: Measuring Emissions from Soil and Water.* Blackwell Science. Cambridge, Massachusetts, USA.

Matson, P. A., and R. C. Harriss, editors. 1995. *Biogenic Trace Gases: Measuring Emissions from Soil and Water. Methods in Ecology.* Blackwell Science, Cambridge, Massachusetts, USA.

Matson, P. A., P. M. Vitousek, G. P. Livingston, and N. A. Swanberg. 1990. Sources of variation in nitrous oxide flux from Amazonian ecosystems. *Journal of Geophysical Research* 95:16789–16798.

Matthias, A. D., D. N. Yarger, and R. S. Weinbeck. 1978. A numerical evaluation of chamber methods for determining gas fluxes. *Geophysical Research Letters* 5:765–768.

Mosier, A. R., D. Schimel, D. Valentine, K. Bronson, and W. Parton. 1991. Methane and nitrous oxide fluxes in native, fertilized and cultivated grasslands. *Nature* 350:330–332.

Mosier, A. R., and L. Klemedtsson. 1994. Measuring denitrification in the field. Pages 1047–1066 *in* R. W. Weaver et al., editors, *Methods of Soil Analysis. Part 2, Microbiological and Biochemical Properties.* Soil Science Society of America, Madison, Wisconsin, USA.

Nadelhoffer, K. J., and J. W. Raich. 1992. Fine root production estimates and belowground carbon allocation in forest ecosystems. *Ecology* 73:1139–1141.

Nay, S. M., K. G. Mattson, and B. T. Bormann. 1994. Biases of chamber methods for measuring soil CO_2 efflux demonstrated with a laboratory apparatus. *Ecology* 75:2460–2463.

Norman, J. M., R. Garcia, and S. B. Verma. 1992. Soil Surface CO_2 fluxes and the carbon budget of a grassland. *Journal of Geophysical Research* 97D:18,845–18,853.

Parkin, T. B. 1987. Soil microsites as a source of denitrification variability. *Soil Science Society of America Journal* 51:1194–1199.

Robertson, G. P. 1993. Fluxes of nitrous oxide and other nitrogen trace gases from intensively managed landscapes: a global perspective. *In* L. A. Harper, A. R. Mosier, J. M. Duxbury, and D. E. Rolston, editors, *Agricultural Ecosystem Effects on Trace Gases and Global Climate Change*. American Society of Agronomy, Crop Science Society of America, Soil Science Society of America, Madison, Wisconsin, USA.

Robertson, G. P., and J. M. Tiedje. 1984. Denitrification and nitrous oxide production in old growth and successful Michigan forests. *Soil Science Society of America Journal* 48:383–389.

Schimel, D. S., B. H. Braswell, and W. J. Parton. 1997. Equilibration of the terrestrial water, nitrogen and carbon cycles. *Proceedings of the National Academy of Sciences* [USA], 94:8280–8283.

Sitaula, B. K., L. Jiafa, and L. R. Bakken. 1992. Rapid analysis of climate gases by wide bore capillary gas chromatography. *Journal of Environmental Quality* 21:493–496.

Starr, J. L., T. B. Parkin, and J. J. Meisinger. 1995. Influence of sample size on chemical and physical soil measurements. *Soil Science Society of America Journal* 59:713–719.

Tans, P. P., T. J. Conway, and T. Nakazawa. 1989. Latitudinal distribution of the sources and sinks of atmospheric carbon dioxide derived from surface observations and an atmospheric transport model. *Journal of Geophysical Research* 94D:5151–5172.

Wofsy, S. C., M. L. Goulden, and J. W. Munger. 1993. Net exchange of CO_2 in a mid-latitude forest. *Science* 260:1314–1317.

11

Measuring Decomposition, Nutrient Turnover, and Stores in Plant Litter

Mark E. Harmon
Knute J. Nadelhoffer
John M. Blair

Decomposition processes represent a major flux of both fixed carbon (C) and nutrients in most terrestrial ecosystems, and quantifying rates of litter mass loss and the concomitant changes in nutrients bound in the litter are important aspects of evaluating ecosystem function. Plant litter decomposition plays an important role in determining carbon and nutrient accumulation, as well as the rate and timing of nutrient release in forms available for uptake by plants and soil biota. Litter decomposition and nutrient dynamics are controlled to varying degrees by substrate quality (litter morphology and chemistry), abiotic conditions (temperature, moisture, soil texture), and biotic activity (microbial and faunal; Kurcheva 1960; Heath et al. 1964; Bunnell et al. 1977; Bunnell and Tate 1977; Parton et al. 1987). Thus, decomposition processes can serve as "integrating variables" for evaluating ecosystem function, for comparing different ecosystems, and for evaluating management practices or other anthropogenic influences (Coleman and Crossley 1996).

Decomposition involves not only mass loss but also changes in the nutrient content of plant litter and the eventual release of nutrients therefrom. Decomposition involves leaching of soluble organic and inorganic components, catabolic breakdown of organic matter, and comminution or physical fragmentation of litter (Swift et al. 1979). These processes ultimately transform senescent plant material into both labile and stable organic matter both above- and belowground. Methods used for quantifying rates of mass loss often can be used to determine changes in nutrient content as well. The dynamics of nutrients in decomposing litter can be complex, and decomposing litter can alternately act as either a nutrient sink or a source. This varies as a function of the nutrient under consideration, litter quality, biotic activity, exogenous nutrient inputs, and stage of decomposition.

In addition to quantifying rates of mass loss and nutrient dynamics of decomposing litter, it is often desirable to quantify the stores, or standing stocks, of various plant litter pools. Standing stocks of both coarse (i.e., woody) and fine (i.e., leaves, fine roots, etc.) plant litter represent important carbon and nutrient reservoirs in terrestrial ecosystems. The sizes of these reservoirs are influenced by both rates of litter production and decomposition, and are sensitive to changes in either process. Unfortunately, there are relatively few large-scale direct measurements of plant litter stores, and regional, national, and global estimates of these pools are often modeled based on input and decomposition rate data (e.g., Birdsey 1992; Harmon and Chen 1992; Kurz et al. 1992; Turner et al. 1995).

Our understanding of, and ability to model, litter decomposition, soil organic matter formation, and the storage of carbon and nutrients in ecosystems will be much improved if researchers design decomposition experiments and conduct inventories that lend themselves to broader synthesis. Our goals in this chapter are to present standard protocols for quantifying decomposition dynamics and standing stocks of most pools of plant litter. Two important exceptions are soil organic matter and very fine roots (<0.5 mm diameter), which are best studied using methods described in Chapter 5, this volume, and Chapter 20, this volume, respectively. Because methodologies vary for different types of plant litter, our discussion is divided into seven sections.

Available Methods

Fine Litter Decomposition

Many methods have been used to determine rates of fine plant litter decomposition. All have problems, and they serve mainly as indices of decomposition rates. Although we recommend only the litterbag method, we discuss other methods as potential alternatives. In the litterbag method preweighed material is confined within mesh bags and changes in mass, nutrient content, and carbon chemistry are measured over time (Falconer et al. 1933; Lunt 1933, 1935; Gustafson 1943; Bocock and Gilbert 1957; Bocock et al. 1960; Gosz et al. 1973). This method excludes macroinvertebrates, which are important elements of the decomposer community in many ecosystems. The litterbag can also alter the microclimate within the bag by slowing drying rates (Witkamp and Olson 1963) and can reduce rates of fungal hyphal colonization and growth (St. John 1980).

Time-series methods analogous to litterbags are (1) litter baskets that confine materials between a fine mesh bottom and a coarser mesh hardware-cloth top to allow access to macroinvertebrates (Stevenson and Dindal 1981; Blair et al. 1991) and (2) tethers to connect litter material while leaving it completely exposed (Witkamp and Olson 1963; Lang 1974). These methods also have problems. In the case of litter baskets, the potential input of additional material in situ may restrict their interpretation to areas with relatively large leaves (e.g., broadleaf forests). Tethered material, while exposed to invertebrates and natural microclimatological conditions, is subject to high rates of physical fragmentation, which can overestimate decomposition.

A chronosequence approach, in which annual accumulation layers are separated and analyzed for changes in mass, nutrient content, particle size, and type has also been used (Kendrick 1959). This approach (i.e., a substitution of space for time) can yield extremely interesting information. However, there must be clear indicators to mark the annual layers, and one must assume annual rates of litterfall are constant and that mixing of the layers is minimal.

There are several indirect methods for measuring decomposition, including harvesting litter plots, comparing paired plots, and calculating input-output balances. In areas with discrete periods of litterfall, decomposition rates can be calculated by comparing the lowest and highest stores of litter (Tyler 1971; Loomis 1975) or by measuring seasonal changes in stores (Capstick 1962; Weary and Merriam 1978). The paired plot method has had limited success; the concept is to remove inputs from two similar plots, harvest the litter at time zero in one plot and after some period in the second plot (Singh and Gupta 1977). The problems are that the paired plots often differ in the initial stores of litter, and preventing additional litter inputs is very difficult. The best-known indirect method is to calculate decomposition rate constants from litter-input:standing-crop ratios (Olson 1963). This method has been widely used, but it may give incorrect values if the standing stock is not in steady state or if the inputs are not completely accounted for (e.g., fine root inputs) or if the standing stock is difficult to measure.

We have selected the litterbag method as the standard protocol for determining the rate at which fine litter decomposes and accumulates or releases nutrients. Although the method has limitations, it is highly repeatable, relatively inexpensive, and widely used. Several types of litterbag systems have been used in the past, most differing in the size of the mesh used to contain the litter. We recommend a range of sizes depending on the purpose of the study. We strongly recommend that studies be conducted well beyond the traditional time length of 1–2 years as the carbon and nutrient dynamics of the early decomposition stages are relatively well known compared with the transition period from litter to stable soil organic matter (Lousier and Parkinson 1978; Berg et al. 1984; Edmonds 1984; Aber et al. 1990).

Woody Detritus Decomposition

Several methods are available to determine rates at which woody detritus decomposes, forms soil organic matter, and accumulates or releases nutrients. Harmon and Sexton (1996) provide a thorough review of these methods, including their relative merits.

Two frequently used approaches are (1) chronosequences that give a short-term snapshot of processes and (2) a time-series approach that is a long-term effort yielding excellent resolution of temporal patterns and processes. In the chronosequence approach, one ages as many pieces of detritus as possible in various states of decay and examines how a parameter such as density changes through time. Dates can be taken from fall scars, seedlings, living stumps, and records of disturbance (e.g., fire, insect outbreak, windstorm, thinning). This approach has been used extensively for coarse woody detritus (e.g., Graham 1982; Grier 1978; Harmon et al. 1987; Means et al. 1987; Sollins et al. 1987) and also can be used for downed fine woody detritus (Erickson et al. 1985) and dead coarse roots (Fahey et al. 1988). The interpretation of chronosequence data varies depending on its use, typically either the conversion of volume measurements into mass or nutrient stores (see later) or the determination the rate mass is lost or nutrients are accumulated or released. If the aim is to use a decay chronosequence to estimate rates of mass loss or nutrient release, then the data must be adjusted for past fragmentation losses to estimate these rates correctly (Harmon and Sexton 1996).

Although chronosequences produce results quickly, there are serious temporal resolution problems caused by errors in dating and estimates of initial conditions. A time series circumvents these problems by examining how a cohort of pieces progresses through time, thereby avoiding the substitution of space for time. Although the method requires substantial investments in effort and time, it lends itself nicely to process studies. In addition to examining a chronosequence of pieces, one can also indirectly estimate decomposition rates of woody detritus from a chronosequence of different-aged stands (e.g., Gore and William 1986; Spies et al. 1988) by assuming each stand-creating disturbance left a similar amount of material. In many cases this assumption is not justified and can lead to significant uncertainty concerning decomposition rates. Finally, the ratio of input to stores can be used to indirectly estimate decomposition rate constants of woody detritus (Sollins 1982). This is subject to the same errors as for fine litter, compounded by the fact that both inputs and stores of woody detritus are highly variable.

Given the greater precision and site specificity of the time-series approach, we recommend this method to study the decomposition and nutrient dynamics of woody detritus above- and belowground. We recommend the chronosequence method to determine the density, carbon content, and nutrient concentrations of decay classes used to estimate mass and nutrient stores.

Standard Substrate Decomposition

Standard substrates can be used to determine the effect of environment on decomposition (Jenney et al. 1949; Tsarik 1975; Piene and Van Cleve 1978). Although these materials are somewhat artificial, they can provide an index of micro- and macroenvironmental controls on decomposition. Standard substrates low in nitrogen (N) (e.g., cellulose and wood) can indicate local variations in nitrogen availability (Binkley 1984). Standard substrates can be natural litter such as wheat straw, which is low in nitrogen and moderate in lignin content, or artificial substrates such as cellulose pulp or filter paper, cotton cloth strips, various small pieces of wood (including Popsicle sticks and chopsticks), and wooden dowels and blocks.

Our standard protocol uses two substrates: cellulose filter paper and hardwood dowels. We have selected these because they are commonly available and are low in nitrogen, and thus sensitive to nitrogen availability.

Organic Horizon Stores

The most common method to determine stores in organic horizons is to harvest material within a small plot. Plot size has varied from as large as 1 m × 1 m (Grier and

Logan 1977) to the more common 10–30 cm squares (Metz 1954; Youngberg 1966; Federer 1984). Steel corers 3–12 cm in diameter have also been used, but in cases where organic horizons are sparse this methodology is difficult to apply. Other methods, such as recording the depth of organic horizons and converting to mass with a measured bulk density, may be suitable for deep layers such as peat, but for most other situations the variation in bulk density is too high for reliable mass determinations. Therefore, the fuel survey methodology (Brown 1974) is not recommended as a standard protocol.

Our recommended method is to harvest organic horizons within a small square template with the approximate dimensions of 25 × 25 cm. Use of larger templates or corers has not been demonstrated to decrease the variability between samples (Capstick 1962). A better strategy is to use a greater number of smaller samples to reduce variability, which can be considerable (Weary and Merriam 1978; Carter and Lowe 1986).

Fine Woody Detritus Stores

Several alternative methods exist for measuring fine woody detritus stores (Harmon and Sexton 1996). The most straightforward is harvesting and weighing material within small plots (<4 m^2). Downed fine woody detritus can also be estimated using planar transects in which the number of pieces in size classes is recorded and then converted to stores using the mean diameter and bulk density of the size class. Unfortunately, the latter two parameters are rarely measured or reported (see Harmon and Sexton 1996 for available data); therefore, stores estimated by this method have questionable accuracy. Moreover, although the planar transect estimates the volume of downed, surface wood, it does not measure other important forms of fine woody detritus including dead branches or dead coarse roots. By using fixed-area plots the same methods can be used on all forms of woody detritus. This allows a better aggregation of these pools into a total woody detritus store that will integrate more closely with the methods used to estimate live tree mass, the source of woody detritus. Fixed-area plot sampling also has advantages for long-term measurement because the area in which trees are dying matches that of the woody detritus sample.

We recommend weighing downed branches in fixed-area plots for determining fine woody detritus stores. Dead branch and coarse root stores have rarely been estimated, and we recommend using an allometric approach based on the basal diameter of downed logs, stumps, and standing dead trees that is adjusted for decay state.

Coarse Woody Detritus Stores

For coarse woody detritus (>10 cm diameter and >1 m long) it is impractical to remove and weigh pieces. Therefore, for downed logs, standing dead trees, and stumps it is more usual to record piece dimensions within fixed-area plots (Harmon et al. 1987) or along planar transects (Warren and Olsen 1964; Van Wagner 1968; Brown 1974) to estimate volume, which is then converted to mass and nutrient stores using decay class–specific bulk density and nutrient concentration values. We strongly discourage visual estimates from photographic comparisons (Maxwell and Ward 1976a, 1976b; Ottmar et al. 1990) because this method can be very inaccurate.

We recommend that fixed-area plots be used for determining woody detritus stores. Although the planar transect is a good, fast method, it does not measure standing dead trees or stumps. Because these two types of woody detritus often form major pools, a methodology that can be used for all types of woody detritus is preferable. By using fixed-area plots one can use the same methods on all forms of woody detritus and can sample them on the same area. This has the advantages for aggregation and long-term measurements discussed earlier under fine woody detritus stores.

Fine Litter Decomposition

The recommended protocol for examining fine litter decomposition, nutrient release, and formation of stable soil organic matter is to use the litterbag method in a time series. This method may be used for fine roots, leaves, twigs, reproductive parts (including cones), and small bark fragments. Because much less is known about mass loss and changes in litter chemistry during later stages of decay (Aber and Melillo 1980, 1982; Berg et al. 1984; Melillo et al. 1989), we suggest designing decomposition studies to last more than 5 years.

Materials

The materials needed to construct, place, and retrieve litterbags include:

1. A suitable quantity of air-dried litter
2. Litterbags (see procedures, below, for construction guidelines)
3. Nylon thread or Monel staples to seal the litterbag
4. Tags, either aluminum or plastic
5. Flagging to mark location
6. Shovel for burying belowground litterbags
7. Heavy nylon monofilament or braided nylon line to tether litterbags
8. Plastic bags to transport and store retrieved litterbags
9. High-quality paper bags to dry litter
10. Drying oven with a 50–55 °C range

Procedure

1. Litter selection. Decomposition data are most useful when the materials studied span a wide range of litter quality (LIDET 1995; Trofymow 1995). The simplest indicators of litter quality are C:nutrient ratios, most often C:N ratios (Singh and Gupta 1977). However, lignin:N ratios (Melillo et al. 1982), the relative concentrations of lignin and cellulose (ligno-cellulose index, or LCI, as defined by Aber et al. 1990), soluble phenolic content (Palm and Sanchez 1990), and phosporus and calcium contents are also useful indicators of litter quality.

2. Litterbag construction. Litterbags should be made of relatively nondegradable, inert materials. Mesh size can have a major effect on the invertebrate community consuming the litter, the microclimate, and the degree of fragmentation (Heath et al. 1964), and will depend on study objectives and environment. For aboveground placement, 1 mm nylon mesh has often been used and in low-light environments can last several decades. For environments with high levels of UV radiation (i.e., deserts, grassland, harvested forest areas), we recommend using fiberglass mesh (1.5 mm). For extremely small litter (e.g., *Larix* needles), we recommend woven polypropylene swimming pool cover or shade cloth (0.4 mm), a material extremely resistant to UV degradation. To allow access to macro- and megafauna, mesh must be at least 2 mm (see Chapter 7, this volume). Litterbags can be constructed to have the same material on the top and bottom or can have a larger mesh on the top than the bottom. The latter design prevents the loss of small fragments during long-term incubations. The smaller-mesh bottom can be made of Dacron sailcloth (50 μm) in low light environments or woven polypropylene in high UV environments. For belowground placement, litterbags can be constructed of Dacron sailcloth on both sides because UV degradation is not a consideration. Litterbags should be 20 cm $\times$ 20 cm and sewn and double-stitched on three sides using nylon thread (polyester thread is sensitive to UV degradation). Bags made of polypropylene can be heat-sealed effectively. Litterbags should be identified with unique numbers embossed on small aluminum or plastic tags that can be attached using UV-resistant cable ties. We also recommend placing a subset of litterbags partially filled with an inert polymer such as polyester fiberfill to estimate the mass and characterize the chemistry of materials transported into litterbags during the course of field incubation.
3. Litter collection. Although leaf litter is the tissue type most commonly used in decomposition studies, inclusion of root and fine woody materials such as twigs is of particular value, since they often represent large inputs to soils (Vogt et al. 1986). If the intent is to mimic natural litterfall, leaves should be collected from senescent plants in the case of herbaceous species or from branches ready to shed leaves in the case of woody plants. In the latter case it is often possible to "strip" leaves off branches. If this is not possible, then placing a clean drop cloth beneath the tree or branch and shaking will cause leaves to fall. In situations where live plant residues are a major source of litter (e.g., an agricultural field or a harvested forest), cutting green material may be appropriate. Regardless of the method used to gather litter, it is essential to report the source when presenting results.

   We recommend that fine roots be excavated from a site similar to where they eventually will be placed. An alternative is to use roots from ingrowth experiments. One may also grow plants in controlled nutrient conditions and harvest the roots. This method has the advantage of allowing one to label roots with isotopes to enhance the interpretation of decomposition and nutrient dynamics. Finally, one can use fine roots from tree seedlings grown in nurseries that are being either discarded or trimmed prior to storage. Given that the substrate quality may vary with the source, even for a single species, it is essential that lignin, nitrogen, and other measures of substrate quality be determined.

4. Filling litterbags. Litter materials should be air-dried for at least 1 week prior to filling bags. Ideally, each bag should initially contain 10 g (air-dried) material because this leaves a sufficient amount for chemical analysis even after extensive decomposition. Subsamples of each litter should be set aside for oven drying at 55 °C and subsequent chemical analyses. We recommend placing the litter on a pre-tared pan and then placing the litter inside the bag after it has been weighed. Litterbags can be sealed in several ways: (1) by sewing the bag shut with thread, (2) by sealing with Monel staples (a nonreactive alloy) using five to six staples per 20 cm length of bag, or (3) heat sealing if the bag is made of polypropylene. It is important to record any losses from fragmentation during transport. One can use a set of "traveler" bags, which are taken to the field site, handled as the other litterbags, and then retrieved after placement. Reweighing these bags determines the average losses caused by transport and handling.
5. Initial chemistry and moisture content. When filling litterbags, 10 g samples should be periodically taken to determine the moisture content and initial chemistry of the litter. If the material has been properly air-dried, the variation in moisture will be quite small (+/− 1%). If weather conditions change radically over the course of filling the litterbags, it is important to take moisture samples frequently. These moisture samples should be weighed prior to and after oven drying to a constant mass to calculate dry weight conversion factors (air-dry mass:oven-dry mass) for each litter type used. Multiplying the dry-weight conversion factor by the air-dry mass will give the estimated oven-dry mass of each sample. Oven-dried material should be stored in sealed containers for future chemical analysis in a cool, dry environment. We strongly recommend a total mass of 50–100 g be set aside for these purposes.
6. Sampling interval. Uniform recommendations of sampling intervals are difficult to make due to climatic variability among regions. However, because mass loss and both carbon and nutrient dynamics change most rapidly during the early stages of decay, sampling intervals should be geometric. If the intent is to determine early leaching and very labile carbon losses, then a sample 1–4 weeks after placement may be necessary. Otherwise samples should be collected for three seasons (spring, summer, fall) in arctic and temperate ecosystems for the first year, and at 1–2 month intervals for moist tropical ecosystems. After the first year, we suggest increasing sampling intervals to once or twice per year in arctic and temperate ecosystems and 3–6 months in tropical systems.
7. Litterbag placement. It is important to avoid pseudoreplication (Hurlbert 1984). Separate sets of litterbags should be placed either in replicated units (ecosystem types, experimental plots, etc.) or in single plot types located along documented environmental gradients (e.g., fertility, moisture, temperature, or elevation gradients). Sufficient numbers of samples should be set out to allow for retrieving at least four to five litterbags per litter type used per plot at each sampling time.

Normally, litterbags should be placed in locations where the litter type under investigation is most likely to enter the soil system. Leaf and fine woody litter samples should be placed at the surface of the litter layer, whereas fine root material should be inserted into the profile where they normally grow and die. We recommend that above ground litterbags be pinned to the surface to limit movement. The recommended procedure for placing litterbags below ground is to push a shovel into the soil at a 45° angle, prying the resultant slit open until there is enough space to slide the litterbag all the way in, and then extracting the shovel. Good soil contact can then be established by gently tamping the raised portion of the soil.

To aid retrieval it is best to tether sets of litterbags to lines (either heavy-gauge monofilament or braided fishing line) that are flagged at both ends. If more than one litter type is tethered to a single line, samples should be placed in random sequence along the line. Each line should be sufficiently long (typically 5–10 m) to encompass variations in microhabitat at the site. Prepare a sketch map indicating the location of the litterbag "lines" with respect to permanent landmarks.

8. Litterbag retrieval. Utmost care should be taken to ensure that decomposing litter materials do not fragment and fall out of litterbags during retrieval or prior to processing. Litterbags should be cleaned of adhering particles (soil, mosses, rock fragments, etc.) to the extent possible in the field and placed individually into plastic bags immediately after being collected. Samples can be refrigerated for up to 1 week before processing, but if processing is delayed for more than a week, samples should be stored frozen.
9. Sample processing. Several options may be used to process litterbags. In cases where samples are not contaminated with large amounts of sand or soil, process the moist samples by carefully brushing the surface of the litterbag, cutting it open, and carefully turning it inside out onto a clean sheet of paper or into a large tray. If decomposition has been extensive, the inside of the litterbag can be scraped with a spatula to remove adhering particles of organic matter. Any living plant parts (e.g., roots or moss) as well as extraneous matter such as rocks and large soil particles should be removed. Do not remove decomposed organic matter or invertebrate feces (frass) from the litter, as these materials could be derived from the original material. Instead, use organic matter accumulation in unfilled litterbags to estimate the possible contribution of exogenous organic matter to the sample. The fresh weight of the material should then be determined and the sample placed in a paper bag, dried at 55 °C until the mass is stable, and then weighed to determine the dry weight.

   When sand or fine soil contamination is high, obvious extraneous matter should be removed. Then oven dry the sample and finally sieve it to remove the bulk of the sand or finer soil. Because it is often difficult to remove all this material, a subsample of the contaminating soil should be retained to determine the carbon and nutrient content so that litter concentrations can be corrected.

   After sample dry weights have been recorded and checked, grind each sample separately to pass a no. 40 sieve and store the dried, ground samples in sealed glass or polypropylene containers until they are analyzed for chemical constituents (see Chapter 8, this volume). Additional sample preparation and grinding may be required depending on the types of analyses that are planned.

Calculations

Samples obtained from litterbags in contact with the mineral soil often contain a mixture of the decomposing original litter and some soil from the surrounding area. Therefore, litter dry weights need to be corrected for soil contamination before determining mass loss or calculating decomposition rate constants. Often a subsample of the ground litter is ashed for 4 hours at 450 °C, and the mass remaining is expressed based on the percent ash-free dry mass (AFDM) of the initial and final litter samples. This is appropriate when soils are very low in organic matter but is not satisfactory for soils with a relatively high organic matter content, since the organic matter of the soil will contribute to the apparent organic matter mass of the litter. Instead, we recommend the use of the following soil correction equation (Blair 1988a):

$$FLi = (SaAFDM - SlAFDM)/(LiAFDM - SlAFDM)$$

where

FLi = the proportion of litterbag sample mass that is actually litter
$SaAFDM$ = the percent AFDM of the entire litterbag sample
$SlAFDM$ = the percent AFDM of the soil from which the litterbag was retrieved
$LiAFDM$ = the percent AFDM of the initial litter

The underlying assumptions of this equation are that the organic matter content (percent AFDM) of the litter remains constant during decomposition, and that organic matter content of the contaminating soil can be determined. The equation then calculates the proportion of litter and soil that must have been mixed to produce the measured percent AFDM of the entire litterbag sample. The weight of the litterbag sample can then be multiplied by the correction factor (FLi) to obtain the weight of the litter remaining. In soils low in carbonates, the same correction can be applied by using percent carbon in place of percent AFDM. For soils high in carbonates, the concentration of these substances will have to be determined before a correction can be made.

The accumulation of soil in the litterbags also affects apparent nutrient concentrations in the litter. Therefore, the nutrient concentrations of litterbag samples contaminated with soil (as indicated by reductions in percent AFDM) should be corrected using the following equation (Blair 1988b):

$$LiNt = [SaNt - (FSl \times SlNt)]/FLi$$

where

$LiNt$ = the nutrient concentration in the residual litter

SaNt = the nutrient concentration of the entire litterbag sample
FSl = the proportion of the litterbag sample mass that is actually soil (1 − *FLi*)
SlNt = the nutrient concentration of the soil
FLi = the proportion of the litterbag sample mass that is litter (from the above soil correction equation)

Special Considerations

The litterbag approach has limitations that need to be considered. In ecosystems where macroinvertebrates play a major role in decomposing litter, the small mesh sizes proposed here will exclude these organisms and thus underestimate decomposition rates. In this case it may be best to use multiple mesh sizes (see Fig. 17.1 in Chapter 17, this volume). For buried litterbags, the high amounts of residual material typically formed (McClaugherty et al. 1984) may be caused by the artificial environment. In particular, the fact that root litter is separate and not intermingled with the soil may alter the decomposition process (Fahey et al. 1988). The proposed protocol thus precludes studying the effect of soil texture and structure on physically protecting litter and incipient soil organic matter. If the latter is of interest, then incorporation of soil of known characteristics into the root litterbags as they are filled may be the method of choice. Finally, the recommended correction for soil contamination can be problematic for species with high ash contents. To test the underlying assumption that ratio of ash to dry mass for litter remains constant over decomposition, plot AFDM versus the cumulative mass loss from a location where soil contamination is minimal.

Fine Woody Detritus Decomposition

Fine woody detritus takes several forms, including attached and downed dead branches and coarse roots (>1 cm). The methods described in this section are appropriate for all these forms of detritus, regardless of whether the material is suspended off the ground, lying on the soil surface, or within the soil. For small pieces (<1 m long and <10 cm diameter), an approach analogous to the litterbag method can be used, weighing entire pieces before and after a period of incubation. Because of their size and structural integrity, however, woody samples do not need to be confined.

Materials

The materials and equipment required to conduct decomposition and nutrient studies for fine woody detritus include the following:

1. A source of branches or coarse roots
2. Chainsaw to cut large wood pieces
3. Miter saw, hand saws, and clippers to cut small wood pieces
4. Calipers (0–150 mm range) to measure thicknesses of samples
5. Diameter tape to measure piece circumference

6. Tape measure or ruler to measure piece length
7. Aluminum tags to mark samples
8. UV-resistant cable ties to attach tags to pieces
9. Plastic bags to carry samples (1–120 L depending on piece size)
10. Paper bags for drying samples (no. 2 to no. 10 depending on piece size)
11. Portable electronic scale if work is conducted at remote site
12. Electronic scales with ranges of 0–1500 g and 0–6000 g depending on piece size
13. 1 mm nylon mesh to cover buried pieces
14. Nylon braided cord to tether pieces

Procedure

1. Sample interval. As with fine litter, the standardization of intervals between sampling times is not recommended given the dependence of decomposition rates on substrate size, quality, and site conditions. We suggest the sample interval should be approximately 10% of the expected maximum life span. The number of times samples are collected will be study-dependent, but five sample times are recommended as a minimum.
2. Species selection. Species selection for a time-series experiment depends on the degree of species richness and the number of functional classes present in a location. Except for large, diverse genera with resistant heartwood such as *Pinus* or *Quercus*, little precision is gained by sampling below the genus level. The minimal suggested design is to select common species that represent the fast and slow ends of the decay resistance spectrum. This has the advantage of allowing one to compare the range of decomposition rates among sites. If additional resources exist, species with intermediate or extremely high decay resistance should be included.
3. Substrate quality descriptors. Physical variables are more commonly used to describe differences in woody substrates than in fine litter, in part because differences in substrate chemistry are less variable in wood than are chemistry differences in leaves, fine roots, and other small plant parts (Harmon et al. 1986). We recommend that at a minimum the total diameter, length, radial thickness of the major tissue types (i.e., outer bark, inner bark, sapwood, and heartwood), the total volume, and bark cover be measured. Because bark, sapwood, and heartwood decay at different rates, it is crucial to know their proportions. This allows one to understand why sizes and species decompose at different rates and to adjust results based on the proportions of these tissues. In addition to these physical descriptors, measurement of the same chemical descriptors used for fine litter is important.
4. Size selection. Since the decomposition of wood is in part a function of size, it is crucial that size differences be considered. We recommend that a range of diameters be used at each site so that statistical adjustments for size effects can be made. A geometric series of diameters is probably most useful, since the effect of diameter tends to decrease with size (e.g., 1, 2, 4, 8, 15 cm). A tolerance of 20% should be allowed for variation in diameter (other-

wise too many pieces will be rejected). When assigning pieces to size classes, also record the diameter of each piece to use as an independent variable. For intersite comparisons, species or sites can be compared directly where classes overlap, or indirectly by comparing regression slopes of decay rate on size, or by adjusting for size effects.

5. Collection of materials. Unlike fine litter, it is often difficult to find woody material that has recently senesced. A more likely source of material will be recently fallen trees from windstorms, or to fell trees using a chainsaw. To collect coarse roots, examine sites of recent road or house construction for access to recently excavated stumps. As materials for fine woody detritus are collected, we suggest cutting pieces at least 2.5 times the final length because this will greatly reduce the number of moisture samples required to estimate the initial oven-dry weight (see later).
6. Sample preparation. For small pieces of branches and coarse roots, one should determine the initial mass by weighing the entire sample. In addition, the diameter at the midpoint and the total length should be measured on each piece. If one is careful to cut lengths uniformly, a subsample of lengths can be measured. However, it is crucial to measure the diameter of each piece. Because one should not oven dry pieces before they are placed in the field, it is necessary to estimate the initial moisture content. This parameter is highly variable and should be determined for each piece. By cutting material in the initial field phase at least 2.5 times the final length, one can generate two pieces for each moisture sample taken. Either a miter or a radial arm saw can be used to prepare branch or coarse root samples. First trim off the ends that have been exposed to drying for more than a few hours. Then remove a thin piece for moisture determination, cut the first sample length, remove another sample for moisture content, cut the next sample length, and so forth, until the remaining piece is too short to be used. Use a uniquely numbered aluminum tag and a UV-resistant cable tie to identify each piece to be placed in the field. Bags containing the moisture content (labeled with individual piece numbers that they correspond to) should be dried at 55 °C until the weight is stable. Use the ratio of oven-dry weight to fresh weight to estimate the initial moisture content of individual sample pieces. Subsamples of this material should be saved for determining the initial chemical composition and ash content of samples. Extra pieces and odd lengths of material should be saved to determine the thickness of the tissue layers comprising samples of various sizes. At a minimum we suggest that bark and wood thickness be recorded, but one should also note the presence of zones high in resins (i.e., knotlike material), which are particularly decay-resistant. Finally, the source of the material should be noted because branch, bole, and coarse root material can differ substantially in their initial density, bark thickness, and chemistry.
7. Sample length. The length of pieces used in fine woody detritus time-series decomposition studies is a crucial consideration. This is because decomposers will colonize from the ends and when pieces are too short the decomposition rate is elevated. A preliminary guideline would be to have the

length 10 times longer than the mean diameter. To some extent the length problem can be avoided by using an end sealer. This may be required when short lengths have to be used because of shortage of material or extreme taper of pieces (e.g., coarse roots). Paraffin, epoxy sealer, and neoprene paint have been used in the past as physical barriers. End surfaces should be clean and dry before application of the sealer. Unless these conditions are met, most sealants will not adhere to the ends properly and the physical barrier will be incomplete.

8. Placement. Pieces can be placed upon the organic horizon, suspended, or buried in the rooting zone. While it is desirable for all these situations to be studied, we recommend that branches placed upon the organic horizon be used as the reference case for each site. If other positions are of interest, using the following methods would assure some degree of standardization. Fine wood pieces may be suspended off the soil surface using two pieces of braided nylon cord to form a ladderlike arrangement, which can then be hung from a tree. Buried fine wood should be inserted into the soil using a shovel as with buried litterbags; however, we recommend that sheets of 1 mm nylon mesh be wrapped around the pieces to aid recovery. As with litterbags, tethering all fine downed or buried wood pieces is recommended. Since many woody detritus decomposition studies may take decades to complete, it is essential that a sketch map showing the location of the pieces relative to obvious landmarks be made at the time of study initiation.
9. Sample replication. At least four sites should be sampled at each time to avoid pseudoreplication problems (Hurlbert 1984). It is recommended that if the entire branch or root is not harvested and weighed, then at least three cross sections should be taken for moisture content.
10. Sample retrieval. Tethered wood pieces should be located and uncovered before attempting retrieval. Pulling the tether line to retrieve samples may cause breakage and loss of sample. If pieces are shorter than the initial length, then the length recovered should be noted to adjust results for breakage. Clean samples of any adhering pieces of organic or mineral matter and place in a plastic bag.
11. Sample processing. In the laboratory, reclean the samples, then place into paper bags and oven dry at 55 °C until the sample weight is stable. For pieces exceeding 5 cm in diameter, cut the sample into pieces to speed drying. After oven-dry weight has been determined, chop the samples and coarse grind to 2 mm using a large Wiley or hammer mill. Grind further to pass a 40-mesh (0.417 mm) screen using a conventional Wiley mill and then store in glass or polypropylene bottles in a cool, dry environment until chemical analysis.

Calculations

Because wood ash contents are consistently low, there is usually no need to correct for soil contamination. Exceptions are coarse roots or any very decayed material that has come into contact with the mineral soil; for these types of material, corrections for ash content should follow those outlined for fine litter. Calculations used for determining the composition and nutrient dynamics of woody detritus are described later.

Special Considerations

The position of the pieces with respect to the soil (suspended above, resting on, or buried within) can have a major impact on decomposition rates. Therefore, one may wish to test this effect by either burying or suspending pieces in addition to placing them on the surface of the organic horizon.

Coarse Woody Detritus Decomposition

Coarse woody detritus takes several forms, including downed and standing boles, stumps, and very large branches and coarse roots. Decomposition of large pieces (>1 m long and >10 cm diameter) of woody detritus is best studied by recording the volume of the entire piece to determine fragmentation losses and then removing disks to determine changes in density.

Materials

The materials and equipment required to conduct decomposition and nutrient studies for coarse woody detritus include the following:

1. A source of boles
2. Chainsaw
3. Hatchet to remove subsamples
4. Hammer and chisel to trim and remove subsamples
5. Calipers (0–150 mm range) to measure thicknesses of samples
6. Diameter tape to measure piece circumference
7. Tape measure or ruler to measure piece length
8. Aluminum tags to mark samples
9. Aluminum nails to attach tags to large pieces (>10 cm)
10. Plastic bags to carry samples (1–120 L depending on piece size)
11. Paper bags for drying samples (no. 2 to no. 10 depending on piece size)
12. Portable electronic scale if work is conducted at remote site
13. Electronic scales with ranges of 0–1500 g and 0–6000 g depending on piece size

Procedure

1. Sample interval. Recommendations for sample intervals are the same as for fine woody detritus.
2. Species selection. Recommendations for species selection are the same as for fine woody detritus.
3. Substrate quality descriptors. The same physical and chemical descriptors of substrate quality used for fine woody detritus should be used for coarse woody

detritus. In addition, the depth and type of any existing decay (white rot versus brown rot) should be measured. It is also useful to record the depth of the pith because this serves as a useful reference point as the piece fragments. If bark was removed during felling or transport, the total bark cover should be estimated.

4. Collection of materials. A good source of material is recently fallen trees from windstorms, or one can fell trees using a chainsaw. For coarse woody detritus allow an additional 20% to the final length to prevent sample disks from excessive drying during piece preparation.
5. Initial mass. For coarse woody detritus it is impractical to weigh samples to determine their initial mass. It is more practical to remove disks, or "cookies," from the ends of pieces to determine the density and to estimate the initial total volume of the piece. When removing disks trim off a short length (e.g., 5 cm) if the ends have been exposed to drying before cutting the sample disk. As a minimum, the end diameters and the middle diameter as well as total length should be measured for initial volume determinations of each piece (see Newton's formula below). The maximum and minimum diameter at each point should be measured with a caliper or diameter tape. Initial mass is the product of the initial volume and density of the disk.
6. Subsampling. Bark and wood should be the minimum layers that are examined on pieces exceeding 10 cm diameter because the nutrient content and decomposition rates of these materials are very different. It is also very useful to separate the sapwood from the heartwood because heartwood decay resistance is the primary basis for differences in tree species (Harmon et al. 1986). Even in species without decay-resistant heartwood, this layer decays slower than the sapwood due to the time required to colonize the inner layers. Although it is interesting to separate the inner and outer bark (these two layers are usually the fastest- and slowest-decaying layers, respectively), it is often very difficult to separate with any degree of accuracy or safety. It is therefore probably best to treat bark as one tissue and then try to separate the dynamics of the individual layers using the two-component exponential model outlined later.
7. Placement. We recommend that pieces be placed upon the organic horizon as the standard protocol for each site. Moving large pieces of woody detritus can be difficult; therefore, they may have to be left "in place." If pieces exceeding a diameter of 25 cm and a length of 2 m are to be moved, logging machinery may be required. Given that many woody detritus decomposition studies may take decades to complete, it is essential that a sketch map showing the location of the pieces relative to obvious landmarks be made at the time of study initiation.
8. Sample replication. At least three sites should be sampled at each time to avoid pseudoreplication problems (Hurlbert 1984).
9. Mass loss. After a suitable period of decomposition, determine the remaining volume, bark cover, and density of parts of the pieces. As a minimum, the end diameters and the middle diameter, as well as total length, should be measured for total volume determinations (see Newton's formula below). The maxi-

mum and minimum diameter at each point should be measured with a caliper or diameter tape. This current volume should be compared with the estimated original volume to see if a correction for fragmentation losses is required. Total bark cover should also be estimated, using a frame of known size to help determine the total area missing or remaining. To determine the density, moisture, and nutrient content of a piece, a minimum of three disks should be removed per piece, and these should be systematically spaced along the length.

Various methods are used to remove subsamples of decomposing bark and wood for density, moisture, and nutrient determination. These include mapping out and subsampling zones with different appearance (Sollins et al. 1987; Harmon et al. 1987), systematically cutting the disk into pieces (Harmon 1992), removing "typical" subsections, or removing entire tissue layers (e.g., bark). Unless one is interested in studying the internal heterogeneity, we suggest the latter approach. If fragmentation of layers has not occurred, then record the diameter with and without a layer, as well as the longitudinal thickness (along the long axis of the piece) of each layer in a disk. Use a hammer and chisel to separate the layers. The total fresh weight of each layer can then be determined and a subsample used to determine the moisture and nutrient content of each layer. The volume of each layer is calculated as for a cylinder,

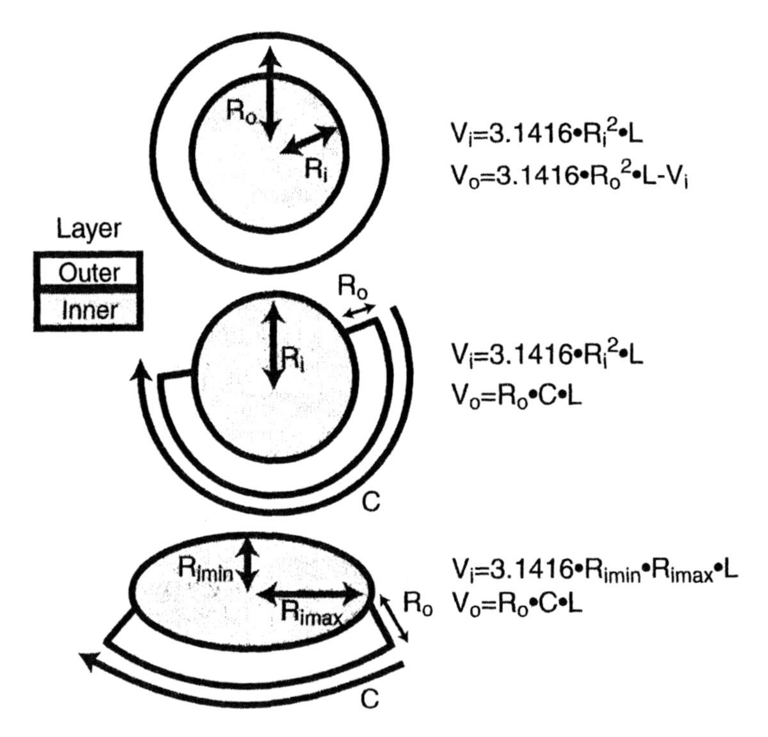

Figure 11.1. Measurements to be taken and appropriate formulas to determine volume of layers within a cross section removed from a piece of large woody detritus. *R* is the radial dimension, and *C* is the length along the circumference. *L* is the longitudinal dimension and is not shown on the cross-section drawings. The subscripts indicate whether the dimension is from the inner (*i*) or outer layer (*o*) or the minimum (*min*) or maximum (*max*) axis.

with the volume of any layer occurring inside it deducted (Fig. 11.1). If fragmentation of a layer has occurred, then record the radial thickness and circumferential length of the layer as well as its longitudinal thickness. The volume of this layer can be computed as a rectangular form. Compute the volume of the remaining unfragmented layers as described earlier for unfragmented disks. Extremely decomposed pieces often have elliptical forms, and it is also difficult to remove intact disks from them. In this case it is best to cut the disk free and then carefully excavate it from the piece. One can then record the maximum and minimum diameters of the disk from the parts that were not removed. The area of the elliptical disk can be computed using the equation for an ellipse. The longitudinal thickness of elliptical pieces can be determined from the pieces of the cross section that are removed. Separation of layers, weight determination, and subsampling for moisture and nutrient contents for these last two cases are the same as for unfragmented disks. Subsamples should be chopped into smaller pieces and placed in paper bags to be oven dried at 55 °C until the weight is stable. The oven-dry weight of the total disk can be computed by multiplying the ratio of oven-dry weight to fresh weight of the subsample by the total fresh weight of the layer. Subsamples for a given layer and pieces may be pooled, coarse ground, fine ground, and stored as for fine wood samples.

Calculations

Because wood ash contents are consistently low, there is usually no need to correct for soil contamination except in locations where insects transport soil into downed wood. Calculations used for determining the composition and nutrient dynamics of woody detritus are described later. Use of the two-component exponential equation is highly recommended when tissue layers with highly different properties (e.g., outer and inner bark) are not physically separated. Density calculations should be based on oven-dry mass divided by green or fresh volume because wood below 30% moisture content (the fiber saturation point) will shrink. Equations for calculating the volume of samples are presented in Fig. 11.2.

Special Considerations

The method proposed to estimate the volume of samples is likely to overestimate the volume of bark for species with rough surfaces. If this is a concern, then displacement measurements should be used to determine volume. If water displacement is used, then separate samples should be used to determine nutrient concentrations.

Standard Substrate Decomposition

The recommended standard substrates to be used in conjunction with fine litter or woody detritus decomposition experiments are cellulose filter paper and hardwood dowels. The advantage of using these materials is that they are more likely to be uni-

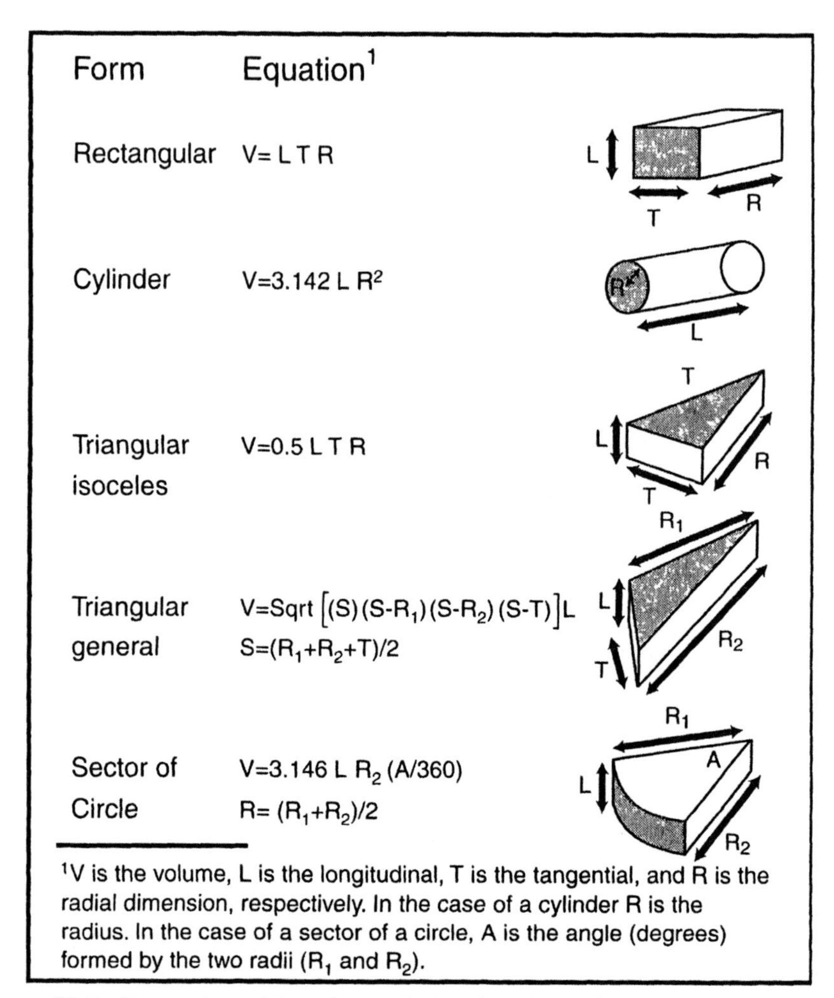

Figure 11.2. Commonly used formulas to calculate the volume of samples used for density.

form and they are very sensitive to nutrient availability. Both characteristics make them ideal for directly comparing the effects of the environment among studies and sites (Binkley 1984; O'Lear et al. 1996).

Materials

The materials needed to construct, place, and retrieve standard substrates include the following:

1. Cellulose filter paper.
2. Litterbags (see earlier for construction guidelines)
3. Nylon thread or Monel staples to seal the litterbags
4. Tags, either aluminum or plastic
5. Flagging to mark location

6. Hardwood dowels, 6 mm diameter; 60 or 120 cm long
7. Dowel sleeves (see later for construction guidelines)
8. Steel rebar, 6 mm diameter, 45 cm long, and hammer to make pilot hole for dowel
9. Heavy nylon monofilament or braided nylon line to tether litterbags
10. Plastic bags to transport and store retrieved litterbags and dowels
11. High-quality paper bags to dry litter
12. Drying oven with a 50–55 °C range
13. Data forms to record the time of recovery, fresh weight, oven-dry weight, and any peculiarities of the samples

Required materials and construction of litterbags for incubating cellulose filter paper are the same as for those containing natural fine litters. For hardwood dowels, some modification is required. If the dowels are to be placed belowground, they should be encased in a sleeve of 1 mm nylon mesh. This can be constructed by sewing a narrow strip of nylon mesh (4 cm wide and 30 cm long) into a sleeve that can be slipped over the portion that is placed below ground. This greatly aids in the recovery of the decomposed dowels from soil.

For dowel studies we recommend using a hardwood species that does not have a decay-resistant heartwood because this reduces variation both within and between species. Species commonly available with this characteristic include ramin (*Gonystylus bancanus*), birch (*Betula* spp.), and basswood (*Tilia* spp.).

Procedure

1. Cellulose standard substrates. Procedures for filling litterbags with cellulose filter papers are the same as for litterbags using natural litters. We recommend using 5–10 g of paper. Placement should be similar to that of the natural litter that is being placed. In addition, it may be of interest to place filter paper filled bags at several depths within the soil. Recovery and treatment of the decomposed material also follow the procedures for fine litter (see earlier). Because the nitrogen concentration of filter paper varies, it is essential this parameter be reported when results are presented.
2. Dowel standard substrate. Procedures for the hardwood dowels are similar to those for small woody detritus pieces. Use 60 cm lengths of 6 mm diameter dowel. Our recommended protocol is to place 30 cm of the dowel belowground and 30 cm aboveground so that these two environments can be compared. UV-resistant cable ties should be used to attach tags to the aboveground portion of the dowels. For the belowground portions attach a tag (with the same number as that on the upper portion) to the nylon mesh sleeve that will eventually encase it. Weigh the entire dowel after taring out the weight of the cable tie and tag attached to the upper portion. As with fine litter, periodically save subsamples to determine the oven-dry weight to air-dry weight conversion factor and to have materials for initial chemical analysis. After the dowel is weighed, slip the nylon mesh sleeve over the portion that is to be placed belowground.

3. Dowel placement. Dowel placement will largely depend on the design of other experiments being conducted. Once a location is selected, we recommend driving a 6 mm diameter by 45 cm long piece of steel rebar into the soil to form a pilot hole in which to place the dowel. Mark the 30 cm depth on the rebar so that the hole is the correct depth. In very rocky soils, it may be necessary to search for a "rock-free" zone or to place the dowels at a shallower depth. Slide the dowel into the pilot hole and note the length of the dowel remaining aboveground.
4. Dowel retrieval and processing. Retrieving the dowel involves finding the aboveground portion (it may no longer be attached to the belowground part) and placing it in a plastic bag. For the belowground portion, locate the tag attached to the nylon sleeve, excavate the dowel using a shovel, and place it in a plastic bag. As the dowel parts are recovered it is important to record the lengths of the above- and belowground parts that are found, as well as noting any obvious insect damage. In the laboratory, brush off any soil still adhering to the dowel with a moist paper towel and clip the above- and belowground portions into short (2–5 cm) sections so they will fit in a small paper bag and dry faster. Dry at 55 °C until the mass is stable (5–7 days) and record the dry weight. To grind the dowel samples in a standard Wiley mill, it may be necessary to first coarse grind to a 2–3 mm particle size by using a larger mill. Store ground samples in closed glass or polypropylene containers in a cool, dry environment until analysis can be conducted.

Calculations

Results for standard substrates should be reported in ash-free values. Calculations of mass loss and adjustments for soil contamination should be the same as those for fine litter. To calculate the initial oven-dry weight of the above- versus belowground portions of dowels, assume that the density is uniform:

$$IODW_{position} = IODW_{total} \times Length_{position} / Length_{total}$$

where

$IODW_{position}$ = the initial oven-dry mass of the position (above- or belowground)
$IODW_{total}$ = the total initial oven-dry mass
$Length_{position}$ = the portion of the dowel in a position
$Length_{total}$ = the total length of the dowel

Special Considerations

There is likely to be some spread of decomposers from the belowground portion of the dowel into the aerial portions. Using separate dowels for above- and belowground measurements will eliminate this effect. One may also suspend dowels in the air to prevent incorporation into the organic horizon. Further subdivision of the dowels beyond the above- and belowground segments recommended here is also possible.

Organic Horizon Stores

The organic horizons to be sampled may be composed of many forms of plant litter. To avoid double counting, one should not include any wood pieces that are greater than 1 cm in diameter and recognizable as branches or boles. Measurements for these materials are described later. Organic horizons should include any thoroughly decomposed wood (usually red-brown in color) that is located in the organic horizon. Although this is often discarded as a nonorganic horizon, it can constitute a considerable fraction of some organic horizons, especially in conifer forests (McFee and Stone 1966; Youngberg 1966; Harvey et al. 1979; Harvey et al. 1981; Little and Ohmann 1988). This material is important to include because it is the wood analog to the humus or O_2 layer in a forest floor and can have high nitrogen availability (Sollins et al. 1987).

The methodology proposed, sampling in 25 cm × 25 cm quadrats, is suitable for most situations where organic horizons are continuous. In situations where organic horizons are sparse or interspersed with rock outcrops, bare soil, logs, or other objects that cover more than 5% of the surface, we recommend a stratified sampling, with line transects being used to determine the area covered in organic horizons versus the other surfaces. Organic horizons that are sampled can then be adjusted to represent the overall area.

A final sampling consideration is the time of year to sample. In ecosystems with distinct pulses of litter inputs, seasonal variation in organic horizon stores can vary by 20–30% (Loomis 1975). This variation can be almost as large as that observed over succession (Federer 1984); it is therefore important to note the season of sampling relative to the peak in litter inputs. Ideally sampling should be conducted before and just after the peak litter inputs so that the annual range in stores would be available for comparative purposes.

Materials

The materials required to sample organic horizons are

1. Wooden or steel sampling template, 25 cm × 25 cm recommended
2. Serrated knife to cut organic horizons
3. Small pruning saw to cut buried branches and coarse roots
4. Pruning shears to cut buried branches and coarse roots
5. Small file to sharpen bottom edge of frame
6. Plastic or plastic-lined paper bags to store samples
7. 30–50 m tape to locate sample points and determine cover of nonorganic surfaces
8. Sorting tray larger than frame to field process sample
9. Random number table

The sampling template can be made out of wood or metal. If a metal template is to be used we recommend it be fashioned as an open frame constructed from stainless steel and welded together. Handles can also be welded on the frame to help push it into the organic horizon. The bottom edge of the metal frame can be sharpened with a file to help it cut through the organic layers. Paper bags for storing samples should be avoided unless they are lined with plastic, since moisture from the samples will weaken even the thickest paper bags.

Procedure

1. Site characterization. Once an ecosystem has been located for sampling, one must decide if the cover of surfaces other than the organic horizon exceeds 5%. If the cover is less than 5%, then proceed to sample the organic horizon as outlined later. If the cover of nonorganic horizon surfaces exceeds 5%, then use line transects to determine the cover of these surfaces. A transect length of at least 100 m should be used to record the length covered by surface rocks and outcroppings, exposed mineral soil, tree roots, logs, stumps, or other surfaces that will be sampled by other means. Ideally the transect or grid used to sample organic horizon cover can also be used for the location of samples. If the ecosystem occurs on sloping ground, it is important to note the average slope steepness because results should be reported on a horizontal and not a slope area basis.
2. Plot placement and replication. Sample plots for organic horizons can be placed either systematically or at randomly spaced locations along the tape measure. The number of samples adequate for an ecosystem will vary. The use of two to three samples (e.g., Metz 1954; Youngberg 1966; Loomis 1975) is strongly discouraged. As a starting point, we recommend 20–50 samples to provide a standard error within 10% of the mean (McFee and Stone 1965; Wallace and Freedman 1986). It is also useful to plot a running mean of samples to determine when additional samples change the mean less than 5%.
3. Sample removal. Once the samples are located along the transect or grid, place the sample template parallel to the surface and press it into the organic horizon until firm resistance is felt. Use a knife to cut the organic layer and pruning shears or saw to cut any roots or buried branches that prevent cutting through to the mineral soil.

   Remove the template, and remove the organic horizon and any mineral soil adhering to the bottom. A spatula can often be used to lift the intact sample off the underlying horizons. Place the sample in a metal sorting tray as intact as possible and remove any adhering mineral soil. It is important to consistently remove the mineral soil from the organic horizon. Remove any branches greater than 1 cm in diameter from the sample and place the remaining sample in a plastic or lined paper bag that is sealed and clearly labeled with the date, location, sample number, and any other critical information. If red-brown, thoroughly decayed wood is found in the sample, separate this from the rest of the material and bag it separately. Further separation of other organic layers is optional (e.g., O1 versus O2), but given the different systems used for each ecosystem, it is unlikely these values could be directly compared outside a given region. The separation of decayed wood is quite important because this material has generally not been measured and is derived from a source different than the rest of the organic horizon.

4. Sample processing. In the laboratory, the samples should be removed from the plastic bags, placed in heavy paper bags or trays, and oven dried at 55 °C until the weight is stable. After determining the dry weight, samples may be pooled to determine chemical properties, since there is a good correlation between pooled samples and the mean of individual samples for most properties (Carter and Lowe 1986). However, if one is interested in the internal variation within a plot or experiment, then we would recommend against sample pooling for determining chemical properties. Samples used for chemical properties should be passed through a screen and homogenized. Subsamples of the material should be ground to 40-mesh sieve and stored in glass or polypropylene containers in a dry, cool location until ash and nutrient contents can be determined (see Chapter 8, this volume).

Calculations

Results should be expressed as ash-free mass using the methods described for fine litter decomposition experiments. For ecosystems where organic horizons cover less than 95% of the surface, the total store in organic horizons should be decreased to represent the average surface:

$$Mass_{corrected} = Mass_{OH} \times Area_{OH}$$

where

$Mass_{correctin}$ = the organic horizon mass corrected for other surfaces
$Mass_{OH}$ = the mass of the surfaces covered by organic horizons
$Area_{OH}$ = the fraction of the ecosystem covered by organic horizons

If the ecosystem occurs on a slope exceeding 10°, then a correction should be made to report results on a horizontal area basis. The equation for this correction is:

$$Mass_{slope\ corr} = \text{cosine}\ (slope) \times Mass_{slope}$$

where

$Mass_{slope\ corr}$ = the slope corrected mass
$slope$ = slope angle in degrees
$Mass_{slope}$ = the mass of organic matter based on slope distance

Special Considerations

Separation of organic horizons from the upper mineral soil is problematic for many soils. Distinctions between organic and mineral horizons are clearer in mor-type layers, but are quite gradual in mull-type layers. In the latter case, close coordination of sampling of the organic and upper mineral horizons is crucial to avoid double counting of stores. Ash content of organic horizons in mull soils is likely to be highly variable; therefore, determining the ash content of each sample is recommended in this case.

Fine Woody Detritus Stores

The forms of fine woody detritus that should be sampled include downed and suspended fine wood (<10 cm diameter and <1 m long), and dead coarse roots. For meaningful comparisons it is extremely important to include all forms of woody detritus in inventories and to report them in the same units.

The recommended protocol is to define a large plot (hopefully the same area in which ongoing experiments or live biomass measurements are being conducted). Use 10–20 1 m × 1 m subplots to estimate the mass of downed fine wood, and use allometric relationships to estimate the stores of dead coarse roots and fine woody detritus attached to standing dead trees. Fine downed woody detritus can be directly harvested, weighed in the field, and subsampled for moisture and nutrient content.

Materials

The materials needed to measure fine woody detritus stores are

1. Diameter tape for measuring diameters of standing dead trees and stumps
2. 30 or 50 m tape for defining plot boundaries
3. Compass to help lay out plot boundaries
4. Flagging to mark boundaries of large plot
5. 1 m × 1 m sample frame to define downed fine woody detritus plot
6. Pruning saw and/or clippers to cut fine downed woody detritus
7. Portable scale, electronic version with accuracy to 1 g preferred
8. Burlap or other large cloth bags to hold fine wood samples
9. Tray to hold fine wood samples while weighing
10. 1–4 L plastic bags to hold fine wood moisture samples

Procedure

1. Sample selection and number. To be most useful, fine woody detritus should be measured in plots or stands that have ongoing experiments or inventories of living biomass and other detrital pools. Because the distribution of fine downed wood is highly variable, we recommend that at least 10–20 subplots be used for each stand sampled (Harmon and Sexton 1996), since this results in estimates with a standard error within 30% and 20% of the mean, respectively (Harmon and Sexton 1996). Alternatively, a running mean may be plotted for a subset of samples to assure adequate replication (i.e., addition or subtraction of a sample does not change the overall mean more than 5%).
2. Downed fine woody detritus. Downed fine wood should be estimated in 1m × 1 m fixed-area subplots. The entire subplot, including the organic horizon, should be searched. Woody material that is <1 cm diameter or decayed to the point its source is not recognizable should not be gathered because such material is considered part of the organic horizon. Once all the fine woody material is harvested, weigh using a portable electronic scale. Spring scales should be avoided due to their general lack of precision. Subsamples of this wood (100–200 g) are then taken and weighed in the field, oven dried at 55 °C, and weighed to determine the moisture content. The total field weight

is then adjusted using the estimated field moisture content. An alternative is to remove the entire sample to be weighed and dry it in the laboratory; given the volume of material this entails, however, subsampling may be preferred.

3. Suspended fine woody debris. It is extremely difficult to estimate suspended fine wood directly unless it is close to the ground, where it should be included in the 1 m × 1 m plots used for downed fine wood. We recommend that an indirect estimate be used, based on the inventory of standing dead tree mass. For standing dead trees that are not broken, one should estimate the volume of branches from allometric equations. For boles that have broken, this volume should be pro rated according to snag height so that only snags with the entire length have the entire branch volume and those that have broken off below the crown do not have any branch volume. The mass of suspended fine wood should then be estimated from the branch volume by multiplying the branch bulk density for the appropriate snag decay class. Unfortunately, there are few estimates of branch density as a function of snag decay class. Lacking such data, assume a branch density.
4. Dead coarse roots. We know of no one who has tried to directly inventory dead coarse roots by excavation, but given the potential mass of material, some estimate should be made. One possible indirect method is to use allometric relationships based on tree diameter to predict the volume of dead roots for each dead tree in the fixed-area plot used to sample coarse woody detritus. Rather than use the diameter at breast height, which would not be available for stumps or many logs, the diameter at the piece base should be measured for this purpose. Equations exist to convert basal diameter to diameter at breast height for many species (Harmon and Sexton 1996). Predict dead coarse root volume for each dead tree inventoried and then use the bulk density of dead roots to estimate the mass. As with suspended fine wood, until bulk density data for decomposing coarse roots become generally available, it may be necessary to assume a value.

Calculations

Corrections for stores measured on plots with a slope greater than 10° are the same as those for organic horizons. Because woody detritus generally has a very low ash content (i.e., <2%), correcting for soil contamination is less of a concern than for organic horizons.

Special Considerations

Other methods are more suited to larger-scale surveys. Planar transects are particularly useful in this context, although other methods will have to be used to estimate suspended fine wood and dead coarse roots.

Coarse Woody Detritus Stores

The forms of coarse woody detritus that should be sampled include stumps (specifically meaning the lower part of trees that were cut by a saw), downed coarse wood (hereafter called logs, >10 cm diameter at the large end and >1 m long), and standing dead trees (including everything from freshly killed trees to extremely decayed, short vertical pieces not cut by saw). For meaningful comparisons it is extremely important to include all forms of woody detritus in inventories and to report them in the same units. Past synthesis efforts have been severely hampered because these two problems have repeatedly not been considered in the primary literature (Harmon et al. 1986; Harmon 1993). For example, reporting standing dead trees as numbers per area and downed wood as volume per area (as is common) makes it impossible to total stores.

The recommended protocol is to define a large plot (preferably the same area in which ongoing experiments or live biomass measurements are being conducted) and to measure the dimensions of all large pieces of woody detritus on the entire plot. The volumes of large woody detritus estimated from the dimensional data can be converted to mass and nutrient stores using decay class–specific bulk density and nutrient concentration data.

Materials

The materials needed to measure coarse woody detritus stores are

1. 1 m caliper for measuring diameters of downed maerial
2. Diameter tape for measuring diameters of standing dead trees
3. 30 or 50 m tape for defining plot boundaries and measuring pieces' lengths
4. Clinometer to determine standing dead tree heights
5. Compass to lay out plot boundaries
6. Flagging to mark boundaries of large plot
7. Sample forms, clipboards, etc., for recording data
8. In addition to these materials, it is also essential that investigators check into the availability of decay classifications and bulk density values for their local species and situations (see Harmon and Sexton 1996 for some compiled values). It is entirely unacceptable to use unrelated genera (e.g., *Pseudotsuga*) to convert volume to mass and nutrient stores; unfortunately, this has occurred in the past. If suitable conversion factors do not exist, then a serious effort needs to be made to create them.

Procedure

1. Plot selection and size. To be most useful, woody detritus should be measured in plots or stands that have ongoing experiments or inventories of living biomass and other detrital pools. This gives a more complete inventory of the ecosystem but also allows coupling of process rates (e.g., mortality) to these "static" measurements. For large pieces of woody detritus, plot size is a crucial consideration. The size of the coarse woody detritus plots may correspond to that of preexisting tree plots. If new plots are being established, a cumulative area of at least 0.1 ha to represent a normally stocked stand is recommended. Even with this plot size, at least 10 replicates may be required to have the standard error range within 10% of the mean (Harmon and Sexton 1996).

2. Dimension measurements. Large or coarse woody detritus assumes at least four forms: standing dead trees (also called snags), stumps, logs, and blobs. The last refers to the piles of decomposed bark and wood that accumulate around the bases of large snags. The variables recorded for each log inventoried include diameters at both ends and at the midpoint, length, species, position, and decay class, and whether the piece is hollow or solid. In many forests it is very important to subtract out the volume associated with hollows by noting the diameter and length of the hollow as well as the exterior diameter. The variables required for all other forms of large woody detritus are similar to those used for logs, with the exception of diameter. For snags, the diameter at breast height is recorded for intact boles, and the diameters at the base and top for boles that have broken. The base and top diameters can also be recorded for stumps. Finally, the diameter at the base is the only dimension required for blobs.
3. Diameters are best measured using 100 cm calipers because it is often impossible to wrap a tape around logs and parallax errors are large if a meter stick is used. When pieces are elliptical, record the maximum and minimum diameters and convert to a round-equivalent diameter using a modified version of the formula for the area of an ellipse. The top diameter of tall snags can usually be accurately measured by simply finding the top. If the top cannot be located, a visual estimate will usually suffice as long as one calibrates one's eye.
4. The length of logs can be measured with a tape or, if available, a sonic tape measure. The height of snags is often difficult to measure or estimate. If the snag is not broken, estimate the snag volume or height from the breast-height diameter by using allometric relationships developed for living trees. If a snag is not intact, then estimate the length or height. For snags less than 4 m, use a 100 cm caliper or meter stick to estimate height; for taller snags, a clinometer and tape can be used.
5. Volume to mass conversion. Regardless of the dimensions measured, these data must first be converted to volume and then to mass to estimate mass and nutrient stores. To convert from volume to mass or nutrient stores, use the density and/or nutrient content of wood and bark in various stages of decay. The latter values can be taken from the literature, although there exists potential for error by not using site-specific values (especially for nutrient stores). It is preferable that decay class conversion factors be site-specific, although this need not include every forest for which dimensional data are gathered. To establish an objective decay class system, it is necessary to correlate the external characteristics to variables of interest such as density, bark cover, and nutrient content. Samples are then removed from three to five logs of each decay class to determine the mean bulk density and nutrient concentration (see the section "Coarse Woody Detritus Decomposition," above, for sampling methods).
6. It is crucial to report the characteristics used to separate decay classes for data to be comparable. Unspecified modifications of another decay class system are not sufficient descriptions. Physical characteristics that have proven use-

ful in the past to distinguish decay classes include presence of leaves, twigs, branches, bark cover on branches and boles, sloughing of wood, collapsing and spreading of log (indicating the transition from round to elliptical form), degree of soil contact, friability or crushability of wood, color of wood, and whether the branch stubs can be moved. Biological indicators such as moss cover, fungal fruiting bodies, or presence of insect galleries seem to be of very little value in separating decay classes because they vary widely even within a limited area. In areas with high species diversity, it will probably be impossible to have decay classes measured for each species. This problem can be addressed by defining larger functional classes, such as decay-resistant and non-decay-resistant species (Harmon et al. 1995).

Calculations

Corrections for stores measured on plots with a slope greater than 10° are the same as those for organic horizons. As woody detritus generally has a very low ash content (i.e., <2%), correcting for soil contamination is less of a concern than for organic horizons.

To convert the maximum and minimum diameters of elliptical pieces to a round diameter equivalent, use a modified version of the formula for the area of an ellipse:

$$A = D_{max} \times D_{min}$$

and backtransform to diameter using a modified version of the area of a circle:

$$D_{round} = \sqrt{A}$$

where

D_{max}, D_{min}, and D_{round} are the maximum, minimum, and round equivalent diameters, respectively

The volume of large woody debris pieces can be calculated by several formulas, depending on the number of diameter measurements taken. For logs in which diameters were measured at three points use Newton's formula:

$$V = L \times [A_b + (4 \times A_m) + A_t]/6$$

where

V = the volume
L = the length
A_b, A_m, and A_t = the areas of the base, middle, and top, respectively

For logs, standing dead trees, or stumps that have two diameter measurements, use the formula for a frustum of a cone to estimate volume:

$$V = L \times [A_b + (A_b A_t)^{0.5} + A_t]/3$$

where

V = the volume
L = the length
A_b and A_t = the areas of the base and top, respectively

For blobs, which only have the basal diameter measured, use a modified version of the formula for a paraboloid to estimate volume:

$$V = L \times (A_b/2)$$

where

V = the volume
L = the length
A_b = the area of the base

Special Considerations

Other methods are more suited to larger-scale surveys. Planar transects are particularly useful in this context, although other methods will have to be used to estimate stumps and standing dead trees. It is essential that decay classes similar to those used in fixed-area plots be used, since the original sound-versus-rotten classification suggested by Brown (1974) is too crude. Methods for other forms of woody detritus, which might work in a large-scale survey context, would include variable-radius plots (Grosenbaugh 1958; Harmon and Sexton 1996) and point-centered quarter sampling (Cottam and Curtis 1956; Mueller-Dombois and Ellenberg 1974). These alternative methods, however, have yet to be tested.

Calculation of Decomposition and Nutrient Mineralization Rates

Time-series data, such as that generated from litterbag studies, can be presented as the percentage of initial mass remaining over time. "Decomposition curves" can also be mathematically described. Once weights have been corrected for mineral soil contamination (see the "Fine Litter Decomposition" section, below), the percent mass remaining can be calculated. Initial air-dry weights should be converted to equivalent oven-dry weights before doing this. Percent mass remaining from individual litterbags can be averaged, and mean percent mass remaining over time can be plotted, by treatment or litter type, and used to calculate decomposition rate constants.

There are several available models to which mass loss data can be fit (Olson 1963; Minderman 1968; Wieder and Lang 1982; Andren and Paustian 1987). The simplest of these is the single negative exponential model (Jenny et al. 1949; Olson 1963) of the form:

$$X_t/X_0 = e^{-kt}$$

where

X_t/X_0 = proportion of litter mass remaining at time t
t = time elapsed, expressed as years or days (see discussion)
e = the base of the natural logarithms
k = the decomposition rate constant

This model is attractive because it produces a single decomposition rate constant (k value), which can be used to compare data from different treatments, species, or studies. Thus, we recommend the calculation of decomposition rate constants using this model when possible. A major disadvantage is that it does not accurately describe litter decomposition kinetics where relative decay rates vary over time, as is the case when there is a rapid loss or an extended lag phase early in decomposition.

The single negative exponential model can be fit to the data by least-squares linear regression of the natural logarithm of mean percent mass remaining over time. To calculate annual decomposition rate constants, time in the field should be expressed as a fraction of 1 year (i.e., 182 days = 0.5 years). For litter types that decompose much faster, such as green crop residues, daily decomposition rates may be more appropriate. Least-squares regression will give values for slope (k), intercept (predicted % mass remaining at t = 0), and coefficients of determination (r^2). Values of k indicate the rate of mass loss; greater k values indicate faster mass loss rates.

Intercept values are often not reported, although they can provide insight into both the appropriateness of the single exponential model and the kinetics of the decomposition process (Witkamp and Olson 1963; Harmon et al. 1990, 1995). Intercepts that are significantly below 100% at t = 0 indicate a more rapid loss of material early in decomposition than would be predicted by the single negative exponential model (Fig. 11.3). Conversely, intercepts significantly above 100% indicate an extended lag phase early in decomposition, which may be due to climate or may indicate a colonization or conditioning phase. We strongly suggest that the mass remaining at each sample time be reported in tabular form and/or included in graphic presentations of the modeled decomposition curves generated, thus allowing reanalysis in future syntheses.

In cases where the single negative exponential model does not fit the data well, a multiple-component exponential model may be appropriate (Wieder and Lang 1982). This model assumes that the litter can be partitioned into two fractions, one labile and one more recalcitrant. This model may be more appropriate for litter types that exhibit a rapid mass loss phase followed by slower decomposition. Another model is the single negative exponential with asymptote, in which mass loss declines to zero and a fixed proportion of recalcitrant litter remains. Although this is not realistic over longer time scales, it may be appropriate for estimating the amount of "stable" organic matter produced as a product of litter decomposition.

In addition to mass loss, changes in litter nutrient concentrations and patterns and amounts of net nutrient accumulation and release should be calculated if nutrient analyses are available. Patterns of net accumulation and/or release of nutrients are typically more complex than patterns of mass loss, since nutrients can accumulate in the litter, microbes, and microbial by-products as decomposition proceeds. The

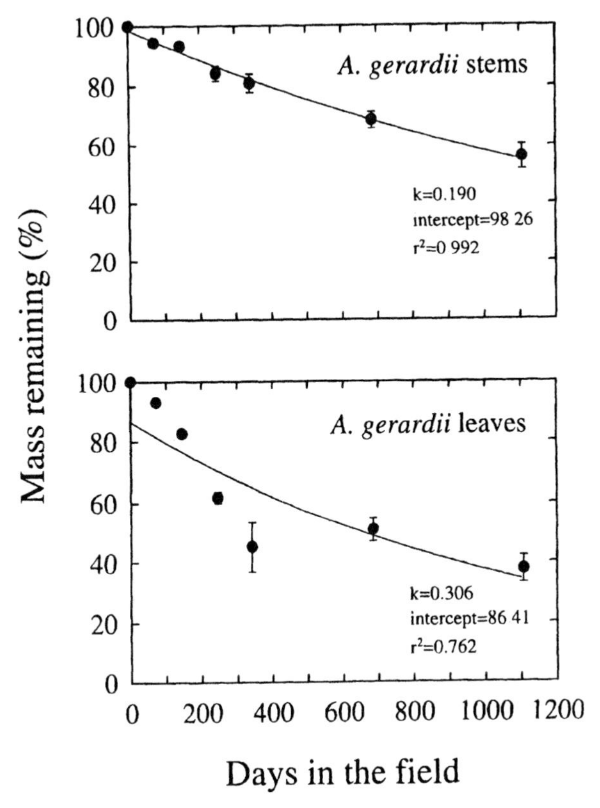

Figure 11.3. Illustration of the fit of a single exponential decomposition model to litter decomposition data. Top: Pattern of mass loss for stems of the C_4 grass big blue stem (*Andropogon gerardii*) decomposing at Konza Prairie. Solid symbols represent the mean percentage of initial mass remaining after 72, 146, 247, 342, 685, and 1107 days in the field. The solid line is the decomposition curve generated by fitting the field data to a single negative exponential decomposition model. Also presented are the annual decomposition rate constant (k), the y intercept, and the coefficient of determination (r^2). Bottom: The same data are presented for leaves of *A. gerardii* decomposing at the Konza Prairie. Note the lower coefficient of determination and lower y-intercept value than in the case for stem decomposition. This is due to a more rapid loss of material early in decomposition and a slower loss of mass later in decomposition than would be predicted by the single negative exponential model. This suggests a two-component model might be more appropriate to describe leaf decomposition.

processes controlling these three mechanisms differ, but they are inseparable. We recommend the terms *net release* and *net accumulation* in place of *net mineralization* and *net immobilization,* respectively, since it is not possible to distinguish actual conversion of organic to inorganic forms of nutrients (mineralization) or microbial uptake (immobilization) with these methods (Berg 1988). In fact, a considerable proportion of the nutrients released from decomposing litter may be in organic form (Yavitt and Fahey 1986; Setala et al. 1990; Qualls et al. 1991), and physical processes as well as microbial uptake may retain nutrients in litter.

It also is important to recognize that changes in nutrient concentration and net accumulation/release are not synonymous. That is, net accumulation/release is a function of both mass loss and changes in the nutrient concentration of the residual litter. Changes in nutrient concentration over time may be presented graphically. However, additional information can be acquired by examining changes in nutrient concentration in relation to mass loss (Aber and Melillo 1980). Net accumulation/release can be calculated as the product of proportion mass remaining at time *t* and the nutrient concentration in the residual litter at *t*, divided by the initial nutrient concentration in that litter type, and then plotted as a function of time. A discussion of the behaviors of various nutrients in decomposing litter is beyond the scope of this chapter. However, many references describe pattens of nutrient accumulation and release in decomposing leaf (Gosz et al. 1973; Berg and Staaf 1981; Blair 1988a, 1988b) and woody litter (Harmon et al. 1986; Sollins et al. 1987; Arthur and Fahey 1990; Harmon and Chen 1992).

Conclusions

Plant litter decomposition is a key ecosystem process that plays major roles in determining carbon and nutrient accumulation in soils, as well as in regulating the rate and timing of nutrient release to plant roots and soil organisms. In addition to understanding the dynamics of plant litter, it is essential to quantify the stores or standing stocks of these pools. Our ability to understand and quantitatively model these processes and pools will be much improved if researchers use comparable methods. This chapter has examined commonly used methods and presented standard protocols to determine the decomposition dynamics and stores for most forms of dead plant matter. These include (1) the litterbag method for determining the rate at which fine litter decomposes and accumulates or releases nutrients; (2) time-series experiments to study the decomposition and nutrient dynamics of woody detritus above- and belowground; (3) using cellulose filter paper and hardwood dowels as two standard substrates; (4) determining organic horizon stores by harvesting in 25- by 25 cm quadrats; (5) determining fine woody detritus stores (<10 cm diameter and <1 m long) by harvesting in 1- by 1-m quadrats; and (5) determining coarse woody detritus stores (>10 cm diameter and >1 m long) by recording piece dimensions in large plots and converting volume to mass or nutrient stores using species-specific and decay class–specific bulk densities and nutrient concentrations.

Acknowledgments We wish to thank David C. Coleman, Charles McClaugherty, and an anonymous reviewer for helpful comments on the manuscript. Funding was provided by grants from the National Science Foundation.

References

Aber, J. D., and J. M. Melillo. 1980. Litter decomposition: measuring the relative contribution of organic matter and nitrogen to forest soils. *Canadian Journal of Botany* 58:416–421.

Aber, J. D., and J. M. Melillo. 1982. Nitrogen immobilization in decaying hardwood leaf litter as a function of initial nitrogen and lignin content. *Canadian Journal of Botany* 60:2263–2269.

Aber, J. D., J. M. Melillo, and C. A. McClaugherty. 1990. Predicting long-term patterns of mass loss, nitrogen dynamics, and soil organic matter formation from initial fine litter chemistry in temperate forest ecosystems. *Canadian Journal of Botany* 68:2201–2208.

Andren, O., and K. Paustian. 1987. Barley straw decomposition in the field: a comparison of models. *Ecology* 68:1190–1200.

Arthur, M. A., and T. J. Fahey. 1990. Mass and nutrient content of decaying boles in an Engelmann spruce–subalpine fir forest, Rocky Mountains National Park, Colorado. *Canadian Journal of Forest Research* 20:730–737.

Berg, B. 1988. Dynamics of nitrogen (^{15}N) in decomposing Scots pine (*Pinus sylvestris*) needle litter. Long-term decomposition in a Scots pine forest. VI. *Canadian Journal of Botany* 66:1539–1546.

Berg, B., G. Ekbohm, and C. A. McClaugherty. 1984. Lignin and hollocellulose relations during long-term decomposition of some forest litters. Long-term decomposition in a Scots pine forest. IV. *Canadian Journal of Botany* 62:2540–2550.

Berg, B., and H. Staaf. 1981. Leaching, accumulation and release of nitrogen in decomposing forest litter. Pages 163–178 *in* F. E. Clark and T. Rosswell, editors, *Terrestrial Nitrogen Cycles. Ecological Bulletin* 33. Swedish Natural Science Research Council, Stockholm, Sweden.

Binkley, D. 1984. Does forest removal increase rates of decomposition and nutrient release? *Forest Ecology and Management* 8:229–233.

Birdsey, R. A. 1992. *Carbon Storage and Accumulation in US Forest Ecosystems.* USDA Forest Service, General Technical Report WO-59. Washington, DC, USA.

Blair, J. M. 1988a. Nitrogen, sulfur and phosphorus dynamics in decomposing deciduous leaf litter in the southern Appalachians. *Soil Biology and Biochemistry* 20:693–701.

Blair, J. M. 1988b. Nutrient release from decomposing foliar litter of three tree species with special reference to calcium, magnesium, and potassium dynamics. *Plant and Soil* 110:49–55.

Blair, J. M., D. A. Crossley Jr., and L. C. Callaham. 1991. A litterbasket technique for measurement of nutrient dynamics in forest floor. *Agriculture, Ecosystems and Environment* 34:465–471.

Blair, J. M., R. W. Parmelee, and M. H. Beare. 1990. Decay rates, nitrogen fluxes and decomposer communities of single- and mixed-species foliar litter. *Ecology* 71:1976–1985.

Bocock, K. L., and O. J. Gilbert. 1957. The disappearance of leaf litter under different woodland conditions. *Plant and Soil* 9:179–185.

Bocock, K. L., O. J. Gilbert, C. K. Capstick. D. W. Twinn, J. S. Waid, and M. J. Woodmann. 1960. Changes in leaf litter when placed on the surface of soils with contrasting humus types. I. Losses in dry weight of oak and ash leaf litter. *Journal of Soil Science* 11:1–9.

Brown, J. K. 1974. *Handbook for Inventorying Downed Woody Material.* USDA Forest Service General Technical Report INT-16. Ogden, Utah, USA.

Bunnell, F. L., and D. E. N. Tait. 1977. Microbial respiration and substrate loss. II. A model of the influences of chemical composition. *Soil Biology and Biochemistry* 9:41–47.

Bunnell, F. L., D. E. N. Tait, P. W. Flanagan, and K. Van Cleve. 1977. Microbial respiration and substrate loss. I. A general model of the influences of abiotic factors. *Soil Biology and Biochemistry* 9:33–40.

Capstick, C. K. 1962. The use of small cylindrical samplers for estimating the weight of forest floor. Pages 353–356 *in* P. W. Murphy and D. Phil, editors, *Progress in Soil Zoology.* Butterworths, London, UK.

Carter, R. E., and L. E. Lowe. 1986. Lateral variability of forest floor properties under second-growth Douglas-fir stands and the usefulness of composite sampling techniques. *Canadian Journal of Forest Research* 16:1128–1132.

Coleman, D. C., and D. A. Crossley Jr. 1996. *Fundamentals of Soil Ecology*. Academic Press, New York, New York, USA.

Cottam, G., and J. T. Curtis. 1956. The use of distance measures in phytosociological sampling. *Ecology* 37:451–460.

Edmonds, R. L. 1984. Long-term decomposition and nutrient dynamics in Pacific silver fir needles in western Washington. *Canadian Journal of Forest Research* 14:395–400.

Erickson, H. E., R. L. Edmonds, and C. E. Peterson. 1985. Decomposition of logging residue in Douglas-fir, western hemlock, Pacific silver fir, and ponderosa pine ecosystems. *Canadian Journal of Forest Research* 15:914–921.

Fahey, T. J., J. W. Hughes, P. Mou, and M. A. Arthur. 1988. Root decomposition and nutrient flux following whole-tree harvest in Northern hardwood forest. *Forest Science* 34:744–768.

Falconer, G. J., J. W. Wright, and H. W. Beall. 1933. The decomposition of certain types of fresh litter under field conditions. *American Journal of Botany* 55:1632–1640.

Federer, C. A. 1984. Organic matter and nitrogen content of the forest floor in even-aged northern hardwood. *Canadian Journal of Forest Research* 14:763–767.

Fogel, R., and K. Cromack Jr. 1977. Effect of habitat and substrate quality on Douglas-fir litter decomposition in western Oregon. *Canadian Journal of Botany* 55:1632–1640.

Gore, J. A., and A. P. William. 1986. Mass of downed wood in north hardwood forests in New Hampshire: potential effects in forest management. *Canadian Journal of Forest Research* 16:335–339.

Gosz, J. R., G. E. Likens, and F. H. Bormann. 1973. Nutrient release from decomposing leaf and branch litter in the Hubbard Brook forest, New Hampshire. *Ecological Monographs* 43:173–191.

Graham, R. L. 1982. Biomass dynamics of dead Douglas-fir and western hemlock fir boles in mid-elevation forests of Cascade Range. Ph.D. dissertation. Oregon State University.

Grier, C. C. 1978. A *Tsuga heterophylla–Picea sitchensis* ecosystem of coastal Oregon: decomposition and nutrient balance of fallen logs. *Canadian Journal of Forest Research* 8:198–206.

Grier, C. C., and R. S. Logan. 1977. Old-growth *Pseudotsuga menzisii* communities of a western Oregon watershed: biomass distribution and production budgets. *Ecological Monographs* 47:373–400.

Grosenbaugh, L. R. 1958. *Point-Sampling and Line-Sampling: Probability Sampling, Geometric Implications, Synthesis*. USDA Forest Service, Southern Forest Experiment Station, Occasional Paper 160. New Orleans, Louisiana, USA.

Gustafson, F. G. 1943. Decomposition of the leaves of some forest trees under field conditions. *Plant Physiology* 18:704–707.

Harmon, M. E. 1992. *Long-Term Experiments on Log Decomposition at the H. J. Andrews Experimental Forest*. USDA Forest Service, General Technical Report PNW-280. Portland, Oregon, USA.

Harmon, M. E. 1993. Woody debris budgets for selected forest types in the U.S. Pages 151–178 *in* D. P. Turner, J. J. Lee, G. J. Koerper, and J. R. Barker, editors, *The Forest Sector Carbon Budget of the United States: Carbon Pools and Flux under Alternative Policy Options*. EPA/600/3–93/093. United States Environmental Protection Agency, Corvallis, Oregon, USA.

Harmon, M. E., G. A. Baker, G. Spycher, and S. E. Greene. 1990. Leaf-litter decomposition in the *Picea-Tsuga* forests of Olympic National Park, Washington, USA. *Forest Ecology and Management* 31:55–66.

Harmon, M. E., and H. Chen. 1992. Coarse woody debris dynamics in two old-growth ecosystems. *Bioscience* 41:604–610.

Harmon, M. E., K. Cromack Jr., and B. G. Smith. 1987. Coarse woody debris in mixed-conifer forests, Sequoia National Park, California. *Canadian Journal of Forest Research* 17:1265–1272.

Harmon, M. E., J. F. Franklin, F. Swanson, P. Sollins, S. V. Gregory, J. D. Lattin, N. H. Anderson, S. P. Cline, N. G. Aumen, J. R. Sedell, G. W. Lienkaemper, K. Cromack Jr., and K. W. Cummins. 1986. Ecology of coarse woody debris in temperate ecosystem. *Advances in Ecological Research* 15:133–302.

Harmon, M. E., and J. Sexton. 1996. *Guidelines for Measurements of Woody Detritus in Forest Ecosystems.* U.S. LTER Network Office Publication, no. 20. University of Washington, Seattle, Washington, USA.

Harmon, M. E., D. F. Whigham, J. Sexton, and I. Olmsted. 1995. Decomposition and mass of woody detritus in the dry tropical forests of the northeastern Yucatan Peninsula, Mexico. *Biotropica* 27:305–316.

Harvey, A. E., M. J. Larsen, and M. F. Jurgensen. 1979. Comparative distribution of ectomycorrhiza in soils of three western Montana forest habitat types. *Forest Science* 25:350–358.

Harvey, A. E., M. J. Larsen, and M. F. Jurgensen. 1981. *Rate of Woody Residue Incorporation into Northern Rocky Mountain Forest Soils.* USDA Forest Service, Research Paper INT-282. Ogden, Utah, USA.

Heath, G. W., C. A. Edwards, and M. K. Arnold. 1964. Some methods for assessing the activity of soil animals in the breakdown of leaves. *Pedobiologica* 4:80–87.

Houghton, R. A., J. E. Hobbie, J. M. Melillo, B. Moore, B. J. Peterson, G. R. Shaver, and G. M. Woodwell. 1983. Changes in carbon content of terrestrial biota and soils between 1860 and 1980: a net release of CO_2 to the atmosphere. *Ecological Monographs* 53:235–262.

Howard, P. J. A., and D. M. Howard. 1974. Microbial decomposition of tree and shrub litter. I. Weight loss and chemical composition of decomposing litter. *Oikos* 25:314–352.

Hurlbert, S. H. 1984. Pseudoreplication and the design of ecological field experiments. *Ecological Monographs* 54:187–211.

Jansson, P. E., and B. Berg. 1985. Temporal variation of litter decomposition in relation to simulated soil climate: long-term decomposition in a Scots pine forest. *Canadian Journal of Botany* 63:1008–1016.

Jenny, H., S. P. Gessel, and F. T. Bingham. 1949. Comparative study of decomposition of organic matter in temperate and tropical regions. *Soil Science* 68:419–432.

Kendrick, W. B. 1959. The time factor in decomposition of coniferous leaf litter. *Canadian Journal of Botany* 37:907–912.

Kurcheva, G. F. 1960. The role of invertebrates in the decomposition of oak litter. *Pedology, Leningrad* 4:16–23.

Kurz, W. A., M. J. Apps, T. M. Webb, and P. J. McNamee. 1992. *The Carbon Budget of the Canadian Forest Sector: Phase I.* Information Report NOR-X-326. Forestry Canada, Edmonton, Alberta, Canada.

Lang, G. E. 1974. Dynamics in a mixed oak forest on the New Jersey Piedmont. *Journal of the Torrey Botanical Club* 101:277–286.

LIDET. 1995. *Meeting the Challenge of Long-Term, Broad-Scale Ecological Experiments.* U.S. LTER Network Office Publication, no. 19. University of Washington, Seattle, Washington, USA.

Little, S. N., and J. L. Ohmann. 1988. Estimating nitrogen lost from forest floor during prescribed fires in Douglas-fir/western hemlock clear cuts. *Forest Science* 34:152–164.

Loomis, R. M. 1975. Annual changes in forest floor weights under a southeastern Missouri oak stand. USDA Forest Service Research Note NC-184. Asheville, North Carolina, USA.

Lousier, J. D., and D. Parkinson. 1978. Chemical element dynamics in decomposing leaf litter. *Canadian Journal of Botany* 56:2795–2812.

Lunt, H. A. 1933. Effects of weathering upon composition of hardwood leaves. *Journal of Forestry* 31:43–45.

Lunt, H. A. 1935. Effect of weathering upon dry matter and composition of hardwood leaves. *Journal of Forestry* 33:607–608.

Maxwell, W. G., and F. R. Ward. 1976a. *Photo Series for Quantifying Forest Residues in the Coastal Douglas-Fir Hemlock Type, Coastal Douglas-Fir-Hardwood Type.* USDA Forest Service, General Technical Report PNW-GTR-51. Portland, Oregon, USA.

Maxwell, W. G., and F. R. Ward. 1976b. *Photo Series for Quantifying Forest Residues in the Ponderosa Pine Type, Ponderosa Pine and Associated Species Type, Lodgepole Pine Type.* USDA Forest Service, General Technical Report PNW-GTR-52. Portland, Oregon, USA.

McClaugherty, C. A., J. D. Aber, and J. M. Melillo. 1984. Decomposition dynamics of fine roots in forested ecosystems. *Oikos* 42:378–386.

McClaugherty, C. A., J. Pastor, J. D. Aber, and J. M. Melillo. 1985. Forest litter decomposition in relation to soil nitrogen dynamics and litter quality. *Ecology* 66:266–275.

McFee, W. W., and E. L. Stone. 1965. Quality, distribution and variability of organic matter and nutrients in a forest podzol in New York. *Soil Science Society of America Proceedings* 29:432–436.

McFee, W. W., and E. L. Stone. 1966. The persistence of decaying wood in humus layers of northern forests. *Soil Science Society of America Proceedings* 30:513–516.

McLellan, T., J. D. Aber, J. M. Melillo, and K. J. Nadelhoffer. 1991a. Determination of nitrogen, lignin and cellulose content of decomposing leaf litter by near infrared reflectance spectroscopy. *Canadian Journal of Forest Research* 21:1684–1688.

McLellan, T., M. E. Martin, J. D. Aber, J. M. Melillo, K. J. Nadelhoffer, B. Dewey, and J. Pastor. 1991b. Comparison of wet chemical and near infrared reflectance measurements of carbon fraction chemistry and nitrogen content of forest foliage. *Canadian Journal of Forest Research* 21:1689–1693.

Means, J. E., K. Cromack Jr., and P. C. Macmillan. 1987. Comparison of decomposition models using wood density of Douglas-fir logs. *Canadian Journal of Forest Research* 15:1092–1098.

Melillo, J. M., J. D. Aber, A. E. Linkins, A. Ricca, B. Fry, and K. Nadelhoffer. 1989. Carbon and nitrogen dynamics along the decay continuum: plant litter to soil organic matter. Pages 53–62 *in* M. Clarholm and L. Bergstrom, editors, *Ecology of Arable Land.* Kluwer Academic Publishers, Dordrecht, Netherlands.

Melillo, J. M., J. D. Aber, and J. F. Murtore. 1982. Nitrogen and lignin control of hardwood leaf litter decomposition dynamics. *Ecology* 63:621–626.

Melillo, J. M., R. J. Naiman, J. D. Aber, and A. E. Linkins. 1984. Factors controlling mass loss and nitrogen dynamics of plant litter decaying in northern streams. *Bulletin of Marine Science* 35:341–356.

Metz, L. J. 1954. Forest floor in the Piedmont region of South Carolina. *Soil Science Society of America Proceedings* 18:335–338.

Minderman, G. 1968. Addition, decomposition, and accumulation of organic matter in forests. *Journal of Ecology* 56:355–362.

Mueller-Dombois, D., and H. Ellenberg. 1974. *Aims and Methods of Vegetation Ecology.* Wiley, New York, New York, USA.

O'Lear, H. A., T. R. Seastedt, J. M. Briggs, J. M. Blair, and R. A. Ramundo. 1996. Fire and topographic effects on decomposition rates and nitrogen dynamics of buried wood in tallgrass prairie. *Soil Biology and Biochemistry* 28:322–329.

Olson, J. S. 1963. Energy stores and the balance of producers and decomposers in ecological systems. *Ecology* 44:322–331.

Ottmar, R. D., C. C. Hardy, and R. E. Vihnanek. 1990. *Stereophoto Series for Quantifying Forest Residues in the Douglas-Fir-Hemlock Type of the Willamette National Forest.* USDA Forest Service, General Technical Report PNW-GTR-258. Portland, Oregon, USA.

Palm, C. A., and P. A. Sanchez. 1990. Decomposition and nutrient release patterns of the leaves of three tropical legumes. *Biotropica* 22:330–338.

Parton, W. J., D. S. Schimel, C. V. Cole, and D. S. Ojima. 1987. Analysis of factors controlling soil organic matter levels in Great Plains Grasslands. *Soil Science Society of America Journal* 51:1173–1179.

Piene, H., and K. Van Cleve. 1978. Weight loss of litter and cellulose bags in a thinned white spruce forest in interior Alaska. *Canadian Journal of Forest Research* 8:42–46.

Qualls, R. G., B. L. Haines, and W. T. Swank. 1991. Fluxes of dissolved organic nutrients and humic substances in a deciduous forest. *Ecology* 72:254–266.

Rustad, L. E. 1994. Element dynamics along a decay continuum in a red spruce ecosystem in Maine, USA. *Ecology* 75:867–879.

Setala, H., E. Martikainene, M. Tyynismaa, and V. Huhta. 1990. Effects of soil fauna on leaching of nitrogen and phosphorus from experimental systems simulating coniferous forest. *Biology and Fertility of Soils* 10:170–177.

Singh, J. S., and S. R. Gupta. 1977. Plant decomposition and soil respiration in terrestrial ecosystems. *Botanical Review* 43:449–528.

Sollins, P. 1982. Input and decay of coarse woody debris in coniferous stands in western Oregon and Washington. *Canadian Journal of Forest Research* 12:18–28.

Sollins, P., S. P. Cline, T. Verhoeven, D. Sachs, and G. Spycher. 1987. Patterns of log decay in old-growth Douglas-fir forest. *Canadian Journal of Forest Research* 17:1585–1595.

Spies, T. A., J. F. Franklin, and T. B. Thomas. 1988. Coarse woody debris in Douglas-fir forests of western Oregon and Washington. *Ecology* 69:1689–1702.

St. John, T. V. 1980. Influence of litterbags on growth of fungal vegetative structures. *Oecologia* 46:130–132.

Stevenson, B. C., and D. L. Dindal. 1981. A litter box method for the study of litter arthropods. *Journal of the Georgia Entomological Society* 16:151–156.

Swift, M. J., O. W. Heal, and J. M. Anderson. 1979. *Decomposition in Terrestrial Ecosystems.* University of California Press, Berkeley and Los Angeles, California, USA.

Trofymow, J. A. 1995. Litter quality and its potential effect on decay rates of materials from Canadian forests. *Water, Air and Soil Pollution* 82:215–226.

Tsarik, I. V. 1975. Decomposition of cellulose in the litter layer and soil of *Pinus mugo* elfin woodland in the Ukrainian Carpathians. *Lesvedenie* 1:88–90.

Turner, D. P., G. J. Koerper, M. E. Harmon, and J. J. Lee. 1995. A carbon budget for forests of the conterminous United States. *Ecological Applications* 5:421–436.

Tyler, G. 1971. Distribution and turnover of organic matter and minerals in a shore meadow ecosystem. *Oikos* 22:265–291.

Van Wagner, C. E. 1968. The line intercept method in forest fuel sampling. *Forest Science* 14:20–26.

Vogt, K. A., C. C. Grier, and D. J. Vogt. 1986. Production, turnover, and nutrient dynamics of above- and below-ground detritus of world forests. *Advances in Ecological Research* 15:303–377.

Wallace, E. S., and B. Freedman. 1986. Forest floor dynamics in a chronosequence of hardwood stands in central Nova Scotia. *Canadian Journal of Forest Research* 16:293–302.

Warren, W. G., and P. F. Olsen. 1964. A line transect technique for assessing logging waste. *Forest Science* 10: 267–276.

Weary, G. C., and H. G. Merriam. 1978 Litter decomposition in a red maple woodlot under natural conditions and under insecticide treatment. *Ecology* 59:180–184.

Wieder, R. R., and G. E. Lang. 1982. A critique of the analytical methods used in examining decomposition data obtained from litter bags. *Ecology* 63:1636–1642

Witkamp, M., and J. S. Olson. 1963. Breakdown of confined and nonconfined oak litter. *Oikos 14*:138–147.

Yavitt, J. B., and T. J. Fahey. 1986. Litter decay and leaching from the forest floor in *Pinus contorta* (lodgepole pine) ecosystems. *Journal of Ecology* 74:525–545.

Youngberg, C. T. 1966. Forest floors in Douglas-fir forests. I. Dry weight and chemical properties. *Soil Science Society of America Proceedings* 30:406–409.

12

Dinitrogen Fixation

David D. Myrold
Roger W. Ruess
Michael J. Klug

Biological N_2 fixation is the conversion of atmospheric N_2 to NH_3 by bacteria. The reduction of the triple bond of N_2 to form NH_3 is performed under anaerobic conditions by the nitrogenase enzyme complex at the expense of 12–16 ATP with H_2 as a by-product. The NH_3 produced by dinitrogen fixation is subsequently assimilated into organic forms of nitrogen by standard metabolic pathways. This input of organic nitrogen from N_2 fixation is often the largest input to the nitrogen cycle of nonfertilized soil ecosystems.

The bacteria that fix N_2 are taxonomically diverse. They differ in their source of energy (some are phototrophs, but most are chemotrophs), carbon (some are autotrophs, but most are heterotrophs), their tolerance for oxygen, and the degree of their association with other organisms. This latter characteristic is the most important with respect to the measurement of N_2 fixation because it largely determines the magnitude of fixation. Free-living N_2-fixing bacteria, such as *Azotobacter* and the cyanobacteria, typically fix <5 kg N ha^{-1} y^{-1}; N_2-fixing bacteria intimately associated with plant roots, such as *Azospirillum,* may fix up to 50 kg N ha^{-1} y^{-1}, although <10 kg N ha^{-1} y^{-1} is more typical; whereas the root nodule–forming symbioses of *Rhizobium* and *Frankia* can fix several hundred kg N ha^{-1} y^{-1}.

Nitrogen fixation is often measured as part of an ecosystem's nitrogen cycle, in which case an annual rate of N_2 fixation is desired. Estimates of N_2 fixation are also made over shorter periods to assess the effects of environmental factors or treatment manipulations. These short-term measurements are often focused on understanding the controls and regulation of N_2 fixation.

Table 12.1. Major Characteristics of Methods Commonly Used to Measure N_2 Fixation

Method	Features
N accretion	Integrative, requires a control or baseline, requires accounting of other imputs and outputs, insensitive
Acetylene reduction	Short-term rate, requires conversion factor, sensitive, susceptible to interferences
^{15}N isotope dilution	Integrative, requires a control, moderately sensitive
^{15}N natural abundance	Integrative, requires a control, insensitive, sometimes only qualitative
$^{15}N_2$ incorporation	Short-term rate, moderately sensitive

Available Methods

Several methods have been devised to measure N_2 fixation (Tab. 12.1). As with all methods, each has advantages and disadvantages. The major considerations for selection are sensitivity, duration, sample type, and whether relative or absolute rates are required.

Nitrogen Balance

This method is also known as the nitrogen accretion or nitrogen difference method. In unmanaged ecosystems the approach has been either to measure the accumulation of nitrogen in ecosystem components over time, often decades (e.g., Youngberg and Wollum 1976), or to measure nitrogen accumulation along a chronosequence (e.g., Newton et al. 1968). Although any accumulation of nitrogen is obviously the net difference between all gains and losses of N, in undisturbed systems the net accumulation is often attributed exclusively to N_2 fixation because other gains and losses are assumed to balance each other.

The nitrogen difference approach has also been used in managed, particularly agricultural, ecosystems over shorter periods, usually one growing season. Most commonly this approach has been used to estimate symbiotic or associative N_2 fixation by comparing the accumulation of plant nitrogen between the fixing plant and a nonfixing control. Rates of N_2 fixation must be more than 20 kg N ha^{-1} y^{-1} to be determined by this approach (Weaver 1986).

A more detailed description of the nitrogen accretion method will not be given because it is relatively insensitive and its inherent assumptions may be unjustified. It should be noted, however, that periodic monitoring of ecosystem nitrogen pools may be useful at many sites for measuring long-term patterns of nitrogen accumulation, particularly sites that contain leguminous or actinorhizal N_2 fixing symbioses. See Chapter 5, this volume, for methods of soil nitrogen analysis.

Acetylene Reduction

The discovery 30 years ago that the nitrogenase enzyme will reduce acetylene (C_2H_2) to ethylene (C_2H_4) quickly led to the widespread use of the acetylene reduction assay as a measure of N_2 fixation. This assay has been used extensively because it has great sensitivity and is inexpensive. Its sensitivity allows short-term (minutes to days) measurements at even low levels of activity. It is less useful as an integrative measure, however, and has been found to have many shortcomings, including the variability of the factor to convert from ethylene production to N_2 fixation and the sensitivity of activity to sample handling. Disturbance effects are especially important when measuring associative or symbiotic N_2 fixation, which limits the utility of this method under field conditions. Despite these limitations, however, it is useful for comparative purposes and for qualitative assessment of N_2 fixation, and it can be used quantitatively under some circumstances if carefully used and calibrated.

^{15}N-Based Methods

Although it has long been recognized that ^{15}N can be used in various ways to assess N_2 fixation, the use of ^{15}N-based methods has expanded in recent years. This is partly because of dissatisfaction with nitrogen balance and acetylene reduction methods but probably more because of the declining cost of ^{15}N-labeled materials, the increased availability and sensitivity of mass spectrometers, and the lower cost of ^{15}N analysis. There are several ^{15}N-based methods, including

- labeling soils with ^{15}N and applying the principles of isotope dilution;
- taking advantage of variations in the natural abundance of ^{15}N in the atmosphere and soils; and
- using $^{15}N_2$ to measure directly ^{15}N incorporation by N_2 fixation.

^{15}N Isotope Dilution

Over the past decade, the ^{15}N isotope dilution method has become increasingly common in field studies of N_2 fixation. It has most often been applied to studies of symbiotic or associative N_2 fixation and is the most sensitive way to measure long-term rates of N_2 fixation.

The principle behind this method is to label soil with sufficient ^{15}N to raise it significantly above ^{15}N natural abundance and subsequently to compare the ^{15}N labeling of a putative N_2-fixing plant-microbe association with that of a non-N_2-fixing reference plant. The ^{15}N abundance of the reference plant is assumed to reflect the ^{15}N labeling of the plant-available soil N. The ^{15}N abundance of the putative N_2-fixing plant-microbe association should then lie somewhere between that of the control plant and that of the atmosphere.

In practice, there are numerous pitfalls to the ^{15}N isotope dilution technique. These include the uniformity with which soil is labeled with ^{15}N, whether the reference plant and N_2-fixing plant are similar in their spatial and temporal nitrogen uptake patterns, and what plant tissues should be sampled. These potential problems can be controlled and minimized, and are normally less severe than the shortcomings of other N_2 fixation methods. A greater difficulty to overcome, particularly with woody perennial plants, is scaling the fraction of nitrogen derived from N_2 fixation to an areal estimate.

^{15}N Natural Abundance

Many soils are enriched in ^{15}N relative to the N_2 in the atmosphere. Thus, plants with associative or symbiotic N_2-fixing bacteria will have a ^{15}N abundance intermediate between that of the atmosphere and that of reference plants taking up nitrogen from the soil, one that reflects the proportions of nitrogen from N_2 fixation and from nitrogen uptake. Multiplying the proportion that came from N_2 fixation by the total nitrogen content of the plant will yield an estimate of N_2 fixed.

This approach is straightforward in principle (Shearer and Kohl 1986) and has the advantage that soils do not have to be disturbed by the addition of ^{15}N-labeled materials. However, it also has many potential drawbacks, the major one being its sensitivity. Although the instrument precision of modern mass spectrometers has improved, replicate plant or soil samples seldom have a precision better than $\pm$ 0.2‰. For a soil that differs from the atmosphere by 4.0‰, this means that the proportion of nitrogen coming from N_2 fixation can be determined within only $\pm 5\%$ (e.g., Unkovich et al. 1994). Because many soils with native vegetation have ^{15}N abundances closer to the atmosphere than this (Hansen and Pate 1987), the use of variations in ^{15}N natural abundance may be limited to agricultural soils. It is also true that soils, and their associated vegetation, can show a high degree of spatial variability; the ^{15}N abundance of plants also varies from tissue to tissue and temporally over the growing season (Shearer et al. 1983; Selles et al. 1986; Bremer and van Kessel 1990). Several recent field studies (Bremer and van Kessel 1990; Stevenson et al. 1995) have shown that, despite this natural variability, the ^{15}N natural abundance method gave similar mean estimates of N_2 fixation as nitrogen balance or ^{15}N isotope dilution studies. It should be noted, however, that these same studies found no significant correlation in the percentage of nitrogen derived from the atmosphere when comparing the ^{15}N natural abundance method with other approaches based on individual measurements.

At the current time, it would appear that the ^{15}N natural abundance method could be used to qualitatively test plant tissue to assess for the presence of symbiotic N_2 fixation and as a semiquantitative measure of N_2 fixation, provided there is at least a 2–4‰ difference between the ^{15}N abundance of the soil and the atmosphere. Greater differences may allow this technique to be used quantitatively.

Because the basic principles for choosing reference plants and performing the isotope dilution calculations are the same for those of the ^{15}N isotope dilution method, a detailed protocol will not be presented for the ^{15}N natural abundance method.

$^{15}N_2$ Incorporation

The most direct way to demonstrate the presence of N_2 fixation is to measure the incorporation of ^{15}N into a sample exposed to $^{15}N_2$ (Warembourg 1993). Like the acetylene reduction assay, $^{15}N_2$ incorporation provides a short-term rate and requires a closed incubation system. Because of the need for a gastight assay system, $^{15}N_2$ incorporation is better suited for laboratory than field studies. Compared with the acetylene reduction assay, $^{15}N_2$ incorporation does not suffer from interferences

Table 12.2. Decision Matrix for Selecting the Most Appropriate Method for Measuring N_2 Fixation

		Type and Rate of Fixation		
Sample Duration	Sample Location	Nonsymbiotic (2 kg $N \cdot ha^{-1} \cdot y^{-1}$ or 0.5 nmol $N_2 \cdot g^{-1} \cdot d^{-1}$)	Associative (20 kg $N \cdot ha^{-1} \cdot y^{-1}$ or 5 nmol $N_2 \cdot g^{-1} \cdot d^{-1}$)	Symbiotic (200 kg $N \cdot ha^{-1} \cdot y^{-1}$ or 50 nmol $N_2 \cdot g^{-1} \cdot d^{-1}$)
Short-term (<4 days)	Laboratory	Acetylene reduction	$^{15}N_2$ incorporation, acetylene reduction	$^{15}N_2$ incorporation, acetylene reduction
	Field	Acetylene reduction	Acetylene reduction	Acetylene reduction
Long-term (>28 days)	Laboratory	—	^{15}N isotope dilution	^{15}N isotope dilution
	Field	—	^{15}N isotope dilution	^{15}N isotope dilution

Notes To select a method, first decide on the type of sample you will be using, followed by the desired duration of the measurement period, and whether the measurement is to be made in the laboratory or in the field. The approximate sensitivity represents the typical activity over an annual or daily period for the type of sample chosen. When more that two methods are given, they are listed in order of preference.

and, of course, does not require a conversion factor. It is slightly less sensitive than the acetylene reduction assay, however. A major use of $^{15}N_2$ incorporation has been to determine appropriate conversion factors for the acetylene reduction assay.

Recommended Protocols

As indicated by the previous discussion, there is no single, universal protocol for measuring N_2 fixation. Table 12.2 has been constructed to assist in the selection of the most appropriate method for a given sample type, duration, and location. For example, if one were interested in assessing N_2 fixation of bacteria associated with the rhizosphere of a Douglas fir seedling, one could perform a short-term assay in the lab using either acetylene reduction or $^{15}N_2$ incorporation. If only relative rates were needed, acetylene reduction would probably be the method of choice because it is less expensive. However, if actual N_2 fixation rates are needed, then $^{15}N_2$ incorporation might be more appropriate because no conversion factor is required. As suggested by Table 12.2, estimates of N_2 fixation in the field are generally best done by using ^{15}N isotope dilution. We will describe all three methods.

N_2 Fixation by Acetylene Reduction

Measurement of ethylene production is easy, inexpensive, and sensitive. It is ideally suited for short-term assays, although the results of several short-term assays made sequentially can be integrated to provide a long-term estimate of N_2 fixation. The acetylene reduction assay is best suited for litter, soil, or samples of woody debris, but it can be adapted for use with root systems or root nodules. As mentioned previously, the lack of a universal factor to convert from ethylene production to N_2 fixed is a drawback; a conversion factor must always be determined empirically for a given system using $^{15}N_2$ incorporation.

The following protocol is designed for use with samples of litter, soil, or woody debris, although modifications are noted for applying acetylene reduction to systems that include plants. This method is also described by Bergersen (1980), Silvester et al. (1989), and Weaver and Danso (1994).

Materials

1. Gas chromatograph. Acetylene and ethylene are measured using a gas chromatograph equipped with a flame ionization detector. A 1.5 m long × 3.2 mm diameter column packed with Porapak Q is commonly used to separate the gases of interest. Nitrogen is used as a carrier gas, and compressed air and H_2 are used to fuel the flame. A strip-chart recorder or integrator is used to measure the signal of the gas chromatograph. Ethylene standards are needed for calibration.
2. Acetylene. Acetylene can be purchased commercially or generated by reacting calcium carbide (CaC_2) with water (Weaver and Danso 1994). Generation from CaC_2 is preferred because the acetylene produced in this manner has

fewer contaminants. In either case, however, it is necessary to purify the acetylene by passing it through traps of concentrated sulfuric acid and water to remove residual contaminants. Mylar balloons or PVC beach balls are convenient for storage of purified acetylene. Remember that acetylene is a flammable gas and a potential explosion hazard.

3. Incubation vessel. Gastight incubation containers are needed for the incubations. These could consist of metal or plastic pipes used to take the sample and subsequently capped at each end with rubber stoppers or septa. Another convenient incubation vessel is a canning jar sealed with a canning lid fitted with a rubber septum. Other more sophisticated cuvettes have been developed for use with plant root systems and flow-through systems (Warembourg 1993; Silvester et al. 1989). The important point is that the system is gastight and sufficiently large so that there is an adequate supply of O_2 for the duration of the experiment but small enough to enhance sensitivity.
4. Miscellaneous equipment. Plastic syringes of various sizes for the addition of acetylene and collection of gas samples, and needles for the syringes are needed. Gas samples can be stored for a short period (<1 hour) in 1 mL plastic syringes fitted with stopcocks or with the needles stuck into a rubber stopper. For longer storage, evacuated, leak-proof vials can be used (see Chapter 10, this volume). With any type of storage container it is important to also store reference standards handled in the same manner as samples to account for contamination, leaks, and absorption of target gases.

Procedure

1. Five to ten replicate samples are normally taken for analysis of litter, soil, or woody debris. To avoid diffusion problems with samples of woody debris, it is best to sample by layer (e.g., bark, sapwood, or decay class) and to cut the wood from each layer into matchstick-sized pieces (Jurgensen et al. 1987; Griffiths et al. 1993). The volume or mass of each layer must also be known to calculate an overall rate.
2. The samples are placed into the incubation vessels, which are sealed gastight.
3. Acetylene is added to a final concentration of 10 kPa (10% v/v), and the headspace is mixed well. An internal standard, such as propane, can be added to estimate total gas volume if sample volume and water content varies. Two types of controls are needed:
   (a) one without a sample and with acetylene to check for contaminating levels of ethylene in the acetylene, and
   (b) one with a sample but without acetylene to measure background ethylene production.
4. Gas samples are taken periodically throughout the incubation for analysis of acetylene and ethylene. Concentrations of acetylene and ethylene are determined from standard curves. Because the ethylene produced is insignificant compared with the acetylene in the system, any decrease in acetylene concentration indicates a gas leak from the incubation vessel.

5. The volume of the headspace and sample dry weight are determined at the end of the incubation. Volume can be calculated:
   (a) by difference from the total volume of the container if the volume of the dry solid sample and water are known,
   (b) from dilution of an internal gas standard, e.g., propane, or
   (c) using a pressure transducer (see Chapter 14, this volume).

Calculations

Ethylene production rates (i.e., acetylene reduction rates) are obtained from the ethylene concentration versus time data. This can be done in four steps:

1. Correct for any contaminating ethylene contained in the acetylene by subtracting the ethylene contaminant concentration (C_c) from the ethylene concentration of each sample incubated in the presence of acetylene (C) to obtain the contaminant-corrected ethylene concentration (C_a) for each sample:

$$C_a = C - C_c$$

2. Calculate the rate of contaminant-corrected ethylene production by linear regression of contaminant-corrected ethylene concentrations with time of sampling. The slope (P_a) is the rate of increase in contaminant-corrected ethylene production. Ideally this should be a linear increase with an r^2 value greater than 0.9.
3. Adjust the contaminant-corrected ethylene production rate for background production of ethylene in the absence of acetylene. This is done by calculating the slope (P_b) of background ethylene production from the linear regression of background ethylene concentration with time of sampling and then subtracting this rate of background ethylene production from the rate of ethylene production in the presence of acetylene to give the background-corrected rate of ethylene production (P):

$$P = P_a - P_b$$

4. The acetylene reduction activity (ARA) is then calculated on a unit weight (e.g., $\text{nmol}\cdot\text{g}^{-1}\cdot\text{d}^{-1}$) or area basis as follows:

$$ARA = P \times H/D$$

where

ARA = acetylene reduction activity
P = ethylene production rate
H = headspace volume
D = sample dry weight or surface area

Conversion to amount of N_2 fixed can be done by dividing ARA by the empirical (using $^{15}N_2$) or theoretical ratios of moles of acetylene reduced per mole of N_2 fixed. Because of the questionable nature of the theoretical stoichiometric conversion factor, it has become standard to simply report *ARA* when the $^{15}N_2$-derived factor is unavailable.

Special Considerations

Although sophisticated flow-through systems have been developed for the measurement of N_2 fixation associated with plant roots or nodules, a cuvette that encloses the root system and is sealed about the stem is probably the most practical (e.g., Warembourg 1993). In field situations, excised root segments containing nodules have been used for short-term (<3 minutes) in situ incubations.

The use of the acetylene reduction method with symbiotic root nodules or root-associated N_2-fixing bacteria has several unique difficulties, however. The most important problems are associated with the negative impact that disturbance of root systems (or shoots, for that matter) have on N_2 fixation (Boddey 1987; Giller 1987) or that acetylene may have on nodule physiology—the so-called acetylene-induced decline (Minchin and Witty 1989). Thus, when applying the acetylene reduction method for plant systems, it is important to minimize physical disturbance. The acetylene-induced decline is more difficult to overcome, although several studies have shown that the initial, high rate of ethylene production before the onset of the decline may give a reasonable estimate of N_2 fixation activity (Minchin and Witty 1989; Schwintzer and Tjepkema 1994). Thus, for excised root/nodule systems, 150 second incubations are used.

Rates of ethylene production in root nodules are often expressed on a per gram nodule basis and must therefore be multiplied by total nodule biomass per unit area when extrapolating to an area basis. Frequent temporal measurements must be made to accurately assess seasonal or annual fixation rates because symbiotic N_2 fixation is sensitive to plant phenology and fluctuations in temperature, moisture, and photoperiod. These extrapolations in space and time can be associated with significant error.

N_2 Fixation by $^{15}N_2$ Incorporation

The use of $^{15}N_2$ gas to directly determine the amount of N_2 fixed is most conveniently performed in the laboratory. As for the acetylene reduction method, the $^{15}N_2$ incorporation method is a short-term assay that requires a closed incubation system. Although it could be applied to samples of litter, soil, or woody debris, it is about one-tenth as sensitive as the acetylene reduction assay. Thus, the details of $^{15}N_2$ incorporation will be described for plant root systems. Additional details can be found in Warembourg (1993) and Weaver and Danso (1994).

Materials

1. Mass spectrometer. ^{15}N abundance is most accurately and precisely measured with a mass spectrometer (Hauck 1982; Mulvaney 1993; Hauck et al. 1994).

If a mass spectrometer is not available locally, a number of laboratories will perform ^{15}N analysis on a fee basis. The most critical aspect of preparing plant or soil samples for ^{15}N analysis is to make certain that the sample is very finely ground (40-mesh) to ensure sample homogeneity.

2. $^{15}N_2$. $^{15}N_2$ is available from several commercial sources, usually at 99 atom % ^{15}N. It can be purchased in lecture bottles or in sealed glass ampules. Although these are the most convenient forms in which to acquire $^{15}N_2$, it can also be made by oxidizing $^{15}NH_4^+$ salts with alkaline hypobromite (Warembourg 1993; Weaver and Danso 1994).
3. Incubation vessel. A closed, gastight system is needed. For litter, soil, and woody debris the same type of incubation system as described earlier for the acetylene reduction assay can be used. More elaborate systems may be needed for plant systems (e.g., Warembourg 1993). Small plants can be enclosed entirely, although care must be taken to minimize any excess heat load from exposure to growth lights. For larger plants simply seal root systems from the external atmosphere. As with the acetylene reduction assay, the gas volume of the incubation system is a compromise between minimizing the gas volume to reduce the amount of $^{15}N_2$ that must be used and having a sufficiently large volume for adequate aeration and dilution of any gaseous products that might interfere with the assay. Weaver and Danso (1994) suggest a ratio of 0.3 L gas phase volume per 1 g of plant tissue.
4. Miscellaneous equipment. Adequate conditions must be available for plant growth (e.g., environmental chamber, greenhouse) because N_2 fixation rates of plant-associated bacteria are dependent on photosynthetic rates. Plastic syringes and needles of various sizes are convenient for transferring and sampling gases. An oven for drying plant tissue and a device for finely grinding the dried tissue are needed. For small quantities, a mortar and pestle will work. As in all ^{15}N methods, care must be taken to avoid cross-contamination of samples.

Procedure

The protocol described is for use with plant-associated N_2 fixation, but it can easily be modified for litter, soil, or woody debris, which will likely require longer incubation times because of lower rates of N_2 fixation.

1. The plant may be grown in water culture, a soilless mix, or soil. It is important to maintain good plant growth conditions. Five replicates are often sufficient.
2. The plant's root system is placed in the gastight incubation vessel and sealed with a nontoxic substance, e.g., Terostat putty or adhesive mastic (Winship and Tjepkema 1990).
3. $^{15}N_2$ is added to the gastight system and the headspace mixed well. A nonlabeled control should also be used to determine the background ^{15}N abundance. Often 10 atom % $^{15}N_2$ is an adequate working concentration, however, this can be adjusted depending on the expected N_2 fixation rate, duration of the experiment, and mass spectrometer precision (Weaver and Danso 1994).

It may also be a good idea to add a small amount of an inert gas, such as He, to serve as an internal standard to check for gas leakage.

4. Shortly after the addition of $^{15}N_2$, a sample should be taken to determine the actual ^{15}N abundance of the headspace atmosphere. As stated earlier, the incubation length depends on the N_2-fixing activity and the $^{15}N_2$ abundance, although when using plants it is probably best to incubate through one or more diurnal cycle because N_2 fixation rates are dependent on photosynthetic activity. It is important that the O_2 concentration not be depleted during the incubation.
5. At the end of the incubation another gas sample should be taken for ^{15}N analysis and averaged with that of the initial gas sample to determine the mean atom % ^{15}N of the N_2 in the headspace.
6. The plant is then sampled for ^{15}N and total N analysis. Plant tissue is dried (70 °C) and ground (40-mesh). Drying and grinding the entire plant for analysis is simplest, although plant parts can be sampled, analyzed separately, and a dry mass weighted average based on nitrogen content used for calculations.

Calculations

It is most convenient to work in terms of atom % ^{15}N excess (AE). For plant tissue (AE_p), this is the difference in atom % ^{15}N between the labeled and control plant tissues. For the ^{15}N-labeled atmosphere (AE_a) this is the difference in the mean atom % ^{15}N of the N_2 in the headspace and that of the unlabeled atmosphere, or natural abundance (0.3663 atom % ^{15}N).

The fraction of nitrogen the plant derived from N_2 fixation (*FNA*) is calculated by dividing the atom % ^{15}N excess of the plant tissue (either whole plant or weighted average based on nitrogen content of various plant parts) by the average atom % ^{15}N excess of the labeled atmosphere:

$$FNA = AE_p/AE_a$$

where

FNA = fraction of plant nitrogen derived from N_2 fixation, which can be expressed as a percentage by dividing by 100 to give the percent nitrogen derived from the atmosphere (commonly known as % Ndfa)
AE_p = atom % ^{15}N excess of the plant tissue
AE_a = mean atom % ^{15}N excess of the N_2 in headspace

If plant parts are analyzed separately, then a weighted atom % ^{15}N excess must be used in place of AE_p.

$$WAE_p = \Sigma AE_{p',i} \times TN_i/\Sigma TN_i$$

where:

WAE_p = total nitrogen weighted average atom % ^{15}N excess of the plant
$AE_{p,i}$ = atom % ^{15}N excess of the ith plant tissue
TN_i = total nitrogen content of the ith plant tissue

Multiplying *FNA* by the total plant nitrogen content gives the total amount of N_2 fixed, which can be expressed as a N_2 fixation rate by dividing by the incubation time and normalizing to the desired plant metric, e.g., nodule dry weight, total plant mass, leaf area index.

N_2 Fixation by ^{15}N Isotope Dilution

The ^{15}N isotope dilution approach labels the soil, or nutrient solution, with ^{15}N instead of the N_2 in the atmosphere. This is usually much simpler than labeling atmosphere, which makes this method amenable to the field as well as the laboratory. It also provides an integrative measure of N_2 fixation, which is difficult with either the $^{15}N_2$ incorporation or the acetylene reduction methods. Although ^{15}N isotope dilution has become the method of choice for field studies of N_2 fixation, there are limitations that must be evaluated carefully (Chalk 1985; Danso 1986; Vose and Victoria 1986; Winship and Tjepkema 1990; Danso et al. 1992; Warembourg 1993; Weaver and Danso 1994).

The following protocol describes the application of ^{15}N isotope dilution to the field measurement of N_2 fixation by associations of N_2-fixing bacteria and non-woody plants, annuals in particular. Modifications of the method for use with perennial woody plants are also given.

Materials

1. ^{15}N -enriched fertilizer. Most commonly ^{15}N-enriched urea, ammonium salts, or nitrate salts are used to enrich the soil solution, with ammonium sulfate often being the least expensive. Many companies sell ^{15}N-labeled materials.
2. Reference plants. The selection of appropriate reference plants as a control for determining N_2 fixation by ^{15}N isotope dilution is critical. The ideal reference plant should not fix N_2 and should have a root system that exploits about the same volume of soil, that grows and take up soil nitrogen in the same temporal pattern, and that shows similar response to environmental factors and cultural manipulations. Nonnodulating varieties of some grain legumes (e.g., soybeans) exist and may be a good choice. If a soil is devoid of the nodulating N_2-fixing bacteria, then inoculated plants can be compared with non-inoculated plants. Sudan grass has been suggested as a good reference plant for grain legumes and perennial ryegrass for pasture legumes. There is no standard reference plant for native plant species, and it is often impractical to introduce a reference plant. Thus, non-N_2-fixing plants already occupying the site are commonly used as reference plants. Careful selection of preexistent reference plants is required with the goals of selecting species with similar rooting habit and temporal patterns of nitrogen uptake.
3. Mass spectrometer. ^{15}N abundance is most accurately and precisely measured with a mass spectrometer (Hauck 1982; Mulvaney 1993; Hauck et al. 1994). If a mass spectrometer is not available locally, several laboratories will perform ^{15}N analysis on a fee basis. The most critical aspect of preparing plant

or soil samples for ^{15}N analysis is to make certain that the sample is very finely ground (40-mesh) to ensure sample homogeneity.

4. Miscellaneous equipment. An oven for drying plant tissue and a device for finely grinding the dried tissue are needed. For small quantities, a mortar and pestle will work. As in all ^{15}N methods, care must be taken to avoid cross-contamination of samples.

Procedure

1. Field plots need to be selected and designed. A randomized complete plot design with six replicates is normally sufficient. The size of the plots is dependent on the type of plant. For grain legumes, 1–5 m^2 plots that encompass four to five rows are adequate; 1–2 m^2 plots are adequate for pasture legumes. Often ^{15}N-fertilized subplots are located within each plot to reduce the cost of ^{15}N fertilizer. The nonfixing reference plants should be located as close as possible to the N_2-fixing plants and can be intermixed in pasture systems.
2. The ^{15}N fertilizer is applied to plots of the fixing and nonfixing plants at the same time and rate. Unlabeled plots containing fixing and nonfixing plants should also be established to determine the background ^{15}N abundance of the fixing and nonfixing plants. For most situations, adding about 1 kg ^{15}N/ha is adequate for a study lasting 1 year (e.g., an enrichment of 5–10 atom % ^{15}N applied at rates of 5–10 kg N/ha). It is important that the total amount of nitrogen added is not high enough to adversely affect N_2 fixation rates. Sometimes ^{15}N-enriched organic materials are used (e.g., plant residues or animal manures). When using organic materials, the suggested 1 kg ^{15}N/ha is probably still a good guide. The ^{15}N-labeled material can be added in many different ways, although application as a solution using a sprayer is often the most convenient and provides for uniform labeling of soil. There may be some advantages to labeling the soil in advance to allow for the added ^{15}N to equilibrate with the native soil nitrogen or for adding the ^{15}N in multiple additions, particularly for perennial plants (Baker et al. 1995). Care should be taken that ^{15}N is applied only where desired to minimize potential contamination between plots.
3. The ^{15}N isotope dilution approach is integrative, although it is possible to sample plants periodically over the growing season to determine N_2 fixation over different times. More commonly, however, plants are harvested at the end of the growing season. Because there are usually some differences in the ^{15}N abundance of different plant parts, it is important to either harvest and process the entire plant or to carefully sample the different tissues to obtain a weighted average.
4. Plant tissues are oven dried at 70 °C and then finely ground (40-mesh) for analysis.

Calculations

It is most convenient to work in terms of atom % ^{15}N excess. The atom % ^{15}N excess of the fixing plant (AE_f) and reference plant (AE_r) are defined as

$$AE_f = A_{fl} - A_{fu}$$
$$AE_r = A_{rl} - A_{ru}$$

where

A_{fl} = atom % ^{15}N of the fixing plant grown on the ^{15}N-labeled plot
A_{fu} = atom % ^{15}N of the fixing plant grown on the unlabeled plot
A_{rl} = atom % ^{15}N of the reference plant grown on the ^{15}N-labeled plot
A_{ru} = atom % ^{15}N of the reference plant grown on the unlabeled plot

The fraction of nitrogen coming from (*FNA*) is calculated as

$$FNA = 1 - AE_f/AE_r$$

If plant tissues are analyzed separately, then *WAE* (calculation shown earlier) can be used in place of *AE*.

Converting *FNA* to an areal N_2 fixation rate requires information about the biomass and N content of the N_2-fixing plant. Typically this is performed by multiplying *FNA* by the total N concentration of the plant and by the biomass of the N_2-fixing plant on an areal basis (e.g., kg/m).

Special Considerations

Adapting the ^{15}N isotope dilution method for use with woody perennial plants requires several modifications (Danso et al. 1992; Parrotta et al. 1994; Baker et al. 1995).

1. Plot size. For N_2-fixing trees, larger plots are needed, with crown size being a reasonable guide to the size of the ^{15}N-labeled area. It is common to use paired-tree plots to conserve the amount of ^{15}N fertilizer needed. This can be facilitated by using paired fixing and reference plants or having N_2-fixing and reference plants located in a checkerboard pattern.
2. ^{15}N labeling. Several (two or three) applications over the growing season will better label the entire rooting volume.
3. There is no standard reference plant for trees or shrubs; therefore, one often has to use the nonfixing plants that are present on the site. This has obvious disadvantages, but if the fixing plant gets most of its nitrogen from N_2 fixation, any deviations from the "ideal" reference plant may not be very important (e.g., Unkovich et al. 1994).
4. An additional problem with perennial plants is that these plants also contain a reserve pool of nitrogen in addition to nitrogen taken up from the soil or N_2 fixed. Baker et al. (1995) have developed an equation that accounts for the influence of plant nitrogen reserves. This calculation requires the measurement of total nitrogen at each time interval as well as ^{15}N abundance.
5. When calculating the amount of N_2 fixed on an areal basis for woody perennials, it is necessary to multiply the fraction of nitrogen derived from N_2 fixation by the nitrogen increment of above- and belowground tissues rather than

the total biomass of these tissues. Accurate estimates of plant nitrogen increment are not easy to obtain and make scaling to an areal basis difficult at best.

As mentioned previously, using differences in the natural ^{15}N abundance of N_2-fixing and nonfixing reference plants to calculate *FNA* has many similarities to the isotope dilution method, except that no ^{15}N fertilizer is used. Reference plants must be chosen with care, and the native soil must be sufficiently different in ^{15}N abundance (at least 2‰) than that of the atmosphere. Additional details can be found in Danso et al. (1992), Shearer and Kohl (1993), and Weaver and Danso (1994).

Conclusions

There are many approaches to measuring N_2 fixation, none of which is universally applicable. It is possible, however, to wisely choose the most appropriate method for a given system and thereby obtain reasonable estimates of N_2 fixation rates. For most field studies, particularly those measuring annual N_2 fixation rates, the ^{15}N isotope dilution method is recommended. For shorter-term assays or measurements of free-living N_2 fixation, the acetylene reduction method calibrated with the $^{15}N_2$ incorporation method, or just the $^{15}N_2$ incorporation method itself, is recommended.

Although a significant amount of research has been performed on N_2 fixation methodology, there are still relatively unexplored areas that need further research. It would be useful to reevaluate whether the acetylene reduction assay can be adapted for use with root nodule symbioses in the field to give similar estimates of N_2 fixation compared with the ^{15}N isotope dilution method. More work is also needed on questions regarding the measurement of N_2 fixation of root-nodulated woody perennials, e.g., differentiating the importance of the reserve pool of plant nitrogen and ways of estimating yearly increments of root and nodule biomass. Further evaluation of the use of variations in ^{15}N natural abundance to quantitatively measure N_2 fixation rates is also warranted because of the approach's potential as a relatively nondisruptive means of estimating annual rates of N_2 fixation.

Acknowledgments We thank Kerstin Huss-Danell, Mark Harmon, Tim Fahey, Xiaoming Zou, and several anonymous reviewers for improving this paper.

References

Baker, D. D., M. Fried, and J. A. Parrotta. 1995. Theoretical implications for the estimation of dinitrogen fixation by large perennial plant species using isotope dilution. Pages 225–236 *in* K. S. Kumarasinghe, editor, *Nuclear Techniques in Soil-Plant Studies for Sustainable Agriculture and Environmental Preservation.* International Atomic Energy Agency, Vienna, Austria.

Bergersen, F. J. 1980. *Methods for Evaluating Biological Nitrogen Fixation.* Wiley, New York, New York, USA.

Boddey, R. M. 1987. Methods for quantification of nitrogen fixation associated with Gramineae. *Critical Reviews in Plant Science* 6:209–266.

Bremer, E., and C. van Kessel. 1990. Appraisal of the nitrogen-15 natural-abundance method for quantifying dinitrogen fixation. *Soil Science Society of America Journal* 54:404–411.

Chalk, P. M. 1985. Estimation of N_2 fixation by isotope dilution: an appraisal of techniques involving ^{15}N enrichment and their application. *Soil Biology and Biochemistry* 17:389–410.

Danso, S. K. A. 1986. Review: Estimation of N_2-fixation by isotope dilution: an appraisal of techniques involving ^{15}N enrichment and their application—comments. *Soil Biology and Biochemistry* 18:243–244.

Danso, S. K. A., G. D. Bowen, and N. Sanginga. 1992. Biological nitrogen fixation in trees in agro-ecosystems. *Plant and Soil* 141:117–196.

Giller, K. E. 1987. Use and abuse of the acetylene reduction assay for measurement of "associative" nitrogen fixation. *Soil Biology and Biochemistry* 19:783–784.

Griffiths, R. P., M. E. Harmon, B. A. Caldwell, and S. E. Carpenter. 1993. Acetylene reduction in conifer logs during early stages of decomposition. *Plant and Soil* 148:53–61.

Hansen, A. P., and J. S. Pate. 1987. Evaluation of the ^{15}N natural abundance method and xylem analysis for assessing N_2 fixation of understorey legumes in Jarrah (*Eucalyptus marginata* Donn ex Sm.) forest in S.W. Australia. *Journal of Experimental Botany* 38:1446–1458.

Hauck, R. D. 1982. Nitrogen: isotope-ratio analysis. Pages 735–779 *in* A. L. Page, R. H. Miller, and D. R. Keeney, editors, *Methods of Soil Analysis. Part 2, Chemical and Microbiological Properties. 2d edition.* American Society of Agronomy Book Series, No. 9. American Society of Agronomy, Madison, Wisconsin, USA.

Hauck, R. D., J. J. Meisinger, and R. L. Mulvaney. 1994. Practical considerations in the use of nitrogen tracers in agricultural and environmental research. Pages 907–950 *in* R. W. Weaver, J. S. Angle, and P. J. Bottomley, editors, *Methods of Soil Analysis. Part 2, Microbiological and Biochemical Properties.* Soil Science Society of America Book Series, No. 5. Soil Science Society of America, Madison, Wisconsin, USA.

Jurgensen, M. F., M. J. Larsen, R. T. Graham, and A. E. Harvey. 1987. Nitrogen fixation in woody residue of northern Rocky Mountain conifer forests. *Canadian Journal of Forest Research* 17:1283–1288.

Minchin, F. R., and J. F. Witty. 1989. Limitations and errors in gas exchange measurements with legume nodules. Pages 79–96 *in* J. G. Torrey and L. J. Winship, editors, *Applications of Continuous and Steady-State Methods to Root Biology.* Kluwer Academic Publishers, Dordrecht, Netherlands.

Mulvaney, R. L. 1993. Mass spectrometry. Pages 11–57 *in* R. Knowles and T. H. Blackburn, editors, *Nitrogen Isotope Techniques.* Academic Press, San Diego, California, USA.

Newton, M., B. A. El Hassan, and J. Zavitkovski. 1968. Role of red alder in western forest succession. Pages 73–83 *in* J. M. Trappe, J. F. Franklin, R. F. Tarrant, and G. M. Hansen, editors, *Biology of Alder.* U.S. Department of Agriculture, Forest Service, Pacific Northwest Forest Range Experimental Station, Portland, Oregon, USA.

Parrotta, J. A., D. D. Baker, and M. Fried. 1994. Application of ^{15}N-enrichment methodologies to estimate nitrogen fixation in *Casuarina equisetifolia. Canadian Journal of Forest Research* 24:201–207.

Schwintzer, C. R., and J. D. Tjepkema. 1994. Factors affecting the acetylene to $^{15}N_2$ conversion ratio in root nodules of *Myrica gale* L. *Plant Physiology* 106:1041–1047.

Selles, F., R. E. Karamanos, and R. G. Kachanoski. 1986. The spatial variability of nitrogen-15 and its relation to the variability of other soil properties. *Soil Science Society of America Journal* 50:105–110.

Shearer, G., and D. Kohl. 1986. N_2-fixation in field settings: estimations based on natural ^{15}N abundance. *Australian Journal of Plant Physiology* 13:699–756.

Shearer, G., and D. Kohl. 1993. Natural abundance of ^{15}N: fractional contribution of two sources to a common sink and use of isotope discrimination. Pages 89–125 *in* R. Knowles, and T. H. Blackburn, editors, *Nitrogen Isotope Techniques.* Academic Press, San Diego, California, USA.

Shearer, G., D. H. Kohl, R. A. Virginia, B. A. Bryan, J. L. Skeeters, E. T. Nilsen, M. R. Sharifi, and P. W. Rundel. 1983. Estimates of N_2-fixation from variation in the natural abundance of ^{15}N in Sonoran Desert ecosystems. *Oecologia* 56:365–373.

Silvester, W. B., R. Parsons, F. R. Minchin, and J. F. Witty. 1989. Simple apparatus for growth of nodulated plants and for continuous nitrogenase assay under defined gas phase. Pages 55–66 *in* J. G. Torrey and L. J. Winship, editors, *Applications of Continuous and Steady-State Methods to Root Biology.* Kluwer Academic Publishers, Dordrecht, Netherlands.

Stevenson, F. C., J. D. Knight, and C. van Kessel. 1995. Dinitrogen fixation in pea: controls at the landscape- and micro-scale. *Soil Science Society of America Journal* 59:1603–1611.

Unkovich, M. J., J. S. Pate, P. Sanford, and E. L. Armstrong. 1994. Potential precision of the ^{15}N natural abundance method in field estimates of nitrogen fixation by crop and pasture legumes in south-west Australia. *Australian Journal of Agricultural Research* 45:119–132.

Vose, P. B., and R. L. Victoria. 1986. Re-examination of the limitations of nitrogen 15 isotope dilution technique for the field measurement of dinitrogen fixation. Pages 23–58 *in* R. D. Hauck and R. W. Weaver, editors, *Field Measurement of Dinitrogen Fixation and Denitrification.* Soil Science Society of America Special Publication No. 18. Soil Science Society of America, Madison, Wisconsin, USA.

Warembourg, F. R. 1993. Nitrogen fixation in soil and plant systems. Pages 127–156 *in* R. Knowles and T. H. Blackburn, editors, *Nitrogen Isotope Techniques.* Academic Press, San Diego, California, USA.

Weaver, R. W. 1986. Measurement of biological dinitrogen fixation in the field. Pages 1–10 *in* R. D. Hauck and R. W. Weaver, editors, *Field Measurement of Dinitrogen Fixation and Denitrification.* Soil Science Society of America Special Publication No. 18. Soil Science Society of America, Madison, Wisconsin, USA.

Weaver, R. W., and S. K. A. Danso. 1994. Dinitrogen fixation. Pages 1019–1045 *in* R. W. Weaver, J. S. Angle, and P. J. Bottomley, editors, *Methods of Soil Analysis. Part 2, Microbiological and Biochemical Properties.* Soil Science Society of America Book Series, No. 5. Soil Science Society of America, Madison, Wisconsin, USA.

Winship, L. J., and J. D. Tjepkema. 1990. Techniques for measuring nitrogenase activity in *Frankia* and actinorhizal plants. Pages 263–280 *in* C. R. Schwintzer and J. D. Tjepkema, editors, *The Biology of* Frankia *and Actinorhizal Plants.* Academic Press, San Diego, California, USA.

Youngberg, C. T., and A. G. Wollum. 1976. Nitrogen accretion in developing *Ceanothus velutinus* stands. *Soil Science Society of America Journal* 40:109–112.

13

Soil Carbon and Nitrogen Availability

Nitrogen Mineralization, Nitrification, and Soil Respiration Potentials

G. Philip Robertson
David Wedin
Peter M. Groffman
John M. Blair
Elisabeth A. Holland
Knute J. Nadelhoffer
David Harris

A soil's capacity to transform organic nitrogen in soil organic matter to inorganic nitrogen—its nitrogen mineralization potential—is often used as an index of the nitrogen available to plants in terrestrial ecosystems. It is perhaps the most common and best means available to assess nitrogen fertility (e.g., Keeney 1980; Binkley and Hart 1989; Palm et al. 1993), related as it is to both the size of the labile soil organic matter (SOM) pool and the activity of the organisms responsible for its oxidation. Mineralization potentials (the net production of inorganic nitrogen under standard conditions) are superior to inorganic soil nitrogen concentrations (pool size) as an indicator of site fertility simply because the supply rate of a limiting nutrient is more important to its availability than is its instantaneous concentration. Most mineralization assays are designed to exclude plant uptake and leaching, and most also ignore immobilization (microbial uptake) and denitrification, so net mineralization potentials provide a good general index of the capacity of a soil to make nitrogen available to plants via the soil solution.

Operationally, N mineralization usually refers to the net increase in both ammonium (NH_4^+) and nitrate (NO_3^-) in soil, since any nitrate formed must first have been ammonium. While other forms of inorganic nitrogen are also produced during mineralization assays (e.g., NO_2^-, N_2O, and NO_x), in most soils their appearance is highly transient and pools are quickly converted to another form (NO_2^-), or their fluxes are inconsequential relative to increases in the NH_4^+ and NO_3^- pools (N_2O, NO_x).

Net nitrification refers specifically to the conversion of ammonium to nitrate by nitrifiers, bacteria that oxidize ammonium to nitrite and then nitrate. Nitrification assays also usually measure the net flux, excluding the nitrate that may be immobilized into microbial biomass (e.g., Davidson et al. 1991) or denitrified to nitrogen gas (e.g., Robertson and Tiedje 1985) during the course of the assay.

While both net N mineralization and net nitrification assays have their limitations as measures of nitrogen availability, they can nevertheless provide substantial insight into soil fertility and ecosystem function at many sites, and they are used widely. Large differences among sites or among experimental treatments, for example, imply large differences in plant-available nitrogen, as well as large differences in the potential loss of nitrogen from an ecosystem. Nitrate, for example, is more readily lost from most ecosystems than is ammonium, so high potential nitrification rates at a site can indicate a higher likelihood of nitrogen loss, all else being equal. It is harder to interpret small differences in mineralization or nitrification among sites or treatments because of artifacts intrinsic to the assays—microbial immobilization may be higher in an incubation than in situ, for example—but relative differences can still provide insight into ecosystem-level processes among sites.

Soil carbon availability is another component of soil fertility that can provide important information about ecosystem status. As for nitrogen availability, carbon availability also can be assessed by measuring a soil's ability to transform organic carbon to an inorganic form—in this case CO_2. A soil's capacity to oxidize fixed carbon to CO_2 under optimal moisture and temperature conditions—its respiration potential—can be a useful index of carbon availability in most ecosystems. The long-term release of CO_2 from SOM under optimal conditions can, in fact, be used mathematically to indicate the functional pools of SOM commonly referred to as active and passive fraction SOM (e.g., Juma and Paul 1981; Hess and Schmidt 1995). Respiration potentials can thus also provide valuable insight into the potentials for carbon storage at a site (e.g., Paul et al. 1998).

Available Protocols

There are many different ways to assess nitrogen and carbon mineralization rates in soil; no single method enjoys universal acceptance, reflecting the complexity of factors that can affect rates of carbon and nitrogen turnover and the diverse nature of the compromises required to produce a reasonable estimate for a specific site. Chief among these compromises is the level of soil disturbance permitted. All available methods require some degree of soil disturbance, and this disturbance can artificially depress or, more commonly, accelerate mineralization rates. A second compromise is the environmental conditions under which mineralization is assessed. It is, for example, a straightforward matter to duplicate the effects of field temperature conditions on mineralization rates but very difficult to duplicate the effects of in situ soil moisture. Thus it is important to recognize that all common mineralization protocols are indices that can be sensitive to assay conditions; consequently, temporal and cross-site comparisons demand that close attention be paid to reproducing comparable disturbance and soil moisture/temperature regimes across experimental units.

Common to all net nitrogen mineralization assays is an incubation period over which accumulated inorganic nitrogen is used to calculate the rate of nitrogen mineralization during that period. As noted earlier, this approach represents the net balance between gross mineralization and microbial immobilization and in vitro losses of nitrogen. The gross rate of mineralization may in fact be more than an order of magnitude higher than net mineralization (Hart et al. 1994); the net rate thus represents the *minimum* amount of nitrogen available for either plant uptake or inorganic nitrogen loss. During incubation a portion of the organic nitrogen mineralized to ammonium will be oxidized to nitrate by nitrifiers; net mineralization is thus calculated as the sum of ammonium plus nitrate nitrogen at the end of the incubation interval. Nitrification itself can be assayed by considering only the net rate of nitrate increase during incubation and can be expressed as a proportion of net mineralization, (NO_3^--N/[NH_4^+−N + NO_3^-−N]).

The variety of techniques available for measuring nitrogen mineralization and nitrification can be classed into three main groupings: (1) in situ incubations of enclosed soils, in which inorganic nitrogen accumulation is measured at the end of a 2–6 week incubation period; (2) laboratory incubations under standard moisture and temperature conditions in which inorganic nitrogen accumulation is monitored at 7–30 day intervals for up to a year or more; and (3) isotopic incubations during which changes in a ^{15}N-labeled inorganic nitrogen pool is measured over the course of a 1–3 day incubation. These techniques all isolate a quantity of soil from its environment during the incubation period; this isolation provides the opportunity for monitoring inorganic N accumulation in the absence of processes that might otherwise affect inorganic N pools. These processes include primarily plant uptake, leaching, and atmospheric N deposition. Other processes that might affect these pools—notably denitrification and NH_3 volatilization—are measured separately, are assumed to be minimal or constant among sites, or are treated like immobilization—implicit factors that effectively reduce net nitrogen mineralization rates.

There are many possible permutations of procedures within the three major groups. These include the amount of soil disturbance (intact cores versus sieved soils), incubation temperatures (in situ temperature fluctuations versus a constant temperature such as 25 °C), incubation moisture (field moisture at the time of sampling versus some proportion of water-filled pore space or field capacity), and—for laboratory incubations—aeration status (ranging from added sand to increase aeration to anaerobic slurries to eliminate oxygen altogether). In the procedures outlined here we suggest specific protocols for each of these three major assay types.

The assays described here should be chosen for use based on experimental objectives and available resources. If one is comparing long-term treatments and can sample only once, then long-term laboratory incubations will provide a reasonable basis for insightful comparisons. If one is attempting to capture differences among sites or treatments with respect to shorter-term in situ dynamics, then the 28 day field incubation is appropriate. Detailed comparisons of specific processes among sites or treatments will require the ^{15}N isotope dilution approach.

Other methods for assessing nitrogen availability have been used to great advantage in many types of ecosystems. These include ion exchange resin bags inside intact cores (e.g., DiStefano and Gholz 1986; Zou et al. 1992), anaerobic slurries

(Waring and Bremner 1964; Keeney 1980), and various enzyme assays (Tabatabai 1994; Schmidt and Belser 1994). Bundy and Meisinger (1994), Hart et al. (1994), and Binkley and Hart (1989) provide excellent reviews of various techniques and their advantages and limitations.

There are also a variety of techniques available for measuring biologically available carbon. While the total pool of soil carbon can be adequately evaluated by chemical means (see Chapter 5, this volume), available carbon—that portion of the total pool actually available to the microbial community—is best assayed biologically. We recommend the laboratory incubation described later because of its simplicity and power (see "Potential Carbon Availability—Respiration Potentials" in this chapter). Measurements of CO_2 flux in the field (see Chapter 10, this volume) can also indicate potential differences in C availability among sites (e.g., Paul et al. 1998), but interpretation can be complicated by the inclusion of in situ root respiration. Likewise, enriched carbon-isotope tracer techniques are complicated by difficulties associated with labeling specific carbon pools. An additional value of the incubation technique described in this chapter is that C availability can be assessed for the same samples (and at the same time) as N availability.

Potential Nitrogen Mineralization—Field Incubations

Field or in situ incubations of intact soil cores provide estimates of net nitrogen mineralization at temperatures typical for a given site without unduly disturbing soil structure. In the technique described here (a modification of methods proposed by Adams and Attiwill 1982; Raison et al. 1987; and Hart et al. 1994), a set of soil cores enclosed in PVC sleeves are loosely capped to minimize moisture changes but not gas exchange, and cores are then allowed to incubate for several weeks in situ. Adjacent cores are taken to provide an estimate of initial nitrate and ammonium levels. At the end of the incubation period the field-incubated cores are removed and analyzed for accumulated nitrate and ammonium. The difference between final and initial levels of total inorganic N (ammonium + nitrate) is the rate of net nitrogen mineralization, best expressed on both a gravimetric ($mg\ N \cdot kg^{-1} \cdot d^{-1}$) basis for within-site comparisons and on an areal ($g\ N \cdot m^{-2} \cdot d^{-1}$) basis to normalize for differences in soil bulk density among different sites. The net nitrification rate is the amount of nitrate nitrogen accumulated over this period. The relative net nitrification rate—the proportion of net mineralized nitrogen that is nitrified—can be expressed on a percentage basis as noted earlier.

Monthly incubations can provide information on seasonal patterns of nitrogen mineralization and, if moisture at the time of sampling is typical for the site over the remainder of each incubation period, can provide a rough estimate of the annual rates of net nitrogen mineralization and net nitrification for a site. If moisture at sampling significantly differs from the site's moisture dynamics over the incubation interval, the method will over- or underestimate actual net nitrogen mineralization rates, but rates may still be valuable for nearby treatment comparisons if moisture dynamics are similar among treatments. For cross-ecosystem comparisons, however, it is extremely important to design sampling intervals that capture local moisture, temperature, and vegetation dynamics. Otherwise a standardized laboratory approach may be more appropriate.

Mineralization rates tend to exhibit high spatial variability in the field. As for many soil properties (see Chapter 1, this volume), net mineralization rates are usually lognormally distributed, and a stratified sampling strategy may be justified. Even so, it is not uncommon for net rates to span an order of magnitude over even a several-meter area in a site that visually appears to be homogeneous. Replication thus becomes very important and is often critical for quantifying significant differences among sites.

How many samples are sufficient? The answer to this question will be site-specific. Rarely are fewer than six cores adequate, and many sites may require a dozen or more to accurately quantify a mean flux. An initial intensive sampling will help to identify the variability present and thus the most efficient sampling strategy for subsequent efforts (see also Chapter 1, this volume).

Materials

1. At least six thin-walled cylinders (schedule 40 PVC or steel) per site, 24 cm long $\times \geq 5$ cm inside diameter (of a sufficient length to penetrate the organic and A horizons and of a sufficient diameter to minimize compaction; see the section "Special Considerations," below), sharpened on one end with a file or bench grinder
2. Loose-fitting caps for each cylinder
3. 1 mol/L potassium chloride solution and other supplies sufficient for the triplicate extraction of initial soil samples equal to the number of cylinders (see Chapter 5, this volume). The same amount of this solution will be needed for the final extraction.
4. Paper bags or soil cans for determination of soil moisture as described in Chapter 3, this volume
5. Other materials and supplies as required for NH_4^+ and NO_3^- analyses of KCl soil extracts

Procedure

1. Hand-drive with a rubber mallet all but 2 cm of each cylinder into the soil in a pattern consistent with the area's vegetation cover (see Chapter 1, this volume); for many sites this pattern will be random. Include the O-horizon layer if present. Cover the top of the cylinder with a loose-fitting cap.
2. Remove a soil sample taken in an equivalent manner (same diameter, depth) from within 20–30 cm of each core and transport back to the laboratory for immediate extraction and inorganic-N analysis. These provide initial N values for each in situ core. If extraction will not take place within 6 hours, put the cores in an ice-cooled container after collecting and extract within 24 hours.
3. In the laboratory, weigh, sieve, and extract in triplicate the initial soil samples, then analyze for inorganic-N as described in Chapter 5, this volume. Analyze

a 50 g subsample for gravimetric moisture as described in Chapter 3, this volume.

4. At the end of 28 days (4 weeks), remove the field-incubated cores and composite, weigh, sieve, extract in triplicate, and analyze for inorganic-N as for the initial cores. Also analyze a 50 g subsample for gravimetric moisture as described earlier. Discard any cores disturbed by animals.

Calculations

The following equations assume NO_3^-–N and NH_4^+–N units that have already been converted to both a gravimetric basis (mg N/kg soil) and an areal basis (g N/m^2) using formulas presented in Chapter 6, this volume. It is sometimes useful to include rates expressed on another basis such as SOM (μg N·kg SOM^{-1}· d^{-1}). Always report soil bulk density values together with turnover rates to allow others to interconvert. A typical mineralization range for a forest or grassland soil sampled during a growing-season period with adequate moisture is 0.1–1.0 g N·m^{-2}·d^{-1} or 0.5–5.0 mg N·kg^{-1}·d^{-1} (Robertson 1982b); low moisture can readily reduce these values by an order of magnitude or more.

Net N Mineralization

$$N_{mineralized} = [(Nitrate_f + Ammonium_f) - (Nitrate_0 + Ammonium_0)]/T_{days}$$

where

$N_{mineralized}$ = net N mineralization rate, expressed as mg N·kg^{-1}·d^{-1} or g N·m^{-2}·d^{-1}

$Nitrate_f$ = final nitrate concentration, expressed as mg NO_3^-–N/g soil or g NO_3^-–N/m^2

$Ammonium_f$ = final ammonium concentration, expressed as mg NH_4^+–N/g soil or g NH_4^+–N/m^2

$Nitrate_0$ = initial nitrate concentration, as mg NO_3^-–N/g soil or g NO_3^-–N/m^2

$Ammonium_0$ = initial ammonium concentration, as mg NH_4^+–N/g soil or g NH_4^+–N/m^2

T_{days} = incubation time, in days

Net Nitrification

$$N_{nitrified} = (Nitrate_f - Nitrate_0)/T_{days}$$

where

$N_{nitrified}$ = net nitrification rate, expressed as mg NO_3^-–N·kg^{-1}·d^{-1} or g NO_3^-–N·m^{-2}·d^{-1}

Relative Nitrification

$$\% \text{ Nitrified} = 100 \times N_{nitrified}/N_{mineralized}$$

Special Considerations

Specific sites may require different soil core dimensions than those described earlier (5 cm inner diameter × 24 cm long). Soils with shallow A horizons may be sampled with shorter cores, and A horizons deeper than 20 cm will require longer cores. Ideally one should analyze soil nitrogen mineralization at least once at multiple depths and then use a core length that captures most of the activity present. Fine-textured soils may compact with cores only 5 cm in diameter, in which case a larger diameter will be necessary. Compaction should be kept to 5% (e.g., 1 cm per 20 cm length). Compacted or rocky soils (e.g., desert soils and some agricultural soils) may require steel rather than PVC cylinders, and in some cases it may be necessary to substitute polyethylene bags for cores, in which case due care should be taken to further minimize soil structural disturbance when sampling.

Many sites may require more than six pairs of cores to adequately characterize nitrogen mineralization within an area. If not already available, it will be important to perform an initial sampling to assess the sample intensity necessary to adequately capture within-site variability (see Chapter 1, this volume).

Potential Nitrogen Mineralization—Laboratory Incubations

Often it is useful to assess the nitrogen mineralization potential of a soil independently of short-term environmental conditions such as temperature and moisture. Such measurements of a soil's potential to mineralize organic nitrogen, usually made under constant moisture and temperature conditions, provide a different means for comparisons of nitrogen availability across systems and treatments.

A variety of approaches are available for measuring potential nitrogen mineralization in incubated soils, ranging from microlysimeter methods (e.g., Stanford and Smith 1972; Robertson 1982a; Nadelhoffer 1990) to simple static incubations of soil in jars or bags. We recommend a static incubation approach because of its simplicity and because with this approach it is easy to assess potentially available carbon at the same time by using incubation vessels that can be tightly sealed (see the section "Potentially Available Carbon," below). Additionally, this approach is suitable for indices of both short-term (28 day) and long-term (252 day) turnover.

Materials

1. Six 125 mL flasks or equivalent per soil sample location
2. Access to a 25 °C incubator or constant-temperature room
3. Paper bags or soil cans for moisture determinations
4. Inorganic N extraction material as per the field incubation technique
5. Access to NO_3^- and NH_4^+ analysis as per the field incubation technique

Procedure

1. Collect soil from each area to be sampled (experimental replicate or subsite of a larger area) and composite by area, weigh, and sieve (4 mm). As for field

incubations, the sample pattern should be consistent with the area's vegetation cover, and the sample number consistent with within-site variability; samples should include the O-horizon.

2. Determine the moisture content of each sample as per Chapter 3, this volume.
3. Adjust the moisture content of >100 g of each sieved soil to approximately 60% water-filled pore space (determined as per Chapter 3, this volume); be certain the moisture-adjusted soil is well mixed.
4. Weigh 10 g of each soil composite into each of the six flasks.
5. Set aside three flasks for immediate inorganic N (NO_3^-–N and NH_4^+–N) analysis.
6. Cap the remaining flasks loosely or cover with polyethylene film (permeable to O_2 and CO_2 but not to H_2O_v) and place in a humidified, darkened, 25 °C incubator or constant-temperature room.
7. Extract the soil in each of the initial flasks in KCl and analyze for inorganic N as described in Chapter 6, this volume.
8. Check for water loss periodically by weighing a subset of flasks at the outset and reweighing at intervals. Replace evaporated soil moisture as needed for each soil. After 28 days, extract soil as in step 7.

Calculations

Convert nitrate and ammonium values to both an areal basis (g N/m^{-2}) and a gravimetric basis (mg N/kg^{-1}) using formulas provided in Chapter 6, this volume. Use the formulas for field incubations (earlier) to determine net nitrogen mineralization potential, potential net nitrification, and relative nitrification.

Special Considerations

See the field incubation procedure for special considerations related to field sampling. To use this method for long-term N mineralization assays, use a sufficient number of flasks per soil composite to allow three extractions per incubation interval; use the same intervals as for long-term respiration potentials, discussed later.

Potential Carbon Availability—Respiration Potentials

Potentially available carbon is best assayed in aerobic incubations and, as for potential nitrogen mineralization, under environmental conditions that are near optimal. The CO_2 produced over the incubation interval is used as an index of the amount of carbon available to microbes. Microbial growth over this interval (C-assimilation) is presumed to be a constant or insignificant alternate sink for metabolized carbon, or is assumed to be similar among experimental areas. Short-term incubations provide information on immediately available carbon. Long-term incubations provide information on those carbon pools that turn over more slowly, and thus provide a means for partitioning SOM into the functionally distinct pools that are an important component of most of today's major soil carbon and nitrogen models (e.g., Juma and Paul 1981; Molina et al. 1983; Jenkinson et al. 1987; Parton et al. 1987; Paustian et al. 1992). Both short- and long-term incubations can provide valuable insight into carbon and nitrogen cycling in a given ecosystem.

Materials

1. Three 125 mL Erlenmeyer flasks (e.g., Corning no. 5020), small canning jars, or equivalent sealable containers per soil composite or site. For flasks, cap with large butyl rubber septa (e.g., Aldrich Chemical Co., Milwaukee, WI; part no. Z10, 145–1 for Corning no. 5020 flasks); for canning jars, lids should be new, boiled before use, and fitted with a rubber septa (e.g., the cap from a disposable blood collection tube) that is inserted snugly into a predrilled hole in the jar lid.
2. Access to a 25 °C incubator or constant-temperature room
3. Paper bags or soil cans for moisture determinations
4. Access to an infrared gas absorption (IRGA) analyzer or gas chromatograph for CO_2 analysis

Procedure

1. Collect soil from each area to be sampled (experimental replicate or subsite of a larger area) and composite by area, weigh, and sieve (4 mm preferred). As for field incubations, the sample pattern should be consistent with the area's vegetation cover, and the sample number consistent with within-site variability; samples should include the O-horizon.
2. Determine the moisture content of the sample as per Chapter 3, this volume.
3. Adjust the moisture content of >100 g of sieved soil to approximately 60% water-filled pore space (determined as per Chapter 4, this volume); be certain the moisture-adjusted soil is well mixed.
4. Weigh 10 g of each composite into each incubation jar.
5. For short-term incubations, sample flasks after 7 days. For long-term incubations, sample flasks at intervals of 7, 14, 28, 42, 63, 84, 105, 140, 196, and 252 days (1, 2, 4, 6, 9, 12, 15, 20, 28, and 36 weeks). At each sample time, vent each flask with a stream of humidified air sufficient to fully aerate the flask headspace and soil macropores (humidify the airstream by bubbling it through a water-filled flask). Recap the flask and at intervals over the next 2–3 hours remove at least three 1 mL headspace samples for CO_2 analysis as per Chapter 10, this volume. The rate of production over this 2–3 hour period (the slope of a linear regression of concentration versus time) represents the respiration potential over that time interval.

Calculations

Short-term Respiration Potential (C Mineralization Rate)

The following calculations assume that CO_2 flux values will be reported in parts per million by volume (ppm_v) per minute or $\mu L \cdot L^{-1}$ $CO_2 \cdot min^{-1}$ (gas standards are usually purchased in ppm_v concentrations). Total CO_2–C in an enclosed soil volume includes CO_2–C in the headspace (including soil pore space) plus the CO_2–C dissolved in soil water. Normally increases in the CO_2–C dissolved in soil water over the incubation interval constitute a minor part of the CO_2 flux and can be ignored. For wet soils (e.g., bog soils or sediments), however, CO_2–C dissolved in water should be included in the calculation and can be calculated based on headspace CO_2 concentration, water content, pH, and a temperature-dependent equilibrium constant termed the *Bunsen coefficient* (see Tiedje 1982).

A. Convert concentrations of CO_2 to mass units and correct for incubation conditions through the application of the Ideal Gas Law:

$$C_m = (C_v \times M \times P)/(R \times T)$$

where

C_m = μg CO_2–C/L headspace
C_v = ppm_v CO_2 or μL CO_2/L headspace
M = molecular weight of CO_2–C (12 μg/μmol)
P = barometric pressure (in atmospheres), e.g., 1 atm
R = universal gas constant (0.0820575 L atm·°K·mole)
T = incubation temperature, in °K (°K = °C + 273.15)

B. Calculate CO_2–C flux for the incubation period:

$$F = C_{rate} \times V/A \text{ or } W$$

where

F = C mineralization rate, expressed as μg CO_2–C·cm^{-2}·d^{-1} or μg CO_2–C·g $soil^{-1}$·d^{-1}
C_{rate} = change in CO_2 concentration over the incubation period, expressed as μg CO_2–C·L $headspace^{-1}$·d^{-1}, calculated by regressing C_m versus incubation time (d)
V = Headspace volume of flask (L), calculated as total flask volume less soil volume; soil volume can be calculated from mass and bulk density (BD) (e.g., mL soil volume = g soil × [1/BD])
A = surface area represented by soil in flask, based on soil bulk density values (see Chapter 4, this volume)
W = dry mass equivalent of soil in flask (g)

Long-term Respiration Potential (SOM Pool Sizes)

Soil organic matter (SOM) pool sizes can be inferred from a graph of long-term carbon mineralization versus time with a two-pool model. The C mineralization rate formula noted earlier should be used to calculate a separate carbon mineralization rate for each incubation interval; these rates are then graphed as per Figure 13.1 and fit to the following model using a nonlinear regression procedure such as that available in SAS (SAS Institute 1985).

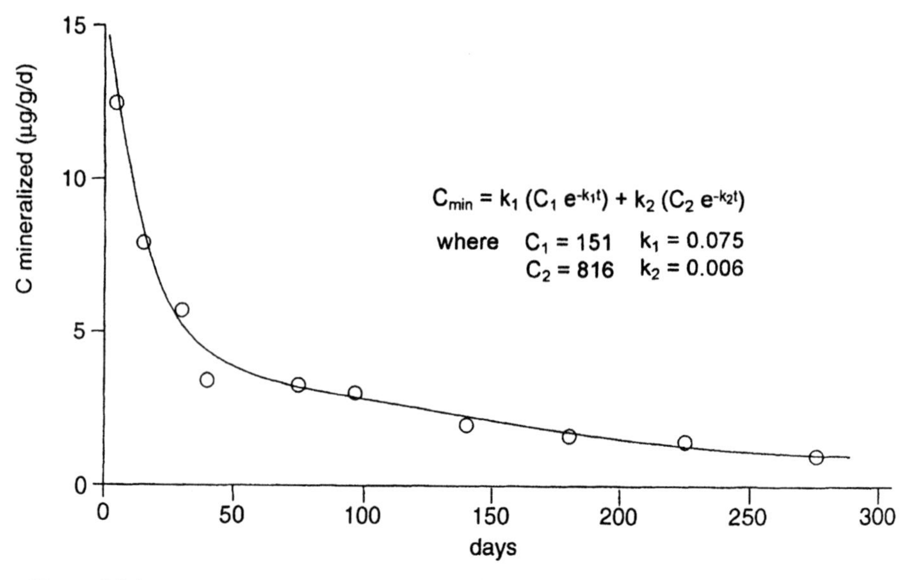

Figure 13.1. Long-term carbon mineralization in a cultivated soil from the KBS LTER site. The solid line is fit with the two-pool nonlinear model as noted in the figure and described in the text ($r^2 = 0.99$).

$$\text{Mineralized C} = k_1(C_1 e^{-k_1 t}) + k_2(C_2 e^{-k_2 t})$$

where

C_1 = carbon content of the active pool
k_1 = rate constant for the active pool
C_2 = carbon content of the intermediate pool
k_2 = rate constant for the intermediate pool
e = the base of the natural logarithms

The third, resistant pool can be very roughly estimated as total C (as measured chemically; see Chapter 5, this volume) less C_1 and C_2. A more accurate estimate of the resistant pool can be made by using acid hydrolysis to define its size (Paul et al. 1998). In this case the intermediate pool (C_2) can then be estimated by difference (C_2 = Total C $- C_3 - C_1$), leaving only three parameters to be estimated by regression. See "HCl-Insoluble Organic Carbon" in Chapter 5 (this volume) to measure the resistant pool by acid hydrolysis.

Special Considerations

If an infrared gas absorption (IRGA) analyzer or gas chromatograph is unavailable for CO_2 determinations, a vial of 1.0 mL of 2.0 mol/L NaOH can be used to trap CO_2 over a 24 hour incubation period. CO_2 captured by the vial can then be determined via titration with 0.1 N HCl as described in Chapter 15, this volume.

^{15}N Isotope Dilution

Where it is important to understand gross rates of nitrogen mineralization or to quantify nitrogen turnover on a relatively short time scale (e.g., following a rain event or other transient disturbances), we recommend the use of ^{15}N isotope dilution techniques as described by Hart et al. (1994). This technique successfully avoids problems of other tracer techniques such as those associated with changes in the size of N-substrate pools (Davidson et al. 1991). The technique is especially useful for exploring links with C cycling and for providing a more accurate picture of the nature and extent of soil N cycling.

In the isotope-dilution technique, a product pool (either NH_4^+ or NO_3^-) is labeled, and the rate of dilution of one or the other of these now-enriched pools with unlabeled NH_4^+ or NO_3^- is taken to indicate the rates of gross nitrogen mineralization or nitrification, respectively. The calculations for these rates—which take into account the simultaneous production and consumption of inorganic-N—were worked out in the early 1950s (Kirkham and Bartholomew 1954).

The specific procedure recommended for LTER sites follows that presented in Hart et al. (1994:990–999). Because the procedure described therein is specific and because this is not a technique that will be commonly used at all sites, we provide no further details here. We do, however, note that a number of laboratories have found spatial variability within individual soil cores to be a major impediment to the interpretation of results using the Hart et al. (1994) protocol, a problem that is commonly reduced by mixing the soils instead of using intact cores. Results should be expressed in the same units as for other nitrogen mineralization processes, i.e., $g\ N \cdot m^{-2} \cdot d^{-1}$ and $mg\ N \cdot kg\ soil^{-1} \cdot d^{-1}$.

Acknowledgments We thank a number of individuals for their helpful discussions and comments on earlier versions of this manuscript, including E. A. Paul, D. D. Myrold, G. Ponciroli, and two thorough, anonymous reviewers. Financial support was provided in part by the KBS LTER project and the Michigan Agricultural Experiment Station.

References

Adams, M. A., and P. M. Attiwill. 1982. Nitrogen mineralization and nitrate reduction in forests soil. *Soil Biology and Biochemistry* 14:197–202.

Binkley, D. 1984. Ion-exchange resin bags for assessing soil N availability: the importance of ion concentration, water regime, and microbial competition. *Soil Science Society of America Journal* 48:1181–1184.

Binkley, D., and S. C. Hart. 1989. The components of nitrogen availability assessments in forest soils. *Advances in Soil Science* 10:57–112.

Bundy, L. G., and J. J. Meisinger. 1994. Nitrogen availability indices. Pages 951–984 *in* R. W. Weaver, J. S. Angle, P. J. Bottomley, D. F. Bezdicek, M. S. Smith, M. A. Tabatabai, and A. G. Wollum, editors, *Methods of Soil Analysis. Part 2, Microbiological and Biochemical Properties.* Soil Science Society of America, Madison, Wisconsin, USA.

Davidson, E. A., S. C. Hart, and M. K. Firestone. 1992. Internal cycling of nitrate in soils of a mature coniferous forest. *Ecology* 73:1148–1156.

Davidson, E. A., S. C. Hart, C. A. Shanks, and M. K. Firestone. 1991. Measuring gross nitrogen mineralization, immobilization, and nitrification by ^{15}N isotopic pool dilution in intact soil cores. *Journal of Soil Science* 42:335–349.

DiStefano, J. F., and H. L. Gholz. 1986. A proposed use of ion exchange resins to measure nitrogen mineralization and nitrification in intact soil cores. *Communications in Soil Science and Plant Analysis* 17:989–998.

Firestone, M. K. 1982. Biological denitrification. Pages 289–326 *in* F. J. Stevenson, editor, *Nitrogen in Agricultural Soils.* American Society of Agronomy, Madison, Wisconsin, USA.

Hart, S. C., and M. K. Firestone. 1989. Evaluation of three *in situ* soil nitrogen availability assays. *Canadian Journal of Forest Research* 19:185–191.

Hart, S. C., J. M. Stark, E. A. Davidson, and M. K. Firestone. 1994. Nitrogen mineralization, immobilization, and nitrification. Pages 985–1018 *in* R. W. Weaver, J. S. Angle, P. J. Bottomley, D. F. Bezdicek, M. S. Smith, M. A. Tabatabai, and A. G. Wollum, editors, *Methods of Soil Analysis. Part 2, Microbiological and Biochemical Properties.* Soil Science Society of America, Madison, Wisconsin, USA.

Hess, T. F., and S. K. Schmidt. 1995. Improved procedure for obtaining statistically valid parameter estimates from soil respiration data. *Soil Biology and Biochemistry* 27:1–7.

Horwath, W. R., and E. A. Paul. 1994. Microbial biomass. Pages 753–774 *in* R. W. Weaver, J. S. Angle, P. J. Bottomley, D. F. Bezdicek, M. S. Smith, M. A. Tabatabai, and A. G. Wollum, editors, *Methods of Soil Analysis. Part 2, Microbiological and Biochemical Properties.* Soil Science Society of America, Madison, Wisconsin, USA.

Jenkinson, D. S., P. B. S. Hart, J. H. Rayner, and L. C. Parry. 1987. Modeling the turnover of organic matter in long-term experiments at Rothamsted. *INTECOL Bulletin* 15:1–8.

Juma, N. G., and E. A. Paul. 1981. Use of tracers and computer simulation techniques to assess mineralization and immobilization of soil nitrogen. Pages 145–154 *in* M. J. Frissel and D. A. vanVeen, editors, *Simulation of Nitrogen Behavior in Soil-Plant Systems.* PUDOC, Wageningen, Netherlands.

Keeney, D. R. 1980. Prediction of soil nitrogen availability in forest ecosystems: a literature review. *Forest Science* 26:159–171.

Kirkham, D., and W. V. Bartholomew. 1954. Equations for following nutrient transformations in soil, utilizing trace data. *Proceedings of the Soil Science Society of America* 18:33–34.

Molina, J. A. E., C. E. Clapp, M. J. Shaffer, F. W. Chichester, and W. E. Larson. 1983. NCSOIL, a model of nitrogen and carbon transformations in soil: description, calibration and behavior. *Soil Science Society of America Journal* 47:85–91.

Myrold, D. D., and J. M. Tiedje. 1986. Simultaneous estimation of several nitrogen cycle rates using 15N: theory and application. *Soil Biology and Biochemistry* 18:559–568.

Nadelhoffer, K. J. 1990. A microlysimeter for measuring nitrogen mineralization and microbial respiration in aerobic soil incubations. *Soil Science Society of America Journal* 54:411–415.

Palm, C., G. P. Robertson, and P. M. Vitousek. 1993. Nitrogen availability. Pages 158–163 *in* J. M. Anderson and J. S. I. Ingram, editors, *Tropical Soil Biology and Fertility: A Handbook of Methods. 2d edition.* CAB International, Oxford, UK.

Parton, W. J., D. S. Schimel, C. V. Cole, and D. S. Ojima. 1987. Analysis of factors controlling soil organic matter levels in Great Plains grasslands. *Soil Science Society of America Journal* 51:1173–1179.

Paul, E. A., H. P. Collins, D. Harris, U. Schulthess, and G. P. Robertson. 1998. The influence of biological management inputs on carbon mineralization in ecosystems. *Applied Soil Ecology* 3:1–13.

Paustian, K., W. J. Parton, and J. Persson. 1992. Modeling soil organic matter in organic-amended and nitrogen-fertilized long-term plots. *Soil Science Society of America Journal* 56:476–488.

Raison, R., M. Connell, and P. Khanna. 1987. Methodology for studying fluxes of soil mineral-N *in situ*. *Soil Biology and Biochemistry* 19:521–530.

Robertson, G. P. 1982a. Factors regulating nitrification in primary and secondary succession. *Ecology* 63:1561–1573.

Robertson, G. P. 1982b. Nitrification in forested ecosystems. *Philosophical Transactions of the Royal Society of London* 296:445–457.

Robertson, G. P., and J. M. Tiedje. 1985. Denitrification and nitrous oxide production in successional and old growth Michigan forests. *Soil Science Society of America Journal* 48:383–389.

SAS-Institute. 1985. *SAS User's Guide: Statistics, Version 5*. SAS Institute, Cary, North Carolina, USA.

Schmidt, E. L., and L. W. Belser. 1994. Autotrophic nitrifying bacteria. Pages 159–177 *in* R. W. Weaver, J. S. Angle, P. J. Bottomley, D. F. Bezdicek, M. S. Smith, M. A. Tabatabai, and A. G. Wollum, editors, *Methods of Soil Analysis. Part 2, Microbiological and Biochemical Properties*. Soil Science Society of America, Madison, Wisconsin, USA.

Stanford, G., and S. J. Smith. 1972. Nitrogen mineralization potentials of soils. *Soil Science Society of America Journal* 36:465–472.

Tabatabai, M. A. 1994. Soil enzymes. Pages 775–826 *in* R. W. Weaver, J. S. Angle, P. J. Bottomley, D. F. Bezdicek, M. S. Smith, M. A. Tabatabai, and A. G. Wollum, editors, *Methods of Soil Analysis. Part 2, Microbiological and Biochemical Properties*. Soil Science Society of America, Madison, Wisconsin, USA.

Tiedje, J. M. 1982. Denitrification. Pages 1011–1026 *in* A. L. Page, editor, *Methods of Soil Analysis. Part 2, Chemical and Microbiological Properties*. American Society of Agronomy, Madison, Wisconsin, USA.

Waring, S. A., and J. M. Bremner. 1964. Effect of soil mesh-size on the estimation of mineralizable nitrogen. *Nature* 202:1141.

Zou, X., D. W. Valentine, R. J. J. Stanford, and D. Binkley. 1992. Resin-core and buried-bag estimates of nitrogen transformations in Costa Rica lowland forests. *Plant and Soil* 139:275–283.

14

Denitrification

Peter M. Groffman
Elisabeth A. Holland
David D. Myrold
G. Philip Robertson
Xiaoming Zou

Denitrification is the reduction of the nitrogen oxides, nitrate (NO_3^-) and nitrite (NO_2^-), to the gases nitric oxide (NO), nitrous oxide (N_2O), and dinitrogen (N_2). The process is carried out mainly by facultative anaerobes, i.e., organisms that normally use oxygen (O_2) to accept electrons during respiration but in its absence can use nitrogen oxides as electron acceptors. Most denitrifying bacteria are heterotrophs, using organic carbon compounds as a source of energy.

Denitrification is important in ecosystems for several reasons. First, removal of inorganic nitrogen by denitrification can influence the productivity of plants because their growth is frequently limited by nitrogen (Vitousek and Howarth 1991). Second, denitrification is important to water quality. Nitrate is a federally listed drinking water pollutant (Keeney 1987) and is an agent of eutrophication in marine ecosystems (Ryther and Dunstan 1971). Denitrification in soils, wetlands, streams, and groundwater can prevent movement of NO_3^- from intensive upland land uses into aquatic ecosystems. Third, N_2O, one of the gaseous products of denitrification, is a "greenhouse" gas that can influence the earth's radiative budget and plays a role in stratospheric ozone destruction (Prather et al. 1995). Finally, anaerobic metabolism is responsible for a significant portion of energy flow in many soils and wetlands. Denitrification is the most energetically favorable form of anaerobic metabolism, allowing for rates of energy generation close to those in aerobic metabolism (Thauer et al. 1977), and thus is essential to the overall microbial function of anaerobic (i.e., wet) soils.

Denitrification is a difficult process to measure. Methods for measuring denitrification are flawed because they either change substrate concentrations, disturb the soil physical environment, lack sensitivity, or are prohibitively costly in time and expense (Tiedje et al. 1989). The quantification of denitrification has also been hindered by high spatial and temporal variation in the field. This variation is especially problematic given the lack of methods amenable to the collection of large numbers of samples with reasonable expenditures of time and money.

Available Protocols

The difficulty with measuring denitrification stems from the fact that it is hard to quantify either the production of the terminal end product of denitrification (N_2) or the specific depletion of the substrate (i.e., NO_3^-). It is difficult to measure production of N_2 because of the already high atmospheric concentration of this gas. It is difficult to quantify denitrification by measuring decreases in NO_3^- because this ion is also consumed by plants, heterotrophic microbes, dissimilatory reduction to NH_4^+, leaching, and runoff, and is produced by nitrification.

Like many biological processes, denitrification exhibits high spatial and temporal variability. Rates of denitrification in the field frequently vary over two or three orders of magnitude in both time and space in a wide variety of environments (e.g., Foloronuso and Rolston 1984; Robertson and Tiedje 1985, 1988; Burton and Beauchamp 1985; Robertson et al. 1988; Parkin 1987; Starr et al. 1995). Total soil denitrification is often dominated by very high rates of activity in very small activity centers (hot spots) where O_2 is low and NO_3^- and carbon availability are high (Parkin 1987; Christensen et al. 1990; Murray et al. 1995). High variability hinders quantification of field rates, comparisons of treatments, and evaluation of different methods.

The Acetylene (C_2H_2) Inhibition Method

A major development in denitrification research was the discovery that acetylene inhibits the reduction of N_2O to N_2 (Balderston et al. 1976; Yoshinari and Knowles 1976), making N_2O the terminal product of denitrification. Quantifying denitrification by measuring production of N_2O in the presence of acetylene is relatively easy because of the low atmospheric concentration of N_2O and the availability of sensitive detectors for this gas. Since 1980 acetylene inhibition has been the most common method used to quantify denitrification (Tiedje et al. 1982; Keeney 1986; Tiedje et al. 1989; Nieder et al. 1989; von Rheinbaben 1990; Payne 1991; Aulakh et al. 1992).

Although acetylene-based methods have been widely applied, they have serious drawbacks. Perhaps the most critical problem is that acetylene inhibits the production of NO_3^- via nitrification (Hynes and Knowles 1978; Walter et al. 1979; Mosier 1980). Inhibition of nitrification can lead to underestimation of denitrification rates as NO_3^- pools become depleted during incubations in the presence of acetylene. This problem is especially critical in natural ecosystems, where NO_3^- pools are often inherently low.

Other (less critical) problems with acetylene methods arise from the difficulty of getting acetylene to diffuse to active denitrification sites in soil (Ryden et al. 1979;

Jury et al. 1982; Parkin et al. 1984); from the effects of acetylene on soil carbon metabolism (Yeomans and Beauchamp 1982; Terry and Duxbury 1985; Topp and Germon 1986; Flather and Beauchamp 1992); from the inhibition of chemoautotrophic oxidations (in addition to nitrification) that can provide energy to denitrifiers (Payne 1984); and from the contamination of acetylene with other gases that can affect denitrification (Hyman and Arp 1987; Gross and Bremner 1992). A more critical (but quantifiable) problem is the failure of the inhibition of N_2O reduction at low NO_3^- concentrations (Oremland et al. 1984; Slater and Capone 1989; Seitzinger et al. 1993).

Direct Flux Methods

Although it is difficult to directly measure the fluxes of denitrification substrates or products as discussed earlier, direct flux techniques have application in certain cases. These techniques are particularly useful in situations where the use of acetylene is inappropriate.

Measuring depletion of NO_3^- can be used as a quantification of denitrification when other possible fates of NO_3^- have been either measured or eliminated. These techniques are thus limited to specific laboratory applications.

Production of nitrogen gases has been measured to quantify denitrification in laboratory studies with artificial atmospheres (e.g., without N_2; Seitzinger et al. 1980, 1993; Swerts et al. 1995). The problem inherent in all direct N_2 flux methods is reducing the background level of N_2 gas sufficiently that rates of N_2 production by denitrification are detectable. This can require unacceptably long preincubation times and/or complex laboratory equipment (Aulakh et al. 1991).

Devol (1991) developed a technique for direct field measurement of N_2 production from marine sediments. This technique is based on measuring the accumulation of N_2 gas dissolved in water within a field chamber placed on the sediment surface and may be applicable to flooded soils.

A final approach to direct flux measurement of denitrification is the quantification of changes in Ar:N_2 ratios, i.e., a decrease in this ratio is used as evidence of denitrification (Wilson et al. 1990; Martin et al. 1995). These methods are not very sensitive unless a mass spectrometer is used to quantify the Ar:N_2 ratio. A mass spectrometer can also be used for very sensitive direct measurement of N_2 production (Thomas and Lloyd 1995).

It is important to note that all direct N_2 flux methods are rather cumbersome, limiting the number of replicate samples that can be run at any one time. This limitation is important given the high spatial and temporal variability of denitrification.

^{15}N Balance Methods

Balance methods are based on tracing the movement of $^{15}NH_4^+$ or $^{15}NO_3^-$ into different ecosystem pools and processes (plants, volatilization, leaching, runoff, soil inorganic and organic pools). In these methods, denitrification is quantified as the ^{15}N unaccounted for at the end of the experiment (Rolston et al. 1979; Parkin et al. 1985; Mosier et al. 1986). This estimate of "unaccounted for N" includes the accumulated errors associated with estimates of the other pools and processes and is thus not very accurate or precise. The precision of this estimate is also limited by how well other loss processes (leaching, runoff, volatilization) are controlled or quantified.

In addition to accumulated error problems, there are also questions about how well added ^{15}N simulates the behavior of soil N. Although added inorganic ^{15}N is likely a good surrogate for fertilizer N, it may not be a good tracer of nitrogen in soil organic matter and microbial biomass. A final concern, common to ^{15}N methods in general, is that addition of ^{15}N can significantly enrich the nitrogen pools under study, leading to artificially high rates of activity.

$^{15}N_2$ Flux Techniques

Techniques have been developed to trace the movement of ^{15}N added to soil into gaseous denitrification products. These techniques generally require high enrichment of soil inorganic nitrogen pools with ^{15}N and thus have primarily been used in situations where soil nitrogen levels are already relatively high (Siegel et al. 1982; Mulvaney 1984; Mosier et al. 1986). However, recent improvements in mass spectrometer techniques have made it possible to make measurements with very low, tracer-level additions of ^{15}N (Brooks et al. 1993). As with ^{15}N balance approaches, although $^{15}N_2$ flux methods can reliably trace fertilizer-derived fluxes, their ability to depict fluxes of nitrogen associated with soil organic matter turnover is less certain (Nielsen 1992).

Perhaps the most important constraint on the use of $^{15}N_2$ flux techniques is that they are expensive and time-consuming. Despite active research in this area (Arah et al. 1993; Avalakki et al. 1995), costs of ^{15}N and for mass spectrometer analysis are high, and sample preparation and collection techniques are time-consuming. These cost and time constraints limit the number of flux measurements that can be made, which is a serious problem given the high spatial and temporal variability of denitrification.

Sample Type—Cores Versus Chambers

The need to add acetylene to soil in a controlled atmosphere motivated the use of extracted soil cores in denitrification research. However, the use of cores creates disturbance effects that are difficult to quantify. The alternative to extracted cores is a chamber method, where chambers are placed over the soil surface and the accumulation of N_2O is measured in the air space under the chamber or in a stream of air circulating through the chamber. A variety of methods have been developed for introducing acetylene to infield chambers (Ryden et al. 1979; Burton and Beauchamp 1984; Hallmark and Terry 1985). The main advantage of chamber methods is that they allow for infield measurement of actual fluxes of nitrogen gases from soil to the atmosphere.

There are several problems with chamber methods for measuring denitrification. Physical processes that inhibit diffusion (e.g., wet and/or fine-textured soils) inhibit the movement of acetylene and N_2O into and out of sites of denitrification activity in soil. Jury et al. (1982) reported that several weeks of monitoring may be required to accurately assess production of nitrogen gases associated with a particular rainfall or irrigation event. Gas diffusion problems can be easily overcome with core methods, however, either by using forced-air-flow recirculation systems (Parkin et al. 1984) or by thorough mixing of the air space of the soil core, e.g., with a large syringe (Robertson et al. 1987; Groffman and Tiedje 1989). Other problems with chambers relating to pressure, concentration, and temperature changes within the chamber can be accounted for with proper chamber design (Mosier 1989; see Chapter 10, this volume).

Detailed comparisons of core versus chamber approaches have shown that cores produce accurate measurements of soil-atmosphere gas fluxes, except when cores are held for long periods (several days) before incubation (Burton and Beauchamp 1984; Ryden et al. 1987; Aulakh et al. 1991; Dunfield et al. 1995). Ryden et al. (1987) found a very strong relationship between denitrification rates in cores versus chambers, over a wide range of denitrification rates, during 24 hour incubations. In very wet soils, cores were superior to chambers due to the difficulty of introducing acetylene into, and slow diffusion of N_2O out of, these soils. An additional advantage of cores is that it is possible to run numerous core incubations cheaply and quickly, whereas chamber measurements can be expensive and time-consuming, limiting the number of replicates and/or sites that can be analyzed. Dunfield et al. (1995) found that extracted cores produced very similar estimates of soil-atmosphere N_2O and CH_4 fluxes as in-field chamber and soil gas concentration/diffusion flux methods.

Measurement of Denitrification Potentials

The high variability and methodological problems associated with measuring denitrification have led many investigators to resort to measures of denitrification potential. A variety of measures of denitrification potential have been made, where amendments are used, frequently under slurried, laboratory conditions, to increase rates of denitrification above those occurring in nature and to reduce the variability of the process.

Of all measurements of denitrification potential, the assay of denitrification enzyme activity (DEA) developed by Smith and Tiedje (1979) is the most common. In this assay, all limiting factors of denitrification (O_2, NO_3^-, C) are present in excess, growth is inhibited (by addition of chloramphenicol), and the nitrogen gas production measured (usually N_2O in the presence of acetylene) is a function only of the level of enzyme present in the sample.

It was originally hoped that DEA would be strongly related to actual denitrification activity because, in culture at least, the denitrifying enzymes are strictly inducible (Payne 1981). However, DEA has been found to be poorly related to hourly or daily denitrification rates due to the persistence of viable but inactive enzymes in soil (Smith and Parsons 1985; Groffman 1987; Martin et al. 1988; Parsons et al. 1991). However, the DEA assay has proven very useful for comparison of soils, ecosystems, and treatments because it responds well to longer-term variation in the factors that control denitrification (soil water, NO_3^- availability, carbon availability).

Recommended Protocols

Although many studies have compared different methods of measuring denitrification, there have been few conclusive results (Tiedje et al. 1982; Keeney 1986; Tiedje et al. 1989; Nieder et al. 1989; Payne 1991; Aulakh et al. 1991, 1992; Beauchamp and Bergstrom 1993; Tiedje 1994; Mosier and Klemedtsson 1994). In most cases, high variability has made it difficult to determine differences among techniques. As a result, it is difficult to produce a "consensus" recommended protocol.

We recommend two approaches for assessing denitrification, one for quantification of denitrification potential (DEA) and one for measurement of actual denitrification nitrogen flux (an acetylene-based, static core method). Although we have a high degree of confidence and consensus about the DEA method for quantifying denitrification potential, our recommendation for quantification of actual denitrification nitrogen flux comes with considerable reservations given the problems with acetylene-based methods described earlier. Our recommendation is based on the fact that many studies have used this core method, in a wide range of ecosystems, and several methodological comparisons/validations have been performed (Burton and Beauchamp 1985; Tiedje et al. 1989; Christensen et al. 1991; Aulakh et al. 1991; Groffman et al. 1993b). The method was designed to allow for large numbers of samples to be run simultaneously, and it is thus suitable for ecosystem and landscape-scale studies. However, investigators should be aware of the problems with this method and should be alert for new methodological developments. We did not select a chamber-based method because the problems with chambers (described previously), especially the fact that the number of chamber incubations that can be run at one time is relatively small, outweigh their advantages.

Denitrification Potentials—Denitrification Enzyme Activity

The objective of the denitrification enzyme assay is to measure the maximum activity of the biomass of enzymes present in soil at the time of sampling. In this assay all limiting factors of denitrification (O_2, NO_3^-, C) are removed, growth is inhibited (by the addition of chloramphenicol), and the nitrogen gas produced is measured as the accumulation of N_2O in the presence of acetylene.

Materials

1. Flasks that can be sealed with airtight stoppers, e.g., Corning no. 5020 125 mL Erlenmeyer flasks with an Aldrich no. z12468-0 25.5 mm rubber septa
2. Media capable of providing NO_3^- (100 mg N kg^{-1}), dextrose (40 mg kg^{-1}) and chloramphenicol (10 mg kg^{-1}). The concentration of the media will vary depending on expected activity (see later).

3. Purified acetylene. Commercially available "laboratory" or "welding" grade acetylene can be purified to remove acetone and other contaminants by passing it through two concentrated H_2SO_4 traps and a distilled water trap in sequence. Protocols for this purification are described in detail in Hyman and Arp (1987) and in Gross and Bremner (1992). Relatively clean acetylene can also be produced by adding water to calcium carbide (CaC_2) in an evacuated flask. Water reacts with the CaC_2 to produce acetylene. Caution must be taken to avoid adding too much water to a large amount of CaC_2 because the reaction can be explosive.
4. A gas manifold capable of evacuation (700 mm Hg) and flushing with an O_2-free gas such as N_2
5. A rotary shaker table capable of maintaining 125 rpm
6. Syringes (disposable, 1, 5, or 10 mL) to add acetylene to flasks and to take gas samples from flasks
7. Airtight storage vials for gas samples and standards. Investigators have used a variety of vials to store gas samples, including commercially available blood collection tubes (e.g., Vacutainer or Venoject), headspace autosampler vials, and polypropylene syringes. With any vials, there can be contamination, leakage, or absorption problems that should always be checked for with blanks and spikes. See Chapter 10, this volume, for more detail on these problems.
8. A gas chromatograph equipped with an electron capture detector

Procedure

1. Weigh sieved field moist soil samples (two to three analytical replicates per sample) into flasks and add media (e.g., 25 g soil, 25 mL media). The weight of soil and the amount of media are varied (by trial and error) with the expected activity of the samples. The objective of this variation is to ensure that N_2O concentrations in the headspace of the flask stay within the range of the standard curve used in the gas chromatographic analysis (e.g., 0.3–50 ppm).
2. Seal flasks with stoppers and make soils anaerobic by repeated evacuation and flushing with oxygen-free gas (e.g., N_2 or Ar). We recommend at least three cycles of flushing (1 minute) followed by evacuation to 700 mm Hg vacuum. Flasks should then be brought to atmospheric pressure.
3. Add acetylene to 10% of the volume of the headspace of the flask. Incubating slightly pressurized flasks prevents contamination with laboratory air during sampling and the development of negative pressure in the flasks from sample removal.
4. Incubate the flasks at 125 rpm on a rotary shaker at constant temperature.
5. Take gas samples at 30 and 90 minutes and store them in evacuated, airtight storage vials. A 60 minute sample is recommended but optional.
6. Analyze gas samples for N_2O by gas chromatography. The most common method is to use an electron capture detector at 350 °C with a Porapak Q, 80/100-mesh column (2 m × 0.32 cm), with a carrier gas of 95% Ar/5% CH_4 at a flow rate of between 10 and 40 mL m^{-1}, with an oven temperature of be-

tween 25 and 50 °C. See Chapter 10, this volume, for more details on N_2O analysis.

Calculations

The basic calculation to quantify the amount of N_2O produced by the soil involves multiplying the concentration of N_2O in the headspace of the flask at 30 and 90 minutes by the volume of the headspace and then dividing by the dry weight of soil:

$$DR = [(C_{90} \times H) - (C_{30} \times H)]/(D \times T)$$

where

DR = denitrification rate, expressed as μg N·kg soil^{-1}·h^{-1}
C_{30} = N_2O concentration at 30 minutes, expressed as μg N_2O–N/L headspace (see Chapters 10 or 13, this volume, for formula to convert ppm$_v$ or μL N_2O/L headspace to μg N_2O–N/L headspace)
C_{90} = N_2O concentration at 90 minutes, expressed in same way as C_{30}
H = flask headspace volume (it is necessary to account for removal of air by sampling). Volume (L) can be calculated as total flask volume less added media volume less soil volume. Soil volume can be calculated based on bulk density.
D = soil dry weight
T = time (duration) of incubation, expressed as h, e.g., 1 h for samples taken at 30 and 90 minutes

It is necessary to account for N_2O dissolved in solution using Bunsen coefficients that predict the amount of gas dissolved in the liquid phase from the concentration in the gas phase (Moraghan and Buresh 1977; Wilhelm et al. 1977):

$$M = C_g \times (V_g + V_l \times \beta)$$

where:

M = total amount of N_2O in the water plus gas phase
C_g = concentration of N_2O in the gas phase
V_g = volume of the gas phase
V_l = volume of liquid phase
β = Bunsen coefficient (1.06 at 05 °C; 0.882 at 10 °C; 0.743 at 15 °C; 0.632 at 20 °C; 0.544 at 25 °C; 0.472 at 30 °C)

In a shaken assay such as this, it is safe to assume that liquid- and gas-phase N_2O are in equilibrium (i.e., that the Bunsen coefficients are accurate). Total N_2O production values can be converted to an areal basis using bulk density values (see Chapter 4, this volume).

Special Considerations

1. Sampling depth varies with site and experimental objectives. For site comparison work, it is important to sample the soil profile to encompass the most

biologically active zone of the soil (e.g., 0–20 cm). As with all biological activities, activity can be highly stratified in the soil profile, with the 0–2 cm or 0–5 cm depth having much higher activity than lower depths. On the other hand, low, but significant, activity can occur to relatively great depth in the soil profile (e.g., 2 or 3 m in some tropical soils), which in aggregate can be more important than surface soil activity.

2. Recently, there has been concern that chloramphenicol may inhibit the activity of existing denitrification enzymes (Brooks et al. 1992); the effect varies with soil type (Wu and Knowles 1995; Pell et al. 1996). We recommend periodically testing for this effect by running very short term (30 minute) assays with and without chloramphenicol. This testing is especially important for comparisons across different experimental sites.
3. Analytical variability (i.e., variation of samples taken from the same bag of well-mixed soil) for the DEA assay ranges from 10% to 20%. Field variability (i.e., variation of different samples from the same plot) ranges from 25% to 75%.
4. Temporal variability of DEA is much less than for actual denitrification rate. In north temperate forest ecosystems with well-distributed rainfall, six to eight sample dates during the snow-free season are sufficient to characterize this variability. In ecosystems with more marked seasonal changes in moisture (e.g., tropical dry forests), sampling should be stratified by season.
5. Sampling should not be done within 3–5 days of drying and rewetting events if possible (Groffman and Tiedje 1988).

Actual Denitrification Rate

Our recommended "static core" method has been used for ecosystem and landscape-scale denitrification studies for over 10 years (Robertson and Tiedje 1984; Groffman 1985; Robertson et al. 1987, 1988; Myrold 1988; Tiedje et al. 1989; Groffman and Tiedje 1989; Groffman et al. 1993a; Hanson et al. 1994). In this method, 2 cm diameter × 15 cm long intact soil cores are taken in acrylic sleeves and sealed with rubber serum stoppers at both ends. The headspace of the cores is sampled at various time intervals to quantify the accumulation of gases. A pressure transducer is used to quantify headspace volume and to check for leakage of each core. This sampling design allows for highly replicated measurement of denitrification rates and related variables (water content, NO_3^- levels, porosity) on the same samples.

Materials

1. A 2 cm diameter punch auger capable of holding acrylic tube inserts. Several companies manufacture punch augers that hold 2.54 cm diameter acrylic tubing, although the tubing they sell with these samplers is often very thin-walled and not gastight. Gastight, more durable acrylic tubing usually can be purchased from local suppliers. Custom-made samplers, or 2 cm diameter "tube" or "Oakfield" samplers can also be used. Larger-diameter cores may produce

less variable estimates of denitrification rate in some cases (Parkin et al. 1987; Starr et al. 1995). However, sampling with large cores is much more labor-intensive.

2. Rubber stoppers capable of providing an airtight seal in the acrylic tubes. We use Aldrich no. z12468-0 25.5 mm rubber septa.
3. Purified acetylene (as described earlier)
4. Syringes for adding acetylene to core tubes (5 or 10 mL), for mixing acetylene into the soil core (30 or 60 mL), and for removing gas samples from the core tubes (5 or 10 mL).
5. Airtight storage vials for gas samples and standards (as described earlier)
6. A pressure transducer capable of measuring pressure changes induced by a 5 or 10 mL addition of air to the headspace of the core/tubes
7. A gas chromatograph equipped with an electron capture detector

Procedure

1. Intact soil cores (0–15 cm depth) are taken directly into, or are inserted into, the acrylic tubes. We recommend taking 10–20 replicate core samples from each field plot for a maximum of 200 cores per sampling date. Core tubes should be stoppered at the bottom only and stored upright. Incubations should be initiated within 24 hours, although some studies (Breitenbeck and Bremner 1987; Parkin et al. 1984) have shown that intact soil cores can be stored at 4 °C for up to 30 days without a significant effect on denitrification. Such stability cannot be assumed for any given soil, however.
2. To begin the incubation, cores should be sealed with rubber stoppers. Acetylene (to at least 10% of the volume of the headspace) should be added to the headspace of each core and mixed into the soil pores by repeated pumping with a 30 or 60 mL syringe.
3. The cores should be incubated at constant, field soil temperatures and sampled at least twice. For example, cores can be incubated for 6 hours, with duplicate gas samples removed from the headspace after 2 and 6 hours, or single samples removed at different times over the 6 hour incubation period. The headspace of the core should be mixed by repeated pumping with a syringe prior to each sampling. Note that it is important to account for the amount of air removed by each sampling. If the headspace is small relative to the sample, negative pressure develops in the headspace.

   The rate of N_2O production between 2 and 6 hours is taken as the rate of denitrification. The 2 hour lag period before initial sampling ensures that acetylene has diffused into soil pores. It is necessary to run time-course experiments to determine that rates of gas production between the initial and final samples are linear (Fig. 14.1). The final length of the incubation should be chosen based on consideration of depletion of soil O_2 levels or the NO_3^- pool (which motivates a shorter incubation), the detection of low rates of activity (which motivates a longer incubation), and convenience (e.g., time of day). Depletion of the NO_3^- pool results in a decrease in denitrification rate, and depletion of soil O_2 levels can result in an increase in rate (Fig. 14.1).

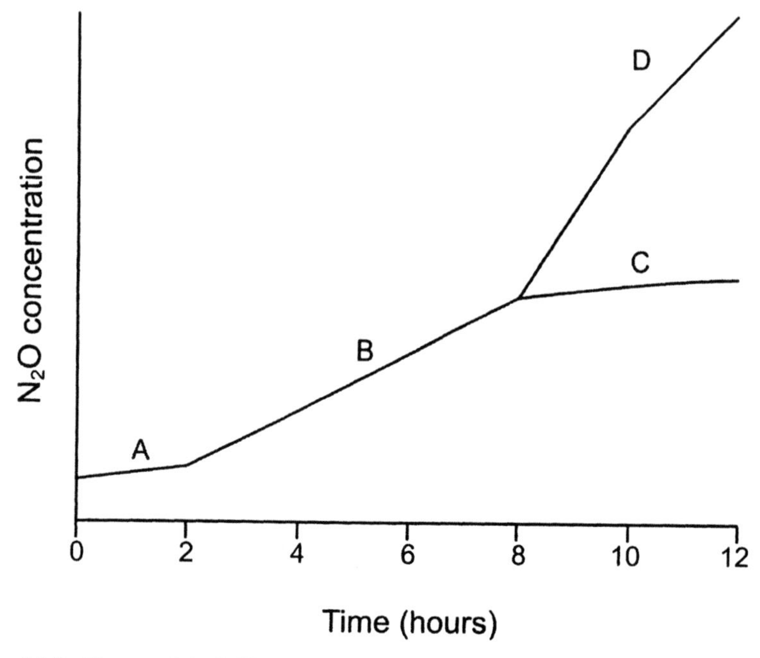

Figure 14.1. Phases of denitrification during intact, static core incubation: A—lag phase while C_2H_2 diffuses into soil pores; B—linear phase; C—NO_3^- depletion phase; D—O_2 depletion phase. Time-course experiments must be performed on subset of samples to ensure that denitrification rates are calculated using data from the linear phase.

4. Store gas samples, blanks, and standards in airtight storage vials and analyze for N_2O as described earlier.
5. Following incubation, cores should be weighed and measured for area and bulk density calculations. The internal headspace volume of each core can be measured with a pressure transducer (Parkin et al. 1984) calibrated to produce volume estimates from pressure changes induced by an addition of 5 or 10 mL of air to the headspace of the core tube. This procedure also facilitates testing for leaks. Alternatively, headspace can be calculated by calculating the volume of the empty tube and subtracting the volume of the soil core (accounting for its pore space and water content).
6. Cores should be processed for soil water content and inorganic nitrogen (see Chapters 3 and 5, this volume).

Calculations

The basic calculation involves quantifying the amount of N_2O produced by the soil by multiplying the concentration of N_2O in the headspace of the core at 2 and 6 hours (or whatever sampling times are used) by the volume of the headspace and then dividing by the dry weight of soil or the surface area of the core. Results are commonly expressed as $\mu g\ N\ kg^{-1}\ d^{-1}$ or as $\mu g\ N\ ha^{-1}\ d^{-1}$:

$$DR = [(C_2 \times H) - (C_1 \times H)]/(D \times T)$$

where

DR = denitrification rate, expressed as $\mu g\ N \cdot kg^{-1} \cdot d^{-1}$ or $\mu\ N \cdot ha^{-1} \cdot d^{-1}$
C_1 = N_2O concentration at the first sampling time, expressed as $\mu g\ N_2O$–N/L headspace (see Chapters 10 or 13, this volume, for formula to convert ppm_v or $\mu L\ N_2O$/L headspace to $\mu g\ N_2O$–N/L headspace)
C_2 = N_2O concentration at the second sampling time
H = core headspace volume (L) (it is necessary to account for removal of air by sampling)
D = soil dry mass equivalent (kg) or core surface area (ha)
T = time between sampling points (d), e.g., 0.17 d for samples removed at 2 and 6 hours

If the headspace is sampled at multiple times during the incubation, the numerator of the equation can be replaced by a regression of N_2O concentration with time ($\mu g\ N_2O$–$N \cdot L$ headspace$^{-1} \cdot d^{-1}$). It is necessary to account for N_2O dissolved in solution using Bunsen coefficients as described earlier for DEA. In intact cores, it is not always safe to assume that liquid- and gas-phase N_2O are in equilibrium (e.g., N_2O is often supersaturated in soil water), but this is usually a small error unless soils are very wet.

Results can be expressed on an areal basis either by using the bulk density or the surface area of the cores. It is also possible to calculate water or air-filled pore space on each core using bulk density and soil water values (see Chapter 4, this volume).

Estimates of annual or seasonal denitrification nitrogen flux can be produced by extrapolating measured rates over the intervals between sampling dates, i.e., assuming that rates at a sampling date are representative of some period before and/or after that date. The validity of these extrapolations is controlled by sampling frequency and the spatial and temporal variability of the measured rates.

Special Considerations

1. Given the earlier discussion about the depth distribution of activity above (see the section "Special Considerations" for the denitrification potential protocol, above), it may be important to take cores from depths greater than 0–15 cm in some cases. It may also be appropriate to take shallower cores as well.
2. It is impossible to quantify analytical variability of an "intact core" method because cores cannot be subdivided. However, taking multiple samples of the headspace during the incubation allows for evaluation of the analytical variability of the headspace N_2O analysis, which ranges from 5% to 15%. Field variability with this method ranges from 50% to 200%.
3. Knowledge about spatial and temporal dynamics of water, nitrogen, and carbon fluxes in a particular system should be used to design optimal sampling strategies for denitrification. Transitions between cold and warm or between dry and wet seasons are often periods of high denitrification because plants do not dominate water and nitrogen dynamics during these periods. In many

ecosystems, denitrification is most vigorous outside of the plant growing season. Drying and rewetting and freezing and thawing events have also been found to stimulate denitrification. Activity may be significant in unfrozen soils under a snowpack.

Ancillary Data

Ancillary data valuable for interpreting spatial and temporal variation in denitrification include soil temperature, moisture, and NO_3^- content, air-filled pore space, soil respiration, texture, organic matter content, NH_4^+ content and pH, vegetation type and productivity, microbial biomass, and mineralization and nitrification rates. Soil moisture and NO_3^- content are essential ancillary data for the intact core method. Denitrification data are frequently lognormally distributed. Approaches for analyzing such data are described by Parkin and Robinson (1992).

Acknowledgments The authors wish to thank William Peterjohn, Timothy Parkin, and two anonymous reviewers for helpful suggestions to improve this manuscript.

References

Arah, J. R. M., I. J. Crichton, and K. A. Smith. 1993. Denitrification measured directly using a single-inlet mass spectrometer and by acetylene inhibition. *Soil Biology and Biochemistry* 25:233–238.

Aulakh, M. S., J. W. Doran, and A. R. Mosier. 1991. Field evaluation of four methods for measuring denitrification. *Soil Science Society of America Journal* 55:1332–1338.

Aulakh, M. S., J. W. Doran, and A. R. Mosier. 1992. Soil denitrification: significance, measurement, and effects of management. *Advances in Soil Science* 18:1–52.

Avalakki U. K., W. M. Strong, and P. G. Saffigna. 1995. Measurement of gaseous emissions from denitrification of applied nitrogen-15. I. Effect of cover duration. *Australian Journal of Soil Research* 33:77–87.

Balderston, W. L., B. Sherr, and W. J. Payne. 1976. Blockage by acetylene of nitrous oxide reduction in *Pseudomonas perfectomarinus. Applied and Environmental Microbiology* 31:504–508.

Beauchamp, E. G., and D. W. Bergstrom. 1993. Denitrification. Pages 351–357 *in* M. R. Carter, editor, *Soil Sampling and Methods of Analysis.* Lewis Publishers, Boca Raton, Florida, USA.

Breitenbeck, G., and J. M. Bremner. 1987. Effects of storing soils at various temperatures on their capacity for denitrification. *Soil Biology and Biochemistry* 19:377–380.

Brooks, M. H., R. L. Smith, and D. L. Macalady. 1992. Inhibition of existing denitrification enzyme activity by chloramphenicol. *Applied and Environmental Microbiology* 58: 1746–1753.

Brooks, P. D., D. J. Herman, G. J. Atkins, S. J. Prosser, and A. Barrie. 1993. Rapid, isotopic analysis of selected soil gases at atmospheric concentrations. Pages 193–202 *in Agricultural Ecosystem Effects on Trace Gases and Global Climate Change.* Soil Science Society of America, Madison, Wisconsin, USA.

Burton, D. L., and E. G. Beauchamp. 1984. Field techniques using the acetylene blockage of nitrous oxide reduction to measure denitrification. *Canadian Journal of Soil Science* 64:555–562.

Burton, D. L., and E. G. Beauchamp. 1985. Denitrification rate relationships with soil parameters in the field. *Communications in Soil Science and Plant Analysis* 16:539–549.

Christensen, S., P. Groffman, A. Mosier, and D. R. Zak. 1991. Rhizosphere denitrification: a minor process but indicator of decomposition activity. Pages 199–211 *in* N. P. Revsbech and J. Sorenson, editors, *Denitrification in Soils and Sediments.* Plenum Press, New York, New York, USA.

Christensen, S., S. Simkins, and J. M. Tiedje. 1990. Spatial variation in denitrification: dependence of activity centers on the soil environment. *Soil Science Society of America Journal* 54:1608–1613.

Devol, A. H. 1991. Direct measurement of nitrogen gas fluxes from continental shelf sediments. *Nature* 349:319–321.

Dunfield, P. F., E. Topp, C. Archambault, and R. Knowles. 1995. Effect of nitrogen fertilizers and moisture content on CH_4 and N_2O fluxes in a humisol: measurements in the field and intact soil cores. *Biogeochemistry* 29:199–222.

Flather, D. H., and E. G. Beauchamp. 1992. Inhibition of the fermentation process in soil by acetylene. *Soil Biology and Biochemistry* 24:905–911.

Foloronuso, O. A., and D. E. Rolston. 1984. Spatial variability of field-measured denitrification gas fluxes and soil properties. *Soil Science Society of America Journal* 49:1087–1093.

Groffman, P. M. 1985. Nitrification and denitrification in conventional and no-tillage soils. *Soil Science Society of America Journal* 49:329–334.

Groffman, P. M. 1987. Nitrification and denitrification in soil: a comparison of incubation, enzyme assay and enumeration techniques. *Plant and Soil* 97:445–450.

Groffman, P. M., C. W. Rice, and J. M. Tiedje. 1993a. Denitrification in a tallgrass prairie landscape. *Ecology* 74:855–862.

Groffman, P. M., and J. M. Tiedje. 1988. Denitrification hysteresis during wetting and drying cycles in soil. *Soil Science Society of America Journal* 52:1626–1629.

Groffman, P. M., and J. M. Tiedje. 1989. Denitrification in north temperate forest soils: relationships between denitrification and environmental factors at the landscape scale. *Soil Biology and Biochemistry* 21:621–626.

Groffman, P. M., D. R. Zak, S. Christensen, A. R. Mosier, and J. M. Tiedje. 1993b. Early spring nitrogen dynamics in a temperate forest landscape. *Ecology* 74:1579–1585.

Gross, P. J., and J. M. Bremner. 1992. Acetone problem in use of the acetylene blockage method for assessment of denitrifying activity in soil. *Communications in Soil Science and Plant Analysis* 23:1345–1358.

Hallmark, S. L., and R. E. Terry. 1985. Field measurement of denitrification in irrigated soils. *Soil Science* 140:35–44.

Hanson, G. C., P. M. Groffman, and A. J. Gold. 1994. Denitrification in riparian wetlands receiving high and low groundwater nitrate inputs. *Journal of Environmental Quality* 23:917–922.

Hyman, M. R., and D. J. Arp. 1987. Quantification and removal of some contaminating gases from acetylene used to study gasutilizing enzymes and microorganisms. *Applied and Environmental Microbiology* 53:298–303.

Hynes, R. K., and R. Knowles. 1978. Inhibition by acetylene of ammonia oxidation in *Nitrosomonas europaea. FEMS Microbiology Letters* 4:319–321.

Jury, W. A., J. Letey, and T. Collins. 1982. Analysis of chamber methods used for measuring nitrous oxide production in the field. *Soil Science Society of America Journal* 46:250–255.

Keeney, D. R. 1986. Critique of the acetylene blockage technique for field measurement of denitrification. Pages 103–115 *in* R. D. Hauck and R. W. Weaver, editors, *Field Measurement of Dinitrogen Fixation and Denitrification.* Soil Science Society of America, Madison, Wisconsin, USA.

Keeney, D. 1987. Sources of nitrate to ground water. *CRC Critical Reviews in Environmental Control* 16:257–304.

Martin, G. E., D. D. Snow, E. Kim, and R. F. Spalding. 1995. Simultaneous determination of argon and nitrogen. *Groundwater* 33:781–785.

Martin, K., L. L. Parsons, R. E. Murray, and M. S. Smith. 1988. Dynamics of soil denitrifier populations: relationships between enzyme activity, most-probable-number counts, and actual N gas loss. *Applied and Environmental Microbiology* 54:2711–2716.

Moraghan, J. T., and R. Buresh. 1977. Correction for dissolved nitrous oxide in nitrogen studies. *Soil Science Society of America Journal* 41:1201–1202.

Mosier, A. R. 1980. Acetylene inhibition of ammonium oxidation in soil. *Soil Biology and Biochemistry* 12:443–444.

Mosier, A. R. 1989. Chamber and isotope techniques. Pages 175–187 *in* M. O. Andreae and D. S. Schimel, editors, *Exchange of Trace Gases between Terrestrial Ecosystems and the Atmosphere.* Wiley, New York, USA.

Mosier, A. R., W. D. Guenzi, and E. E. Schweizer. 1986. Field denitrification estimation by nitrogen-15 and acetylene inhibition techniques. *Soil Science Society of America Journal* 50:831–833.

Mosier, A. R., and L. Klemedtsson. 1994. Measuring denitrification in the field. Pages 1047–1066 *in* R. W. Weaver, J. S. Angle, and P. S. Bottomley, editors, *Methods of Soil Analysis. Part 2, Microbiological and Biochemical Properties.* Soil Science Society of America, Madison, Wisconsin, USA.

Mulvaney, R. L. 1984. Determination of ^{15}N-labeled dinitrogen and nitrous oxide with triple-collector mass spectrometers. *Soil Science Society of America Journal* 46:1178–1184.

Murray, R. E., Y. S. Feig, and J. M. Tiedje. 1995. Spatial heterogeneity in the distribution of denitrifying bacteria associated with denitrification activity zones. *Applied and Environmental Microbiology* 61:2791–2793.

Myrold, D. D. 1988. Denitrification in ryegrass and winter wheat cropping systems of western Oregon. *Soil Science Society of America Journal* 52:412–415.

Nieder, R., G. Schollmayer, and J. Richter. 1989. Denitrification in the rooting zone of cropped soils with regard to methodology and climate: a review. *Biology and Fertility of Soils* 8:219–226.

Nielsen, L. P. 1992. Denitrification in sediment determined from nitrogen isotope pairing. *FEMS Microbiology Ecology* 86:357–362.

Oremland, R. S., C. Umberger, C. W. Culbertson, and R. L. Smith. 1984. Denitrification in San Francisco Bay intertidal sediments. *Applied and Environmental Microbiology* 47:1106–1112.

Parkin, T. B. 1987. Soil microsites as a source of denitrification variability. *Soil Science Society of America Journal* 51:1194–1199.

Parkin, T. B., H. F. Kaspar, A. J. Sexstone, and J. M. Tiedje. 1984. A gasflow soil core method to measure field denitrification rates. *Soil Biology and Biochemistry* 16:323–330.

Parkin, T. B., and J. M. Robinson. 1992. Analysis of lognormal data. *Advances in Soil Science* 20:193–236.

Parkin, T. B., A. J. Sexstone, and J. M. Tiedje. 1985. Comparison of field denitrification rates determined by acetylenebased soil core and nitrogen15 methods. *Soil Science Society of America Journal* 49:94–99.

Parkin, T. B., J. L. Starr, and J. J. Meisinger. 1987. Influence of sample size on measurements of soil denitrification rates. *Soil Science Society of America Journal* 51:1492–1501.

Parsons, L. L., R. E. Murray, and M. S. Smith. 1991. Soil denitrification dynamics: spatial and temporal variations of enzyme activity, populations, and nitrogen gas loss. *Soil Science Society of America Journal* 55:90–95.

Payne, W. J. 1981. *Denitrification.* Wiley, New York, New York, USA.

Payne, W. J. 1984. Influence of acetylene on microbial and enzymatic assays. *Journal of Microbiological Methods* 2:117–133.

Payne, W. J. 1991. A review of methods for field measurements of denitrification. *Forest Ecology and Management* 44:5–14.

Pell, M., B. Stenberg, J. Stemstrom, and L. Torstensson. 1996. Potential denitrification activity assay in soil—with or without chloramphenicol. *Soil Biology and Biochemistry* 28:393–398.

Prather, M., R. Derwent, D. Ehhalt, P. Fraser, E. Sanhueza, and X. Zhou. 1995. Other trace gases and atmospheric chemistry. Pages 77–126 *in* J. Houghton, L. G. Meira, E. Haites, N. Harris, and K. Maskell, editors, *Climate Change 1994: Radiative Forcing of Climate Changes and an Evaluation of the IPCC IS92 Emission Scenarios.* Cambridge University Press, New York, New York, USA.

Robertson, G. P., M. A. Huston, F. C. Evans, and J. M. Tiedje. 1988. Spatial variability in a successional plant community: patterns of nitrogen availability. *Ecology* 69:1517–1524.

Robertson, G. P., and J. M. Tiedje. 1984. Denitrification and nitrous oxide production in old growth and successional Michigan forests. *Soil Science Society of America Journal* 48:383–389.

Robertson, G. P., and J. M. Tiedje. 1988. Denitrification in a humid tropical rainforest. *Nature* 336:756–759.

Robertson, G. P., P. M. Vitousek, P. A. Matson, and J. M. Tiedje. 1987. Denitrification in a clearcut Loblolly pine (*Pinus taeda* L.) plantation in the southeastern U.S. *Plant and Soil* 97:119–129.

Rolston, D. E., F. E. Broadbent, and D. A. Goldhammer. 1979. Field measurements of denitrification. II. Mass balance and sampling uncertainty. *Soil Science Society of America Journal* 43:703–708.

Ryden J. C., L. J. Lund, J. Letey, and D. D. Focht. 1979. Direct measurement of denitrification loss from soils. II. Development and application of field methods. *Soil Science Society of America Journal* 43:110–118.

Ryden, J. C., J. H. Skinner, and D. J. Nixon. 1987. Soil core incubation system for the field measurement of denitrification using acetylene-inhibition. *Soil Biology and Biochemistry* 19:753–757.

Ryther, J. H., and W. M. Dunstan. 1971. Nitrogen, phosphorus and eutrophication in the coastal marine environment. *Science* 171:1008–1013.

Seitzinger, S. P., L. P. Nielsen, J. Caffrey, and P. B. Christensen. 1993. Denitrification measurements in aquatic sediments: a comparison of three methods. *Biogeochemistry* 23:147–167.

Seitzinger, S. P., S. Nixon, M. E. Q. Pilson, and S. Burke. 1980. Denitrification and N_2O production in nearshore marine sediments. *Geochimica Cosmochimica Acta* 44:1853–1860.

Siegel, R. S., R. D. Hauck, and L. T. Kurtz. 1982. Determination of $^{30}N_2$ and application to measurement of N_2 evolution during denitrification. *Soil Science Society of America Journal* 48:99–103.

Slater, J. M., and D. G. Capone. 1989. Nitrate requirement for acetylene inhibition of nitrous oxide reduction in marine sediments. *Microbial Ecology* 17:143–157.

Smith, M. S., and L. L. Parsons. 1985. Persistence of denitrifying enzyme activity in dried soils. *Applied and Environmental Microbiology* 49:316–320.

Smith, M. S., and J. M. Tiedje. 1979. Phases of denitrification following oxygen depletion in soil. *Soil Biology and Biochemistry* 11:262–267.

Starr, J. L., T. B. Parkin, and J. J. Meisinger. 1995. Influence of sample size on chemical and physical soil measurements. *Soil Science Society of America Journal* 59:713–719.

Swerts, M., G. Uytterhoeven, R. Merckx, and K. Vlassak. 1995. Semicontinuous measurement of soil atmosphere gases with gas-flow soil core method. *Soil Science Society of America Journal* 59:1336–1342.

Terry, R. E., and J. M. Duxbury. 1985. Acetylene decomposition in soils. *Soil Science Society of America Journal* 49:90–94.

Thauer, R. K., K. Jungermann, and K. Decker. 1977. Energy conservation in chemotropic anaerobic bacteria. *Bacteriology Reviews* 41:100–180.

Thomas, K. L., and D. Lloyd. 1995. Measurement of denitrification in estuarine sediment using membrane inlet mass spectrometry. *FEMS Microbiology Ecology* 16:103–114.

Tiedje, J. M. 1994. Denitrifiers. Pages 245–268 *in* R. W. Weaver, J. S. Angle, and P. S. Bottomley, editors, *Methods of Soil Analysis. Part 2, Microbiological and Biochemical Properties.* Soil Science Society of America, Madison, Wisconsin, USA.

Tiedje, J. M., A. J. Sexstone, D. D. Myrold, and J. A. Robinson. 1982. Denitrification: ecological niches, competition and survival. *Antonie van Leeuwenhoek* 48:569–583.

Tiedje, J. M., S. Simkins, and P. M. Groffman. 1989. Perspectives on measurement of denitrification in the field including recommended protocols for acetylene based methods. *Plant and Soil* 115:261–284.

Topp, E., and J. C. Germon. 1986. Acetylene metabolism and stimulation of denitrification in an agricultural soil. *Applied and Environmental Microbiology* 52:802–806.

Vitousek, P. M., and R. W. Howarth. 1991. Nitrogen limitation on land and in the sea: how can it occur. *Biogeochemistry* 13:87–115.

von Rheinbaben, W. 1990. Nitrogen losses from agricultural soils through denitrification: a critical evaluation. *Zeitschrift fur Pflanzenernahrung und Bodenkunde* 153:157–166.

Walter, H. M., D. R. Keeney, and I. R. Fillery. 1979. Inhibition of nitrification by acetylene. *Soil Science Society of America Journal* 43:195–196.

Wilhelm, E., R. Battino, and R. J. Wilcock. 1977. Low-pressure solubility of gases in liquid water. *Chemical Reviews* 77:219–262.

Wilson, G. B., J. N. Andrews, and A. H. Bath. 1990. Dissolved gas evidence for denitrification in the Lincolnshire limestone groundwaters. *Eastern England Journal of Hydrology* 113:51–60.

Wu, Q., and R. Knowles. 1995. Effect of chloramphenicol on denitrification in *Flexibacter canadensis* and *Pseudomonas denitrificans. Applied and Environmental Microbiology* 61:434–437.

Yeomans, J. C., and E. G. Beauchamp. 1982. Acetylene as a possible substrate in the denitrification process. *Canadian Journal of Soil Science* 62:137–144.

Yoshinari, T., and R. Knowles. 1976. Acetylene inhibition of nitrous oxide reduction by denitrifying bacteria. *Biochemical and Biophysical Research Communications* 69:705–710.

Part IV

Soil Organisms

15

The Determination of Microbial Biomass

Eldor A. Paul
David Harris
Michael J. Klug
Roger W. Ruess

The soil biota represents 1–3% of the soil carbon and 3–5% of soil nitrogen. The microbial component, often ranging from 100 to >1000 $\mu g\ C\ g^{-1}$, is usually considered to contain those soil organisms not visible without magnification. It includes nematodes, protozoa, filamentous and yeast forms of the fungi, microalgae, and bacteria. The bacteria, in turn, contain great diversity, including the actinomycetes, archae, and chemo- and photolithotrophs plus many still uncultured forms. Often the effects of climate, landscape, soil type, and management on ecosystem functioning and nutrient dynamics in the environment are most easily interpreted through the highly responsive microbial component. Microbial biomass acts both as a nutrient reservoir and as a catalytic force in decomposition, and is crucial for understanding nutrient fluxes within and between ecosystems (Smith and Paul 1990). Microbial biomass can also be a sensitive indicator of environmental toxicity attributable to pesticides, metals, and other anthropogenic pollutants. Because microbial biomass is a sensitive indicator of many belowground components and interactions, it is an important measurement for cross-site comparisons in ecosystem studies.

Microbial biomass can be used to describe steady-state soil characteristics. It is more useful, however, when the dynamic nature of the microbiota is taken into account. Changes in size and composition represent the influence of above- and belowground inputs, climate, and disturbance. Although the MB is usually greatest near the soil surface, it occurs throughout the rooting zone and often at great depths in the vadose layer and in deeper subsurface sediments such as oil deposits.

The turnover rate of the biomass cannot be readily measured by changes with time because of concurrent growth. Tracers can be used to show the movement and turnover of nutrients in this active pool, especially when combined with mathematical modeling. Growth studies also can be conducted by measuring the incorporation of specifically labeled substrates such as ^{3}H thymidine or ^{14}C leucine (Christensen and Christensen 1995; Harris and Paul 1994). A knowledge of the growth rate makes it possible to determine values for maintenance energy. The amount of substrate utilized for maintenance energy is of significance in many soils because of the slow turnover rate of the microbial biomass in these soils. Available methods include microscopy (Fry 1990) and the measurement of cell constituents released on fumigation (Horwath and Paul 1994; Joergensen 1995). Other methods include component analyses such as ATP (Inubushi et al. 1989), phospholipids (Tunlid and White 1992), ergosterol (Newell 1992), and ninhydrin-reactive materials (Amato and Ladd 1988). Substrate-induced respiration (SIR) is an activity-based method that has been applied to the measurement of biomass (Anderson and Domsch 1978). These techniques have not been conducted on a sufficient number of soils for us to be able to recommend them for routine measurements.

Microscopy facilitates the determination of size and shape, as well as the numbers of organisms such as bacteria, fungi, yeasts, protozoa, and diatoms. Fluorescent stains (Bloem et al. 1995b) make it possible to identify and measure the biota in the soil matrix. Analysis of digitized images by computer software can eliminate much of the operator error in cell identification and measurement. However, the high cost of the microscope and associated image analysis equipment and the lengthy time involved in analysis are deterrents to the use of this method. The relation of biovolume to biomass also is confounded by the fact that microorganisms can have variable C, N, S, and P contents.

Chloroform fumigation kills most soil organisms and destroys their membranes and cell walls. Fumigation can be combined with incubation (chloroform-fumigation-incubation, or CFI) to release the C as CO_2 and the N as NH_4. Alternatively, fumigation can be followed by extraction of the cell constituents (chloroform-fumigation-extraction, or CFE). Both approaches recover elemental cell constituents, making it possible to utilize tracers to determine pool sizes and turnover rates. A direct estimate of the pool size and turnover rate of the microbial biomass is a prerequisite for studies involving the role of the microbiota in organic matter dynamics and nutrient cycling. The tracers ^{14}C, ^{13}C, ^{15}N, and ^{32}P can be directly determined after lysis with $CHCl_3$ and incubation (CFI) or extraction (CFE).

Microscopy has been used to standardize fumigation methods. Although slow and tedious, it provides a number of advantages that include estimates of cell size and shape and bacterial-to-fungal ratios, as well as providing an overall estimate of biovolume. It is applicable to a wide variety of samples such as peats, sediments, litter, and contaminated sites where there can be problems using chemical approaches. Both fumigation methods can be set up to handle larger numbers of samples than microscopy within restricted time periods. The CFE method is faster than CFI and is now used on many sites where it gives reproducible results. The CFI method requires an incubation period but directly gives the products of decomposition as CO_2 and NH_4, which are readily measured. In some soils we have found CFI to give better reproducibility than CFE. Here we present methods for microscopy, CFI, and CFE. No one method is best suited for all purposes. All need to be calibrated and standardized before use. There is no simple approach to standardization of biomass measurements. Samples fixed on slides for microscopy are transportable between laboratories and should be used for comparison if possible. The interchange, between laboratories, of soil with a living biomass is not easy. Preincubating soil samples at constant moisture and temperature for 7 days results in a fairly stable biomass. Such samples could be sent to colleagues for cross-comparison. Sample standardization should be a prerequisite for funding and publication. When properly standardized, the different approaches give estimates that are reasonably well correlated with each other (Smith and Paul 1990; see also Wardle and Ghani 1995).

Soil Sampling and Handling

Sampling, in interdisciplinary studies, nearly always involves a number of analyses from one set of samples. As long as it is realized that one is working with living organisms, the sample coring and handling techniques used for other analyses are satisfactory for biomass determinations. Because biota are very site-specific and change during the season, compositing, replication, and recognition of site variability are most important and must be incorporated into the sampling design. The samples, stored in bags that retain moisture but allow respiration (thin polyethylene), should be placed in a temperature-controlled environment immediately upon sampling. A portable cooler has been found satisfactory for transportation. Immediate processing, although most desirable, is often not possible. Overnight storage at 5 °C or 15 °C depending on the original soil temperature is often used. The recognition that a delay in processing of field samples is often impossible to avoid has led Joergensen (1995) to recommend that all samples be preincubated for 5–7 days under laboratory conditions before analysis. This attenuates some of the disturbance associated with sampling but does not allow measurement of dynamic situations.

Sieving removes plant debris and large solids and provides mixing to decrease sample heterogeneity. Small soil samples (20 g) are attractive because of economy in containers, extractant volumes, and overall sample handling capacity, but they increase sample variance compared with larger samples. Fifty gram samples obtained from well-mixed composite subsamples are recommended as a compromise. Either 2 or 4 mm mesh sieve sizes are recommended, depending on the needs for other analyses also being conducted. Sieving removes much of the litter fraction and its associated microbiota. This should be analyzed separately.

Adjustment of the water content of medium-textured soils to 50% of water-holding capacity (WHC) is recommended before proceeding with either analysis. Water-holding capacity (WHC) is defined as the gravimetric water content of sieved soil that has been saturated and allowed to drain for over 24 hours in a filter funnel at 100% humidity. Some sandy or high clay soils will require adjustment based on water potential (−5 to 10 kPa) rather than WHC (Voroney et al. 1993). The CFI method is not satisfactory for waterlogged soils due to methane production (Inubushi et al.

1984), but CFE can be used (Inubushi et al. 1991). Dryness affects CFE more than it does CFI. In CFE, moist soils are required for good fumigation and for the required proteolysis thereafter (Sparling and West 1989).

Microscopy

Fluorescence microscopes combined with a skilled observer or image analysis software can differentiate between the soil biota and other similar-sized particles such as clay minerals. Fluorescent stains that have been tested for soil microbial biomass measurement (Kepner and Pratt 1994; Bloem et al. 1995a) include protein stains such as fluorescein isothiocyanate (FITC) and DTAF (5-(4,6-dichlorotriazin-2-yl)aminofluorescein)). Acridine orange, europium chelate, and DAPI (4′,6-diamidino-2-phenylindole) stain nucleic acids. Research on fungal stains has concentrated on cell walls because many fungi have large sections free of cytoplasm. The stain now most often used for fungi is fluorescent brightener calcofluor white M2R (Bloem et al. 1995a).

Other available staining techniques include those that measure cell activity such as fluorescein diacetate (FDA) activated by esterase (Söderström 1979). The redox-sensitive dye INT (tetrazolium chloride), when reduced by microbial electron transfer, produces formazan visible within the cell (Dutton et al. 1983). The fluorescent redox probe, 5-cyano-2,3 ditolyl tetrazolium chloride, allows one to utilize the superior differential viewing of the fluorescence microscope to separate active from inactive cells (Rodriguez et al. 1992). Techniques that allow the use of molecular probes for direct analyses under the microscope open the field for phylogenetic analyses of that huge population of soil biota that can be microscopically seen but are not culturable. They are not yet available for routine analysis. Herein we report on the use of fluorescence microscopy using DTAF staining for bacteria and calcofluor M2R for fungi.

For microbial biomass measurements, the soil must be adequately dispersed and placed onto a microscopic slide or appropriate filter as a thin film. Smears produce flatter preparations with better contrast then membrane filters and are recommended for bacteria. Fungi in some soils do not adequately stick to smears; if so, filters instead of smears should be used for fungi. The individual fields must be selected and the biovolume in each field determined through measurement of the size, shape, and number of organisms.

Materials

1. Buffer: 0.05 mol/L Na_2HPO_4 (7.8 g/L) in 0.15 mol/L NaCl (8.8 g/L) adjusted to pH 9.0
2. DTAF stain: 2 mg (5-(4,6-dichlorotriazin-2-yl) aminofluorescein) in 10 mL buffer. Prepare fresh daily.
3. Calcofluor M2R (fluorescent brightener) stain: 2 g/L in water. Stain is stable at room temperature for 1 month.
4. Filter stains, buffer, and wash water filtered through a 0.2 μm filter

Procedure

Slide Preparation

1. Homogenize 10 g soil in 190 mL filtered water for 1 minute at full speed in a Waring blender.
2. Allow coarse particles to settle for 30 seconds. With a wide-bore pipette remove a bulk (5 mL) sample to a capped test tube.
3. Add 0.1 mL formalin (40% aqueous formaldehyde) as a preservative.
4. Vortex this bulk sample before subsampling to prevent further sedimentation.
5. Place 4 μL drops of soil suspension onto each 6 mm diameter "well" of printed slides (e.g., Bellco toxoplasmolysis slides). Spread the suspension uniformly across the well with the pipette tip without touching the tip to the slide surface.
6. Allow the smears to air dry completely; this fixes the organisms to the slide. Prepare two slides for each sample; one will be stained for bacteria, the other for fungi.

Staining Bacteria

1. Flood each smear (well) with 8 μL DTAF stain.
2. Store slides in a container with wet tissue (100% relative humidity) for 30 minutes.
3. Remove excess stain by placing slides in a series of staining jars containing buffer; make three changes of buffer (30 minutes each) and finally a change of water (30 minutes).
4. Air-dry slides flat.
5. Add a small drop of low-fluorescence immersion oil (Cargille type FF recommended) to each smear and cover the slide with a 50 × 25 mm coverslip.

Staining Fungi

1. Flood each smear with 10 μL Calcafluor M2R stain.
2. Stain in a covered container with wet tissue (100% relative humidity) for 2 hours.
3. Rinse by soaking slides in water in a staining jar for 30 minutes, three times.
4. Air-dry slides flat.
5. Add small drops of low-fluorescence immersion oil (Cargille type FF recommended) to each well and cover slide with a 50 × 25 mm coverslip.

Bacterial Counting and Biovolume Estimation

Observe the green bacteria using a fluorescence microscope with a 63× or 100× oil-immersion objective. Suitable filters are excitation 450–490 nm, dichroic 510 nm, and suppression <515 nm. The bacteria may be counted by image analysis or by a human observer. Counting by humans is not recommended because it is difficult to achieve consistency between individuals and by the same individual over time. It is also extremely difficult to estimate the dimensions of individual cells by direct observation of fluorescent images because of their small size and unresolved edges. Important advantages of image analysis over human counting are consistent application of rules so that the same features in the image are counted on any occasion. It is possible to estimate the dimensions and volume of each cell counted. Collection of images typically is much faster than human counting, so photobleaching of the fluorochrome is minimized. Analysis of the images can be automated and run as a batch process separate from image acquisition. Comparisons of automatic image analysis counts with those made by skilled human observers show that average numbers of bacteria counted are similar but that the variance of counts on replicate slides by the image analysis system is greater (Bloem et al. 1995b; Harris, unpublished data). This effect is probably the result of the unconscious effort of the human counter to find more bacteria in slides where there are few and to miss bacteria where there are many.

Image analysis procedures are presented in the Appendix. Because relatively few laboratories will be suitably equipped, we describe a manual counting method that can be used in the absence of image analysis equipment. An ocular grid (10×10 squares) that encloses approximately 50% of the field of view is used to define a counting area in each field of view. The dimensions of the grid can be determined with a stage micrometer. A mechanical counter or simple computer program (Bloem et al. 1992) is used to tally the counted bacteria in each field. The observer should work methodically through the array of 100 squares within the grid, counting each cell recognized as a bacteria. Any cells crossed by the upper and left limits of the grid should be excluded, while those crossed by the lower and right lines are counted. Fields should be prearranged on equatorial transects of the smear such that at least 300 cells per smear are counted. Four to six replicate smears are recommended. Assuming a Poisson distribution, the 95% confidence interval is approximately twice the square root of the count. If $4 \times 300 = 1200$ cells are counted, then the 95% confidence interval is about $\pm$ 5%, equivalent to 60 cells.

It is too slow to manually measure the dimensions of individual cells in fluorescent images because of photobleaching. A subsample of fields should be photographed on 35 mm transparency film (ISO 400 or faster), and the images should include a size marker or grid. The dimensions of bacteria can be measured more easily and accurately in projected images calibrated with the known scale. Two perpendicular diameters represent the maximum and minimum dimensions of the cell. The transparencies can also be used to train counters and to test interoperator calibration, and they may be archived for comparison between dates, experiments, and sites.

The bacterial biovolume is calculated from measurements of the large and small diameters of each cell by assuming that the cells are prolate spheroids (Sieracki et al. 1989):

$$bv = (\pi/6) \times d^2 \times l$$

where

bv = bivolume
d = small diameter
l = large diameter

The average bacterial volume from the measured sample can be multiplied by the count of bacteria per smear to give bacterial biovolume per smear.

Fungal Counting and Biovolume Estimation

Observe the Calcofluor M2R (blue)–stained fungi under epifluorescence. Suitable light filters are excitation 340–380 nm, dichroic 400 nm, and suppression <430 nm. Fluorescent dyes usually fail to stain heavily melanized hyphea, which may be common or prevalent in some soils. Low magnification (10 × objective), to include a relatively large field, can be used with imaging systems. A method for measurement of fungal hyphae by image analysis is given in the Appendix (see also Morgan et al. 1991; Daniel et al. 1995). Higher magnification objectives (25–40×) may be needed for manual measurements, especially for hyphal diameter estimation. Increased magnification decreases the frequency of hyphal fragments per field; use the lowest magnification that allows clear discrimination of hyphae.

Manual measurements of hyphal length can be made by the grid intercept method (Newman 1966) where intersections between hyphae and a set of lines of known length distributed in a fixed area are counted. Hyphal length (*hl*) is estimated by:

$$hl = (\pi \times n \times a)/(2 \times l)$$

where

hl = length of hyphae, in μm
n = the number of intersections
a = the defined area, in μm^2
l = the length of all the lines, in μm

The 10 × 10 ocular grid used for bacteria counting is suitable. Scan the grid and count each contact between hyphae and each of the lines of the grid. In soils with relatively low hyphal populations (<100 m hyphae/g soil), many microscope fields will contain no hyphae. For example, in a smear prepared as described earlier, from a soil with 100 m hyphae/g, a 1 mm^2 grid with a total line length of 22 mm would contain about 6 μg soil and an average of about 600 μm hyphae corresponding to an average of 8.4 grid intersections. In our experience 1 minute homogenization at full speed in a Waring blender yields hyphal fragments with a median length of about 40 μm, which is close to the optimum fragment size recommended by Sundman and Sivelä (1978). The error of the measured hyphal length is proportional to the number of hyphal fragments measured (Hanssen et al. 1974); therefore, preparations with larger hyphal fragments require a larger counting area (more fields) for equal precision.

Hyphal diameters can be estimated by visual comparison with the lines in the grid, by using an ocular scale, or, better, by measurement of projected photographic images. The volume (*hv*) of each hyphal fragment can be calculated as a cylinder:

$$hv = \pi \times (d/2)^2 \times l$$

where

hv = hyphal volume, in μm^3
d = diameter, in μm
l = length, in μm

The hyphal biovolume per image is the sum of the volumes of the fragments.

Many, perhaps most, hyphae in soil are apparently dead or empty. This poses the greatest problem in the estimation of fungal biomass by microscopic means. Vital staining (activity detecting) with FDA (Söderström 1979; Ingham and Klein 1984) has been tried but fails to penetrate many hyphae. Other vital stains such as sulphofluorescein diacetate (SFDA; Tsuji et al. 1995) or 5-cyano-2,3-ditolyl tetrazolium chloride (CTC; Rodriguez et al. 1992) may prove useful in this regard. However, in preliminary work with CTC many regions of soil hyphae where dye reduction occurred were colonized by superficial or intralumenar bacteria; these appeared to be the active agents in dye reduction (D. Harris, unpublished data).

Calculations

Mass of Soil Per Image (field of view)

The initial suspension (200 mL) contains 10 g moist soil, and each smear contains 4 μL of the suspension. Thus each smear contains the following mass of soil:

$$\text{g soil per smear} = \{10\text{g soil}/[(\%H_2O/100) + 1]\} \times (1/200 \text{ mL}) \times (4/1000\text{mL})$$

The mass of soil per field is given by

$$s = [(x \times y)/(\pi \times r^2)] \times \{10/[(\%H_2O/100) + 1]\} \times (2/10^5)$$

where

s = soil mass per field
x = length of the counting area, in mm
y = width of the counting area, in mm
r = radius of the smear, in mm

The sum of the volumes of bacteria (bv) or fungal hyphae (hv), divided by the appropriate mass of soil (s), represents biovolume per gram soil (μm^3/g). If filters are used, instead of smears, for fungi the calculations will need to be modified.

Calculation of Biomass

The biovolume is converted to biomass C using a conversion factor for the C content per unit volume. Conversion factors for C have been estimated by a variety of methods using different preparation techniques, organisms, and growth conditions (van Veen and Paul 1979; Bakken and Olsen 1983; Bratbak and Dundas 1984; Fry 1990; Bloem et al. 1995b). These factors have shown a range for the C content of bacteria of 150–310 fg/μm^3. Norland (1993) compared four models of variation in the C content-to-volume ratio with cell size in marine bacteria. In these models the

C content (fg C/μm^3) increases as biovolume decreases. The variation in C content of bacteria with cell size may help to explain some of the differences in reported conversion factors. Three of these models, applied to our measurements of soil bacteria, gave biomass C values that agreed well (±5%) with those obtained using the factor 200 fg C/μm^3 (Bloem et al. 1995b). Carbon contents of cultured fungi were estimated by van Veen and Paul (1979) and correspond to 150 fg/μm^3.

Chloroform-Fumigation-Incubation

We describe both chloroform fumigation techniques: (1) the release of CO_2 and NH_4 by mineralization in a 10 day incubation following fumigation (CFI) and (2) the extraction of cell constituents by K_2SO_4 after fumigation (CFE). The CFI method has been calibrated for temperate soils against direct microscopy and by the addition of cells with known C and N contents (Jenkinson 1976; Voroney and Paul 1984). Calibration factors for soils of nontemperate regions have not yet been developed. The major problem with CFI is the question of what proportion of the C mineralization flush following fumigation is due to continued mineralization of soil organic matter rather than mineralization of the killed biomass; this is still controversial (Wu et al. 1996). Studies using ^{14}C-labeled straw additions to soil indicate that fumigation reduces but does not eliminate soil organic matter mineralization (Smith et al. 1995; Horwath et al. 1996). The respiration of the unfumigated control soil has been used to estimate this background mineralization rate. But many soils have respiration rates comparable to the CO_2 flush following fumigation. This, if subtracted from the mineralization in the fumigated sample, can result in a low or even negative microbial biomass C estimate. The opposite assumption, that all of the 10 day postfumigation CO_2 flush arises from mineralization of the killed biomass, can result in excessively high microbial biomass C estimates.

Development of a method for estimating the proportion of the unfumigated control mineralization to subtract for each sample (Horwath et al. 1996) has overcome the control problem in CFI. The proportion (p) of the control to subtract is calculated by using the ratio of CO_2 mineralized in the control (C_C) relative to that in the fumigated sample (C_F). Two parameters, k_1 and k_2, are used to relate C_F/C_C to p. These have been obtained by fitting the data obtained by fumigation in a series of soils to the biomass estimated by microscopy (Horwath et al. 1996).

The control problem is not as serious with biomass N determinations by CFI because the unfumigated control values for N mineralization are typically low relative to those in fumigated samples (Jenkinson 1988). However, the N mineralized in CFI is affected by the C:N ratio of the biomass before and after fumigation and by the growth yield of the biomass that develops after fumigation. This results in the mineralization of variable proportions of the biomass N and may lead to net immobilization of N, especially where fungi are prevalent. Harris et al. (1998) adjusted for this variation in biomass C:N by relating the mineralization of N and C after fumigation to the N:C ratios of organisms added to soil. The parameters of this relationship were found applicable to other soils; this approach allows calculation of biomass N even for soils where CFI results in net immobilization of N.

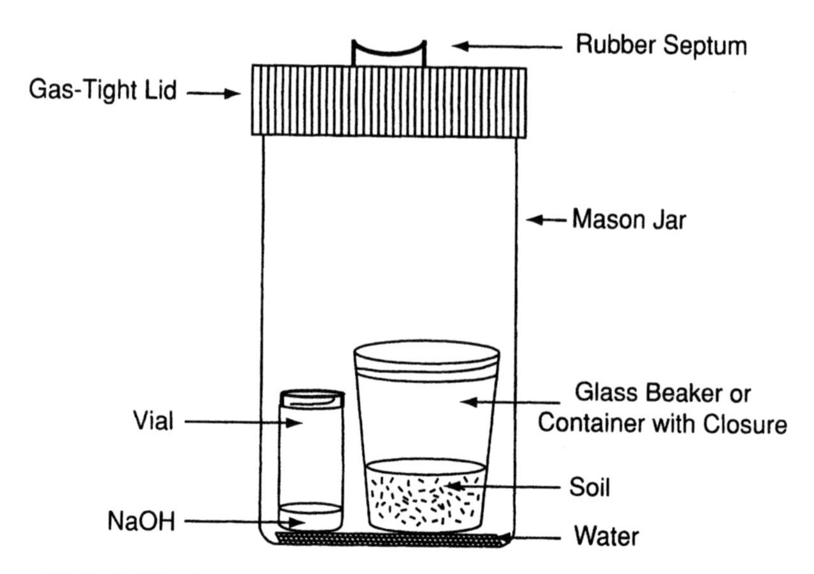

Figure 15.1. Soil incubation jar. The respirometer (vial + NaOH) is not used when mineralized CO_2 is sampled using the gas chromatography (GC) method. The rubber septum is not needed when using the respirometer method to assay CO_2.

Materials

1. Commercial chloroform stabilized with ethanol (0.75%)
2. Anhydrous K_2CO_3
3. 2 mol/L NaOH solution
4. 1 L separatory funnel
5. 100 mL glass beakers for fumigated samples. Containers for controls need not be glass.
6. Incubation containers of approximately 1 L (Fig. 15.1). Home canning jars, such as Mason jars, are readily available and work well if new lids are used.
7. 20 mL scintillation vials. (The dissolution of silica in base traps can be a problem with soda glass vials; high-density polyethylene vials have good base resistance but relatively high permeability to CO_2 and so should be protected from atmospheric CO_2 during any storage before titration.)
8. Large glass vacuum desiccators (20 L desiccators will hold about 20 soil samples in 100 mL beakers)
9. Aspirator or rotary vacuum pump with cold trap

Procedure

Soil Preparation

1. Mix and composite soil samples by sieving to 4 mm. Do this at field moisture (unless partial drying is necessary) to maintain microaggregate stability.

2. Determine the soil moisture content by oven drying samples of each composite.
3. Determine the WHC for each soil type (see Chapter 3, this volume).
4. Weigh 50 g (dry weight equivalent) samples into 100 mL beakers; three samples of each soil are required, one for initial mineral N determination plus one for control and one for fumigated samples. Label beakers to be fumigated with pencil (chloroform will dissolve most markers). Smaller samples (e.g., 20 g) have many advantages, such as the use of smaller containers and extractant volumes, but they may increase sample variability.
5. Add water to bring each sample to 50% of its WHC (or to −5 to −10 kPa).
6. Preincubate the samples for 5 days at 25 °C in closed incubation jars that have a few milliliters of water in the bottom to provide 100% humidity. This preincubation should be omitted in tracer experiments where nutrient dynamics are being measured.

Fumigation

In a fume hood:

1. Line the vacuum desiccator(s) with wet paper towels.
2. Put fumigation samples (approx. 20 in a 20 L desiccator) and a conical flask containing 75 mL ethanol-free chloroform and a few boiling chips in each desiccator.
3. Connect the desiccator(s) to a vacuum manifold that includes a vacuum gauge. If a rotary vacuum pump is used, it should be protected from excessive chloroform vapor with a cold trap (−20 °C or better) and should vent to a fume hood. Change the pump oil frequently. Alternatively, an aspirator pump may be used.
4. Evacuate the desiccators until the chloroform boils for 1 minute.
5. Incubate the samples in chloroform vapor in the dark at 25 °C for 24 hours.
6. After fumigation, open the desiccators in a fume hood and remove the paper towel and the flask of chloroform.
7. Remove the residual chloroform from the soil by evacuating the desiccator to 250 Pa and venting to atmosphere. Repeat this step six times. The meticulous removal of residual chloroform is essential.
8. During the fumigation the unfumigated control samples are incubated in closed canning jars.

Incubation

1. Check the sample weights and readjust the moisture content if necessary.
2. Place the fumigated and control samples in canning jars with an open 20 mL scintillation vial containing 2 mL of 2 mol/L NaOH (base trap) to trap CO_2 (for soils with very high organic matter content the volume or molarity of the trap can be adjusted to be equivalent to about twice the expected CO_2 flush). A few milliliters of water in the bottom of the jar will maintain 100% humidity. Open the canning jars of the unfumigated controls for at least 5 minutes

to equilibrate with atmospheric air. Breathing directly into the jars or preparing the samples in a small, poorly ventilated space (such as a temperature control room) can significantly increase the initial CO_2 concentration in the jars and should be avoided. The O_2 content of the canning jar should be adequate for the 10 day incubation (50 g soil producing 1 mg C/g as CO_2 would consume approximately 40 of the 200 mL of O_2 in the jar).

3. Include several blank jars that contain everything except soil.
4. Tightly close the jars and incubate for 10 days at 25 °C.
5. After 10 days remove the base traps and close with gastight screwcaps (Poly Seal).

Measurement of CO_2 Mineralized—Double End-point Titration

When no stable isotopic analysis of the CO_2 is required, the double end-point titration is preferred because it is a direct analysis of HCO_3^-.

Materials

1. Dual end-point automatic titrator (or pH meter with semi-micro combination electrode, magnetic stirrer and burette)
2. Carbonic anhydrase solution (1 mg/mL)
3. 2 mol/L HCl, not standardized
4. 0.3 mol/L HCl, standardized
5. Thymolpthalein indicator solution (Sigma Chemicals)

Procedure

1. Add 1 drop of indicator and a small stir bar to the base trap.
2. While stirring add 2 mol/L HCl from a pipette (or burette) until the indicator is very pale blue (pH 9.0–9.5; use a white surface under the stirrer).
3. Add 50 µL carbonic anhydrase solution.
4. Transfer the vial to the titrator and add 0.3 mol/L HCl to the first end point of pH 8.35.
5. Titrate to the second end point of pH 3.5.
6. Record the volume of acid added between the two end points.
7. Denatured carbonic anhydrase and indicator will accumulate on the pH electrode. It can be removed, as necessary, with 0.5 mol/L NaOH.
8. The sample titre (T) minus the blank (B) is equivalent to $(T - B) \times M \times 12$ mg C where M is the molarity of the HCl in µmol/L.

Measurement of CO_2 Mineralized—Single End-point Titration

Back titration of excess NaOH after precipitation of CO_3 with $SrCl_2$ is the preferred method when ^{13}C analyses are required because it enables collection of the C as $SrCO_3$, that can be collected for ^{13}C analysis in automatic mass spectrometers (Harris et al. 1997).

Materials

1. Automatic titrator (or pH meter with semi-micro combination electrode, magnetic stirrer and burette)
2. Standardized 0.3 mol/L HCl
3. 2 mol/L $SrCl_2$

Procedure

1. Add 2 mL 2 mol/L $SrCl_2$ to the base trap.
2. Titrate to pH 7.0 with 0.3 M HCl.
3. The blank titre (B) minus the sample titre (T) is equivalent to the CO_2 in the trap ($[B - T] \times M \times 6$ mg C when M is the molarity of the HCl in mol/L).

N Mineralized during Incubation

Materials

1. 1 mol/L KCl (use 2 mol/L for soils with high CEC)
2. 500 mL flasks or bottles
3. Filter funnels and multiposition rack
4. 15 cm Whatman no. 1 filter paper (a finer grade [no. 5] may be necessary for clay soils) washed in 0.1 mol/L HCl and rinsed with water

Procedure

There are three extractions for each soil sample, an initial (day 0) extract (N_0), at the start of the fumigation, and control (N_C) and fumigated (N_F) samples after the 10 day incubation.

1. Measure 250 mL 1 mol/L KCl.
2. Transfer the soil from beakers to the extraction flasks, use some of the KCl to wash residual soil from the beakers, and add the remainder to the flask.
3. Shake vigorously (keep the soil suspended) on a reciprocating or orbital shaker for 30 minutes. It may be necessary to use ultrasound to disperse strongly aggregated clay soils.
4. Filter the extract, and collect the filtrate, which may be stored frozen prior to analysis.
5. Analyze the extract for NH_4^+ and NO_3^-. Automated colorimetric analyzers are preferred.

Calculation of Biomass C and N from CFI

Five measurements are required: the 10-day C mineralization in fumigated and unfumigated soils (C_F, C_C), initial soil mineral N (ammonium plus nitrate), and mineral N in the fumigated and unfumigated samples after 10 days incubation. Four of these measurements are required for biomass C and N determination by previous methods (Jenkinson 1976, 1988). The fifth measurement, that of initial mineral N, is often routinely conducted when taking such samples.

The amounts of CO_2–C evolved in the fumigated (C_F) and unfumigated (C_C) samples are used in the following equation to calculate the biomass C when using a corrected control (Horwath et al. 1996):

$$MBC = [C_F - (p \times C_c)]/k_c$$

where

MBC = biomass C in μg C/g soil
C_F = C mineralized from the fumigated sample (μg C/g soil)
C_C = C mineralized from the unfumigated sample (μg C/g soil)

$$p = [k_1 \times (C_F/C_c)] + k_2$$

where

k_1 = 0.29 (see below)
k_2 = 0.23 (see below)
k_c = 0.41 (see below)

The value 0.41 for the decomposition factor k_c is the average value determined by a number of laboratories for temperate soils incubated at 25 °C. Values for the parameters k_1 and k_2 have been estimated by minimizing the sums of squares of the differences between *MBC* and microscopic biomass of a number of soils. This equation has given meaningful biomass estimates in a number of soils using parameter values of k_1 = 0.29 and k_2 = 0.23 (Horwath et al. 1996). Substituting these values, the above equation can be simplified to

$$MBC = 1.73C_F - 0.56C_c$$

The calculation of biomass N is based on the determination of biomass C and the N:C ratio. The addition of organisms to soil was used to determine the parameters involved (Harris et al. 1998). Calculation of biomass N:C ratio (*MBN*/*MBC*) can be made without correction for mineralization of nonbiomass material in the fumigated sample:

$$MBN/MBC = (0.56 \times N_F/C_F) + 0.095$$

where

MBN = biomass N in μg N/g soil
N_F = N mineralized from the fumigated sample (μg N/g soil)
N_C = N mineralized from the control sample (μg N/g soil)

The preceding uses only the measured C and N mineralization following fumigation-incubation and parameters determined from added organism experiments

(Harris et al. 1998). Biomass N is obtained by multiplying biomass N:C by biomass C:

$$MBN = MBC \times \{[0.56 \times (N_F/C_F)] + 0.095\}$$

The biomass N:C estimate best reflects soil conditions when it includes the contribution of nonbiomass materials to N mineralization in the fumigated soil by using the following relationships:

$$MBN/MBC = \{0.56 \times [(N_F - qN_C)/(C_F - pC_C)]\} + 0.095$$

where

qN_C and pC_C = N and C mineralization, respectively, in the fumigated soil from sources other than chloroform-killed biomass
N_C = mineral N after 10 day incubation of unfumigated soil minus soil mineral N at day 0
N_F = mineral N after 10 day incubation of fumigated soil minus soil mineral N at day 0 and
$q = p = [0.29 \times (C_F/C_C)] + 0.23$

This calculation includes measurements of C and N mineralization in the unfumigated sample and a fraction p (Horwath et al. 1996) determined by calibration of CFI biomass C using microscopy. We found that the calculation with partial control correction yields values 20% higher than the uncorrected version. The corrected values should be more representative of the populations in nature than uncorrected ones because differences in the N:C ratio of biomass and nonbiomass materials decomposed in the fumigated soil are considered.

Chloroform-Fumigation-Extraction

Like the CFI procedure, the CFE procedure also has problems, although less severe, with controls. Here the considerations are the more difficult analysis of the low C and N contents of the extracts and the more involved preparations required when working with stable isotopic tracers. Some investigators have noted a greater inherent variability in the measurements (Sparling and Zhu 1993; Horwath and Paul 1994). Horwath and Paul (1994) also reported a lack of agreement in specific activity of ^{14}C obtained by CFE and CFI even though the overall correlation in total MBC between the two measurements was good. The radioactivity of the C extracted by CFE was one-half that of CFI. This verified findings by Merckx and Martin (1987) and Badalucco et al. (1990), who found anthrone and ninhydrin reactive material of nonbiomass origin in the CFE extract.

We give procedures for determination of C by persulfate digestion and N by Kjeldahl digestion. Other measurement methods are available. Dichromate digestion has been used for C (Jenkinson and Powlson 1976) and ninhydrin analyses for

N (Amato and Ladd 1988). Automated colorimetric methods for both C and N are also available (Voroney et al. 1993).

Materials

1. Ethanol-free chloroform prepared as in CFI procedure
2. Anhydrous K_2CO_3
3. 100 mL glass beakers for fumigated samples
4. Large glass vacuum desiccators (20L; each desiccator will hold about 20 soil samples in 100 mL beakers)
5. Aspirator or rotary vacuum pump with cold trap
6. 0.5 mol/L K_2SO_4
7. 500 mL flasks or bottles

Procedure

Fumigation

In a fume hood:

1. Line the vacuum desiccator(s) with wet paper towels.
2. Arrange fumigation samples (approx. 20) and a conical flask containing 75 mL ethanol-free chloroform and a few boiling chips in each desiccator.
3. Connect the desiccator(s) to a vacuum manifold that includes a vacuum gauge. If a rotary vacuum pump is used it should be protected from excessive chloroform vapor with a cold trap (−20 °C or better) and should vent to a fume hood. Change the pump oil frequently. Alternatively, a water (aspirator) pump may be used.
4. Evacuate the desiccators until the chloroform boils for 1 minute.
5. Seal the desiccator(s) and incubate at 25 °C for 24 hours in the dark. A 24 hour fumigation is widely used (Voroney et al. 1993), but a longer fumigation of 120 hour is recommended for full release of microbial C and N (Horwath and Paul 1994).
6. After fumigation, open the desiccators in a fume hood and remove the paper towel and the flask of chloroform.
7. Remove the residual chloroform from the soil by evacuating the desiccator to 250 Pa and venting to atmosphere six times.

Extraction of Microbial C and N

1. Add 250 mL 0.5 mol/L K_2SO_4 to the control samples; these should be extracted at the time the fumigation is initiated.
2. Measure 250 mL 0.5 mol/L K_2SO_4.
3. Transfer the fumigated soil from beakers to the extraction flasks; use some of the K_2SO_4 to wash residual soil from the beakers, and add the remainder to the flask.
4. Shake vigorously (keep the soil suspended) on a reciprocating or orbital shaker for 1 hour. It may be necessary to use ultrasound to disperse strongly aggregated, clay soils.

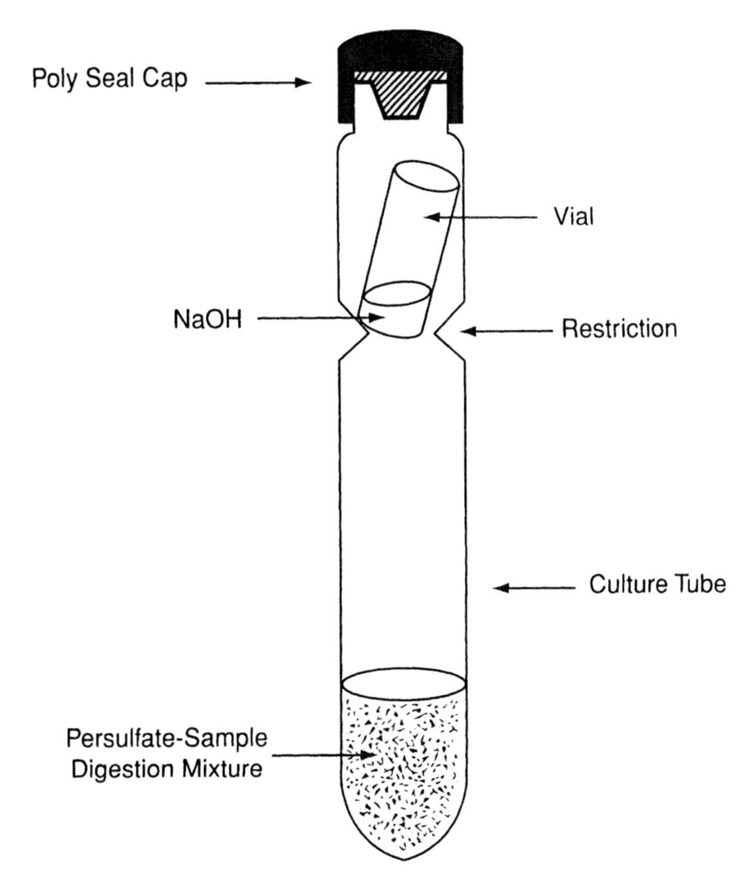

Figure 15.2. Components of the persulfate digestion. The sample, digestion chemicals, and CO_2 trap are placed into the modified culture tube and sealed with a Poly Seal cap. Avoid heating the cap excessively during digestion to avoid distorting the seal.

5. Filter the extracts through Whatman no. 5 filter paper and collect the filtrate; 5–30 mL is required depending on the method of analysis, although more may be needed for some isotopic analyses. The extracts should be stored frozen prior to analysis.

Analysis of Extract Persulfate Digestion (Total C)

Materials

1. 25 × 200 mm screw-cap culture tubes with Poly Seal caps. The tubes may be modified by forming a restriction at about 60 mm from the neck to support a vial containing alkali; otherwise the vial can be supported at about 150 mm on a bent glass rod (Fig. 15.2).
2. 15 × 45 mm glass vials
3. Heater block at 120 °C
4. 0.1 mol/L NaOH

5. Potassium persulfate ($K_2O_8S_2$)
6. 0.05 mol/L H_2SO_4
7. 0.05 mol/L HCl (standardized)

Procedure

1. Prepare an alkali trap containing 1 mL 0.1 mol/L NaOH.
2. Add 15 mL extract, 1 g $K_2O_8S_2$, and 1 mL 0.05 mol/L H_2SO_4 to the culture tube.
3. Insert the alkali trap, supported in the headspace of the tube by the restriction or by a glass rod.
4. Firmly cap the tube.
5. Heat the digests in the heater block at 120 °C for 2 hours. Do not allow the cap to get hot (it may leak or melt).
6. Remove the digests from the heater block and let stand overnight at room temperature to allow complete absorption of the CO_2.
7. Blanks should be prepared with 15 mL K_2SO_4.
8. Titrate the alkali traps by either the double or single end point (if ^{13}C is to be collected) methods (see earlier) except the titrator should be charged with 0.05 mol/L HCl.

Analysis of Extract—Kjeldahl Digestion (Total N)

Materials

1. 25 × 250 mm, Pyrex, graduated (100 mL), constricted-neck digestion tubes
2. Block heater capable of achieving >375 °C
3. Concentrated H_2SO_4
4. $CuSO_4 \cdot 5H_2O$
5. Selenized antibumping granules (Hengar)

Procedure

In an acid-resistant fume hood:

1. Add 3 mL concentrated H_2SO_4 followed by 100 mg $CuSO_4$ and 25 mL extract and 2–3 antibumping granules to the digestion tubes.
2. Heat the tubes to 150 °C to evaporate most of the water, then increase the temperature until the acid refluxes.
3. Digest at acid reflux temperature (about 375 °C) for 1 hour.
4. Allow the tubes to cool, then carefully add water, mix, and adjust to the 100 mL graduation mark. If the digest solidifies on cooling it may be necessary to reheat to dissolve the salts.
5. Analyze the digest for NH_4^+, preferably by a colorimetric autoanalyzer method.

Special Considerations

See also the alkaline persulfate digest for total N described by Cabrera and Beare (1993), which could replace the TKN method, described earlier, and probably could be adapted for simultaneous measurement of C and N in K_2SO_4 extracts.

Calculation of Biomass C and N from CFE

1. Volume of solution in extracted soil:

$$V = FW - DW + EV$$

where

V = volume of solution in the extracted soil (mL)
FW = soil fresh weight, as g
DW = soil dry weight, as g
EV = extractant volume, as mL

2. Mass of extractable C and N in the fumigated and control samples:

$$C_F = EC_F \times V/DW$$

$$C_C = EC_C \times V/DW$$

$$N_F = EN_F \times V/DW$$

$$N_C = EN_C \times V/DW$$

where

C_F = extractable C in fumigated sample in μg/g soil
EC_F = extractable C in fumigated sample in μg/mL extractant
C_C = extractable C in control sample in μg/g soil
EC_C = extractable C in control sample in μg/mL extractant
N_F = extracted N in fumigated sample in μg/g soil
EN_F = extractable N in fumigated sample in μg/mL extractant
N_C = extractable N in control sample in μg/g soil
EN_C = extractable N in control sample in μg/mL extractant

3. Microbial biomass C and N in soil:

$$MBC = (C_F - C_C)/k_{EC}$$

$$MBN = (N_F - N_C)/k_{EN}$$

where

kEC = extraction coefficient for extractable carbon as described below
kEN = extraction coefficient for extractable nitrogen as described below

Voroney et al. (1993) proposed values of 0.25 for k_{EC} and 0.18 for k_{EN} based on in situ calibrations (Bremer and van Kessel 1990). These values are substantially lower than other estimates and will give correspondingly larger biomass estimates. Vance et al. (1987) obtained a k_{EC} of 0.38 by correlating the data for CFE with that obtained by CFI in 10 Rothamsted soils. A k_{EC} of 0.35 was obtained by Sparling and West (1988) using in situ labeling techniques. In later work Sparling and Zhu (1993) found a range of values for k_{EC} (0.30 $\pm$ 0.18) and k_{EN} (0.38 $\pm$ 0.14) for a number of Western Australia soils by comparing CFE to CFI. We recommend the use of factors derived directly from the soils under study either by in situ calibration or by comparison with other methodologies. The values used for all calculations should be reported, and reference soils from other laboratories should be included in the study.

Acknowledgments We thank the National Science Foundation for support through grants to the KBS-LTER project, and the Center for Microbial Ecology at Michigan State University.

Appendix: Measurement of Bacterial and Fungal Biovolume by Fluorescence Microscopy and Image Analysis

General System Specifications

Hardware

1. Fluorescent images are very dim compared with those produced by most other forms of light microscopy. The camera should be sufficiently sensitive so that the bright features in the image have intensity values close to the maximum. This is commonly achieved by summing several frames in video-based systems or by variable shutter speeds in charge-coupled diode (CCD) digital cameras. A monochrome camera is usually preferred. Slow scan, cooled CCD cameras offer the best performance in terms of sensitivity and signal-to-noise ratio. Our system (Harris, Paul) is a Princeton Instruments cooled CCD camera with a Kodak 768 × 512 chip; each CCD is 9 × 9 μm.
2. The effective size of pixels in the image should be small so that each bacterial cell is represented by many pixels. Current systems use a pixel size of 0.1 μm or less. Pixel size is a function of the size of the detector elements and the magnification of the optical system. The detector elements should be small ($<$10 μm) to minimize the required magnification. High magnification, by reducing depth of field, increases the problem of analyzing those parts of the image that are out of focus. Confocal microscopy can be used to solve this problem (Bloem et al. 1995b) but is expensive and slow, since several images from different focal planes are required. The signal-to-noise ratio of photomultipliers typically used in confocal microscopes is also lower than that of cooled CCD detectors.

Software

Bacteria

Analysis of the images to detect bacteria will probably include these steps:

1. *Noise elimination:* Random noise can be removed by smoothing procedures, including averaging and median filters, and by morphological operations such as opening and closing (Russ 1990). Cooled CCD cameras produce less noisy images than other systems and may not require this step.
2. *Background subtraction:* Fluorescent images of soil smears contain a highly variable, nebulous background that must be removed before the bacteria can be detected. Morphological transforms called "top hat" are effective means to isolate small bright features (Bloem et al. 1995b; Bright and Steel 1987) and to effectively eliminate the background.
3. *Edge detection:* The intensity profile along a chord passing through an image of a bacterial cell approximates a Gaussian peak. The true edge is estimated either as the maximum gradient in the intensity profile (Sieracki et al. 1989) or as the half-height of the peak (Bloem et al. 1995b).
4. *Segmentation:* The gray-scale image is segmented at some threshold intensity; those parts of the image above the threshold are retained (presumed bacteria), and the rest are set to zero. Features outside a set range of area may also be rejected at this stage.
5. *Group decomposition:* Some bacteria occur either as dividing cells or as small groups. Image segmentation will frequently fail to separate these groups into individual cells. Groups can be found by searching for local maxima in the gray-scale image of the detected features. More than one local maximum suggests a group. This test is sensitive to noise and may require further image smoothing before use. Individual cells in groups can be reconstructed by isotropically expanding the local maxima until the segment boundary is reached or until further expansion would cause the "cells" again to merge (Bloem et al. 1995b). A related technique replaces pixel values with the distance to the nearest edge and then divides the feature along the watershed(s) of the distance values (Russ 1990); this technique has the advantage of preserving the size relationships of dissimilar structures like budding yeast cells. Another approach is to look for and connect inflections or turn points in the boundaries of groups (Dubuisson et al. 1994). Counts of dividing cells can be used as an estimate of the growth rate of the population (Bloem et al. 1995b).
6. *Cell measurement and biovolume estimation:* Most image analysis software readily calculates the area and perimeter of detected features and also the major and minor diameters of the minimum ellipse that will contain the cell. Cell volumes (*prolatev*) can be calculated from the major (l) and minor (d) diameters by assuming the bacterial cells to be prolate spheroids

$$prolatev = \pi/6 \times d^2 \times l$$

These assumptions work well for cocci and short rods but overestimate the volume of long rods and more complex shapes. Baldwin and Bankston (1988) de-

Table Appendix 1. Examples of Scripts Used to Count Bacteria and Fungi

Action	Comment
Bacteria	
Set variable	Width of "*Tophat brim*" 1
Set variable	Width of "*Tophat*" 15
Set variable	Height of "*Tophat*" 100
Start	Label to mark start of looping section
Open	Opens image file from supplied list
Show image	
Duplicate window	Copies original image; we work on the copy
Show image	Duplicate
Linear filter	Convolution with the 3×3 kernel 0,−1,0,4,−1,0,−1,0 (approx. Laplacian)
Linear filter	Convolution with the 3×3 kernel 3,5,3,5,8,5,3,5,3 (approx. Gaussian $s = 1$)
Tophat	"Horizontal" pass detects peaks in intensity profile, subtracts background; result in new window
Rename window	Renames result of "*Tophat*"—peaks "*h*"
Change window	Back to copy of original
Rotate and scale	Rotates copy of original 90°
Tophat	"Vertical" pass detects peaks in intensity profile, subtracts background; result in new window
Rename window	Renames results of "*Tophat*" peaks "*v*"
Rotate and scale	Rotates peaks v − 90°
Image arithmetic	Maximum value of peaks *h* and *v;* result to peaks *h,* which now contain the net result of peaks detected by both "horizontal" and "vertical" passes of *Tophat;* background is zero
Dispose window	Peaks *v*
Dispose window	Copy of original
Change window	Peaks *h*
Segmentation	Segments image at threshold intensity value 1
Change window	Original image
Transfer attributes	Copies overlay of segmented image (detected features) onto original image
Measurement options	Labels each measurement with image name, erases segments excluded from measurement, fills holes in segments
Set measurements	Collect area, perimeter, major and minor axes for each detected feature; excludes those not meeting set criteria" area $> 5 < 200$ pixels
Measure segments	Measure features matching set criteria
Dispose window	Clean up
Dispose window	Clean up again
Loop	Go to start and get another image from list
End	
Fungi	
Set variable	Width of "*Tophat brim*" 1
Set variable	Width of "*Tophat*" 10
Set variable	Height of "*Tophat*" 100
Start	Label to mark start of looping section
Open	Opens image file from supplied list
Show image	
Duplicate window	Copies original image; we work on the copy
Show image	Show copy
Tophat	"Horizontal" pass detects peaks in intensity profile, subtracts background; result in new window

(continued)

Table Appendix 1 *(continued)*

Action	Comment
MiniMax	"Horizontal" pass retains the value of pixels > ½ maximum peak height, sets others to zero
Rename window	MiniMax "*h*"
Dispose window	*Tophat h*
Change window	Copy of original
Rotate and scale	Rotates copy of original 90°
Tophat	"Vertical" pass detects peaks in intensity profile, subtracts background; result in new window
MiniMax	"vertical" pass retains the value of pixels > ½ maximum peak height, sets others to zero
Rename window	MiniMax *v*
Dispose window	*Tophat v*
Rotate and scale	Rotates MiniMax *v* − 90°
Image arithmetic	Maximum value of MiniMax *h* and MiniMax *v*; result to MiniMax *h*, which now contain the net result of peaks detected by both "horizontal" and "vertical" passes of *Tophat* Background is zero
Dispose window	MiniMax *v*
Dispose window	Copy of original
Change window	MiniMax *h*
Median filter	Sets each pixel to the median value of the 9 pixels in its 3 3 3 neighborhood. Noise reduction
Segmentation	Segments image at threshold intensity value 1
Change window	Original image
Transfer attributes	Copies overlay of segmented image (detected features) onto original image
Measurement options	Labels each measurement with image name, erases segments excluded from measurement, fills holes in segments
Set measurements	Set criteria: area > 200 pixels, radial standard deviation > 35
Measure segments	Measures area of each detected feature matching criteria
Set measurements	No criteria
Modify segments	Erode each segment to a single connected line of pixels (skeletonize)
Measure segments	Area of skeleton estimates length
Dispose window	Original
Loop	Return to start
End	

scribe a stereological calculation of cell volume based on area and perimeter measurements. This avoids the difficult measurement of the diameter of the cell but is sensitive to overestimation of the perimeter due to "pixelation," resulting in underestimation of cell volumes. Sieracki et al. (1987) describe a method in which cell volume is estimated by rotating the cell outline about its long axis and summing the volumes of the disks swept by each edge pixel. This method is accurate for a wide range of cell shapes but is computationally complex.

Fungal Hyphae

Image-processing requirements are similar to those used for bacteria with some modification to detect filaments. The segmented image represents the plan area (*a*) of the hyphae in the image. Length (l) can be estimated by eroding the image of each hypha to a single connected line of pixels (skeletonization). The diameter (d) of each hyphal fragment is estimated from the plan area (a) divided by length (l):

$$d = a/l$$

Alternatively, the logical term AND of the segmented image with an image of a grid can be used to estimate the total length of hyphae in the image by the grid intercept method. If this is used, an average hyphal diameter can be calculated for all the hyphae in the image by dividing the sum of the plan areas by the total length.

Specific Examples of Image-Processing Procedures

Image-processing steps to measure bacterial and fungal hyphal biovolume in soil smears using IPLab Spectrum v 3.0 image analysis software (Macintosh PPC) are shown later in the form of "scripts," which are executable lists of processing steps (see Tab. Appendix.1). The software has the capability to repeat the script for each of a list of image files and in this way can automatically analyze numerous images as a batch process. Two externally programmed functions are used; these are the filters "Tophat" and "MinMax," which were written in the computer programming language C and incorporated into the IPLab Spectrum software as external functions.

References

Amato, M., and J. N. Ladd. 1988. Assay for microbial biomass based on ninhydrin-reactive nitrogen in extracts of fumigated soil. *Soil Biology and Biochemistry* 20:107–114.

Anderson, J. P. E., and K. H. Domsch. 1978. A physiological method for the quantitative measurement of microbial biomass in soils. *Soil Biology and Biochemistry* 10:215–221.

Badalucco, L. P., P. Nannipieri, and S. Grego. 1990. Microbial biomass and anthrone-reactive carbon in soils with different organic matter contents. *Soil Biology and Biochemistry* 22:899–904.

Bakken, L. R., and R. A. Olsen. 1983. Buoyant densities and dry-matter contents of microorganisms: conversion of a measured biovolume to biomass. *Applied and Environmental Microbiology* 45:1188–1195.

Baldwin, W. W., and P. W. Bankston. 1988. Measurement of live bacteria by Nomarski interference microscopy and stereologic methods as tested with macroscopic rod-shaped models. *Applied and Environmental Microbiology* 54:105–109.

Bloem, J., P. R. Bolhuis, R. M. Vininga, and J. Wierenga. 1995a. Microscopic methods to estimate the biomass and activity of soil bacteria and fungi. Pages 162–174 *in* K. Alef and P. Nannipieri, editors, *Methods in Soil Microbiology and Biochemistry.* Academic Press, New York, New York, USA.

Bloem, J., D. K. van Mullem, and P. R. Bolhuis. 1992. Microscopic counting and calculation of species abundances and statistics in real time with an MS DOS personal computer, applied to bacteria in soil smears. *Journal of Microbial Methods* 16:203–213.

Bloem, J., M. Veninga, and J. Sheperd. 1995b. Fully automatic determination of soil bacterium numbers, cell volumes, and frequencies of dividing cells by confocal laser scanning microscopy and image analysis. *Applied and Environmental Microbiology* 61:926–936.

Bratbak, G., and I. Dundas. 1984. Bacterial dry matter content and biomass estimations. *Applied and Environmental Microbiology* 48:755–757.

Bremer, E., and C. van Kessel. 1990. Extractability of microbial ^{14}C and ^{15}N following addition of variable rates of labeled glucose and $(NH_4)_2SO_4$ to soil. *Soil Biology and Biochemistry* 22:707–713.

Bright, D. S., and E. B. Steel. 1987. Two-dimensional top-hat filter for extracting spots and spheres from digital images. *Journal of Microscopy* 146:191–200.

Cabrera, M. L., and M. H. Beare. 1993. Alkaline persulfate oxidation for determining total nitrogen in microbial biomass extracts. *Soil Science Society of America Journal* 57:1007–1012.

Christensen, H., and S. Christensen. 1995. ^{3}H thymidine incorporation technique to determine soil bacterial growth rate. Pages 258–261 *in* K. Alef and P. Nannipieri, editors, *Methods in Applied Soil Microbiology and Biochemistry.* Academic Press, London, UK.

Daniel, O., F. Schönholzer, and J. Zeyer. 1995. Quantification of fungal hyphae in leaves of deciduous trees by automated image analysis. *Applied and Environmental Microbiology* 61:3910–3918.

Dubuisson, M., A. K. Jain, and M. K. Jain. 1994. Segmentation and classification of bacterial culture images. *Journal of Microbiological Methods* 19:279–295.

Dutton, R. J., G. Bitton, and B. Koopman. 1983. Malachite green–INT (MINT) method for determining active bacteria in sewage. *Applied and Environmental Microbiology* 46:1263–1267.

Fry, J. C. 1990. Direct methods and biomass estimation. *Methods in Microbiology* 22:41–85.

Hanssen, J. F., T. F. Thingstad, and J. Goksyr. 1974. Evaluation of hyphal lengths and fungal biomass in soil by a membrane filter technique. *Oikos* 25:102–107.

Harris, D., and E. A. Paul. 1994. Measurement of bacterial growth rates in soil. *Applied Soil Ecology* 1:227–290.

Harris, D., L. K. Porter, and E. A. Paul. 1997. Continuous flow isotope ratio mass spectrometry of $^{13}CO_2$ trapped as strontium carbonate. *Communications in Soil Science Plant Analyses* 28:747–757.

Harris, D., R. P. Voroney, and E. A. Paul. 1998. Measurement of microbial biomass N:C ratio by chloroform fumigation-incubation. *Canadian Journal of Soil Science.* 77:507–514.

Horwath, W. R., and E. A. Paul. 1994. Microbial biomass. Pages 753–773 *in* R. W. Weaver, J. S. Angle, and P. S. Bottomley, editors, *Methods of Soil Analysis. Part 2, Microbiological and Biochemical Properties.* Soil Science Society of America, Madison, Wisconsin, USA.

Horwath, W. R., E. A. Paul, D. Harris, J. Norton, L. Jagger, and K. A. Horton. 1996. Defining a realistic control for the chloroform-fumigation incubation method using microscopic counting and ^{14}C-substrates. *Canadian Journal of Soil Science* 76:459–467.

Ingham, E. R., and D. A. Klein. 1984. Soil fungi: relationships between hyphal activity and staining with fluorescein diacetate. *Soil Biology and Biochemistry* 16:273–278.

Inubushi, K., P. C. Brookes, and D. S. Jenkinson. 1989. Influence of paraquat on the extraction of adenosine triphosphate from soil by trichloracetic acid. *Soil Biology and Biochemistry* 21:741–742.

Inubushi, K., P. C. Brookes, and D. S. Jenkinson. 1991. Soil microbial biomass C, N and ninhydrin-N in aerobic and anaerobic soils measured by the fumigation-extraction method. *Soil Biology Biochemistry* 23:737–741.

Inubushi, K., R. Wada, and Y. Takai. 1984. Determination of microbial biomass nitrogen in submerged soil. *Soil Science Plant and Nutrition* 30:455–459.

Jenkinson, D. S. 1976. The effects of biocidal treatments on metabolism in soil. IV. The decomposition of fumigated organisms in soil. *Soil Biology and Biochemistry* 8:203–208.

Jenkinson, D. S. 1988. Determination of microbial biomass carbon and nitrogen in soil. Pages 368–386 *in* J. R. Wilson editor, *Advances in Nitrogen Cycling in Agricultural Ecosystems.* CAB International, Wallingford, UK.

Jenkinson, D. S., and D. S. Powlson. 1976. The effects of biocidal treatments on metabolism in soil. V. A method for measuring the soil biomass. *Soil Biology and Biochemistry* 8:209–213.

Joergensen, R. G. 1995. Microbial biomass: the fumigation method. Pages 377–396 *in* K. Alef and P. Nannipieri, editors, *Methods in Applied Soil Microbiology and Biochemistry.* Academic Press, London, UK.

Kepner, R. L., Jr., and J. R. Pratt. 1994. Use of fluorochromes for direct enumeration of total bacteria in environmental samples: past and present. *Microbiological Reviews* 58:603–615.

Merckx, R., and J. K. Martin. 1987. Extraction of microbial biomass components from rhizosphere soils. *Soil Biology and Biochemistry* 19:371–376.

Morgan, P., C. J. Cooper, N. S. Battersby, S. A. Lee, S. T. Lewis, T. M. Machin, S. C. Graham, and R. J. Watkinson. 1991. Automated image analysis method to determine fungal biomass in soils and on solid matrices. *Soil Biology and Biochemistry* 23:609–616.

Newell, S. Y. 1992. Estimating fungal biomass and productivity in decomposing litter. Pages 521–561 *in* G. C. Carroll and D. T. Wicklow, editors, *The Fungal Community. 2d edition.* Marcel Dekker, New York, New York, USA.

Newman, E. I. 1966. A method for estimating total length of root in a sample. *Journal of Applied Ecology* 3:139–145.

Norland, S. 1993. The relationship between biomass and volume of bacteria. Pages 303–307 *in* P. F. Kemp, editor, *Handbook of Methods in Aquatic Microbial Ecology.* Lewis Publishers, Boca Raton, Florida, USA.

Rodriguez, G. G., D. Phipps, K. Ishiguro, and H. F. Ridgway. 1992. Use of a fluorescent redox probe for direct visualization of actively respiring bacteria. *Applied and Environmental Microbiology* 58:1801–1808.

Russ, J. 1990. *Computer-Assisted Microscopy: The Measurement and Analysis of Images.* Plenum Press, New York, New York, USA.

Sieracki, M. E., S. E. Reichenbach, and K. E. Webb. 1989. Evaluation of automated threshold selection methods for accurately sizing microscopic fluorescent cells by image analysis. *Applied and Environmental Microbiology* 55:2762–2772.

Sieracki, M. E., C. L. Viles, and K. L. Webb. 1987. Algorithm to estimate cell biovolume using image analyzed microscopy. *Cytometry* 10:551–557.

Smith, J. L., J. J. Halvorson, and H. Bolton Jr. 1995. Determination and use of a corrected control factor in the chloroform fumigation method of estimating microbial biomass. *Biology and Fertility of Soils* 19:287–291.

Smith, J. L., and E. A. Paul. 1990. The significance of soil microbial biomass estimations. Pages 357–396 *in* J. M. Bollag and G. Stotzky, editors, *Soil Biochemistry.* Volume 6, Marcel Dekker, New York, New York, USA.

Söderström, B. E. 1979. Some problems in assessing the fluorescein diacetate–active fungal biomass in the soil. *Soil Biology and Biochemistry* 11:147–148.

Sparling, G. P., and A. W. West. 1988. A direct extraction method to estimate soil microbial C: calibration *in situ* using microbial respiration and ^{14}C-labeled cells. *Soil Biology and Biochemistry* 20:337–343.

Sparling, G. P., and A. W. West. 1989. Importance of soil water content when estimating soil microbial C, N and P by the fumigation-extraction method. *Soil Biology and Biochemistry* 21:245–253.

Sparling, G., and C. Zhu. 1993. Evaluation and calibration of biochemical methods to measure microbial biomass C and N in soils from western Australia. *Soil Biology and Biochemistry* 25:1793–1801.

Sundman, V., and S. Sivelä. 1978. A comment on the membrane filter technique for estimation of length of fungal hyphae in soil. *Soil Biology and Biochemistry* 10:399–401.

Tsuji, T., Y. Kawasaki, S. Takeshima, T. Sekiya, and S. Tanaka. 1995. A new fluorescence staining assay for visualizing living microorganisms in soil. *Applied and Environmental Microbiology* 61:3415–3421.

Tunlid, A., and D. C. White. 1992. Biochemical analysis of biomass, community structure, nutritional status and metabolic activity of microbial communities in soil. Pages 229–262 *in* J. M. Bollag and G. Stotzky, editors, *Soil Biochemistry.* Vol. 6. Marcel Dekker, New York, New York, USA.

Vance, E. D., P. C. Brookes, and D. S. Jenkinson. 1987. An extraction method for measuring soil microbial biomass C. *Soil Biology and Biochemistry* 19:703–707.

van Veen, J. A., and E. A. Paul. 1979. Conversion of biovolume measurements of microorganisms grown under various moisture tensions, to biomass and their nutrient content. *Applied and Environmental Microbiology* 37:686–692.

Voroney, R. P., and E. A. Paul. 1984. Determination of k_c and k_N *in situ* for calibration of the chloroform fumigation incubation method. *Soil Biology and Biochemistry* 16:9–14.

Voroney, R. P., J. P. Winter, and R. P. Bayaert. 1993. Soil microbial biomass C and N. Pages 277–286 *in* M. R. Carter, editor, *Soil Sampling and Methods of Analysis.* Lewis Publishers, Boca Raton, Florida, USA.

Wardle, D. A., and A. Ghani. 1995. Why is the strength of relationships between pairs of methods for estimating soil microbial biomass often so variable? *Soil Biology and Biochemistry* 27:821–828.

Wu, J., D. S. Jenkinson, and P. C. Brookes. 1996. Evidence for the use of a control in the fumigation-incubation method for measuring microbial biomass carbon in soil. *Soil Biology and Biochemistry* 28:511–518.

16

Characterizing Soil Microbial Communities

Robert L. Sinsabaugh
Michael J. Klug
Harold P. Collins
Phillip E. Yeager
Søren O. Petersen

The soil microbial community is intimately linked to primary production, soil structure, and biogeochemical cycling. The analyses described in this chapter provide structural or functional information about the soil microbiota beyond that provided by integrative measures of biomass or respiration. These methods of community analysis can be separated into two categories: those that generate phenotypic data and those that generate genotypic data (Fig. 16.1). The former includes lipid analysis, substrate utilization profiles, and extracellular enzyme activity assays. The latter includes a variety of techniques for analyzing extracted nucleic acids. With all these approaches, the objective is to acquire an unbiased "snapshot" of the whole microbial community, bypassing the well-known problems of culture-based methods. Unfortunately, whole-community analyses also have limitations; these are discussed within each section.

Lipid analysis provides information on the taxonomic composition of the microbiota. It is based on the extraction and quantification of membrane fatty acids, some of which are "signature" molecules for specific taxa, principally bacteria. This approach is the most resource-intensive of the phenotypic methods described.

Substrate utilization profiles provide information on the functional diversity of soil bacterial communities. Suspended organisms are dispensed into the wells of microplates. Each well contains a different growth substrate plus an indicator dye. The pattern of growth over the matrix of potential substrates provides comparative data on community organization. This approach requires fewer resources than fatty acid analyses, but its ecological significance is more difficult to interpret.

Enzyme assays provide functional information about the microbiota. Bacteria

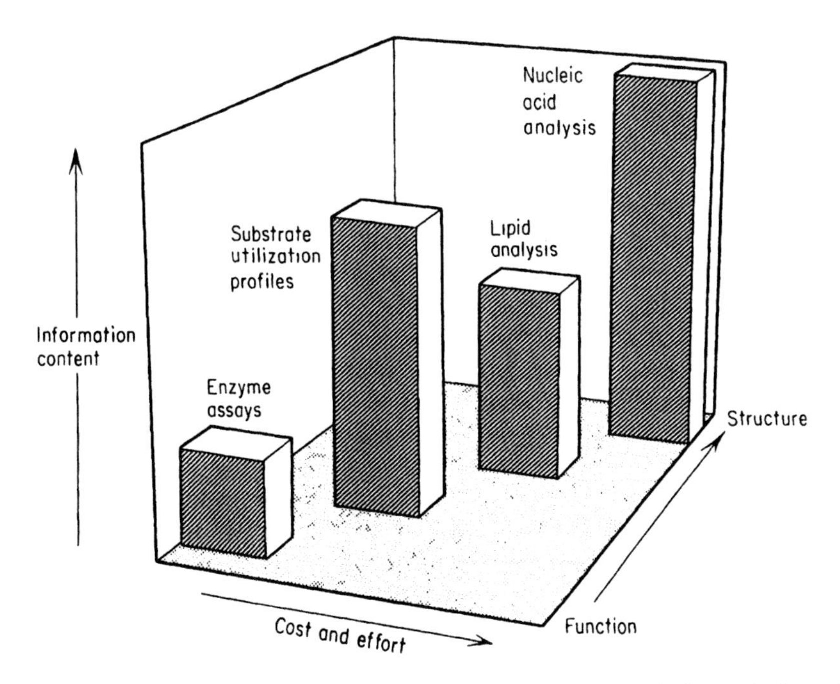

Figure 16.1. Categorical comparison of "whole-community" methods for analyzing soil microbiota with respect to the relative quantity of information obtained, the nature of the information, and the cost and effort required to conduct the analyses.

and fungi deploy enzymes extracellularly to break down organic matter into assimilable forms. By quantifying the potential activity of these enzymes, it is possible to make inferences about the relative effort directed by microorganisms toward obtaining carbon, nitrogen, or phosphorus from specific sources. This approach requires the fewest resources and provides the least resolution.

Nucleic acid analyses provide information on the structural composition of the microbial community. A variety of techniques can be used to assess diversity within a sample, to compare similarities among samples, or to estimate the relative abundance of specific taxa. In general, these methods are more fastidious than phenotypic methods and require more specialized equipment and training.

All these approaches can be applied to demonstrate spatial and temporal variability in microbial communities on virtually any scale. However, the ecological significance of this variability may be difficult to assess. Primarily these methods are used to investigate the organization and dynamics of microbial communities in relation to other ecological variables. Because current knowledge of these relationships is fragmentary, it may be difficult to mechanistically connect these changes in community patterns with other soil properties.

Overview

Lipid Analyses and Substrate Utilization Profiles

Descriptions of soil microbial community structure and function have been largely based upon measures of species composition using isolate-based methods (Mills and Wassel 1980; Tate and Mills 1983; Kinkel et al. 1992). This approach has been shown to underestimate diversity, since less than 10% of the total soil microflora observed by microscopy is isolated (Brock 1987; Torsvik et al. 1990b). Sorheim et al. (1989) concluded that there was no direct correlation between plate isolation and bacterial diversity. Measures of microbial diversity rely not only on the types of organisms present but also on their relative abundance. The latter has rarely, if ever, been accurately determined. Due to the historical difficulties inherent in measuring soil microbial community structure, few efforts have been made to link microbial community structure and function. One link that has been made is a positive correlation between C and N mineralization rates and the size of the microbial biomass following disturbance (Insam et al. 1989; Collins et al. 1992; Henrot and Robertson 1994). Recent evidence, however, suggests that the composition of the biomass may also be critical in determining the rate and efficiency of particular processes. It is also generally assumed that the functional redundancy found among soil microbial populations acts as a buffer against shifts in environmental conditions, i.e., that populations within a group have different environmental optima expressing them at different levels of environmental stress.

Recent advancements in biochemical methods now make it possible to study in situ microbial communities without the problems associated with the isolation or removal of cells from the environment (Tunlid and White 1992; Klug and Tiedje 1993; Petersen and Klug 1994). These biochemical methods examine the entire community and therefore provide a quantitative description of structure and function of the microflora within a particular environment. Two methods that historically were isolate-based are now being adapted to identify differences in the structure and function of microbial communities from diverse ecosystems (e.g., soils, sediments, biofilms). The first characterizes the composition of microbial biomass by identifying extracted cellular phospholipid ester–linked fatty acids (PLFA; White 1986; Guckert et al. 1985; Findlay et al. 1990; Zelles et al. 1992). Results from field studies have also shown that microbial communities under differing agricultural management profiles can be distinguished using a simplified extraction of cellular fatty acid methyl esters (FAME; Cavigelli et al. 1995). The second approach, developed by Garland and Mills (1991), uses substrate utilization profiles of soil communities to assess community structure.

Enzyme Assays

The first paper on enzyme activity in soils was published in 1899, but it was not a topic of extensive investigation until the 1960s (Skujins 1978). During the 1960s and 1970s, much of the work on soil enzymes was directed toward classifying and diagnosing agricultural soils. As information on microbial physiology and the biochemistry of plant fiber decomposition accumulated, enzymic analyses became increasingly useful for functional comparisons of microbial communities and for development of models for microbial processes (Sinsabaugh et al. 1994; Sinsabaugh and Moorhead 1994).

Substrate utilization by microheterotrophs is mediated by enzyme systems arrayed extracellularly. These systems include those involved in breaking covalent bonds (hydrolases and oxidases) and those involved in transporting molecules across the cell membrane (permeases). In natural systems it is generally considered that the rate of generation of utilizable substrates from polymeric or condensed molecules limits microbial metabolism.

The production of extracellular enzymes is presumably expensive, especially in terms of nitrogen, and once deployed the cell has no control over their activity or turnover. However, there are feedback mechanisms that promote the efficient use of cellular resources. When an adequate supply of directly utilizable substrates exists, the enzyme production is repressed at the level of transcription. But even under repression conditions, low-level, constitutive synthesis may occur. When the supply of utilizable substrate dwindles, the end products of constitutive activity may induce the production of new enzymes by triggering a signal cascade that ultimately derepresses transcription. Derepression results in the secretion of additional enzyme until the point at which excess utilizable substrate triggers a repression feedback. Once transported outside the cell, the activity and persistence of enzymes are subject to environmental control by physicochemical variables such as temperature, moisture, and pH and by the relative abundance of substrates and inhibitors. Most enzymes are subject to competitive inhibition by the end products of their activity, which may accumulate if products are generated in excess of utilization. Extracellular enzymes are also subject to other forms of inhibition. The most ubiquitous inhibitors are humic substances, which typically affect enzymes by enhancing their stability (extending turnover time) and reducing their activity (in Michaelis-Menton terms, by increasing K_m and decreasing V_{max}); the sorption of enzymes by clays can exert similar effects (Burns 1983; Sinsabaugh and Linkins 1987; Lahdesmaki and Piispanen 1992).

In practice, assays for enzyme activity in soil and litter are simple. The difficulties lie in choosing assay conditions and in interpreting the results. The basic decision is whether to measure activities under conditions as close as possible to ambient or whether to make standardized comparisons among samples. In the protocols that follow, the latter strategy is presumed. Under that strategy, perhaps the best way to interpret the results is to consider them a measure of the effort or resources that a microbial community is allocating toward decomposing or acquiring a category of substrates.

The enzymes of interest are generally those involved in the decomposition of lignocellulose and its derivatives and in the acquisition of organic nitrogen and phosphorus. These include cellulases, hemicellulases, pectinases, phenol oxidases and peroxidases, chitinases, peptidases, and phosphatases. Each of these functional categories includes multiple, specific enzymes. In general, a different enzyme is required for each type of linkage and each type of monomer. Also enzymes that act along the interior of polymers are usually distinct from those that attack free ends.

Reviews of these enzyme systems include Biely (1985); Blanchette (1991); Burns (1983); Chrst (1991); Dekker (1985); Eriksson and Wood (1985); Eriksson et al. (1990); Higuchi (1990); Kirk and Farrell (1987); Ljungdahl and Eriksson (1985); Marsden and Gray (1986); Rayner and Boddy (1988); Sakai et al. (1993); Skujins (1976); Viikari et al. (1993); and Wong et al. (1988).

Like enzymes, assays have varying degrees of specificity. It is possible to assay the activity of high enzyme categories by measuring the end product of their collective action, e.g., measure cellulase activity as the rate of glucose generation from cellulose, or to assay the activity of a specific hydrolase, e.g., cellobiohydrolase. The protocols described in the following are suggested starting points, selected because of their simplicity and suitability for soils. They are not comprehensive and may not be useful, depending on specific objectives.

Nucleic Acid Analyses

Nucleic acid or "molecular" analyses provide the most comprehensive information on microbial community composition. It is the only approach available for estimating the absolute diversity of soil microbiota (Torsvik et al. 1990a,b) and in many cases is the only available approach for monitoring the dynamics of specific taxa. Molecular ecology is developing rapidly; new techniques, equipment, and reagents for existing techniques are introduced continually.

Lipid Analysis—Direct Extraction of FAMEs from Soil

Lipids are essential components of every cell, representing less than 5% of the dry mass (Kennedy 1994). Lipids have distinct structure and form that are characteristic of specific genera and species (Tab. 16.1), and since lipid composition is highly conserved within taxa, lipids can provide an estimate of in situ community structure. The study of soil microbial communities by lipid (FAME) analysis involves the extraction of lipids from a sample with organic solvents followed by analysis of the extracted material. Extracted lipids are analyzed by high-resolution capillary gas chromatography. This technique is rapid and can be run for a large number of samples.

Materials

1. 25–35 mL glass Teflon-capped centrifuge tubes
2. 13 × 100 mm Teflon-capped test tubes (precombusted at 550 °C)
3. Optional: acid bath for cleaning 13 mm screw-cap test tubes
4. Optional: muffle furnace
5. Two water baths, variable temperature: 80–100 °C
6. Vortex mixer
7. End-over-end mixer
8. Pasteur pipettes
9. Centrifuge

Table 16.1. Marker Fatty Acids

Eubacteria*			
14:0 OH	15:1 iso 7 trans mono	17:0 cyc	
15:0 br	saturated branched and 17.1 at 6	17:1 at 6	
15:1 at 6	straight 16 and 17C	17:1 at 8	
15:1 at 8	16:0 br 10	18:1 cis 11	
15:1 iso 3	17:0 br	19:0 cyc	
Gram-negative eubacteria			
fatty acids (usually 3 OH)			
Gram-positive eubacteria			
branched fatty acids (iso, anteiso) also found in gram-negative eubacteria			
Eukaryotes			
12:0	16:1 at 7	18:1 at 9	18:2 cis 9, cis 12
0-18:3 cis 9, cis 12, cis 15 polyunsaturated fatty acids with > 20 C			
Protozoa			
20:3 at 6	20:4 at 6		
Microfauna			
FE-18:3			
All Organisms			
14:0	16:0	18:0	

*Eubacteria do not, in general, contain polyunsaturated fatty acids.

10. Brown gas chromatography (CG) vials
11. Optional: −20 °C freezer
12. Gas chromatograph: programmable 170–270 °C at 5 °C per minute equipped with an HP Ultra 2 capillary column (cross-linked 5% Ph Me silicone, 25 m × 0.2 mm × 0.33 mm film thickness), a flame ionization detector, and integrator
13. 6 N H_2SO_4 (used in the acid bath)
14. Saponification reagent: 3.25 N NaOH in 50% methanol
15. Methylation reagent: 6.0 N hydrochloric acid (HCL) in 50% methanol
16. Extraction solvent: hexane and methyl tert-butyl ether (MTBE) (1:1 vol:vol)
17. Base wash: 0.33 N sSodium hydroxide
18. Hexane
19. Highly purified methyl ester standards

Procedure

The procedure outlined here is modified from a commercially available method (MIDI; Microbial ID, Inc., Newark, DE) for fatty acid analysis of bacterial cell extracts. The method is also similar to that described by Kennedy (1994).

1. Prepare the four reagents in clean, brown, 1 L bottles. Place a Teflon-coated stir bar in each bottle to facilitate mixing. Only Teflon and glass should come in contact with the solutions; rubber and plastic will contaminate the reagents. The 13 mm tubes can be either acid washed overnight in 6 N H_2SO_4 or ashed in a muffle furnace at 550 °C. Centrifuge tubes should only be acid washed and not ashed, since high temperature will weaken the glass and cause breakage during centrifugation.
2. Weigh 1.0 g of soil directly into an acid-washed 25–35 mL glass centrifuge tube with a Teflon-coated cap.
3. Saponify soil samples by adding 5 mL of 3.25 mol/L NaOH in 50% methanol, vortex, and place in a 100 °C water bath for 30 minutes. Remove the tubes from the water bath and place in cool water.
4. After cooling, add 3 mL of 6 N HCL in 50% methanol (vol:vol), vortex, and place tubes in an 80 °C water bath for 10 minutes. This step methylates the free fatty acids. Do not allow the temperature to vary more than 1° during this step. Remove the tubes and place again in the room temperature water bath to cool.
5. Add 1.25 mL of a 1:1 mixture of MTBE and hexane (vol:vol), vortex, and place on an end-over-end mixer for 10 minutes.
6. Centrifuge the tubes for 3 minutes at about 3000 × G to separate the phases.
7. Transfer the organic (top) phase to an acid-washed or precombusted 13 × 100 mm Teflon-capped tube and store in a 4 °C refrigerator.
8. Repeat steps 4 through 6. Add the second organic phase to the tube containing the first phase.
9. To the combined organic phases add 3 mL of 0.33 N NaOH to wash the phase, and mix.
10. Allow the samples to stand for a clear separation or centrifuge at a low speed.
11. Remove the organic (top) phase to a clean 13 mm test tube and dry the sample under a stream of nitrogen. The samples can be stored in the −20 °C freezer until run on the GC.
12. Before analysis, resuspend the FAMEs in 200–500 μL of hexane/MTBE and transfer to a GC vial. It is best to store the samples in a freezer as the dried lipid if the sample will not be analyzed by GC promptly.

Lipid Analysis—Direct Extraction of Phospholipid FAME's from Soil

This assay is based on a modified Bligh–Dyer single phase extraction (Bligh and Dyer 1959) in dichloromethane (DCM)–methanol-phosphate buffer (Petersen and Klug 1994). Following the extraction, additional DCM and sodium bromide (NaBr) are added to separate the extract into an organic and a water-soluble phase. The primary difference between this modification and other extraction procedures (e.g., White et al. 1979) is the use of DCM instead of chloroform. The lower density of DCM in combination with the heavy NaBr solution places the organic phase, containing the lipids, on top and separated from the residue. Phospholipids in the dried extracts are isolated by solid-phase extraction. The lipid material is transferred to silicic acid columns, where lipids of low or intermediate polarity are eluted with chloroform and acetone, respectively, and subsequently polar lipids (mainly phospholipids) are eluted with methanol. Finally, phospholipid fatty acids are converted to the corresponding methyl esters by a mild alkaline transesterification and analyzed by gas chromatography.

Materials

1. 25–35 mL glass Teflon-capped centrifuge tubes
2. 13 × 100 mm Teflon-capped test tubes (precombusted at 550 °C)
3. Acid bath for cleaning the centrifuge tubes
4. Muffle furnace for cleaning of test tubes, Pasteur pipettes, GC vials, and inserts
5. Vortex mixer
6. Stirring hot plate
7. Pasteur pipettes
8. Centrifuge
9. Side-arm Erlenmeyer flasks
10. SPE columns (100 mg silicic acid)
11. Water bath
12. GC vials with 100 μL inserts
13. Gas chromatograph: programmable 170–270 °C at 5 °C per minute equipped with an HP Ultra 2 capillary column (cross-linked 5% Ph Me silicone, 25 m × 0.2 mm × 0.33 mm film thickness), a flame ionization detector and integrator
14. 6 N analytical-grade Sulfuric acid (H_2SO_4), used in the acid bath
15. 50 mmol/L potassium phosphate (K_2HPO_4) buffer, pH 7.4
16. High performance liquid chromatograph (HPLC)–grade methanol: dichlormethane (DCM) (2:1 vol:vol)
17. Sodium bromide (NaBr) (0.8 g/mL)
18. HPLC-grade chloroform
19. HPLC-grade acetone
20. HPLC-grade methanol
21. HPLC-grade methanol:toluene (1:1 vol:vol)
22. 0.2 N potassium hydroxide (KOH) in 50% methanol
23. n-Hexane, HPLC-grade
24. 1mmolAVEacetic acid, HPLC-grade
25. Nonadecanoate methyl ester standard (100–200 μM) in n-hexane

Procedure

Extraction of Lipids

1. Weigh 1–2 g of soil directly into an acid-washed 25–35 mL glass centrifuge tube with a Teflon-coated cap. Add phosphate buffer to bring the final aqueous volume to 1.9–2.0 mL.
2. Add 5.0 mL methanol and 2.5 mL DCM to each centrifuge tube. The final ra-

tio of DCM, methanol, and phosphate buffer/water per sample should be 1:2:0.8.

3. The tubes are capped, vortexed for 1 minute, and extracted at room temperature for 2–3 hours.
4. Phase separation: Add 2.5 mL DCM followed by 10 mL NaBr solution. The NaBr solution is heated (≤40 °C) to dissolve the NaBr, cooled to 40 °C before use; the solution is supersaturated and will crystallize, so constant stirring is required to obtain a uniform solution. The tubes are vortexed for 1 minute and left at room temperature overnight.
5. The next morning the tubes are centrifuged for 10 minutes at 3000 × G. If the organic phase is not on top of the aqueous phase, 5 mL of the aqueous phase may be replaced with fresh NaBr (heated and cooled); centrifuge without vortexing.
6. With a Pasteur pipette, transfer a portion of the organic phase to a precombusted test tube and cover with a Teflon-lined screw cap. Weigh each test tube, then remove the DCM under a stream of nitrogen in a fume hood while gently heating (using a sand-filled heating plate). Reweigh the tubes after drying. The lipid mass of the total DCM phase can be calculated from the weight difference.

Separation of Phospholipids

1. SPE columns (up to 10 at a time) are mounted on a vacuum manifold. The outlet from the vacuum chamber is connected to a vacuum pump via a sidearm Erlenmeyer flask for collection of solvents other than the methanol containing the lipids.
2. The columns are conditioned with 1.5 mL chloroform (with 1 mL reservoirs 2 × 0.75 mL). The solvent is pulled as slowly as possible through the columns by applying a small vacuum. Between additions the vacuum is turned off to prevent the column from drying.
3. The dried lipid extracts are then transferred to individual columns in 3 × 100 μL chloroform. Draw the chloroform slowly through the columns.
4. After the chloroform/lipid extract has passed through the column, wash each column with 1.5 mL chloroform, followed by 6 mL acetone. Only polar lipids are now retained within the column matrix.
5. After the acetone is gone, turn off the vacuum, remove the lid of the vacuum chamber, and mount the rack with appropriately marked precombusted test tubes to collect the phospholipids which are eluted with 1.5 mL methanol under vacuum.
6. The methanol is evaporated under a stream of nitrogen in a fume hood.
7. Before analysis, resuspend the phospholipids in 200–500 μL of hexane/MTBE and transfer to a GC vial. If the samples will not be analyzed promptly, it is best to store them as dried lipid under nitrogen in a freezer.

Methylation of Phospholipid Fatty Acids

1. To the test tubes containing the dried phospholipid fraction, add 1 mL methanol:toluene, 1 mL KOH/methanol, and 100 μL nonadecanoate methyl

ester as an internal standard. Vortex briefly, then incubate for 15 minutes in a 37 °C water bath.

2. Add sequentially: 2 mL hexane, 0.3 mL 1 mol/L acetic acid, and 2 mL deionized water. Vortex for 1 minute and centrifuge at 1500 × G for 5 minutes.
3. Transfer as much as possible of the upper (organic) phase to another test tube, then add 2 mL hexane to the first tube and repeat the extraction of the fatty acid methyl esters.
4. The combined organic phase is evaporated to dryness (<25 °C) and redissolved in 100 μL hexane, which is transferred to a 2 mL GC vial with insert. Samples should be stored in a freezer as the dried extract under nitrogen if samples cannot be analyzed immediately.

Special Considerations

Identification of the extracted fatty acids and phospholipids described earlier may be based on their retention time compared with retention times of highly purified FAME standards. Concentrations of a given compound in standard and sample should be comparable. Results are reported using standard FAME nomenclature as described by Harwood and Russell (1984). Interpretation of soil community FAME profiles can be aided by the use of fatty acid (FA) markers. Fatty acids that have been found in particular groups of organisms are shown in Table 16.1 (White 1983; Harwood and Russell 1984; Vestal and White 1989; Cavigelli et al. 1995). Abiotic and biotic conditions influence the amount and types of lipids extracted. Lipid concentration varies with organism age, food resource, habitat, and predation pressure.

Clay and soil organic matter concentrations have a significant influence on the extraction of lipids. We recommend using the modified Bligh-Dyer procedure to reduce the concentration of materials that will interfere with lipid identification during GC analysis. Because of differences in extraction efficiencies among soil types, we further suggest that additional soil samples containing known concentrations of lipid standards be included within the group of soils to be extracted.

When making comparisons among different soil types or experimental treatments, it is important that the amount of extracted lipid be similar among samples. Samples containing low microbial biomass produce lipid profiles of limited diversity that can confound interpretation. The absence of particular lipids may be a function of the extraction efficiency and the detection limits of the GC rather than environmental or treatment differences.

Table 16.2. Phosphate Standards for Phospholipid-P Determination

P concentration (μmol/L)	Volume of 50 μmol/L standard (mL)	Volume of $K_2S_2O_8$ (mL)
0	0	5.0
1	0.1	4.9
2.5	0.25	4.75
5	0.5	4.5
10	1	4
15	1.5	3.5

Note: Each standard is diluted to 100 mL in a volumetric flask.

Substrate Utilization Profiles

The use of Biolog gram negative (GN) and gram positive (GP) microplates (Biolog, Inc., Hayward, CA) is becoming more prevalent in studies characterizing bacterial isolates from natural environments. The Biolog system consists of a 96-well microtiter plate containing 95 different carbon substrates plus one negative control with no carbon substrate. In addition to the substrate, the wells contain a complex of growth factors at low concentrations. Each well contains a redox dye (tetrozolium violet) that is reduced to formazan during respiratory activity. Mineralization of the substrate results in cellular accumulation of insoluble formazan. The degree to which substrates are utilized is dependent on the complexity of the microbial community inoculated into the plates and their ability to utilize and compete for the substrate under the conditions provided.

Materials

1. GN or GP microtiter plates
2. 8-channel repeating pipetter; adjustable to deliver 150 μL
3. 1000 μL pipette tips
4. 100 mL dilution bottles
5. Sterilized 100 mL disposable dispensing well
6. Optional: microtiter plate reader
7. Incubator
8. 10 mM potassium phosphate (K_2HPO_4) buffer, pH 7.4
9. Calcium chloride ($CaCl_2$)
10. Magnesium carbonate ($MgCO_3$)

Procedure

1. Add 1 g soil to a 99 mL dilution bottle containing 10 mM phosphate buffer and shake for 20 minutes.
2. After shaking, allow soil particles to settle for 30 minutes at 4 °C, or chemically flocculate the supernatant with 1 g $CaCO_3$ and $CaCl_2$ to remove suspended clay particles.
3. Dispense 150 μL of the supernatant into each well of the GN or GP microplates and incubate at 25 °C for up to 120 hours. Readings of optical density are made at 24 hour periods using a microtiter plate reader set to a wavelength of 590 nm.

Special Considerations

Use of soil extracts as inoculum has raised questions regarding the reliability of results due to the potential of nutrient carryover and interference of optical density measurements by soil particles and organic debris. Turbidity of the inoculant can also create false-positive results. Calcium chloride and magnesium carbonate can be added to flocculate soil particles and reduce interference during optical density readings. Garchow et al. (1993) reported that for a loam soil direct inoculation without flocculation resulted in a high number of false-positive wells as early as the 24 hour reading, which they attributed to interference of soil particles. Visual inspection confirmed that few wells exhibited the characteristic color change associated with respiratory activity. Direct microscopic counts showed little difference in the inoculum level between flocculated and nonflocculated samples. Allowing soil particulates to settle for 1 hour at 4 °C followed by dispensing (Vahjen et al. 1995) and serial dilution (Bossio and Scow 1995) reduces interference. Some organisms have intracellular stores of compounds that may be used instead of the substrate contained in the well. Use of these compounds can also lead to false-positive readings.

Biolog does not provide an estimate of the functional diversity of soil microbial communities but only information about members of the community that can be cultured under the conditions provided. Haack et al. (1995) constructed model communities to test the repeatability of the Biolog assay and to determine whether Biolog profiles of whole communities reflect taxonomic structure. Although the Biolog assay is easy and rapid, it should not be relied upon as a single measure describing microbial functions in the soil environment because of the difficulties in interpreting assay results. Growth of some members of the community and the failure of others to grow in the wells of Biolog microplates have important implications for the interpretation of whole-community substrate utilization profiles. First, they imply that microplates are similar to plate culture in providing selective conditions suitable for some but not all members of the community. Whole-community substrate utilization profiles are probably not, in fact, based on the whole community. Second, Garland and Mills (1991) noted that the utilization profiles obtained represent potential metabolic capabilities of community members, and not the in situ activity. There is a temptation to correlate substrate utilization profiles with community function. In the case where one community utilizes a substrate that another does not, there may be some basis to the assumption that this reflects community function, although consideration should be given to all problems addressed earlier. The response level of different communities does not suggest anything about the number of potential utilizers of a substrate, nor about the likelihood that the substrate would actually be used by the community in situ. In addition, many important organisms (e.g., actinomycetes and fungi) are excluded by the method.

Enzyme Assays for Hydrolase Activity

The most convenient enzyme assays involve hydrolysis of side groups linked to a chromogenic or fluorogenic moiety. Three families of these compounds are commercially available: methylumbelliferyl (MUF), β-naphthyl (βN), and p-nitrophenyl (pNP). Hydrolysis of the first two yield fluorogenic methylumbelliferone and β-naphthol; the latter yields p-nitrophenol, which has an intense yellow color at basic pH. The fluorogenic substrates are more sensitive by at least an order of magnitude, but the pNP-linked substrates are easier to work with in soils because of their greater aqueous solubility and because humic substances often exert severe quenching effects on fluorescence. Thus the protocols presented here are based on pNP-linked substrates.

pNP-linked substrates are commercially available for many esterases, glycosidases, and peptidases. Commonly measured enzymes include phosphatase (pNP-phosphate), 1,4-β-glucosidase (pNP-β-glucopyranoside), cellobiohydrolase (pNP-cellobioside), 1,4-α-glucosidase (pNP-α-glucoside), 1,4-β-xylosidase (pNP-β- xylopyranoside), 1,4-β-N-acetylglucosaminidase (pNP-β-N-acetylglucosaminide), 1,4-β-exochitinase (pNP-β-N,N′-diacetylchitobiose), leucine aminopeptidase (leucine p-nitroanilide), glycine aminopeptidase (glycine p-nitroanilide), and fatty acid esterase (pNP-acetate).

Materials

1. Magnetic stirrer
2. pH meter
3. Spectrophotometer
4. Pipetters, for volumes ranging from 0.2 to 8.0 mL
5. 5 mL polypropylene test tubes with caps
6. Low-speed centrifuge
7. Vortex stirrer
8. Test tube racks
9. Blender or tissue homogenizer
10. 125 mL Nalgene screw-cap bottles
11. Drying oven
12. Muffle furnace
13. 50 mM acetate buffer, pH 5.0. Prepare a 1.0 mol/L stock solution by dissolving 0.5 mole of sodium acetate and 0.5 mole of acetic acid in deionized water; dilute to a final volume of 1.00 L. Make 50 mmol/L buffer solution by diluting the stock 1 in 20 with deionized water and adjusting the pH to 5.0.
14. Substrate solutions (described in procedure, below)
15. 1.0 N NaOH

Procedure

1. Prepare substrate solutions. Most fungal and bacterial glycosidases have pH optima from 4–6. For these assays, dissolve pNP-substrates in 50 mmol/L, pH 5.0, acetate buffer. Most peptidases have pH optima around 8.0; for these assays dissolve substrates in 50 mmol/L, pH 8.0, Tris (tris-hydroxymethyl-aminomethane) buffer. For acid phosphatase use the acetate buffer; for alkaline phosphatase use a pH 9 carbonate buffer. Verify the pH of the substrate solutions before mixing them with samples and adjust if necessary. Check the pH again 15 minutes after mixing substrate solution and samples; if the pH has shifted more than 0.5 units away from the initial value of the substrate solution, it may be necessary to increase the concentration of the buffer used to prepare substrate solutions.

   The goal is to conduct assays under conditions of substrate saturation without consuming excess quantities of expensive substrate; excessive substrate concentrations may also repress enzyme activity. In general, buffer solutions with a substrate concentration of 5 mmol/L are sufficient. However, some substrates are not soluble enough to make 5 mmol/L solutions. The substrate concentration needed to achieve enzyme saturation is closely linked to the size of the sample particles and the mixing energy used during incubation. The 5 mmol/L recommendation is based on well-homogenized and well-mixed sample suspensions. A static mixture of largely intact soil will require a far higher substrate concentration to attain apparent saturation. If uncontaminated, substrate solutions can be made up in batches and stored in the refrigerator for up to a few weeks; they also can be frozen.

2. Prepare samples. Assays for the "total" (i.e., not resolved into extractable and bound fractions) activity associated with soil and litter samples are made on homogenized suspensions. Sieving may be a preliminary step to remove rocks, roots, or other unwanted components from samples. Samples should be homogenized in the same buffer used to make the substrate solution. However, if performing multiple assays that have varying pH requirements, it may be simpler to homogenize samples in water and rely on the substrate solutions to buffer the pH appropriately.

   Samples are homogenized in a blender or tissue homogenizer. For litter, it is usually necessary to cut the sample into small pieces with scissors before blending. For litter samples use the equivalent of 0.2–0.5 g dry mass (DM) per 100 mL; for soil use the equivalent of 5–10 g DM. The primary consideration is handling; one must be able to dispense the resulting slurry in reproducible aliquots. For future calculation of enzyme activity, it is necessary to know either the total quantity of sample homogenized and the total volume of the suspension or the average quantity of sample material per milliliter of suspension. Nalgene 125 mL screw-cap bottles work well for holding sample homogenates.

   Sample preparation and subsequent assays should be conducted as soon as possible after collection. If processing must be delayed for up to a few days, it is best to store samples cold but not frozen. If long delays are unavoidable, samples can be sealed and frozen for indefinite periods. However, freezing and thawing will change the outcome of the assays; in general, activity tends to increase due to the fracture of cells and aggregates (Sinsabaugh and Linkins 1989). Thus, we do not recommend directly comparing the activities of never-frozen samples to those of frozen samples. Drying soils prior to assay generally results in the loss of a significant fraction of enzyme activity. However, drying has been used as a way of "stabilizing" soil samples, i.e., reducing variation within a soil unit to facilitate comparisons between units.

3. Mix 2.0 mL of sample homogenate with 2.0 mL of substrate solution in a 5 mL polypropylene test tube. To dispense the sample homogenates, place the bottle on a magnetic stir plate, mix vigorously to get a uniform suspension, and withdraw 2 mL aliquots using a pipetter with a wide aperture (e.g., a Gilson P5000 pipetter). To prevent clogging of the tip, snip off the end to make an opening about 0.5 cm in diameter. It is possible to scale down the assay mixture to 1.0 mL of sample and 1.0 mL of substrate solution if either sample or substrate quantities are limited. (For some soils, such as those with high

sand content, it may not be possible to reproducibly dispense suspensions. In such cases, it may be necessary to skip the sample homogenization procedure described earlier and weigh out analytical subsamples individually and place them into tubes.) Do four analytical replicates per sample per assay. Prepare sample controls by mixing 2.0 mL of sample homogenate and 2.0 mL of buffer in a tube and incubate it concurrently with samples. Prepare substrate controls by mixing 2 mL of buffer with 2 mL of substrate solution. Prepare both types of controls in duplicate.

4. After all tubes have been prepared and capped, place them on a mixer and incubate for 1–6 hours at 20 °C. A procedure that works well is to place the tubes into Ziplock bags, then place the bags in a blood platelet mixer that tumbles the tubes end over end at about 3 rpm. Reciprocating or orbital shakers can also be used. The mixers can be placed in an incubator or environmental room for better temperature control.

   Minimum incubation times will vary with enzymatic activity, which will vary with the quantity and quality of organic matter in the samples. Litter samples tend to have much higher activities than soils. Among the assays we have used, the highest activities have been found in the phosphatase and (-glucosidase activity assays. For litter, incubation times of an hour or less often suffice. Other assays typically incubate for several hours. We have incubated samples for as long as 24 hours and observed a constant rate of hydrolysis per hour.
5. After incubation, centrifuge the reaction suspensions for 2–3 minutes in a table-top centrifuge and transfer 2.0 mL of the supernatant into a 15 mL glass test tube. Soils with a high clay content may not yield clear supernatants; it may be necessary to spin such samples at higher G for longer times, or to filter residual particulates from the supernatants.
6. Add 0.2 mL of 1.0 N NaOH to each test tube to terminate the reaction and develop the color.
7. Add 8.0 mL of distilled water to each test tube and vortex. The dilution of the supernatants is somewhat arbitrary. A 10 mL volume is sufficient for repeated measurements of absorbance and dilutes the color into the range at which spectrophotometers are most accurate (i.e., abs < 1.000). However, if activity is exceptionally low, sensitivity can be improved by adding less water.
8. Read absorbance of each tube, samples and controls, at 410 nm. If any of the sample absorbencies exceed 1.500, the assay should be repeated using a shorter incubation time or a higher sample dilution.

Calculations

Activity is expressed as μmol of substrate hydrolyzed per gram organic mass (OM) (or dry mass) per hour as follows:

$$\text{Activity } (\mu\text{mol}\cdot\text{h}^{-1}\cdot\text{g}^{-1}\ OM \text{ or } DM) = OD\ /\ (EC \times T_h \times OM \text{ or } DM)$$

where

OD = sample Absorbance − [substrate control absorbance + sample control absorbance]
T_h = incubation period in hours
EC = micromolar extinction coefficient ($EC/\mu mol$) for p-nitrophenol (about 1.6). To calculate EC, make a standard curve by making dilutions of a 1.00 μmole/mL solution of p-nitrophenol in buffer. Mix 1 mL of standard, 0.1 mL of 1.0 N NaOH, and 4 mL of distilled water. Read absorbencies. Perform a linear regression of OD versus concentration. The slope of the line is the extinction coefficient ($EC/\mu mol$). Absorbance is linear with p-nitrophenol concentration up to an OD of about 2.000.
OM = organic matter, as g OM/mL sample homogenate
DM = dry mass of soil, as g dry soil/mL sample homogenate

Special Considerations

There are potential operational and interpretational difficulties with this protocol. With sandy soils, it may be difficult to achieve and dispense uniform soil suspensions, making it necessary to weigh out individual subsamples for each assay. In clay soils, low-speed centrifugation may not remove suspended clay from supernatants, making it necessary to filter supernatants through disposable syringe filters before taking absorbance readings. Soils with high humus content may yield supernatants with high background color, but little can be done about this.

The interpretation of enzyme assay results is complicated by the heterogeneity of soils and by the disturbances associated with the assay procedures. Among the potential difficulties are nonspecific substrate or product binding to colloids, the presence of competing substrates, and the presence of inhibitors (for additional discussion see Sinsabaugh et al. 1991 and Sinsabaugh 1994). Because of the artificial nature of the measurements, one should be cautious about concluding that differences among soils in the rate of a particular biotic process are directly correlated with differences in the apparent activity of a particular enzyme. Such correlations may exist, but they require validation. The factors regulating soil processes are complex, and the activity of the enzyme one assays may not be limiting the process of interest. The potential activities of soil or litter enzymes, especially when expressed per unit of microbial biomass, are perhaps best interpreted as measures of the relative effort expended by a microbial community to acquire a particular category of substrates (Sinsabaugh and Moorhead 1994).

Enzyme Assays for Oxidative Enzyme Activity

In litter and soil, oxidative enzymes, variously classified as phenol oxidases, polyphenol oxidases, laccases, or peroxidases, mediate the formation and degradation of lignin and humus. These enzymes generally have much less substrate specificity than hydrolases. Some of the enzymes involved in the degradation of lignin have been isolated and characterized, e.g., the lignin peroxidase and Mn peroxidase of white rot fungi. However, these enzymes are not universal even among white rot fungi. The mechanisms of lignin and humus decomposition are poorly known for other groups of decomposers. Most of the substrates used in vitro to measure the activity of these enzymes are not suitable for use in environmental samples because of low solubility or the need to measure absorbance at ultraviolet wavelengths. The assay protocol presented here is based on the oxidation of a phenolic amino acid known as L-3,4-dihydroxyphenylalanine (DOPA), which can be measured either with or without the addition of hydrogen peroxide. Without added peroxide, the assay measures the activity of phenol oxidases and oxygenases, i.e., enzymes that use organic molecules or molecular oxygen as electron acceptors; with added peroxide, the activity of peroxidases is captured too.

Materials

1. Magnetic stirrer
2. pH meter
3. Spectrophotometer
4. Pipetters, for volumes ranging from 0.2 to 8.0 mL
5. 5 mL polypropylene test tubes with caps
6. Low-speed centrifuge
7. Vortex stirrer
8. Test tube racks
9. Blender or tissue homogenizer
10. 125 mL Nalgene screw-cap bottles
11. Drying oven
12. Muffle furnace
13. 50 mM acetate buffer, pH 5.0. Prepare a 1.0 mol/L stock solution by dissolving 0.5 mole of sodium acetate and 0.5 mole of acetic acid in deionized water; dilute to a final volume of 1.00 L. Make 50 mmol/L buffer solution by diluting the stock 1 in 20 with deionized water and adjusting the pH to 5.0.
14. 5 mmol/L DOPA in 50 mmol/L, pH 5.0, acetate buffer
15. 0.3% H_2O_2 in deionized water, diluted from commercially obtained 30% H_2O_2 stock

Procedure

1. Prepare a 5 mmol/L substrate solution of DOPA in 50 mmol/L, pH 5.0, acetate buffer. The substrate is unstable; the solution will slowly darken over several hours as the DOPA reacts with O_2; therefore, it must be made fresh before each set of assays.
2. Prepare samples as for the hydrolase enzyme protocol, described earlier.
3. Phenol oxidase assay (measures oxidase and oxygenase activities): Place 2.0 mL of sample homogenate and 2.0 mL of L-DOPA solution in a 5 mL polypropylene tube. Dispense homogenates as described in the hydrolase activity procedure, described earlier. Use 2.0 mL of sample homogenate and 2.0 mL of acetate buffer as a control. Perform four analytical replicates for the sample and two replicates for the control.

4. Peroxidase assay (measures peroxidase activities): Place 2.0 mL of sample homogenate, 2.0 mL of L-DOPA solution. and 0.2 mL of 0.3% hydrogen peroxide in a 5 mL tube. Do four analytical replicates for each sample. Use 2.0 mL of sample homogenate, 2.0 mL of acetate buffer, and 0.2 mL of 0.3% hydrogen peroxide as a background control. Do two replicates of each control.
5. Cap and vortex all tubes and place them on a mixer at 20 °C for 1 hour as per the hydrolase activity procedure, described earlier. Because DOPA is not stable in the presence of oxygen, it is best to limit the incubation period to <1 hour. For longer incubation periods, it is necessary to run a substrate control (2 mL of 5 mmol/L DOPA solution + 2 mL of water). The DOPA will react slowly, and more or less linearly, with oxygen even in the absence of enzymatic activity.
6. Centrifuge all the tubes and withdraw 2.0 mL of supernatant.
7. Measure the absorbance of the supernatants at 460 nm, using distilled water to zero the spectrophotometer.

Calculations

Calculate activity as μmol substrate converted per hour per gram organic mass (or dry mass) of sample as follows:

Phenol Oxidase

$$\text{Activity } (\mu\text{mol} \cdot \text{h}^{-1} \cdot \text{g}^{-1}\ OM \text{ or } DM) = OD \,/\, (EC \times T_h \times OM \text{ or } DM)$$

where

OD = sample absorbance—control absorbance
EC = micromolar extinction coefficient (EC/μmol) for DOPA (about 1.6 under these assay conditions). EC can be determined by mixing 1.0 mL of 1.00 μmol/mL DOPA (in acetate buffer) with 3 mL of horseradish peroxidase (obtain from a commercial source, dissolve lyophilate in acetate buffer) then monitoring optical density at 460 nm until density peaks.
T_h = incubation time in hours

Peroxidase

Note that peroxidase activity is the *increment* in activity between samples incubated with and without hydrogen peroxide.

$$\text{Activity } (\mu\text{mol} \cdot \text{h}^{-1} \cdot \text{g}^{-1}\ OM \text{ or } DM) = OD \,/\, (EC = OD/(EC \times T_h \times OM \text{ or } DM)$$

where

OD = (sample absorbance) − (control absorbance) − (optical density for phenol oxidase)
EC, T_h, OM, and DM as defined above

Special Considerations

The same caveats described in the hydrolase enzyme protocol apply here. Additional difficulties are the instability of the substrate and reaction rates that may not be linear through time. These problems can be minimized by keeping incubation times as short as possible. Because the mechanisms of lignin and humus degradation are far from understood, the interpretation of data from these assays is more tenuous than that from most hydrolase assays. Oxidative activity tends to increase during the later stages of litter decomposition and, unlike hydrolytic activity, is often higher in the upper mineral horizons than in the organic horizons.

Viscometric Assay for Endocellulase Enzymes

1,4-β-endoglucanase (endocellulase) is one of the classes of enzymes involved in the degradation of cellulose. Endocellulase randomly cleaves glycosyl bonds along the interior of cellulose molecules. The viscometric assay for this enzyme was developed by Almin and Eriksson (1967) and Almin et al. (1967). The principal reasons for including this protocol are its simplicity and the observation that endocellulase activity appears to show the consistent relationship with decomposition across systems (Sinsabaugh et al. 1994; Jackson et al. 1995).

Materials

1. Magnetic stirrer
2. pH meter
3. Pipetters, for volumes ranging from 0.2 to 2.0 mL
4. 5 mL polypropylene test tubes with caps
5. Low-speed centrifuge
6. Vortex stirrer
7. Test tube racks
8. Blender or tissue homogenizer
9. 125 mL Nalgene screw-cap bottles
10. Drying oven
11. Muffle furnace
12. Stopwatch
13. 0.1 mL glass pipette mounted vertically on a ring stand
14. 50 mM acetate buffer, pH 5.0. Prepare a 1.0 mol/L stock solution by dissolving 0.5 mole of sodium acetate and 0.5 mole of acetic acid in deionized water; dilute to a final volume of 1.00 L. Make 50 mmol/L buffer solution by diluting the stock 1 in 20 with deionized water and adjusting the pH to 5.0.
15. 1.25% solution (wt/vol) of carboxymethyl cellulose in 50 mM acetate buffer. Preparation details given in the next section.

Procedure

1. Prepare substrate. Prepare a 1.25% solution of sodium carboxymethylcellulose (CMC) (Sigma C5013, High Viscosity) in 50 mmol/L, pH 5, acetate

buffer. Make 1–2 L batches. Stir continuously; it will take at least 24 hours for the CMC to dissolve. Dissolution can be speeded by heating the flask on a stirring hot plate. Dissolution can be further speeded by homogenizing the mixture with a blender or tissue homogenizer and/or by autoclaving. If the CMC solution is to be kept for an extended time, it is a good idea to autoclave it. If autoclaving, CMC should be added to a final concentration of 1.5% because autoclaving causes some polymer degradation, resulting in lower viscosity. When properly prepared, a mixture of 2 mL of CMC solution and 1 mL of acetate buffer will have an efflux time of 45 seconds (see later). If initial efflux is greater than 1 minute, dilute the CMC with buffer. For easiest handling, place finished CMC solution in a bottle capped with a positive displacement pipette.

2. Prepare samples as described in the protocol for hydrolase activity, described earlier.
3. Add 1 mL of sample homogenate and 2 mL of CMC solution to a 5 mL polypropylene tube. Vortex thoroughly; mixture is viscous. Do four analytical replicates per sample. It is advisable to add extra tubes that can be sacrificed during incubation to check the status of the assay.
4. Cap tubes and place on a mixer. Incubate for 1 to several hours (depending on activity) at 20 °C. To determine how long to incubate, periodically remove a tube from the mixer and proceed through steps 5 and 6. Longer incubations provide faster efflux times, which means results can be read faster. However, for best reproducibility, do not allow the efflux time to drop below 7–8 seconds. Tubes with efflux times less than 5 seconds should not be used to calculate results. See the section "Special Considerations," below.
5. Centrifuge the tubes.
6. Draw supernatant into a 0.1 mL glass pipette. With a stopwatch, measure the time required for the meniscus to fall from the 0 to the 0.03 mark or between two marks about 5 cm apart. Record as the final sample efflux time.
7. Prepare a reference tube by thoroughly mixing 1 mL of acetate buffer with 2 mL of CMC solution. Measure its fall time in the pipette (initial efflux time).

Calculations

Initial Viscosity

$$[n]_0 = \{[8 / C_s] \times [(T_i / T_s)^{0.125} - 1]\}$$

where

$[n]_0$ = initial viscosity
C_s = substrate concentration = (approx. 11 mg/mL) × 0.667 (Note: 0.667 is dilution factor)
T_i = initial efflux time, in sec
T_s = solvent efflux (distilled water) = 0.6 sec

Final Viscosity

$$[n]_t = \{[8 / C_s] \times [(T_f / T_s)^{0.125} - 1]\}$$

where

$[n]_t$ = final viscosity
C_s = substrate conc. = (11 mg/mL) × 0.667
T_f = final efflux time, in sec
T_s = solvent efflux time = 0.6 sec

Activity

$$B_v = [([n]_t^{-a} - [n]_o^{-a}) \times C_s] / t$$

where

B_v = activity by volume, in units·mL^{-1}·L^{-1}
t = incubation time, in hr
a = a constant that is specific to the substrate; use a value of 3.25

Final Activity

$$B_g = B_v \times (V / OM)$$

where

B_g = gravimetric activity, in units·g OM^{-1}·h^{-1}
V = total volume of sample extract or homogenate, in mL
OM = total quantity of organic matter used to prepare extract or homogenate, in g

Special Considerations

One problem with this assay is that the results are expressed in relative units rather than as a quantity of substrate converted, which makes it difficult to directly compare rates with those from other assays or processes. Another complication is that viscosity changes are inherently nonlinear with time: the number of bonds that need to be broken to attain a given change in viscosity increases exponentially as the mean size of the residual CMC molecules decreases. The result is that activity appears excessive at efflux times under 5 seconds. For this reason, adjust incubation periods so that efflux times do not go below 7–8 seconds. Some samples may contain surfactants that interfere with this assay. If not severe, surfactants can be dealt with by measuring sample efflux times immediately after step 3 and using these values, rather than the reference tube efflux (step 7), for T_t.

Nucleic Acid Analyses

The first and most critical step for most molecular or nucleic acid techniques remains the isolation of DNA or RNA from soil and its separation from other materials. Although several methods are available (e.g., Holben 1994; Ausubel et al. 1990; Sambrook et al. 1989), we present two methods in this section that we consider the most efficient. However, depending on soil type, these methods may require modification as outlined by Zhou et al. (1996).

Both methods described here use bead beating and lysis solutions to liberate DNA/RNA from soil microbes. While some concern has been raised about the shearing of DNA using this method (Zhou et al. 1996; Liesack et al. 1991) and its subsequent effect on polymerase chain reaction (PCR) analysis, we have found DNA resulting from this procedure to range in size from 5 to 25 kb, with the majority being greater than 10 kb in length.

The first method by More et al. (1994) is limited to the extraction of DNA from soils. The second method by Purdy et al. (1996) allows for the extraction of DNA, RNA, or DNA and RNA from soils, but it requires more sophisticated laboratory techniques. A concluding discussion references some of the available analytical options. The fundamental premise of these analyses is that DNA preparations are an unbiased representation of microbial diversity. Unfortunately, there is good evidence that this is not the case. All of the major steps in the acquisition procedure—lysis, extraction, and purification—have selectivities that distort this ideal (Borneman et al. 1996; Zhou et al. 1996; Holben and Harris 1995).

DNA-Only Extraction

Materials

1. Mini-bead mill (BioSpec Products, Bartlesville, OK)
2. 0.1 mm zirconia/silica beads (available through BioSpec Products), washed and sterilized
3. 2 mL screw-cap microcentrifuge tubes
4. 5 mL polypropylene tube, sterilized
5. Sterile, 1.5 mL microcentrifuge tubes, siliconized
6. Microcentrifuge
7. Refrigerated centrifuge
8. Pipetters, sterilized
9. Latex gloves
10. 100 mmol/L, pH 8, phosphate buffer
11. Lysis solution: 10% SDS w/v, 500 mmol/L Tris, 100 mmol/L NaCl
12. 7.5 mmol/L ammonium acetate

Procedure

1. To a 2 mL screw-top microcentrifuge tube add 2.5 g of 0.1 mm zirconia/silica beads and about 0.5 g soil (dry mass equivalent, or about 0.5 mL volume), 500 μL phosphate buffer, and 250 μL lysis solution (in order).

The soil component tends to stopper the centrifuge tube, disallowing space for the buffer and lysis solution. To remedy this, press the soil to one side of the tube before introducing the buffer and lysis solutions. The soil and phosphate buffer volumes should be about equal.

2. Cap tubes and mill for 5 minutes at 4 °C. The bead mill creates a large amount of frictional heat, which may denature DNA/RNA. It is best to use it prechilled or for short periods with intermittent icing of samples.
3. Remove tubes from bead mill and place *cap down* in ice for 5 minutes.
4. Remove liquid phase: Heat an insect pin over a burner until red hot. Puncture the bottom of the sample tube, at center, with the pin. Nest the 2.0 mL microcentrifuge tube inside a 5 mL polypropylene tube. Place tubes in ice until centrifugation.
5. Spin the nested tubes for 5 minutes at 10,000 × G and 4 °C. The liquid phase, about 750 μL containing extracted DNA, will collect in the 5 mL tube.
6. Remove two 350 μL aliquots of extract and place each aliquot in a separate sterile 1.5 mL microcentrifuge tubes.
7. Add 140 μL of 7.5 mol/L ammonium acetate to the extracts (ratio of 5:2, i.e., 350 μL supernatant: 140 μL 7.5 mol/L ammonium acetate) and cap.
8. Refrigerate tubes at 4 °C for 5 minutes to precipitate proteins.
9. Centrifuge tubes for 3 minutes at 12,000 × G.
10. Pipette the supernatant (approx. 500 μL) into a sterile siliconized 1.5 mL microcentrifuge tube.
11. Purifying the DNA: Although several DNA extraction and cleanup protocols have been published (Kemp et al. 1993; Trevors and van Elsas 1995), we have found bead mill extraction of DNA followed by cleanup with commercially available resin kits such as Wizard (Promega, Madison, WI) works well.

DNA + RNA Extraction

Materials

1. Mini-bead mill (BioSpec Products, Bartlesville, OK)
2. 0.1 mm zirconia/silica beads (available through BioSpec Products): washed and sterilized
3. 2 mL screw-cap microcentrifuge tubes
4. Sterile 1.5 mL microcentrifuge tubes, siliconized
5. 1 and 3 mL sterile disposable syringes
6. Glass wool (sterile)
7. Microcentrifuge (sterile)
8. Refrigerated centrifuge
9. Pipetters, sterilized
10. Latex gloves
11. 120 mmol/L sodium phosphate buffer, pH 8.0
12. 120 mmol/L sodium phosphate buffer, pH 7.2

13. 300 mmol/L potassium phosphate buffer (K_2HPO_4), pH 7.2
14. 140 mmol/L Potassium phosphate buffer (K_2HPO_4), pH 7.2
15. 70% EtOH
16. Tris EDTA (TE) pH 8.0
17. Sephadex G-75
18. Polyethylene glucol 8000
19. Polyvinylpolypyrrolidone (PVPP)
20. Tris equilibrated phenol, pH 8.0
21. Sodium dodecyl sulfate (SDS)
22. Hydroxyapatite (HTP), (Bio-Gel HTP, from BioRad). Do not use DNA grade; the pore size is too restrictive.

Procedure

For Total Nucleic Acid Elution

1. Remove extracellular nucleic acids from the soil matrix by an initial rinse: Wash sediment (v/v) with 120 mmol/L sodium phosphate, pH 8.0 for 15 minutes at 150 rpm, then spin at 6000 × G for 10 minutes, remove supernatant, and store pellet at −70 °C.
2. To extract nucleic acids, place 0.5 g of washed sediment into a 2 mL screw-cap microcentrifuge tube containing 0.5 g zirconia glass beads (0.1 mm diameter). Add 700 μL of 120 mmol/L sodium phosphate buffer, pH 8.0, containing 1% (w/v) acid-washed PVPP; 500 μL Tris equilibrated phenol, pH 8.0; and 50 μL 20% (w/v) SDS.
3. Place the tubes in the bead beater and mill for three 30 second periods at 2000 rpm. Cool the tubes between each cycle by placing them in an ice bath for at least 30 seconds between each cycle. Then centrifuge the tubes at 12,000 × G for 2 minutes, remove supernatant, and store on ice.
4. Add a second 700 μL aliquot of 120 mmol/L sodium phosphate buffer (pH 8.0) to the original soil sample. Place the tube back into the bead beater and mill the sample for 30 seconds; centrifuge the tubes at 12,000 × G for 2 minutes and combine the supernatant with that collected above. Store at −70 °C.
5. Prepare hydroxyapatite (HTP) spin column by placing 0.6–0.7 mL Bio-Gel HTP into a 1 mL plastic syringe plugged with sterile glass wool. Load the prepared HTP column by placing 700 μL of nucleic acid extract onto the column. Spin the column at 100 × G for 2–4 minutes (until sample has eluted). Repeat until all of the supernatant containing the nucleic acid has been eluted. Wash column to remove proteins by adding 500 μL 120 mmol/L sodium phosphate buffer (pH 7.2) and centrifuging at 100 × G. Repeat this step three times.
6. Collect the nucleic acid by inserting the washed column into a sterile 1.5 mL microcentrifuge tube and adding 400 μL, 300 mmol/L K_2HPO_4 (pH 7.2) to the column and spinning at 100 × G for 1 minute.
7. Desalt the eluted nucleic acid using a G-75 Sephadex column: Insert a 2.5 mL G-75 Sephadex spin column into a sterile 1.5 mL microcentrifuge tube. Load

the nucleic acid eluted above onto the column and centrifuge at 3000 × G for 1 minute.

8. Precipitate nucleic acids from the Sephadex eluate by adding 2.5 volumes of ice-cold EtOH. Place the tubes in a −70 °C freezer for 2 hours. Decant the EtOH and resuspend the pellet in 50 μL TE, pH 8.0, and store at −70 °C.
9. If the pellet is brown, indicating humic contamination, resuspend the pellet in 200 μL sterile Milli-Q water. Precipitate the nucleic acids with 200 μL of 1.6 M NaCl containing 13% polyethylene glycol 8000 (w/v) at 4 °C for 1 hour. Then centrifuge at 12,000 × G for 15 minutes. Aspirate and discard the supernatant. Wash the pellet with ice-cold 70% EtOH. Resuspend the pellet in 50 μL TE, pH 8.0. Store at −70 °C.

For Separate Elution of rRNA and DNA

1. rRNA must be eluted from the spin column prior to DNA. To accomplish this, add 700 μL of 140 mmol/L K_2HPO_4 (pH 7.2) to the HTP column containing adsorbed nucleic acids. Elute rRNA into a sterile 1.5 mL microcentrfuge tube by spinning at 100 × G. Repeat this step two more times, collecting each eluant in a separate microcentrifuge tube. Note: Each batch of HTP elutes nucleic acids at different rates and must be checked empirically.
2. To remove proteins from the rRNA eluant, add 700 μL sterile Milli-Q water to each of the three aliquots (this allows rRNA to rebind to the HTP spin column), then reload the rRNA in 700 μL aliquots onto a new HTP column by spinning at 100 × G. Wash this column once with 500 μL of 120 mmol/L sodium phosphate (pH 7.2) to desorb proteins, centrifuge at 100 × G, discard eluant.
3. Elute the rRNA from the column by adding 400 μL of 300 mmol/L K_2HPO_4 (pH 7.2). Desalt and precipitate as in the total nucleic acid elution procedure, described earlier.
4. Elute the DNA sorbed to the original HTP column as described in the total nucleic acid elution procedure, above.

Analysis Options

For DNA quantification the PicoGreen protocol from Molecular Probes (www.probes.com) gives greater sensitivity and precision than the 260/280 absorbance ratio of Sambrook et al. (1989). RNA quantification can be carried out using the fluorometric method of Schmidt and Ernst (1995).

DNA obtained through these protocols may also be used directly in whole-community DNA-DNA hybridization (Yeager and Sinsabaugh 1997; Trevors and van Elsas 1995; Sinsabaugh et al. 1992; Lee and Fuhrman 1990; Sambrook et al. 1989) or used in analyses with higher resolution after further manipulation.

Both DNA and RNA extracted by these protocols may be amplified using the polymerase chain reaction (PCR; Holben 1994) to detect specific DNA/RNA sequences from soil samples. These sequences may be signature sequences that allow one to detect the presence of a specific taxon or a specific gene, e.g., nitrogenase

(Schleper et al. 1997; Borneman et al. 1996; Amann et al. 1995; Lee 1994; Herrick et al. 1993; Tsai et al. 1991; Giovannone et al. 1990; Torsvik et al. 1990, 1990a, b).

Special Considerations

As noted previously, DNA extraction and cleanup procedures for microbial communities in soils are undergoing constant change. The techniques described here are good general approaches to the extraction of nucleic acids. These techniques are somewhat expensive, but samples can be turned over in 3–4 hours. The technique chosen will depend on soil characteristics, time, monetary constraints, and desired outcome.

In general, radioactive probes are simpler, cheaper, and more sensitive than their nonradioactive counterparts. But these techniques require licensing for their use, as well as designated space, storage, and disposal facilities. Nonradioactive techniques such as Dioxygenin (DIG) analysis may be carried out either colormetrically or photometrically and require no special permits. However, the protocols demand more steps, and the colormetric protocol is not readily quantifiable. Using the photometric technique is fine with high levels of hybridization, but at low hybridization efficiencies the chemical luminescence lacks the longevity needed for film exposure.

Analysis of Substrate Utilization and Nucleic Acid Data Sets

The phenotypic or genotypic data collected using any of the approaches described in this chapter can be scrutinized using a variety of multivariate statistical methods. In general, statistical analysis is intended either to reveal internal patterns within the microbial community or to relate microbial community characteristics with external variables. Internal patterns are often examined using factor analyses that examine correlations within the data set. These correlations allow a large number of observed variables (e.g., FAME concentrations, enzyme activities, substrate utilization rates, hybridization scores) to be condensed into a smaller number of new nonobserved variables (factors) while preserving most of the variance of the original data. The results indicate which measured elements of the community tend to covary. These factors can be plotted against other ecological variables to assess whether community characteristics change in predictable ways.

Cluster analyses are commonly used to visualize the relative similarity or difference among spatially or temporally displaced microbial communities. In these analyses, a coordinate axis is created for each measured parameter of the microbial community. An iterative hierarchical process is used to sequentially group samples based on their distance from one another within hyperdimensional space. The results of cluster analysis are commonly shown as dendrograms. The nodes or branches or these dendrograms may be analyzed for statistically significant differences using bootstrapping (Nemec and Brinkhurst 1988). A direct comparison between distance matrices used to form these dendrograms may be made using

Mantel analysis allowing direct correlations between biotic/abiotic and biotic/ biotic matrices.

References

Almin, K. E., and K.-E. Eriksson. 1967. Enzymic degradation of polymers. I: Viscometric method for the determination of enzymic activity. *Biochimica et Physica Acta* 139:238–247.

Almin, K. E., K.-L. Eriksson, and C. Jansson. 1967. Enzymic degradation of polymers. II: Viscometric determination of cellulose activity in absolute terms. *Biochimica et Physica Acta* 139:248–253.

Amann, R. I., W. Ludwig, and K.-L. Schleifer. 1995. Identification and *in situ* detection of individual microbial cells without cultivation. *Microbiological Reviews* 59:143–169.

Ausubel, F. M., R. Brent, R. E. Kingston, D. D. Moore, J. G. Seidman, J., A. Smith, and K. Struhl. 1990. *Current Protocols in Molecular Biology.* Greene Publisher Associates and Wiley-Interscience, New York, New York, USA.

Biely, P. 1985. Microbial xylanolytic systems. *Trends in Biotechnology* 3:286–290.

Blanchette, R. A. 1991. Delignification by wood-decay fungi. *Annual Review of Phytopathology* 29:381–398.

Bligh, E. G., and W. J. Dyer. 1959. A rapid method of total phospholipid extraction and purification. *Canadian Journal of Biochemistry and Physiology* 37:911–917.

Borneman, J., P. W. Skroch, K. M. O'Sullivan, J. A. Palus, N. G. Rumjanek, J. L. Jansen, J. Nienhuis, and E. W. Triplett. 1996. Molecular microbial diversity of an agricultural soil in Wisconsin. *Applied and Environmental Microbiology* 62:1935–1943.

Bossio, D. A., and K. M. Scow. 1995. Impact of carbon and flooding on the metabolic diversity of microbial communities in soils. *Applied Environmental Microbiology* 61:4043–4050.

Brock, T. D. 1987. The study of microorganisms *in situ:* Progress and problems. *In* M. Fletcher, T. R. Gray, and J. C. Jones, editors, *Ecology of Microbial Communities.* Cambridge University Press, London, UK.

Burns, R. G. 1983. Extracellular enzyme-substrate interactions in soil. Pages 249–298 *in* J. H. Slater, R. Whittenbury, and J. W. T. Wimpenny, editors, *Microbes in Their Natural Environments.* Cambridge University Press, Cambridge, UK.

Cavigelli M. A., G. P. Robertson, and M. J. Klug. 1995. Fatty acid methyl ester (FAME) profiles as measures of soil microbial community structure. *Plant and Soil* 70:99–113.

Chróst, R. J. 1991. Environmental control of the synthesis and activity of aquatic microbial ectoenzymes. Pages 25–59 *in* R. J. Chróst, editor, *Microbial Enzymes in Aquatic Environments.* Springer-Verlag, New York, USA.

Collins, H. P., P. E. Rasmussen, and C. L. Douglas. 1992. Crop rotation and residue management effects on soil carbon and microbial dynamics. *Soil Science Society of America Journal* 56:783–787.

Dekker, R. F. H. 1985. Biodegradation of the hemicelluloses. Pages 503–533 *in* T. Higuchi, ed. *Biosynthesis and Biodegradation of Wood Components.* Academic Press, New York, USA.

Eriksson, K. E., R. A. Blanchette, and P. Ander. 1990. *Microbial and Enzymatic Degradation of Wood Components.* Springer-Verlag, Berlin, Germany.

Eriksson, K. E., and T. M. Wood. 1985. Biodegradation of cellulose. Pages 469–503 *in* T. Higuchi, editor, *Biosynthesis and Biodegradation of Wood Components.* Academic Press, New York, USA.

Findlay, R. H., M. B. Trexler, J. B. Guckert, and D. C. White. 1990. Laboratory study of disturbance in marine sediments: response of a microbial community. *Marine Ecology Progress Series* 62:121–133.
Garchow, H., H. P. Collins, and M. J. Klug. 1993. Characterizing whole soil microbial communities using Biolog GN microplates. *In Research Findings Fall 1993.* Center for Microbial Ecology, Michigan State University, East Lansing, Michigan, USA.
Garland, J. L., and A. L. Mills. 1991. Classification and characterization of heterotrophic microbial communities on the basis of patterns of community-level sole-carbon-source utilization. *Applied and Environmental Microbiology* 57:2351–2359.
Giovannone, S. J., T. B. Britschgi, C. L. Moyer, and K. G. Field. 1990. Genetic diversity in Sargasso Sea bacterioplankton. *Nature* 345:60–65.
Guckert, J. B., C. P. Antworth, P. D. Nichols, and D. C. White. 1985. Phospholipid, ester-linked fatty acid profiles as reproducible assays for changes in prokaryotic community structure of estuarine sediments. *FEMS Microbial Ecol*ogy 31:147–158.
Haack, S. K., H. Garchow, M. J. Klug, and L. J. Forney. 1995. Analysis of factors affecting the accuracy, reproducibility, and interpretation of microbial community carbon source utilization patterns. *Applied and Environmental Microbiology* 61:1458–1468.
Harwood, J. L., and N. J. Russell. 1984. *Lipids in Plant and Microbes.* Allen and Unwin, London, UK.
Henrot, J., and G. P. Robertson. 1994. Vegetation removal in two soils of the humid tropics: effect on microbial biomass. *Soil Biology and Biochemistry* 26:111–116.
Herrick, J. B., E. L. Madsen, C. A. Batt, and W. C. Ghiorse. 1993. Polymerase chain reaction amplification of naphthalene-catabolic and 16s rRNA gene sequences from indigenous sediment bacteria. *Applied and Environmental Microbiology* 59:687–694.
Higuchi, T. 1990. Lignin biochemistry: biosynthesis and biodegradation. *Wood Science and Technology* 24:23–63.
Holben, W. E. 1994. Isolation and purification of DNA from soil. Pages 727–751 *in* R. W. Weaver, editor, *Methods of Soil Analysis. Part 2, Microbiological and Biochemical Properties.* Soil Science Society of America, Madison, Wisconsin, USA.
Holben, W. E., and D. Harris. 1995. DNA-based monitoring of total bacterial community structure in environmental samples. *Molecular Ecology* 45:627–631.
Insam, H., D. Parkinson, and K. H. Domsch. 1989. Influence of macroclimate on soil microbial biomass. *Soil Biology and Biochemistry* 21:211–216.
Jackson, C., C. Foreman, and R. L. Sinsabaugh. 1995. Microbial enzyme activities as indicators of organic matter processing rates in a Lake Erie coastal wetland. *Freshwater Biology* 34:329–342.
Kemp, P. F., S. Lee, and J. LaRoche. 1993. Evaluating bacterial activity from cell-specific ribosomal RNA content measured with oligonucleotide probes. Pages 415–422 *in* P. F. Kemp, B. F. Sherr, E. B. Sherr, and J. J. Cole, editors, *Handbook of Methods in Aquatic Microbial Ecology.* Lewis Publishers, Boca Raton, Florida, USA.
Kennedy, A. C. 1994. Carbon utilization and fatty acid profiles for characterization of bacteria. Pages 543–556 *in* R. W. Weaver, S. Angle, P. Bottomley, D. Bezdecek, S. Smith, A. Tabatabai, and A. Wollum, editors, *Methods of Soil Analysis. Part 2, Microbiological and Biochemical Properties.* Soil Science Society of America, Madison, Wisconsin, USA.
Kinkel, L. L., E. V. Nordheim, and J. H. Andrews. 1992. Microbial community analysis in incompletely or destructively sampled systems. *Microbial Ecology* 24:227–242.
Kirk, T. K., and R. L. Farrell. 1987. Enzymatic "combustion": the microbial degradation of lignin. *Annual Review of Microbiology* 41:465–505.
Klug, M. J., and J. M. Tiedje. 1993. Response of microbial communities to changing environmental conditions: chemical and physiological approaches. Pages 371–375 *in* R.

Guernero and C. Pedros-Alio, editors, *Trends in Microbial Ecology.* Spanish Society for Microbiology, Barcelona, Spain.

Lahdesmaki, P., and R. Piispanen. 1992. Soil enzymology: role of protective colloid systems in the preservation of exoenzyme activities in soil. *Soil Biology and Biochemistry* 24:1173–1177.

Lee, K. E. 1994. The biodiversity of soil organisms. *Applied Soil Ecology* 1:251–254.

Lee, S., and J. A. Fuhrman. 1990. DNA hybridization to compare species compositions of natural baterioplankton assemblages. *Applied and Environmental Microbiology* 56: 739–746.

Liesack, W., H. Weyland, and E. Stackebrandt. 1991. Potential risks of gene amplification by PCR as determined by 16s DNA analysis of a mixed culture of strict barophilic bacteria. *Microbial Ecology* 21:191–195.

Ljungdahl, L. G., and K.-E. Eriksson. 1985. Ecology of microbial cellulose degradation. *Advances in Microbial Ecology* 8:237–299.

Marsden, W. L., and P. P. Gray. 1986. Enzymatic hydrolysis of cellulose in lignocellulosic materials. *CRC Critical Reviews in Biotechnology* 3:235–276.

Mills, A. L., and R. A. Wassel. 1980. Aspects of diversity measurement for microbial communities. *Applied and Environmental Microbiology* 40:578–586.

More, M. I., J. B. Herrick, M. C. Silva, W. C. Ghoirse, and E. L. Madsen. 1994. Quantitative cell lysis of indigenous microorganisms and rapid extraction of microbial DNA from sediment. *Applied and Environmental Microbiology* 60:1572–1580.

Nemec, A. F. L., and R. O. Brinkhurst. 1988. Using bootstrap to assess statistical significance in the cluster analysis of species abundance data. *Canadian Journal of Fisheries and Aquatic Sciences* 45:965–970.

Petersen, S. O., and M. J. Klug. 1994. Effects of sieving, storage, and incubation temperature on the phospholipid fatty acid profile of a soil microbial community. *Applied and Environmental Microbiology* 60:2421–2430.

Purdy, K. J., T. M. Embley, S. Takii, and D. B. Nedwell. 1996. Rapid extraction method of DNA and rRNA from sediments by a novel hydroxyapatite spin-column method. *Applied and Environmental Microbiology* 62:3905–3907.

Rayner, A. D. M., and L. Boddy. 1988. *Fungal Decomposition of Wood: Its Biology and Ecology.* Wiley, Chichester, UK.

Sakai, T., T. Sakamoto, J. Hallaert, and E. J. Vandamme. 1993. Pectin, pectinase, and protopectinase: production, properties, and applications. *Advances in Applied Microbiology* 39:213–294.

Sambrook, J., E. F. Fritsch, and T. Maniatis. 1989. *Molecular Cloning: Laboratory Manual.* Cold Spring Harbor Laboratory Press, Cold Spring, New York, USA.

Schleper, C., W. Holben, and H. Klenk. 1997. Recovery of crenarchaeota ribosomal DNA sequences from freshwater-lake sediments. *Applied and Environmental Microbiology* 63:321–323.

Schmidt, D. M., and J. D. Ernst. 1995. A fluorometric assay for the quantification of RNA in solution with nanogram sensitivity. *Analytical Biochemistry* 232:144–146.

Sinsabaugh, R. L. 1994. Enzymic analysis of microbial pattern and process. *Biology and Fertility of Soils* 17:69–74.

Sinsabaugh, R. L., R. K. Antibus, and A. E. Linkins. 1991. An enzymic approach to the analysis of microbial activity during plant litter decomposition. *Agriculture, Ecosystems and Environment* 34:43–54.

Sinsabaugh, R. L., R. K. Antibus, A. E. Linkins, C. A. McClaugherty, L. Rayburn, D. Repert, and T. Weiland. 1992. Wood decomposition over a first-order watershed: mass loss as a function of lignocellulase activity. *Soil Biology and Biochemistry* 24:743–749.

Sinsabaugh, R. L., and A. E. Linkins. 1987. Inhibition of the *Trichoderma viride* cellulase complex by leaf litter extracts. *Soil Biology and Biochemistry* 19:719–725.

Sinsabaugh, R. L., and A. E. Linkins. 1989. Natural disturbance and the activity of *Trichoderma viride* cellulase complexes. *Soil Biology and Biochemistry* 21:835–839.

Sinsabaugh, R. L., and D. L. Moorhead. 1994. Resource allocation to extracellular enzyme production: a model for nitrogen and phosphorus control of litter decomposition. *Soil Biology and Biochemistry* 26:1305–1311.

Sinsabaugh, R. L., D. L. Moorhead, and A. E. Linkins. 1994. The enzymic basis of plant litter decomposition: emergence of an ecological process. *Applied Soil Ecology* 1:97–111.

Skujins, J. J. 1976. Extracellular enzymes in soil. *CRC Critical Reviews in Microbiology* 4:383–421.

Skujins, J. J. 1978. History of abiontic soil enzyme research. Pages 1–49 *in* R. G. Burns, editor, *Soil Enzymes.* Academic Press, London, UK.

Sorheim, R., V. L. Torsvik, and J. Goksoyr. 1989. Phenotypic divergences between populations of soil bacteria isolated on different media. *Microbial Ecology* 17:181–192.

Tate, R. L., and A. L. Mills. 1983. Cropping and the diversity and function of bacteria in a Pahokee muck. *Soil Biology and Biochemistry* 15:175–179.

Torsvik, V., J. Goksoyr, and F. L. Daae. 1990a. High diversity in DNA of soil bacteria. *Applied and Environmental Microbiology* 56:782–787.

Torsvik, V., K. Salte, R. Sorheim, and J. Goksoyr. 1990b. Comparison of phenotypic diversity and heterogeneity in populations of soil bacteria. *Applied and Environmental Microbiology* 56:776–781.

Trevors, J. T., and J. D. Van Elsas, editors. 1995. *Nucleic Acids in the Environment: Methods and Applications.* Springer-Verlag, New York, New York, USA.

Tsai, Y.-L., M. J. Park, and B. H. Olson. 1991. Rapid method for direct extraction of mRNA from seeded soils. *Applied and Environmental Microbiology* 57:765–768.

Tunlid, A., and D. C. White. 1992. Biochemical analysis of biomass, community structure, nutritional status, and metabolic activity of microbial communities in soil. *Soil Biochemistry* 7:229–262.

Vahjen, W., J. C. Munch, and C. C. Tebbe. 1995. Carbon source utilization of soil extracted microorganisms as a tool to detect the effects of soil supplemented with genetically engineered and non-engineered Corynebacterium glutamicum and a recombinant peptide at the community level. *FEMS Microbial Ecology* 18:317–328.

Vestal, J. R., and D. C. White. 1989. Lipid analysis in microbial ecology. *BioScience* 39:535–541.

Viikari, L., M. Tenkanen, J. Buchert, M. RêttÖ, M. Bailey, M. Siika-aho, and M. Linko. 1993. Hemicellulases for industrial applications. Pages 131–182 *in* J. N. Saddler, editor, *Bioconversion of Forest and Agricultural Plant Residues.* CAB International, Wallingford, UK.

White, D. C. 1983. Analysis of microorganisms in terms of quantity and activity in natural environments. Pages 37–66 *in* J. H. Slater, R. Whittenbury, and J. W. T. Wimpenny, editors, *Microbes in Their Natural Environments.* Cambridge University Press, Cambridge, UK.

White, D. C. 1986. Validation of quantitative analysis for microbial biomass, community structure, and metabolic activity. *Ergebnisse der Limnologie* 31:1–18.

White, D. C., W. M. Davis, J. C. Nickels, J. D. King, and R. J. Bobbie. 1979. Determination of the sedimentary microbial biomass by extractable lipid phosphate. *Oecologia* 40:51–63.

Wong, K. K. Y., L. U. L. Tan, and J. N. Saddler. 1988. Multiplicity of β-1,4-xylanase in microorganisms: functions and applications. *Microbiological Reviews* 52:305–317.

Yeager, P., and R. L. Sinsabaugh. 1998. Microbial community organization along a detitral particle size gradient. *Aquatic Ecology.* In press.

Zelles, L., Q. Y. Bai, and F. Beese. 1992. Signature fatty acids in phospholipids and lipopolysaccharides as indicators of microbial biomass and community structure in agricultural soils. *Soil Biology and Biochemistry* 24:317–323.

Zhou, J., M. A. Bruns, and J. M. Tiedje. 1996. DNA recovery from soils of diverse composition. *Applied and Environmental Microbiology* 62:316–322.

17

Soil Invertebrates

David C. Coleman
John M. Blair
Edward T. Elliott
Diana H. Wall

The array of species constituting soil invertebrates is very large, encompassing virtually all terrestrial invertebrate phyla. Communities of soil fauna offer opportunities for the study of such ecological phenomena as species interactions, resource utilization, and responses of ecosystems to perturbations to ecosystems. Soil invertebrates are also important for processing energy in soil systems. It is possible to determine community- and system-level energetics using standard assumptions of consumption, assimilation, respiration, and excretion rates (Petersen and Luxton 1982). Several invertebrate groups such as millipedes, oribatid mites, and earthworms and snails are important accumulators and processors of labile calcium and perhaps other elements in numerous forested and desert ecosystems (Cromack et al. 1988).

The soil invertebrate macrofauna include a variety of surface- and soil-dwelling arthropods (insects, arachnids, millipedes, centipedes, isopods), terrestrial mollusks and crustaceans, and earthworms. The presence and relative abundance of these groups vary tremendously in different biomes, as do the methods used for quantitative sampling of different groups. Coverage of methods for all these groups is beyond the scope of this chapter. Instead, we focus on sampling litter and soil microfauna, mesofauna, and earthworms. However, some of the methods contained in this chapter, e.g., handsorting and flotation, can also provide information on populations on other surface- (mollusks) and soil-dwelling (insect larvae) invertebrates. Colonial insects, e.g., ants and termites, require specialized sampling procedures (see Lee and Wood [1971] and Brian [1965]). Similarly, there are numerous fossorial vertebrates (moles, skinks, some snake species) that are important participants in ecosystem functions. These are beyond the purview of this chapter.

In this chapter we consider fauna that live in the soil portions of terrestrial ecosystems. This, by definition, includes litter and soil, with the litter derived from either aboveground or belowground sources. For more details on measurement of litter standing stocks and decomposition, see Chapter 11, this volume. As is true for all measurements in this volume, we encourage researchers on soil invertebrate numbers, biomass, and functions to report their data on an areal basis, namely per square meter (see Chapter 1, this volume).

As is true for other biota, the use of molecular techniques as tools to aid in taxonomic identification for biodiversity or phylogeny research, geographic distribution, and presence or absence following disturbance is accelerating. Methods exist and are being used particularly for invertebrates such as nematodes, microarthropods (collembola, mites), and some of the higher macrofauna. These techniques are not, at present, sufficiently common to be recommended as a first priority for identification or delineation of groups of biota, and thus are not described in detail here. However, we realize that within the next few years the maturation and standardization of these techniques will enable them to be more widely used and accepted.

Image analysis and the Internet have made the identification of microscopic biota more available for all to see. An image captured on an image analysis software such as the National Institutes of Health (NIH) free ware (http://rsb.info.nih.gov/nih-image) is easily sent via the Internet to an expert for confirmation or for help. Such a system also allows others in a lab to view taxonomic characteristics and come to an educated conclusion. Many Internet and WWW servers have images of the soil biota and keys to identification (e.g., nematodes of the Konza Prairie at http://ianr www.unl.edu/ianr/plntpath/nematode/NNGPROP/Konza—nematodes.htm).

Size Characteristics of the Soil Fauna

Body width of soil fauna is related to soil microhabitats (Fig. 17.1; Swift et al. 1979). The microfauna (protozoa, rotifers, tardigrades) inhabit water films. The mesofauna, such as collembola and mites, inhabit air-filled pore spaces and are mostly restricted to preexisting spaces. Nematodes inhabit water films and move across pore spaces as well. Nematodes and tardigrades can undergo desiccation and enter into an anhydrobiotic ametabolic state at any time in their life cycle when soils dry. The macrofauna such as earthworms and termites, in comparison, have the ability to create their own spaces in the soil matrix by burrowing through it, and they have the capacity to change aspects of the soil structure (Lee 1985; Hendrix 1995). Methods for studying these faunal groups are largely size-dependent and rely on the regions of the soil that the fauna inhabit. Methods for sampling the microfauna have a close parallel to methods used in soil microbiology. Mesofauna are obtained using specialized equipment for extraction, using heat, water-density gradients, or flotation. Macrofauna are often collected by hand or with some mechanical assistance. Faunal groups and extraction techniques are presented in the following pages.

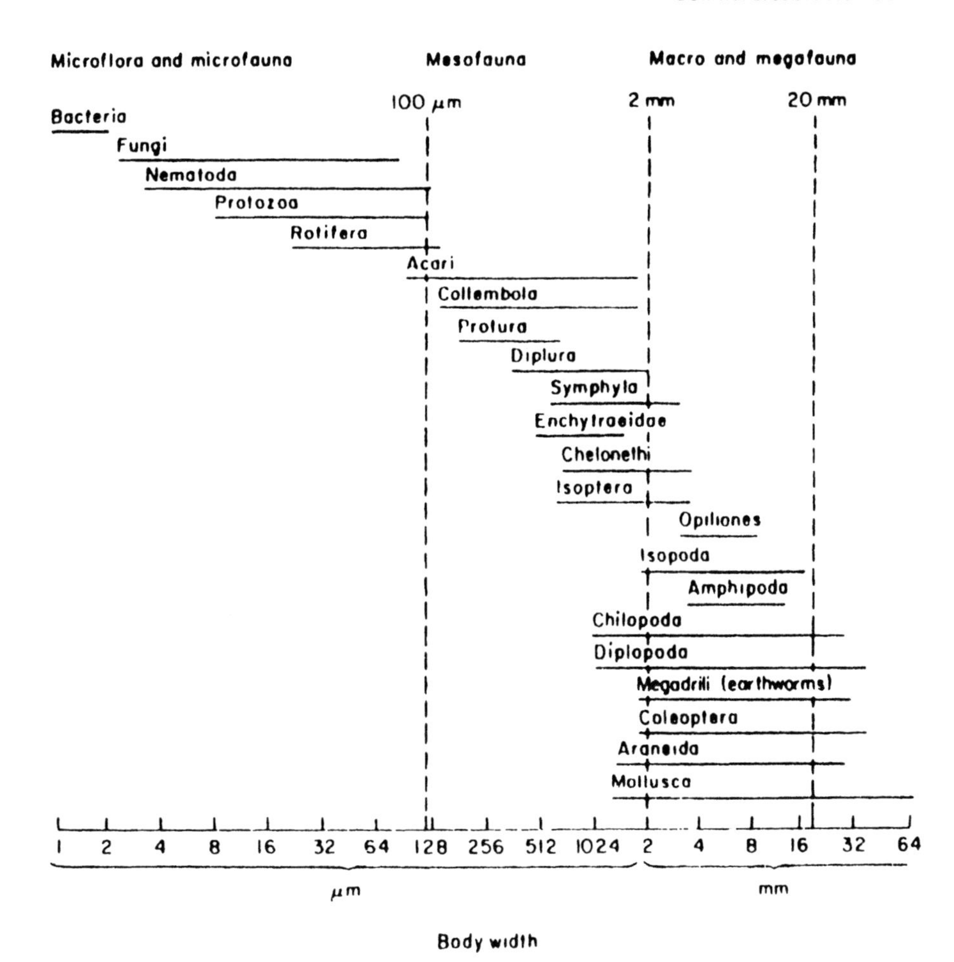

Figure 17.1. Size classification of organisms in decomposer food webs by body width (Swift et al. 1979).

General Sampling Precautions

Biodiversity associations with habitat require an awareness of possible contamination when designing sampling strategies. Soil corers or other collection tools that are used across a diverse number of habitats without thorough cleaning can transfer soil, with accompanying species, to the next-sampled habitat. To reduce organism transfer, clean the instrument appropriately. For microfauna this may mean sterilizing before use. For meso- and macrofauna, simply cleaning with a brush or paper towels should be sufficient. Presterilized hand trowels, which can be purchased at a reasonable cost, eliminate the time factor needed for cleaning, although soil corers are preferred.

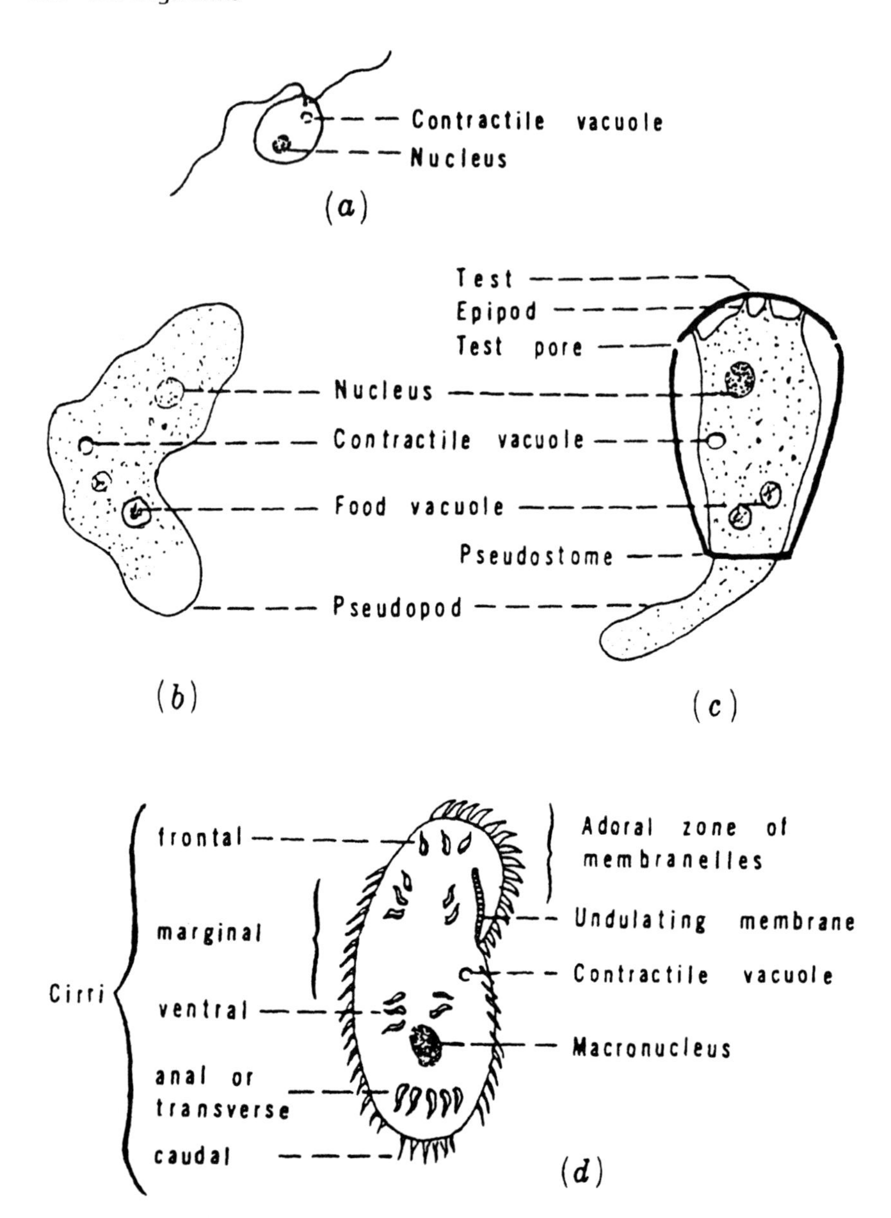

Figure 17.2. Morphology of four types of soil Protozoa: (a) flagellate (*Bodo*); (b) naked amoeba (*Naegleria*); (c) testacean (*Hyalosphenia*); and (d) ciliate (*Oxytricha*) (Lousier and Bamforth 1990).

Microfauna—Protozoa

Protozoa may be considered in four ecological groups: the flagellates, naked amoebae, testacea, and ciliates (Fig. 17.2; Lousier and Bamforth 1990). Protozoa are abundant in all known biomes of the LTER, including the Antarctic LTER sites, at times reaching abundances of 10^6 per gram soil or sediment. Protozoa have several feeding modes, from predation to primary production (if they contain algal symbionts), but they are primarily microbivorous, feeding directly on bacteria, fungal hyphae, or occasionally on fungal spores. Details of biologies and life histories of these fascinating unicellular eukaryotes are presented in Coleman and Crossley (1996) and more extensively in Darbyshire (1994). Protozoa are considered to have the greatest impact, mass for mass, of any of the fauna in many ecosystems because of their very rapid turnover rate (up to 10 generations in a growing season). This is attributed to their voracious feeding patterns in microsites such as rhizospheres, and in accumulations of organic matter (OM) such as aggregates, in a wide variety of ecosystems (Clarholm 1994; Coleman 1994).

Protozoa are concentrated mostly in the upper 2–5 cm of the soil horizon but may be found at great depths associated with phreatophytic vegetation, or even in groundwater at depths up to 200 m (Sinclair and Ghiorse 1989). Most of the protozoa are difficult to observe in soil samples and require culture techniques to quantify their presence, including the use of culture wells or microtiter plates (Darbyshire et al. 1974). A presence-absence (most probable number) procedure is often used to assess protozoan numbers. Briefly, this procedure entails placing small, known quantities of soil or soil suspensions from dilution series in wells and inoculating with a single species of bacteria as a food source for the protozoa. After several days' time (usually 1 week), "positive" wells, i.e., those with protozoa present, are counted by microscopy. By adding 2% HCl to a duplicate set of wells overnight and washing with dilute NaCl afterward to remove the acid, one can assess for the trophozoite (actively feeding forms) versus the encysted (inactive) forms. The assumption is that the dilute hydrochloric acid will kill the trophozoites and allow the encysted forms to excyst in the dilute saline and bacteria afterward. Unfortunately, this treatment method is considered unreliable (Cowling 1994). Alternative methods include heat treatment (45 °C. for approx. 20 minutes; Rutherford and Juma 1992), but comprehensive tests have yet to be made of this technique.

Perhaps more important, different most probable number (MPN) estimates can be obtained if different bacteria are supplied as food. For purposes of comparisons across sites, it is best to use the same quantitative methods throughout the comparisons (Cowling 1994).

In many forested soils, the testate or thecate amoebae are often numerous; for example, over half of all protozoans in an aspen site in the Canadian Rockies were testate amoebae (Lousier and Parkinson 1981), and calculations suggested that they consumed more than the entire standing crop biomass of bacteria during a growing season. These amoebae can be counted in stained samples of 5–30 mg soil, on membrane filters (Lousier and Parkinson 1981).

Materials

1. Distilled deionized water (DDW), agar, fresh soil
2. DDW 0.2 μm filtered
3. 125 mL flask (one per sample)
4. 15 mL plastic centrifuge tube (one per sample)

5. 15 mL glass screw-top tubes (five per sample)
6. "Multiwell" 24 well, tissue culture plates (one per sample)

Procedure

Pre-sample Preparation (Protozoan Plates)

1. From 500 g of fresh soil make an extract with enough DDW to make 1 L volume. Prepare agar using 5 g agar, 25 mL of the soil extract, and 200 mL DDW and autoclave the mixture for 15 minutes at 121 °C.
2. Using a clean bench (positive flow hood), pipet 0.5 mL of soil extract agar into each well of the plates. Place the plate together with two moist paper towels under it (to maintain moisture) in a plastic bag. Tape the bag shut, label it, and store at 3 °C until needed.
3. For each sample, 40 mL of DDW that has been filtered through 0.2 μm Nuclepore filters is dispensed into a 125 mL flask, and 9 mL into an autoclavable centrifuge tube. Five screw-top glass tubes, one with 5 mL and the rest with 9 mL of filtered physiological saline (0.9% NaCl), are also prepared for each sample.

Dilutions

1. Add 10 g of sample soil to the 125 mL flask and mix for 5 minutes at 200 rpm (1:5 dilution). Take a 5 mL subsample of the 1:5 dilution and add it to a test tube with 5 mL of sterile filtered DDW; this is the 1:10 dilution.
2. Take a 1 mL sample of the 1:10 dilution and add it to the 9 mL of sterile filtered DDW in the centrifuge tube; this is the 10^{-2} dilution.
3. Repeat the procedure using the screw-cap test tubes, taking 1 mL from each previous dilution, to make the further dilutions (10^{-3}, 10^{-4}, 10^{-5}, and 10^{-6}).
4. Working in the clean bench, use the 10^{-1}–10^{-6} dilutions to set up the protozoan plates. Using one plate per sample, put 0.5 mL of the 10^{-6} dilution into each of the plate's four replicate wells. Repeat this step using the other dilutions in order (10^{-5}, 10^{-4}, etc., down to 10^{-1}). The same pipette tip can be used for more concentrated dilutions but not for more dilute ones. Incubate the well plates in the dark at 27 °C for 6–9 days.

Bacterial Broth for Feeding Protozoans

Three days prior to counting the protozoan samples, start a bacterial broth to feed the protozoans. Dissolve 2 g "Bacto" dehydrated nutrient broth in 250 mL DDW and autoclave. Using sterile techniques add two loopfulls of bacteria (*Pseudomonas* sp.) to the flask. Allow it to incubate at 35 °C for at least 24 hours. One day prior to counting the samples, add 0.1 mL of broth to all wells of each culture plate.

Counting Protozoan Plates

Beginning with the highest dilution (10^{-6}), take an aliquot and observe under phase-contrast microscopy on a slide at approximately 400–600×. A single pipette can be used per replication if the highest dilution is transferred first. Up to three dilutions can be placed on a single slide under separate coverslips.

Note the presence or absence of protozoans (flagellates, ciliates, and amoebae). Larger protozoans tend to move toward the sides of the slide, so it may be necessary to scan three sides. Another technique is to scan two random transects. Whichever technique is adopted, it should be performed consistently for all samples.

For a given sample, once all three types of protozoans have been observed, the dilutions above that point need not be counted. This data are then entered into the computer (on a spreadsheet) and analyzed using a MPN program (e.g., Bamforth 1995).

Special Considerations

The preceding approach is generally recommended for enumeration of soil protozoa because it is reasonably accurate for soil ciliates in addition to the more abundant flagellates and naked amoebae, and it works well for distinguishing between these two latter forms. It may also allow some coarse taxonomic distinction among types of amoebae or flagellates because microscope slides are made from each well and high-resolution microscopy may be used if needed. However, situations may arise where study objectives may be focused on obtaining intensive estimates of numbers—for example, of the potential for protozoans to consume bacteria and/or produce nutrients. If this is the case, one may prefer a method more similar to that described by Darbyshire et al. (1974) using microtiter plates. This method is faster, but one is unlikely to obtain useful values for ciliated protozoa and it is less easy to discern among protozoan types (although flagellates and amoebae are distinguishable under some circumstances). Also, the method works best where protozoan numbers are relatively high (10^4–10^5 per gram soil), although soils with lower concentrations may be enumerated with practice in viewing the plates.

In this alternative technique, clear plastic microtiter plates are used, preferably those with flat bottoms. Each plate contains 12 rows of eight wells. Sterile physiological saline is quantitatively added to each well (0.05 mL). To the first row of wells, 0.05 mL of an appropriate dilution is added (approx. 10^2–10^3); weaker dilutions are better if one knows that the soils being enumerated have high levels of protozoa because there are fewer soil particles that may interfere with viewing. A set of eight 0.05 mL microtiter pipettes, that work by capillarity, is placed in the first row and rotated to mix the soil suspension and diluent. The pipettes are then raised and placed in the next set of dilution wells, and the process is successively repeated for all 12 rows, thus yielding a series of 12 twofold dilutions. This dilution series gives a more precise estimate of numbers than the 10-fold dilution method described earlier. A weak bacterial suspension (e.g., 0.05 mL of *Pseudomonas* grown on soil extract agar) is added to each well and left to incubate before viewing as per the standard method described earlier.

A key timesaving benefit of this method is the ability to view the wells from the bottom using an inverted microscope and thereby avoiding the need to make microscope slides for each dilution. However, the trade-off is a poorer-quality image and the difficulty of discerning types of protozoa. Wells may be enumerated as positive or negative and a twofold MPN table should be used to calculate an estimate of the number of protozoa in the original soil sample. Another problem with the microtiter method is that the small volume used to incubate the sample may result in the dominance of either amoebae or flagellates as competition for food resources occurs during population growth.

The microtiter method is particularly useful where one is interested in tracking the relationship between protozoan and bacterial populations or nutrient dynamics because these processes usually occur so rapidly that the first described method may be prohibitively time-consuming. Also, for bacterial/nutrient studies, one may be less interested in the kinds of protozoa and more interested in their dynamics. The first-described method is certainly recommended for exploratory and/or characterization studies of new sites within an LTER-like setting, and researchers may find that as they become involved in more process-oriented and sampling-intensive experiments, the latter method proves more practical.

Microfauna—Rotifers

Rotifers, like nematodes and tardigrades, are aquatic animals (inhabiting soil water films), occurring in many of the same habitats where tardigrades are found, and in environments where they would not be expected, i.e., hot North American and cold Antarctic desert soils. Although they are abundant in aquatic habitats, in soils they are primarily confined to the upper layers and in mosses and litter. Rotifers are similar to the other microfauna in their ability to survive desiccation by entering into an anhydrobiotic survival state and, in their shriveled state, being windblown with soil particles. In moist organic habitats, densities can exceed $10^5/m^2$ (Wallwork 1970).

Procedure

Sampling

It is best to sample soils in the top few centimeters because this reflects the habitat for rotifers. Soils should not be dried because it is best to have live specimens for identification. Sampling procedures for planktonic rotifers are more well developed (Green 1977).

Extraction

Vegetation can be agitated in water and the suspension washed through a sieve, similar to sizes used for nematodes. Rotifers are easily extracted using centrifugation techniques as described for nematodes.

Identification

A dissecting microscope at 400× magnification or higher is used for identifying living rotifers. The majority of identified soil and litter rotifers are in one order, the Bdelloidea, and are characterized by the cirri, or "wheel(s)" on their anterior end.

Time required for identification varies with densities, but it will usually be 10–15 minutes per sample if there are fewer than 100 specimens in the sample dish.

Microfauna—Tardigrades

Tardigrades are one of the favorite small soil fauna. Their nickname, "water bears," accurately describes this invertebrate. Small (100–500 μm in length), with a cylindrical body that is flattened on the ventral side, and with four pairs of clawed legs, this aquatic invertebrate is fascinating to watch lumbering slowly about under a microscope. Yet, like many of the soil fauna, little is known about it. Kinchin (1994) presents a recent and comprehensive description of these taxa; for general sampling, extracting, and identification, see Nelson and Higgins (1990). A species distribution list has been compiled by McInnes (1994). Species have been previously described based on habitat (aquatic versus terrestrial), but many species inhabit both. For example, moss habitats on land frequently have "limno" species, although there is no evidence of bryophagy. Tardigrades are polyphagous, including carnivores, bacterial feeders, algal feeders, and unknown feeding habits.

Procedure

Sampling

Tardigrades are found in vegetation and in soils, primarily in wet or moist habitats. There is little information on numbers of samples needed to adequately estimate population densities or species diversity of tardigrades.

For plants and other vegetation (mosses, lichens, etc.), remove the vegetation gently from the soil or rock surface and dry in a paper bag until ready for processing. For soils, sample with a core sampler or handle leaf and organic matter gently when sampling with a trowel. Soils can also be dried. The tardigrades enter anhydrobiosis, if gradually dried over 12–24 hours, and can be extracted later.

Extraction from Vegetation

1. If in a field situation, place the dried sample into water in a container for several hours; then, squeezing gently, stirring and agitating the vegetation will allow the tardigrades (adults, eggs, and cysts) to separate from the vegetation and settle into the sediment in the container. The excess water is then decanted (Nelson and Higgins 1990).
2. The Kinchin technique (1987) uses a 70 μm sieve, placed on a support in a sink. After soaking the dried sample in distilled water for 24 hours, the sample is placed on the sieve, and a high column (Plexiglass) is placed around the sample and on top of the sieve. Water slowly runs down the column for a minimum of 1 hour. Contents of the sieve are then viewed under a dissecting microscope (30× or greater).

3. A third method is soaking the vegetation for many hours and pouring through sieves to catch the tardigrades on a 50 μm sieve.

Extraction from Soil

1. Wash dried soil for several hours, as for vegetation, and then directly examine the sediment in thin films of water. This method can be extremely laborious and time-intensive. For sandy soils, stirring the sand and then quickly decanting the water-tardigrade solution is a sufficient method.
2. Alternatively, sugar and water centrifugation methods have been used by some (see Nelson and Higgins 1990) to acquire a sediment within the size range of the tardigrades. The method is similar to the one described in the section "Extractions of Nematodes in the Lab" but a 325-mesh sieve (45–40 μm) is used to collect the tardigrades (Martin and Yeates 1975).
3. In special situations it may be necessary to employ other techniques. Other methods include one for soil and moss (Hallas 1975) and one for soil, sand, and leaf litter (Greaves 1989).

Preservation

Boiling water or boiling alcohol (85% or greater) in equal amounts to the tardigrade-water solution is used for tardigrade preservation. As is true for nematodes, the average sorting time for tardigrades is significant, but with few of these animals in many soils, the total counting time may approach 10 minutes per sample.

Mesofauna—Nematodes

Prior to beginning a study of nematodes, it is always safest to test at least two extraction methods, since no nematode extraction method is best for all soil types and textures. All nematode extraction methods require water (except for entomophagous nematodes; see later). As an example of the difficulties in recovering nematodes, results from an extraction of known quantities of nematodes suspended in water using a sieving technique varied from 51% to 88% depending on the nematode species (Viglierchio and Schmitt 1983). Differences in nematode density and size, and losses of nematodes not collected on sieves accounted for some of the variability. The choice of extraction method varies with knowledge of the soil type and the nematode species, but in general nematologists recommend elutriation, sugar centrifugation, sieving, or misting as priority techniques. As with most soil biota, extraction should be from fresh soil (or from soil stored a maximum of 2 weeks at 4 °C). The optimum method is one that extracts a high diversity and abundance of living nematodes. A preliminary trial will help to determine whether samples should be field or laboratory processed.

Sampling strategies for nematodes will be dependent on the questions or hypotheses to be addressed. Nematodes are not distributed uniformly across a landscape but are aggregated near their food base (plant roots, microbes, fungi, protozoa, insects; e.g., Robertson and Freckman 1995). Densities generally decrease with depth, but in some situations nematodes can be found to great depths, e.g., at maximum rooting depth in desert ecosystems (Freckman and Virginia 1989).

The frequency distribution of nematodes generally follows a negative binomial distribution (McSorley 1987), which adds a degree of uncertainty to estimates of population densities. For example, Robertson and Freckman (1995) found densities varied from one to three orders of magnitude and 99% of the sample variance of free-living nematodes was spatially dependent at scales of <80 m in an agricultural field.

Nematode abundance and diversity are heavily influenced by the soil type, texture, and other physical and chemical properties. Determining the species diversity of nematodes in soil is quite daunting, with recent estimates of more than 72 species per sample of 200 nematodes in the Cameroon tropical forest (Lawton et al. 1996). If sampling for species biodiversity, fewer samples may be needed, primarily because there are so many species that it may take a taxonomist 2–3 years to identify the nematodes to species, and many species are new and will remain unidentified. If sampling for nutrient cycling or energetics questions, determining trophic group abundance may be a better way to proceed. The impact of a soil disturbance or a manipulation on nematodes might be addressed by quantifying abundance at the family level (which also provides information on trophic group), and using Bongers's Maturity Index (1990). Analyses based on total number of nematodes would be the least desirable type of information, since the density fluctuates vastly with climate, season, and life history of the nematode species.

For sampling a large area, the more replicate samples, the better the estimate of density and species. Sampling even "homogeneous" soils such as agroecosystems, which may vary in soil texture and chemical properties across a field, should include at least five samples of several bulked cores. Guidelines established for plant parasitic nematodes by the Society of Nematologists (Barker 1978) recommend that the number of cores (core size = 2.0 cm diam.) to be composited into one sample should be 10 cores for plots <5 m^2; 20 cores for plots 5–100 m^2; and 30 cores for plots >100 m^2. In a forest, 16 cores/sample represented 96% of the nematode species (Johnson et al. 1972). More recently, Prot and Ferris (1992) found that for extensive nematode surveys of large areas, one sample of 10 composite cores in a 7 ha alfalfa field reliably estimated abundance, but they cautioned that the cores should be spaced evenly across the entire area. In ecological studies, one core is sometimes used, since there are other time-consuming techniques (microbial biomass, fungi, other invertebrates, soil chemistry, soil physical factors) that may be analyzed and related to the nematodes. Perhaps nothing is more important than obtaining representative cores, choosing the better extraction method, and handling the samples consistently from the beginning to the end of the study.

Immediately following sampling, place the plastic bag with soil in an insulated cooler. Exposure to sun and heat, or to extreme cold if the ground was warm, can kill certain species. Mix the soil in the bag gently, breaking up any clumps. Measure out three subsamples for soil moisture determination (see Chapter 3, this volume) and for two extraction methods. Nematode distribution varies with microhabitat;

thus, relating a sample to soil moisture determinations taken at a nearby distance by a data logger or time domain reflectometry (TDR) may not reflect the actual moisture level in the soil sample.

Extractions of Nematodes in the Field

In a field situation with no power source and with a limited water source, the Baermann funnel technique (Freckman and Baldwin 1990; McSorley 1987) is the only extraction choice. The number of samples one can extract is dependent on the number of funnels available. Where electricity and water are available, the sugar centrifugation technique (described later) is preferred. The Baermann funnel technique is used for small soil samples (about 100 cm^3 volume or 40–50 g soil) and for root samples for endoparasites (roots are cut into small 2–5 cm pieces).

Materials

1. Plastic or wire window screen cut larger than the diameter of the funnel
2. Glass funnels
3. Ring stand or support for funnel
4. Kleenex or other brand tissue
5. Rubber hose to fit on bottom of funnel, about 5 cm long
6. Pinch clamp to go on bottom of funnel

Procedure

1. Fill the funnel with water to within about 5 cm from the top. Add the screen and place a tissue over the screen.
2. Place the soil uniformly over the tissue.
3. Add sufficient water to make the soil moist (not covered).
4. Leave for 24 hours, draw off nematodes and water (about 30 mL) into a beaker or dish by squeezing the pinch clamp.
5. Refrigerate or otherwise cool the extracted nematodes.
6. Refill water from the top by moving the screen aside gently.
7. At 48 hours, draw off an additional 30 mL into the same container.
8. Repeat at 72 hours, then remove soil and clean the funnel carefully (nematodes adhere to the sides of the funnel and may contaminate your next sample). Nematodes are drawn off at 24 hour intervals because large densities collecting in the bottom of the tube can result in oxygen depletion and death.

Special Considerations

Higher extraction efficiencies are obtained from sandier soils than from clay soils. Efficiency decreases with a variety of small details, including dirty and pitted glassware, quality of the tissue paper, how tissue paper is layered on the funnel, excess sample volume, and soil spread (lumpy or cone-shaped soil on the paper rather than evenly spread out).

Longer extraction times (>3 days) allow the eggs of bacterial-feeding species to hatch, which may overestimate their importance in decomposition food webs or in soil quality indices.

A modification of the Baermann funnel is the Baermann tray, which uses screening supported on metal pie pans. If the screens do not touch the bottom of the pan, this is a good method, since it has more surface area, allowing a larger soil sample to be placed on the screens. However, samples are frequently dirtier than with the Baermann funnel.

Extractions of Nematodes in the Lab

The sugar centrifugation technique, or modifications of it, is widely used in the United States, as is some form of elutriation when 500–1000 g samples are available (Freckman and Baldwin 1990; McSorley 1987). In the centrifugation technique, nematodes in soil are mixed with water and are centrifuged to concentrate the nematodes, then remixed in a sugar solution of a density in which they will float; this suspension is then centrifuged, and the nematodes are left in a clean solution of sucrose, washed into water. The method is fast and collects both living and dead nematodes.

Materials

1. Table centrifuge
2. Sucrose solution (454 g sucrose/L water)
3. Centrifuge tubes (50 mL)
4. Sieves (400-mesh [38 μm], 500-mesh [26 μm], 40-mesh [380 μm])
5. 1000 mL plastic beakers
6. 150 mL beakers or test tubes, ring stand with funnel
7. If preserving samples, 5% formalin solution

Procedure

1. Place 100–200 cm^3 of soil (approximately 150 g) in a large plastic beaker and obtain a weight of the soil.
2. Remove rocks with forceps.
3. Add water to 800 mL.
4. Stir carefully for 30 seconds using a spatula, spoon handle, or stirring stick (something you can rinse easily), or use motorized stirrer at 1550 rpm.
5. Immediately pour into wet sieves—a 40-mesh on top of a 400-mesh.
6. Rinse gently through the top of the stack, keeping the sieves at an angle as the water filters through.
7. Remove top sieve.
8. Rinse from the top of the sieve down, never directly on top of nematodes that are collecting at the bottom angle of the sieve. Let the water cascade down and carry the nematodes into the bottom wedge of the angled screen.
9. Tap the side of the screen gently to filter all the water through.

10. Rinse from the front and the back, keeping the screen at an angle and not allowing the water in it to overflow the edge of the screen.
11. Backwash the nematodes into a funnel with a 50 mL centrifuge tube beneath, to catch the nematodes, tipping the screen into the funnel above the tube and rinsing gently.
12. Rinse the funnel with water.
13. Put in the centrifuge, making sure to balance the load, for 5 minutes at 1750 rpm.
14. Decant off all liquid.
15. Fill with the sugar solution, which should be made ahead of time and cooled.
16. Stir gently with spatula until the pellet at the bottom of the tube is broken up and suspended.
17. Centrifuge for 1 minute at 1750 rpm (faster slices nematodes!).
18. Decant tube onto wet 500-mesh screen.
19. Rinse well with water and backwash into a beaker.
20. As necessary, preserve samples using a hot 5% formalin solution added to an equal volume of nematode-water solution.

Nematode Identification and Calculations

Trophic groups (bacterial feeders, fungal feeders, omnivores, plant parasites, algal feeders) can be identified based primarily on morphology and known feeding habits (see Freckman and Baldwin 1990; Yeates et al. 1993) using an inverted microscope (100× or 400×). Usually, nematologists count all the sample that has been extracted, but if the densities are too high, dilutions are made. However, if identifying to species, most species are rare, so dilutions may result in an underestimation of species richness.

Once soil moisture is known, calculate the amount of nematodes/per kilogram of dry soil. If bulk density is known, volume measurements (numbers/m^2 to a 10 cm depth) can be calculated.

Biomass can be determined quickly for nematodes using the free NIH image analysis program cited earlier in this chapter. Once the level of knowledge is specified (adults, juveniles, trophic groups, species, etc.), we recommend measuring the length and width of a minimum of 100 nematodes (Freckman 1982; or see Freckman and Mankau 1986). The nematode is displayed on the monitor, and the mouse is used to determine nematode length and width at the widest point (not at the vulva). This information is placed into Andrassy's formula (1956),

$$\mu\text{g nematodes} = (L \times W^2)/(1.6 \times 10^6)$$

where

L = length in μm
W = width in μm

This figure is multiplied by 0.25 to estimate the amount of carbon in nematodes (Freckman 1982).

Special Considerations

A well-trained worker can set up and extract 50–60 soil samples in a day. The identification and enumeration phase is longer, with a mixed sample of soil nematodes requiring 45–50 minutes, if examined in a dish on an inverted microscope. One needs to allow time for operator eye fatigue, so the total number of samples that can be handled reasonably in an 8 hour day may be only six or seven.

Microarthropods

The term *microarthropod* refers to an artificial grouping of the small arthropods that make up a significant part of the belowground food web in most terrestrial ecosystems. Microarthropods range in size from about 100 μm to a few millimeters and include the mites (Acari), collembola, symphyla, protura, diplura, pauropoda, small centipedes and millipedes, and small insects from several orders. Microarthropods include representatives of most trophic groups within the belowground food web (Walter et al. 1987a; Moore et al. 1988). The Acari alone contain species that are chewing fungivores, piercing fungivores, bacterivores, predators, and generalized saprovores. For a more complete discussion of the role of microarthropods on soil processes see reviews by Seastedt (1984b), Moore et al. (1988), Lussenhop (1992), and Moldenke (1994).

The abundance and biomass of different microarthropod groups in various terrestrial biomes have been summarized by Petersen and Luxton (1982). Densities of microarthropods are typically greatest in coniferous forests (up to about 800,000 per m^2), followed by deciduous forests and grasslands. Microarthropod numbers appear to be positively correlated with standing stocks of organic matter, on both regional (McBrayer et al. 1977) and local scales (Blair et al. 1994). The majority of microarthropods occur in the top 5–10 cm of soils. There are, however, few studies of their depth distributions in different biomes (e.g., Silva et al. 1989). A study of the vertical distribution of microarthropods in a mixed hardwood forest at the Coweeta LTER site indicated that about 87% of the total microarthropods collected to a depth of 55 cm occurred in the top 10 cm, with about 80% occurring in the top 5 cm (Seastedt and Crossley 1981). In contrast, microarthropods associated with rooting systems of *Prosopis glandulosa* in a Chihuahuan desert ecosystem were found to depths of 13 m, but a majority occurred in the top meter of soil (Silva et al. 1989).

Mites and collembola generally constitute 90–95% of microarthropod samples, although the relative abundance of different groups varies in different biomes. Oribatid mites (Cryptostigmata or Oribatei) are often the most abundant and diverse mites reported from temperate deciduous and coniferous forests. For example, Abbott et al. (1980) recovered 72 genera of oribatid mites from a mixed hardwood forest at the Coweeta site, with over 160 species recovered in a more extensive study (R. A. Hansen, personal communication). Moldenke and Fichter (1988) reported 93 genera from the H. J. Andrews Forest in western Oregon. Prostigmatid mites appear to be relatively more abundant in grasslands (Seastedt 1984a; Leetham and Milchunas 1985), alpine tundra (O'Lear and Seastedt 1994), and desert ecosystems (Santos et al. 1978).

Microarthropod Sampling

A variety of techniques have been proposed for sampling and extracting microarthropods (Edwards 1991; Moldenke 1994), and the utility of these methods can vary in different biomes and soil types. Virtually all methods begin with collection of appropriate soil or soil/litter samples. Microarthropods can be recovered from either intact or mixed soil or litter samples. Although grab samples from different habitats or litter/soil types are useful for initial qualitative surveys, intact soil cores of known diameter and depth are necessary for expressing population numbers on a per unit area or unit volume basis. We recommend a core diameter of 5 cm, preferably collected with a split corer containing an inner sleeve to keep the cores intact. This is especially useful if microarthropods are to be extracted from the core using a high-gradient type extractor (Crossley and Blair 1991). The depth of sampling, or of individual core segments, depends on the habitat, the goals of the study, and the type of extraction to be used. For less abundant forms, such as Mesostigmata, a larger-diameter core, or 25 cm × 25 cm samples, stratified for depth, is preferable for diversity estimates (A. R. Moldenke, personal communication).

For high-gradient extraction of intact cores (discussed later), a maximum depth of 5 cm per core segment is recommended. The typical total depth is 10 cm, with increments of 0–5 and 5–10 cm. Deeper horizons can be sampled by successively coring the same area. This results in clearly stratified sample depths and minimizes the soil compaction likely to occur when deeper cores are taken. Because microarthropod distributions are typically highly aggregated (Nef 1962; Edwards 1991), a large number of samples per treatment or habitat usually is needed to characterize population densities. While random sampling is generally the best strategy for comparing particular habitats or treatments, it may be desirable to further stratify sampling in some highly heterogeneous environments (e.g., under and away from vegetation in deserts or arid grasslands).

Intact cores should be wrapped with aluminum foil and placed in a cooler as they are collected. Cores that will be mixed before extraction can be placed into plastic bags. In both cases, soil cores should be kept cool and may be refrigerated in the lab until they are processed.

Extraction by Physical Methods

Methods for extracting microarthropods are generally based on behavior of the live organisms (dynamic methods) or on the use of various solutions to float microarthropods out of dispersed soil (physical methods). Each approach has unique advantages and shortcomings, and there is little consensus on a best method for all situations. Therefore, we recommend two methods that cover a range of soil conditions in a variety of biomes. However, this does complicate direct comparisons of data obtained from different studies.

Physical extraction techniques are usually based on dispersing soil samples in a solvent with the subsequent flotation of soil invertebrates based on density differences (salt or sugar solutions) or on the affinity of the arthropod cuticle for organic solvents. Many different solutions and modifications have been proposed, including a heptane flotation technique proposed by Walter et al. (1987b) and further modified by Geurs et al. (1991). This flotation method appears to be very efficient in the soils in which it has been tested, and it is recommended for soil types in which flotation would be appropriate. The main advantages of flotation methods are their efficiency at recovering both active and quiescent life stages (better for detailed studies of population structure), as well as specific taxa that would not respond behaviorally to heat and desiccation, i.e., members of the Collembolan family Onychiuridae, which are notoriously difficult to extract with gradient funnels (R. J. Snider, personal communication). Thus, flotation methods are particularly well adapted for use in arid ecosystems. They are also often considered a standard against which to measure the efficiency of other methods.

A major disadvantage of flotation is that it is exceptionally time-consuming, limiting the number of replicate samples that can be processed in a given time. Soft-bodied arthropods may be badly damaged during sample processing, and dead arthropods are included, since they, as well other organic debris, float out in these samples. Therefore, this method usually is not suitable for organic-rich soils. The need for adequate soil dispersion also limits its utility in soils with a high clay content.

A detailed description of the recommended heptane flotation method is provided by Walter et al. (1987b) and is not repeated here. While the extraction efficiency of this method appears to be quite high, it is very labor-intensive, with estimated sample processing times of 1–1.5 hours per core. Geurs et al. (1991) have suggested some modifications to reduce processing times somewhat. Another disadvantage is the use of organic solvents, which requires the availability of a fume hood and proper organic waste disposal facilities. Thus, it may not be feasible to use this method at some remote field sites. Other flotation methods that utilize sugar (1 mol/L sucrose) or salt (1 mol/L $MgCl_2$) solutions to create density gradients have been proposed to eliminate the problem of dealing with organic solvents. These methods can be substituted for heptane flotation, although their relative efficiency needs to be documented.

Extraction by High-Gradient Extraction

Dynamic methods such as high-gradient extraction depend on the behavioral responses of microarthropods to artificially induced temperature and moisture gradients to drive them out of soil cores and into collection funnels. Perhaps the best known of these is the high-gradient extractor proposed by Macfadyen (1962) and subsequently modified by Merchant and Crossley (1970), Crossley and Blair (1991), and Moldenke (1994). One of the main advantages of this method is that a large number of samples can be processed simultaneously and relatively inexpensively. The method seems especially well suited for highly organic soils where flotation techniques would be problematic. The major disadvantage with high-gradient extraction is its variable, and often unknown, efficiency, which is low for certain Collembolan families, as noted in the preceding section.

Many variations of the high-gradient extraction method have been proposed (Edwards 1991), some of which require elaborate heating and/or cooling systems.

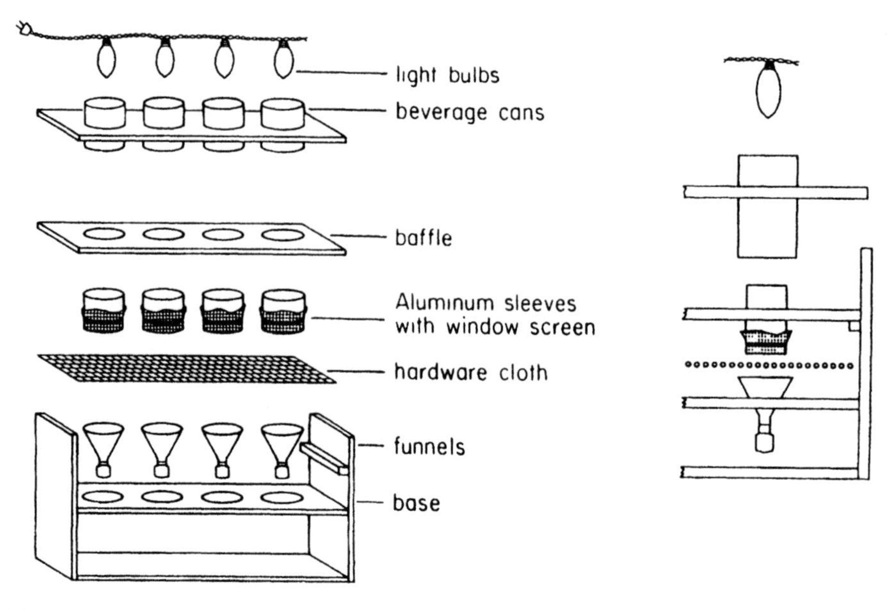

Figure 17.3. Design and assembly of the high-efficiency extractor for soil microarthropods (Crossley and Blair 1991).

Here we summarize a relatively simple and inexpensive extraction system (Crossley and Blair 1991), which is portable and can easily be adapted to a variety of situations (cold room, refrigerator, or bench top extraction).

Materials

The basic extractor (Fig. 17.3) is constructed of the following:

1. Plywood
2. Hardware cloth (6 mm mesh)
3. Polyethylene funnels (65 mm diameter fitted with polyethylene vial caps on the stems)
4. Aluminum tubes (350 mL beverage cans with top and bottom removed)
5. 7 watt lightbulbs
6. Aluminum sleeves in which soil cores are collected and held (5 cm diameter × 5 cm deep)
7. Window screen material (1 mm mesh), which can be attached to the cores with rubber bands
8. Vials

Scoring the funnel stems makes the attachment of vial caps to the funnel stems easier. For sandy and drier soils, double-layered cheesecloth works better than window screen material. The following is a brief description of the assembly and operation of the extractor, which is described in greater detail in Crossley and Blair (1991).

Procedure

1. Insert the funnels, with lids attached, into the plywood base assembly (Fig. 17.3).
2. Fill vials approximately half full with 70% ethanol and attach to the funnels.
3. Insert a collection label, with sample date and ID information written in pencil or alcohol-resistant ink, into each vial.
4. Place the wire hardware cloth (to support the soil cores) over the funnels and put in place the plywood baffle (to separate the top and bottom halves of the cores).
5. Unwrap the aluminum sleeves with soil cores to expose the upper surface of the core. If the soils are dry, the surface can be wetted with deionized water to minimize soil loss when the core is inverted.
6. Cover the upper surface of the core with a 10 × 10 cm square of window screen held in place with a rubber band.
7. For sandy soils, or those with little surface litter, a square of double-layered cheesecloth can be attached over the window screen to minimize soil contamination of the extracted samples.
8. Invert and place soil cores screen down through the baffle to be supported on the hardware cloth. An air space should remain between the bottom of the core and the top of the funnel.
9. Once all the soil cores are in place, cover with the beverage can sleeves, which increase the efficiency of heating the upper portion of the cores.
10. Place individual lightbulbs, attached to a wooden support bar over each can, and place a piece of aluminum foil over the light bar to act as a reflector.
11. Set the entire unit into a refrigerator or cold room at about 10 °C. Although we recommend this for development of a large temperature gradient, it is possible to extract soil microarthropods at room temperatures if a cold room or refrigerator is not available (D. A. Crossley Jr., personal communication).
12. An extraction time of 5 days is usually sufficient for most soil types. Check the adequacy of this period by removing the collection vials after 5 days, replacing them with new vials, and then checking for additional specimens after additional days.
13. At the end of the extraction period, carefully disassemble the extractor and remove the soil cores. The dried cores must be handled carefully to minimize soil contamination at this point.
14. Rinse the inside of the funnels with a small amount of 70% ethanol so that any microarthropods adhering to the funnel walls are washed into the sample vials. Once capped, the samples can be stored.

Special Considerations

The enumeration and identification of extracted microarthropods will vary, depending on the goals of the study and the degree of taxonomic expertise available. Many microarthropods can be identified to suborder or family under a dissecting microscope. Finer taxonomic resolution typically requires mounting individual specimens on microscope slides, or use of an inverted compound microscope

(Wright 1988). There are several keys available for identification of specific groups of microarthropods, such as those in Dindal (1990).

The total time required for extraction and enumeration varies, depending on numbers of microarthropods present and the amount of contaminant soil particles, litter, etc. It is often preferable to do an initial sorting of microarthropods from surrounding contaminating materials, then perform a more careful identification and enumeration using either a binocular dissecting microscope or the inverted compound unit noted above. Time required per sample ranges from 30 minutes to 1 hour.

Enchytraeids

Enchytraeids are small Oligochaetes that are considered important in a number of temperate-to-boreal ecosystems (O'Connor 1955; Ellenberg et al. 1986; Didden 1993) and also in agricultural systems (Didden 1991). Enchytraeids can affect decomposition processes indirectly by comminution and mixing of organic material and soil, and by digesting soil microbes, releasing mineralized nutrients for subsequent plant uptake (van Vliet et al. 1995). Being small, segmented worms, distantly related to earthworms, enchytraeids are very sensitive to a lack of water and prefer to exist in moist water films or moister soils. This trait is employed to facilitate the extraction of these animals from soil samples. Because they require water films to be active, they are more usually found in mesic to moist habitats.

Enchytraeids are commonly sampled with 5 cm diameter soil corers, to a depth of 5 cm. Soil is sliced at depths of 2.5 cm increments and is placed on a modified wet-funnel extractor (O'Connor 1955). The soil, placed on a sieve in a funnel filled with water, is exposed to increasing heat and light. After 4 hours of presoaking (to saturate the soil), the light intensity in 40 watt bulbs is gradually turned up (over approximately 3 hour period) on a rheostat timer until the soil surface reaches a temperature of 45 °C. Enchytraeids respond by moving away from the heat and light and passing through the sieve into the water below. They are then counted and/or preserved in 70% ethanol. To determine the ash-free dry weight, subsamples can be freeze-dried and then ashed in a muffle furnace at 500 °C for 4 hours.

Materials

1. Funnel rack similar to that used for nematode extraction
2. 8 cm diameter plastic funnels attached to collection vials
3. Screening or cheesecloth
4. 40 watt lightbulbs on a single rheostat, positioned 2–4 cm above soil samples

Procedure

1. Place prewetted soil sample on screening or cheesecloth inside the funnel.
2. Add water to barely cover the wet soil sample.
3. Over a 3 hour extraction period gradually increase light intensity to warm the surface soil temperature to 45 °C.

4. Preserve enchytraeids captured in the water beneath the funnels in 70% ethanol.

Special Considerations

Total time required to set up the samples for extracting the Enchytraeids is approximately 4 hours. Enumeration time will require from 40 to 60 minutes for a sample averaging 100–150 individuals. To obtain identifications below the family level, it will be necessary to send specimens to a taxonomic expert.

Macrofauna—Macroarthropods

Surface-dwelling macroarthropods are often censused by pitfall trapping, in which open-top traps are set into the ground level with the surrounding soil surface (Coleman and Crossley 1996). While this method may be useful for surveys, or to assess the relative activities of macroarthropods, it has limited utility as a quantitative sampling technique (Dennison and Hodkinson 1984). Instead, we recommend hand sorting a known area as a standard method for quantifying surface macroarthropod densities on a per unit area basis (Edwards 1991). For most biomes, sample areas of 1 m^2 will be sufficient. Subsurface macroarthropods, such as white grubs and cicada nymphs, can be assessed using the same hand-sorting method recommended in the next section for earthworm sampling (Seastedt et al. 1987).

Macrofauna—Earthworms

Three major ecological groupings of earthworms have been defined based on feeding and burrowing strategies (Bouchê 1977). Epigeic species live in or near the surface litter and feed primarily on coarse particulate organic matter. They are typically small and have high metabolic and reproductive rates as adaptations to the highly variable environmental conditions at the soil surface. Endogeic species live within the soil profile and feed primarily on soil and associated organic matter (geophages). They generally inhabit temporary burrow systems that are filled with cast material as the earthworms move through the soil.

Anecic earthworm species, such as the familiar *Lumbricus terrestris*, live in more or less permanent vertical burrow systems that may extend 2 m into the soil profile. They feed primarily on surface litter, which they pull into their burrows, and also may create "middens" at the burrow entrance, consisting of a mixture of cast soil and partially incorporated surface litter. These broad ecological categories indicate that different sampling methodologies are required for different ecological groupings of earthworms. There also is seasonal variation in depth distribution for many species, with some species adopting resting stages deeper in the soil to escape unfavorable conditions at certain times of year. It is therefore critical to have some knowledge of the species present and their life cycles and seasonal variation before deciding on specific methods for sampling earthworm populations.

There are two main approaches to sampling earthworm populations: physical and behavioral. Physical methods involve removing a known volume of soil and sorting through it manually or mechanically. Behavioral methods rely on the response of earthworms to an irritant, usually a solution of dilute formalin, applied to the soil. The irritant causes the earthworms to come to the surface, where they can be collected manually. Different methods work better for different earthworm species, and the two methods, physical and behavioral, can sometimes be combined successfully to provide more accurate results than could be attained by using either method alone. Before choosing a particular approach, preliminary studies should be performed to assess the relative differences of the various methodologies and their applicability to the particular site under study. Thorough assessments of earthworm populations can be labor-intensive, and the methods selected will necessarily be a compromise between the level of accuracy desired and the labor and time available. There are several excellent reviews of methods for sampling earthworm populations (e.g., Satchell 1969; Bouché 1972; Bouché and Gardner 1984; Lee 1985; Edwards and Bohlen 1995). We provide here a summary of methods that can be adapted successfully to the vast majority of sampling situations.

Extraction by Passive Methods

Passive methods of earthworm sampling include handsorting, washing and sieving, and flotation techniques. Physical methods for assessing earthworm populations involve removing a known volume of soil from which earthworms can be manually or mechanically collected. The soil can simply be dug with a spade or removed with a mechanical device, such as a hydraulic coring tool (e.g., Berry and Karlen 1993). When digging with a spade, it can be difficult to remove replicable volumes of soil, a problem that can be overcome by driving a steel quadrat into the ground to mark the edges of the sampling unit. Use a square-point spade inserted with the back toward the quadrat and keep the spade vertical. Typical dimensions for the surface area of soil removed are 25 × 25 cm. Smaller sampling areas may be inadequate when population densities are low. Also, the proportion of physically damaged worms in the sample is greater when very small sampling units are used. Larger quadrats decrease sampling efficiency when soils are handsorted because of the great volumes of soil that must be processed (Zicsi 1958). The depth of the soil removed must match the vertical distribution of the species present. This can be determined by preliminary sampling and may vary with time of year (e.g., Persson and Lohm 1977; Baker et al. 1992). We recommend sampling depths of 20–30 cm. Soil can be removed incrementally if more detailed data on depth distribution are desired.

Procedure

Earthworms can be separated from the soil either by handsorting or washing and sieving. Handsorting can be performed in the field by placing the soil on plastic sheets and carefully sorting through it or by returning soil to the laboratory. The soil should be handsorted immediately or stored at 4 °C because any worms killed during digging will decay rapidly. Handsorting is the simplest method, but it is laborious, and small worms (<100 mg fresh weight [Satchell 1983; Lohn and Persson 1977] or <2 cm in length [Reynolds 1973]) are easily overlooked during handsorting. In most cases small worms do not constitute a large portion of total earthworm biomass, but they may contribute significantly to earthworm numbers. Handsorting can be supplemented with washing and sieving to correct for very small specimens or to collect earthworm cocoons. Washing and wet sieving can be accomplished using a standard soil sieve (Parmelee and Crossley 1988) or more elaborate mechanical washing sieving devices (e.g., Edwards et al. 1970; Bouché 1972). We recommend standard soil sieves with the mesh size selected to prevent small earthworms or cocoons from washing through. Soils with a large clay content may need to be soaked in a 0.5% metaphosphate solution to facilitate sieving. Formalin may be added to the solution to preserve the worms.

Another situation in which handsorting is ineffective and washing and sieving may be preferred is when sampling turf or pasture grass mats with dense fibrous root systems. Physical sampling methods are by far the best method for an accurate population assessment of shallow-dwelling earthworm species such as those belonging to the lumbricid genera *Aporrectodea* and *Octolasion* or the native American megascolecid genus *Diplocardia.* However, some lumbricid species such as *Lumbricus terrestris,* which form permanent burrows that may extend up to 2 m deep, are difficult to sample by physical methods. Small immature specimens of this species may be recovered by handsorting, but adults, which can constitute the majority of earthworm biomass at some sites, are rarely recovered by digging or coring methods. Behavioral sampling methods must be used for an adequate assessment of populations of species such as *L. terrestris.*

Extraction by Behavioral Methods

The most widely used behavioral sampling method is the formalin expulsion technique, first described by Raw (1959). This method consists of applying a dilute formalin solution to a known area of soil. Earthworms are irritated by the formalin solution and expelled onto the soil surface, where they can be easily collected.

Materials

1. Dilute formalin solution, 0.25%, or approximately 14 mL of 37% formalin in 2 L water
2. Quadrat frames, 50 × 50 cm, made from strips of aluminum sheet metal

Procedure

1. Press quadrat frames 1–2 cm into soil surface.
2. Slowly pour 2 L of dilute formalin onto the soil surface within the quadrat frame.
3. Earthworms will begin to emerge within several minutes and should be collected as they emerge. Continue collection until emergence ceases.

Special Considerations

The formalin expulsion method has some serious limitations. It is ineffective for megascolecid earthworms such as *Diplocardia* spp. (James 1990), and it appears to be reliable for only some species of Lumbricidae. It is most effective for species that have vertical burrows that open to the surface (e.g., *L. terrestris*) and is less effective for horizontally burrowing species. It is totally ineffective at recovering worms in resting stages. The method is not effective at low soil temperatures (below 4–8 °C) or when the soil is either very wet or dry. Because of these limitations, the formalin method works best during times of year when populations are most active, typically spring and fall in temperate regions. Under these conditions, the method is excellent for deep-dwelling lumbricid species and may even provide a good relative measure of populations of those species that are not recovered as well by formalin as by handsorting (Bouché and Gardner 1984; Bohlen et al. 1995a,b).

Another problem with formalin extraction is that it is difficult to determine the exact volume of soil being sampled because there is no way to determine the flow paths of the formalin. Nonetheless, it is by far the best method for sampling *L. terrestris* and similar species. If formalin cannot be used because of objections about its toxicity, a nontoxic alternative is a solution of mustard flour (0.33%; Gunn 1992), although this method is not as well studied or standardized as the formalin method and cannot be recommended for routine scientific sampling.

For estimates of biomass chemistry soil fauna can be sampled into water or into empty vials and then immediately freeze-dried or oven-dried for subsequent analysis.

In situations where both deep-dwelling and shallow-dwelling species coexist, a combined physical-behavioral approach can be used. Formalin solution can be applied to the bottom of the pit dug to obtain samples for handsorting. Worms emerging in the bottom of the pit can be combined with those collected by handsorting (Martin 1976; Barnes and Ellis 1979). This technique has been used successfully in studies of earthworm populations in corn agroecosystems on a silt loam soil in Ohio (Bohlen et al. 1995b).

Interpretation and Analysis of Earthworm Data

Keys for identification of earthworm species are available in Dindal (1990). Data for earthworm populations are expressed as total biomass or number of earthworms per unit area, generally as grams or number $\times m^{-2}$. Biomass should be expressed as ash-free dry mass (AFDM) rather than oven-dry mass because AFDM corrects for varying mass of soil in the earthworms' intestines, which can account for 50–70% of earthworm dry mass. AFDM is determined by combusting oven-dried worms (60 °C to constant weight), or a ground subsample, in a muffle furnace at 500 °C for 4 hours (g AFDM = g dry wt − g ash wt).

A difficulty that arises in analyzing data from physical sampling methods is that each sample usually contains body parts, in addition to whole individuals. This is not a problem for estimates of earthworm biomass per unit area, but it complicates the accurate determination of numbers of individuals in a sample. An arbitrary method for overcoming this problem is to count as an individual any fragment of a whole worm that contains the anterior portion (head). Body fragments lacking a head are included in total biomass but are not counted as individuals.

A cautionary note is needed regarding the evaluation of earthworm communities at a specific site. Glaciation, land-use changes, and the introduction of exotic species have significantly affected the distribution of native North American species of earthworms (Hendrix 1995), and in many localities at least some exotic species will be present (Hendrix et al. 1992). This is significant because introduced and exotic species may have different ecological roles and may respond differently to environmental factors or ecosystem treatments (e.g., James 1988). In many instances the present species composition will be a reflection, in part, of colonization history, which is affected by the proximity of source populations, the presence of barriers to colonization, and various stochastic factors. For this reason it is useful to have some information on the distribution of different species across the landscape, which should be considered before designing a sample plan. In areas where species replacements are recent or ongoing, it may be important to document long-term dynamics of earthworm populations.

References

Abbott, D. T., T. R. Seastedt, and D. A. Crossley Jr. 1980. The abundance, distribution and effects of clear-cutting on oribatid mites (Acari: Cryptostigmata) in the southern Appalachians. *Environmental Entomology* 9:618–623.

Andrassy, I. 1956. Die rauminhalts und gewichtbestimmung der Fadenwurmer (Nematoden). *Acta Zoologica Academy of Sciences Hungary* 2:1–15.

Baker, G. H., V. J. Barrett, R. Grey-Gardner, and J. C. Buckerfield. 1992. The life history and abundance of the introduced earthworms *Aporrectodea trapezoides* and *A. caliginosa* (Annelida: Lumbricidae) in pasture soils in the Mount Lofty Ranges, South Australia. *Australian Journal of Ecology* 17:177–188.

Bamforth, S. S. 1995. Isolation and counting of protozoa. Pages 174–180 *in* K. Alef and P. Nannipieri, editors, *Methods in Applied Soil Microbiology and Biochemistry.* Academic Press, London, UK.

Barker, K. R. 1978. Determining population responses to control agents. Pages 114–125 *in* E. I. Zehr, editor, *Methods for Evaluating Plant Fungicides, Nematicides, and Bactericides.* American Phytopathological Society, St. Paul, Minnesota, USA.

Barnes, B. T., and F. B. Ellis. 1979. Effects of different methods of cultivation and direct drilling and disposal of straw residues on populations of earthworms. *Journal of Soil Science* 30:669–679.

Berry, E. C., and D. L. Karlen. 1993. Comparison of alternative farming systems. II. Earthworm population density and species diversity. *American Journal of Alternative Agriculture* 8:21–26.

Blair, J. M., R. W. Parmelee, and R. L. Wyman. 1994. A comparison of forest floor invertebrate communities of four forest types in the northeastern U.S. *Pedobiologia* 38:146–160.

Bohlen, P. J., W. M. Edwards, and C. A. Edwards. 1995a. Earthworm community structure and diversity in experimental agricultural watersheds in northeastern Ohio. *Plant and Soil* 164:536–543.

Bohlen, P. J., R. W. Parmelee, J. M. Blair, C. A. Edwards, and B. R. Stinner. 1995b. Efficacy of methods for manipulating earthworm populations in large-scale field experiments in agroecosystems. *Soil Biology and Biochemistry* 27:993–999.

Bongers, T. 1990. The Maturity Index: an ecological measure of environmental disturbance based on nematode species composition. *Oecologia* 83:14–19.

Bouché, M. B. 1972. *Lombriciens de France: ecologie et systematique.* INRA Publ. 72-2. Institut Nationale Recherches Agriculturelle, Paris, France.

Bouché, M. B. 1977. Strategies lombriciennes. *Ecological Bulletin* (Stockholm) 25:122–132.

Bouché, M. B., and R. H. Gardner. 1984. Earthworm functions. VIII. Population estimation techniques. *Revue d'Öcologie et de Biologie du Sol* 21:37–63.

Brian, M. V. 1965. *Social Insect Populations.* Academic Press, London, UK.

Clarholm, M. 1994. The microbial loop in soil. Pages 221–230 *in* K. Ritz, J. Dighton, and K. E. Giller, editors, *Beyond the Biomass.* Wiley, Chichester, UK.

Coleman, D. C. 1994. The microbial loop concept as used in terrestrial soil ecology studies. *Microbial Ecology* 28:245–250.

Coleman, D. C., and D. A. Crossley Jr. 1996. *Fundamentals of Soil Ecology.* Academic Press, San Diego, California, USA.

Cowling, A. J. 1994. Protozoan distribution and adaptation. Pages 5–72 *in* J. F. Darbyshire, editor, *Soil Protozoa.* CAB International, Wallingford, UK.

Cromack, K., Jr., B. L. Fichter, A. M. Moldenke, J. A. Entry, and E. R. Ingham. 1988. Interactions between soil animals and ectomycorrhizal fungal mats. *Agriculture, Ecosystems and Environment* 24:161–168.

Crossley, D. A., Jr., and J. M. Blair. 1991. A high-efficiency, low-technology Tullgren-type extractor for soil microarthropods. *Agriculture, Ecosystems and Environment* 34:187–192.

Darbyshire, J. F. 1994. *Soil Protozoa.* CAB International, Wallingford, UK.

Darbyshire, J. F., R. E. Wheatley, M. P. Greaves, and R. H. E. Inkson. 1974. A rapid micromethod for estimating bacterial and protozoan populations in soil. *Revue d'Öcologie et de Biologie du Sol* 11:465–475.

Dennison, D. F., and J. D. Hodkinson. 1984. Structure of the predatory beetle community in a woodland soil ecosystem. IV. Population densities and community composition. *Pedobiologia* 26:157–170.

Didden, W. A. M. 1991. Population ecology and functioning of Enchytraeidae in some arable farming systems. Ph.D. dissertation. Agricultural University Wageningen, Netherlands.

Didden, W. A. M. 1993. Ecology of terrestrial Enchytraeidae. *Pedobiologia* 37:2–29.

Dindal, D. L. 1990. *Soil Biology Guide.* Wiley, New York, New York, USA.

Edwards, C. A. 1991. The assessment of populations of soil-inhabiting invertebrates. *Agriculture, Ecosystems and Environment* 34:145–176.

Edwards, C. A., and P. J. Bohlen. 1995. *Biology and Ecology of Earthworms.* Chapman and Hall, London, UK.

Edwards, C. A., A. E. Whiting, and G. W. Heath. 1970. A mechanized washing method for separation of invertebrates from soil. *Pedobiologia* 10:141–148.

Ellenberg, H., R. Mayer, and J. Schauermann 1986. Ökosystemforschung: Ergebnisse des Sollingprojekts: 1966–1986. Ulmer, Stuttgart, Germany.

Freckman, D. W. 1982. Parameters of nematode contribution to ecosystems. Pages 81–97 *in* D. W. Freckman, editor, *Nematodes in Soil Ecosystems.* University of Texas Press, Austin, Texas, USA.

Freckman, D. W., and J. G. Baldwin. 1990. Soil Nematoda. Pages 155–200 *in* D. L. Dindal, editor, *Soil Biology Guide.* Wiley, New York, New York, USA.

Freckman, D. W., and R. Mankau. 1986. Abundance, distribution, biomass and energetics of soil nematodes in a northern Mojave Desert ecosystem. *Pedobiologia* 29:129–142.

Freckman, D. W., and R. A. Virginia. 1989. Plant feeding nematodes in deep-rooting desert ecosystems. *Ecology* 70:1665–1678.

Geurs, M., T. Bongers, and L. Brussaard. 1991. Improvements of the heptane flotation method for collecting microarthropods from silt loam soil. *Agriculture, Ecosystems and Environment* 34:213–221.

Greaves, P. M. 1989. An introduction to the study of tardigrades. *Microscopy* 36:230–239.

Green, J. 1977. Sampling rotifers. *Archivs fuer Hydrobiologie* 8:9–12.

Gunn, A. 1992. The use of mustard to estimate earthworm populations. *Pedobiologia* 36:65–67.

Hallas, T. E. 1975. A mechanical method for the extraction of Tardigrada. Proceedings of the First International Symposium on Tardigrades. *Memorie dell'Istituto di Idrobiologia* 32:153–158.

Hendrix, P. F. 1995. *Earthworm Ecology and Biogeography in North America.* Lewis Publishers, Boca Raton, Florida, USA.

Hendrix, P. F., B. R. Mueller, R. R. Bruce, G. R. Langdale, and R. W. Parmelee. 1992. Abundance and distribution of earthworms in relation to landscape factors on the Georgia Piedmont, U.S.A. *Soil Biology and Biochemistry* 24:1357–1361.

James, S. W. 1988. The post-fire environment and earthworm populations in tallgrass prairie. *Ecology* 69:476–483.

James, S. W. 1990. Oligochaeta: Megascolecidae and other earthworms from southern and midwestern North America. Pages 379–386 *in* D. Dindal, editor, *Soil Biology Guide.* Wiley, New York, New York, USA.

Johnson, S. R., V. R. Ferris, and J. M. Ferris 1972. Nematode community structure of forest woodlots. I. Relationships based on similarity coefficients of nematode species. *Journal of Nematology* 4:175–182.

Kinchin, I. 1987. The moss ecosystem. *School Science Review* 68:499–503.

Kinchin, I. 1994. *The Biology of Tardigrades.* Portland Press, Chapel Hill, North Carolina, USA.

Lawton, J. H., D. E. Bignell, G. F. Bloemers, P. Eggleton, and M. E. Hodda. 1996. Carbon flux and diversity of nematodes and termites in Cameroon forest soils. *Biodiversity and Conservation* 5:261–273.

Lee, K. E. 1985. *Earthworms: Their Ecology and Relationships with Soils and Land Use.* Academic Press, Sydney, Australia.

Lee, K. E., and T. G. Wood. 1971. *Termites and Soils.* Academic Press, London, UK.

Leetham, J. W., and D. G. Milchunas. 1985. The composition and distribution of soil microarthropods in the shortgrass steppe in relation to soil water, root biomass, and grazing by cattle. *Pedobiologia* 28:311–325.

Lohm, U., and T. Persson, editors. 1977. *Soil Organisms as Components of Ecosystems.* Ecological Bulletins, no. 25. Swedish Natural Science Research Council, Stockholm, Sweden.

Lousier, J. D., and S. S. Bamforth 1990. Soil protozoa. Pages 97–136 *in* D. L. Dindal, editor, *Soil Biology Guide.* Wiley, New York, New York, USA.

Lousier, J. D., and D. Parkinson. 1981. Evaluation of a membrane filter technique to count soil and litter Testacea. *Soil Biology and Biochemistry* 13:209–213.

Lussenhop, J. 1992. Mechanisms of microarthropod-microbial interactions in soil. *Advances in Ecological Research* 23:1–33.

Macfadyen, A. 1962. Soil arthropod sampling. Pages 1–24 *in* J. B. Cragg, editor, *Advances in Ecological Research.* Volume 1. Academic Press, London, UK.

Martin, N. A. 1976. Effect of four insecticides on the pasture ecosystem. V. Earthworms (Oligochaeta: Lumbricidae) and Arthropoda extracted by wet sieving and salt flotation. *New Zealand Journal of Agricultural Research 1*9:111–115.

Martin, N. A., and G. W. Yeates. 1975. Effect of four insecticides on the pasture ecosystem.

III. Nematodes, rotifers and tardigrades. *New Zealand Journal of Agricultural Research* 18:307–312.

McBrayer, J. F., L. J. Metz, C. S. Gist, B. W. Cornaby, Y. Kitazawa, T. Kitazawa, J. G. Werntz, G. W. Krantz, and H. Jensen. 1977. Decomposer invertebrate populations in U.S. forest biomes. *Pedobiologia* 17:89–96.

McInnes, S. J. 1994. Zoogeographic distribution of terrestrial/freshwater tardigrades from current literature. *Journal of Natural History* 28:257–352.

McSorley, R. 1987. Extraction of nematodes and sampling methods. Pages 13–48 *in* R. H. Brown and B. R. Kerry, editors, *Principles and Practices of Nematode Control in Crops.* Academic Press, London, UK.

Merchant, V. A., and D. A. Crossley Jr. 1970. An inexpensive, high-efficiency Tullgren extractor for soil microarthropods. *Journal of the Georgia Entomological Society* 5:83–87.

Moldenke, A. R. 1994. Arthropods. Pages 517–542 in R. W. Weaver, S. Angle, and P. Bottomley, editors, *Methods of Soil Analysis. Part 2, Microbiological and Biochemical Properties.* Soil Science Society of America, Madison, Wisconsin, USA.

Moldenke, A. R., and B. L. Fichter. 1988. *Invertebrates of the H. J. Andrews Experimental Forest, Western Cascade Mountains, Oregon. IV. The Oribatid Mites (Acari: Cryptostigmata).* USDA Forest Service, General Technical Report PNW-GTR-217. Portland, Oregon, USA.

Moore, J. C., D. E. Walter, and H. W. Hunt. 1988. Arthropod regulation of micro- and mesobiota in belowground food webs. *Annual Review of Entomology* 33:419–439.

Nef, L. 1962. The distribution of Acarina in soil. Pages 56–58 *in* P. W. Murphy, editor, *Progress in Soil Zoology.* Butterworths, London, UK.

Nelson, D. R., and R. P. Higgins. 1990. Tardigrada. Pages 393–419 *in* D. L. Dindal, editor, *Soil Biology Guide.* Wiley, New York, New York, USA.

O'Connor, F. B. 1955. Extraction of Enchytraeid worms from a coniferous forest soil. *Nature* 175:815–816.

O'Lear, H. A. and T. R. Seastedt. 1994. Landscape patterns of litter decomposition in alpine tundra. *Oecologia* 99:95–101.

Parmelee, R. W., and D. A. Crossley Jr. 1988. Earthworm production and role in the nitrogen cycle of a no-tillage agroecosystem on the Georgia Piedmont. *Pedobiologia* 32:353–361.

Petersen, H., and M. Luxton. 1982. A comparative analysis of soil fauna populations and their role in decomposition processes. *Oikos* 39:287–388.

Prot, J. C., and H. Ferris. 1992. Sampling approaches for extensive surveys in nematology. *Supplement to the Journal of Nematology* 24S:757–764.

Raw, F. 1959. Estimating earthworm populations by using formalin. *Nature* 184:1661–1662.

Reynolds, J. W. 1973. Earthworm (Annelida: Oligochaeta) ecology and systematics. Pages 95–120 *in* D. L. Dindal, editor, *Proceedings of the First Soil Microcommunities Conference,* Syracuse, New York, 1973. U.S. Atomic Energy Commission, Office of Information Services, Technical Information Center, Washington, DC, USA.

Robertson, G. P., and D. W. Freckman. 1995. The spatial distribution of nematode trophic groups across a cultivated ecosystem. *Ecology* 76:1425–1433.

Rutherford, P. M., and N. G. Juma. 1992. Influence of texture on habitable pore space and bacterial-protozoan populations in soil. *Biology and Fertility* of Soil 12:221–227.

Santos, P. F., E. DePree, and W. G. Whitford. 1978. Spatial distribution of litter and microarthropods in a Chihuahuan desert ecosystem. *Journal of Arid Environments* 1:41–48.

Satchell, J. E. 1969. Methods of sampling earthworm populations. *Pedobiologia* 9:20–25.

Satchell, J. E. 1983. *Earthworm Ecology from Darwin to Vermiculture.* Chapman and Hall, London, UK.

Seastedt, T. R. 1984a. Microarthropods of burned and unburned tallgrass prairie. *Journal of the Kansas Entomological Society* 57:468–476.

Seastedt, T. R. 1984b. The role of microarthropods in decomposition and mineralization processes. *Annual Review of Entomology* 29:25–46.

Seastedt, T. R., and D. A. Crossley Jr. 1981. Microarthropod response following cable logging and clear-cutting in the southern Appalachians. *Ecology* 62:126–135.

Seastedt, T. R., T. C. Todd, and S. W. James. 1987. Experimental manipulations of arthropod, nematode and earthworm communities in a North American grassland. *Pedobiologia* 30:9–18.

Silva, S., W. G. Whitford, W. M. Jarrell, and R. A. Virginia. 1989. The microarthropod fauna associated with a deep-rooted legume, *Prosopis glandulosa,* in the Chihuahuan Desert. *Biology and Fertility of Soils* 7:330–335.

Sinclair, J. L., and W. C. Ghiorse. 1989. Distribution of aerobic bacteria, protozoa, algae, and fungi in deep subsurface sediments. *Geomicrobiological Journal* 7:15–31.

Swift, M. J., O. W. Heal, and J. M. Anderson. 1979. *Decomposition in Terrestrial Ecosystems.* University of California Press, Berkeley, California, USA.

van Vliet, P. C. J., M. H. Beare, and D. C. Coleman. 1995. Population dynamics and functional roles of enchytraeidae (Oligochaeta) in hardwood forest and agricultural ecosystems. *Plant and Soil* 170:199–207.

Viglierchio, D. R., and R. V. Schmitt. 1983. On the methodology of nematode extraction from field samples: comparison of methods for soil extraction. *Journal of Nematology* 15:450–454.

Wallwork, J. A. 1970. *Ecology of Soil Animals.* McGraw-Hill, New York, New York, USA.

Walter, D. E., H. W. Hunt, and E. T. Elliott. 1987a. The influence of prey type on the development and reproduction of some predatory soil mites. *Pedobiologia* 30:419–424.

Walter, D. E., J. Kethley, and J. C. Moore. 1987b. A heptane flotation method for recovering microarthropods from semiarid soils, with comparison to the Merchant-Crossley high-gradient extraction method and estimates of microarthropod biomass. *Pedobiologia* 30:221{endash}232.

Wright, D. H. 1988. Inverted microscope methods for counting soil mesofauna. *Pedobiologia* 31:409–411.

Yeates, G. W., T. Bongers, R. G. M. DeGoede, D. W. Freckman, and S. S. Georgieva. 1993. Feeding habits in soil nematode families and genera: an outline for soil ecologists. *Journal of Nematology* 25:315–331.

Zicsi, A. 1958. Determination of number and size of sampling unit for estimating lumbricid populations of arable soils. Pages 68–71 *in* P. W. Murphy, editor, *Progress in Soil Zoology.* Butterworths, London, UK.

18

Methods for Ecological Studies of Mycorrhizae

Nancy C. Johnson
Thomas E. O'Dell
Caroline S. Bledsoe

Mycorrhizae are generally mutualistic associations in which plant hosts receive soil resources (mineral nutrients and water), while fungal endophytes receive photosynthetically derived carbon compounds (Harley 1989; Allen 1991). Mycorrhizae are inseparable components of soil ecosystems and are important in plant nutrition, nutrient cycling, food webs, and the development of soil structure. The mycorrhizal condition has evolved independently many times. Four distinct types of mycorrhizae are traditionally recognized: arbuscular mycorrhizae, ectomycorrhizae, orchid mycorrhizae, and mycorrhizae in the Ericales (Smith and Read 1997). These mycorrhizal types differ greatly in their plant and fungal associates, morphology, and ecological distributions (Tab. 18.1). Arbuscular mycorrhizae usually predominate in grasslands, agricultural systems, arid and semiarid biomes, certain temperate hardwood forests (including orchards), and many tropical forests (Brundrett 1991; Allen et al. 1995). Ectomycorrhizae tend to occur in forest trees and shrubs in areas where plant growth is seasonal or intermittent. Orchid mycorrhizae are most common in tropical systems where orchids are abundant and diverse. Ericales mycorrhizae are common in habitats suited for ericaceous plants, such as bogs and heathlands in temperate and boreal ecosystems.

No single measurement can adequately assess the structure and function of mycorrhizae in natural ecosystems, but a series of relatively simple measurements can provide a reasonable knowledge of the abundance and diversity of mycorrhizae. In this chapter we focus on the two most common types of mycorrhizae, arbuscular mycorrhizae and ectomycorrhizae. For information about other types of mycorrhizae, refer to Read (1983), Leake (1994), and Smith and Read (1997). Here we describe the plant and fungal components of arbuscular mycorrhizae and ectomycorrhizae, review many of the standard mycorrhizal research techniques, and present a

Table 18.1. Summary Characteristics and Global Distribution of the Main Types of Mycorrhizae

	Plant Hosts	Fungal Symbionts	Ecological Distribution
Arbuscular mycorrhizae	Most plant species (both herbaceous and woody), including pteridophytes, gymnosperms, and angiosperms	About 150 species of Zygomycota in the order Glomales. Six genera are currently recognized: *Acaulospora, Entrophospora, Gigaspora, Glomus, Sclerocystis, and Scutellospora.*	Distributed globally, from the Arctic to the tropics, and are especially abundant in areas with mineral soils
Ectomycorrhizae	Many trees and shrubs, including most taxa in Pinaceae, Betulaceae, Fagaceae, Myrtaceae, Caesalpiniaceae, and Dipterocarpaceae	Over 5,000 species of fungi, including: 25 families of Basidiomycota, 9 families of Ascomycota, and 1 genus of Zygomycota (*Endogone*)	Widely distributed in about 20 families of plants, most taxa are woody.
Orchid mycorrhizae	All members of the Orchidaceae	About 8 genera of Basidiomycota, and possibly some Ascomycota	Widely distributed, but especially common in the tropics where orchids are abundant and diverse.
Ericales mycorrhizae[†]	Families in the order Ericales, including Ericaceae, Epacridaceae, Empertraceae, Pyrolaceae, and Monotropaceae	Not well known, appears to be primarily Ascomycota and some Basidiomycota. Many EM fungi are known to form Ericales mycorrhizae.	Common in regions with acid, peaty soils where nitrogen may be more limiting to plant growth than phosphorus.

Source: Modified from Smith and Douglas (1987).

[†]This category may be further divided into ericoid, arbutoid, and monotropoid mycorrhizae (see Smith and Read 1997)

Table 18.2. Protocols to Assess the Mycorrhizal Status of Plants, the Diversity of Mycorrhizal Fungi, and the Relative Densities of Mycorrhizal Fungal Propagules

	Research Goals		
Protocol	Assess Mycorrhizal Status of Plants	Assess the Diversity of Mycorrhizal Fungi	Assess Relative Densities of Mycorrhizal Fungal Propagules
Quantification of AM colonization	AM associations [1]		
Quantification of EM colonization and identification of EM morphotypes	EM associations [1]	EM associations [2–3]	
Quantification and identification of AM fungal spores		AM associations [1–3]	
Establishment of AM trap cultures		AM associations [2]	
Quantification and identification of EM sporocarps		EM associations [1–3]	
Bait-plant bioassays for both AM and EM			AM and EM associations [1]
DNA fingerprinting to study EM diversity		EM associations [3]	

Note: Numbers refer to the technical difficulty of the protocol (1 is least difficult; 3 is most difficult).

set of recommended protocols to study mycorrhizae in diverse ecosystems. Ecological field studies frequently consider (1) mycorrhizal status of plants, (2) diversity of mycorrhizal fungi, or (3) relative densities of mycorrhizal fungal propagules. Table 18.2 relates these general research goals to the appropriate set of protocols described in detail in the last section of this chapter.

Taxa of the Symbionts

Arbuscular mycorrhizae form in virtually all types of vascular plants: ferns, trees, grasses, and forbs, and in virtually all terrestrial (and many aquatic) ecosystems (Smith and Read 1997; Brundrett 1991; Tab. 18.1). Nearly 90% of all plant species belong to families that can form arbuscular mycorrhizal (AM) associations (Trappe 1987). Notable exceptions are ruderal plants in the families Brassicaceae, Chenopodiaceae, and Cyperaceae, in which the majority of taxa (61–87%) are nonmycorrhizal (Newman and Reddell 1987). In contrast to the high diversity of AM hosts, AM fungi have a remarkably low diversity. All AM taxa belong to a single order of Zygomycota, the Glomales (Morton and Benny 1990). This order contains about 150 species within the genera *Acaulospora, Entrophospora, Gigaspora, Glomus,* and *Scutellospora.* A sixth genus, *Sclerocystis,* currently contains only one species (*S. coremioides*), which is likely to be transferred to the genus *Glomus* (J. B. Morton, personal communication). These fungi are obligate biotrophs and have never been successfully cultured in the absence of a host plant.

Ectomycorrhizal (EM) plants often cover vast areas as canopy-dominant trees (e.g., members of the Betulaceae, Dipterocarpaceae, Fagaceae, Leguminosae, Myrtaceae, Pinaceae). Approximately 50 families of (primarily woody) angiosperms and gymnosperms form ectomycorrhizae (Harley and Smith 1983; Wilcox 1991). Although fewer plant taxa form ectomycorrhizae than arbuscular mycorrhizae, many more fungal taxa are EM than AM. Several thousand species of Basidiomycota, Ascomycota, and a single genus of Zygomycota are known to form ectomycorrhizae (Tab. 18.1). While EM fungi often reproduce sexually and form elaborate sporocarps (mushrooms, truffles, and allies), AM fungi reproduce asexually through vegetative growth and asexual soilborne spores that are multinucleate. Unlike AM fungi, which are obligate biotrophs, some EM fungi have saprotrophic capabilities and can be cultured. Also, EM fungi typically exhibit more host specificity than AM fungi (Molina et al. 1992).

Components of Mycorrhizal Systems

Mycorrhizae are not fungi but rather are symbiotic associations between fungi and plant roots. These symbioses are generally mutualistic; however, environmental conditions mediate mycorrhizal functioning and neutral or parasitic relationships are known to occur (see Johnson et al. 1997). Because mycorrhizae are the interface between roots and soils, many of the effects of mycorrhizae on plant fitness are mediated by soil properties, particularly those related to the availability of limiting resources like nutrients or water. Consequently, it is useful to envision these symbioses as "mycorrhizal systems" composed of plant tissues, fungal tissues, plant-fungal interfaces (intraradical structures), and soil-fungal interfaces (extraradical hyphae). Although AM and EM associations are analogous, the fungal and mycorrhizal structures differ considerably. Therefore, the structure and function of AM and EM associations are described individually.

Arbuscular Mycorrhizae

Although AM fungi can cause subtle changes in root branching patterns (Fitter 1985; Hetrick 1991), they do not alter the gross morphology of plant roots, and their presence can be determined only through microscopic examination of cleared and stained roots (Fig. 18.1). Traditionally these associations have been called *endomy-*

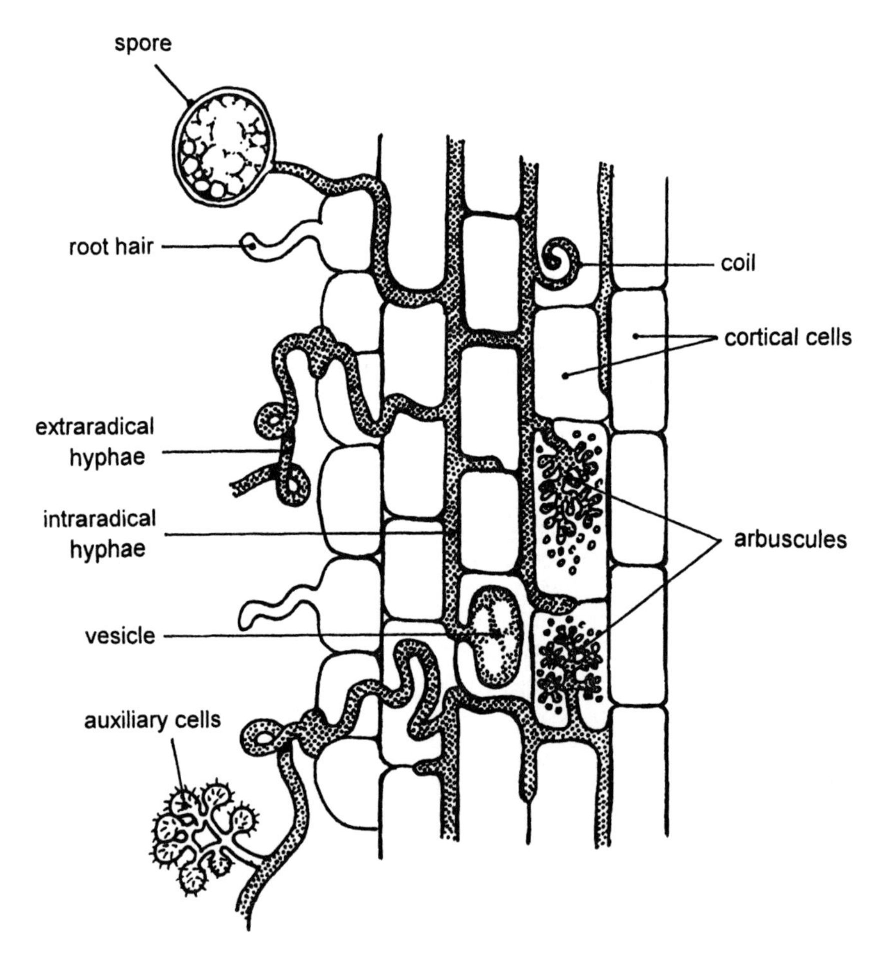

Figure 18.1. Longitudinal section of an arbuscular mycorrhizal root showing characteristic intra- and extraradical structures. Modified from Brundrett et al. (1994).

corrhizae because a large proportion of AM fungal biomass is inside plant cortical cells. These internal organs, or *intraradical fungal organs,* include *arbuscules, vesicles, hyphae, coils,* and occasionally *spores.* The portion of the fungus that is outside the root, or *extraradical,* includes *hyphae, spores,* and in some cases *auxiliary cells.* The structure and function of each of these organs are described in the following.

- *Arbuscules* are intricately branched structures that are sites of carbon, water, and mineral transfer between plant and fungus (Smith and Gianinazzi-Pearson 1988). They are formed inside host cortical cells through repeated dichotomous branching of fine hyphae. Arbuscules are ephemeral, generally forming and then degenerating within 2 weeks (Cox and Tinker 1976; Toth and Miller 1984). By definition, all taxa of AM fungi form arbuscules.
- *Vesicles* are round, elliptical, or irregularly shaped saclike structures that form as hyphal swellings both between and within cortical cells. Because vesicles accumulate lipids, it has been suggested that they function as storage organs in AM fungi (Brown and King 1984). Vesicles are formed by taxa in the genera *Acaulospora, Entrophospora,* and *Glomus,* but not by *Gigaspora* or *Scutellospora* species.
- *Intraradical AM hyphae* often extend between or within cortical cells and connect vesicles and arbuscules. Intraradical hyphae also sometimes form coils or loops inside cortical cells. The morphology of these hyphae may vary considerably among genera of AM fungi and host plants (see Fig. 6.4 in Brundrett et al. 1994; also Abbott 1982; Smith and Smith 1996; Widden 1996).
- *Extraradical hyphae* of arbuscular mycorrhizae absorb soil resources, initiate new colonization sites, and participate in spore formation. The structure of AM hyphae varies with their function. Hyphae involved in nutrient absorption tend to be finer than those involved in the initiation of new colonization sites (Friese and Allen 1991a).
- *Auxiliary cells* are clusters of terminal swellings on external hyphae. They may have a spiny, knobby, or smooth surface. Auxiliary cells are formed only by taxa in the genera *Gigaspora* and *Scutellospora,* the two genera that do not form intraradical vesicles. The function of auxiliary cells is uncertain.
- *Spores* are produced asexually on extraradical hyphae, or occasionally intraradically (e.g., *Glomus intraradices*). Spores may form singly, in clusters (e.g., *G. aggregatum*), or in more highly organized sporocarps (e.g., *G. clavispora*). The taxonomy of AM fungi is based entirely on the morphology and wall structure of the spores (Morton and Benny 1990; Schenck and Pérez 1990).

Ectomycorrhizae

Ectomycorrhizae form on fine roots, generally less than 0.5 mm diameter. Unlike AM roots, EM roots can usually be easily distinguished from nonmycorrhizal roots using no, or low, magnification. Except in the Salicaceae and a few tropical species, EM fungi usually induce swelling and proliferation of root tips. Nonmycorrhizal roots often have root hairs, while EM root tips tend to be smooth or have associated

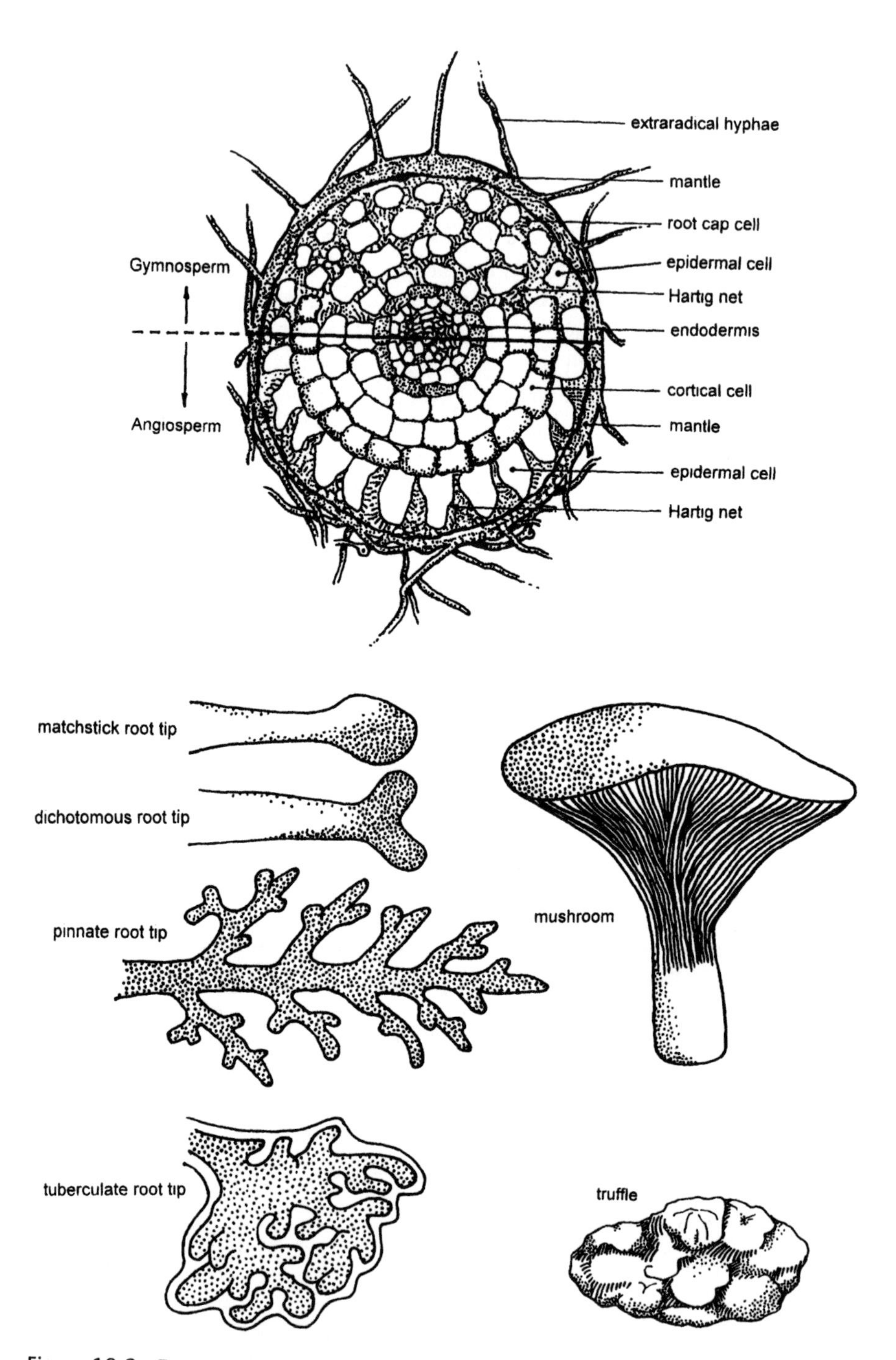

Figure 18.2. Cross section of an ectomycorrhiza showing characteristic structures in a gymnosperm and an agiosperm root (top). Generalized morphologies of root tips (bottom left) and sporocarps (bottom right). Modified from Brundrett et al. (1994).

hyphae, much finer than root hairs. As the name *ectomycorrhizae* implies, fungal biomass is almost entirely outside the root; however, some hyphae penetrate epidermal and outer cortical cells in the formation of *Hartig nets* (Fig. 18.2). The structure and function of these organs are briefly described in the following.

- *Hartig nets* form when fungal hyphae penetrate between epidermal and outer cortical cells of a root. These labyrinthine or digitate structures greatly increase the surface area for exchange of carbon compounds, mineral nutrients, and water between plant and fungus (Massicotte et al. 1987).
- *Mantles* (or sheaths) of fungal hyphae usually cover root tips and often generate a stubby, swollen appearance. Mantles can increase the diameter of the root by 40–80 μm (Harley and Smith 1983). The fungal tissues forming these mantles are often colored and, combined with general root-tip morphology, can help characterize taxa of EM fungi.
- *Extraradical hyphae* extend from EM root tips into the soil, where they absorb nutrients and colonize other roots (Read 1984). They may form extensive, and obvious, mycelial mats on the forest floor. Alternatively, extraradical hyphae may be diffuse and relatively difficult to observe. Extraradical hyphae sometimes aggregate into mycelial strands or cordlike rhizomorphs and may be the EM component with the greatest fungal biomass (Rousseau et al. 1994).
- *Sporocarps,* including mushrooms, truffles, and allies, are sexual reproductive structures of EM fungi. Sporocarps are the most visible component of EM fungi and are often the only structure readily identifiable to species. Some sporocarps are important food sources for certain animals, and many are managed as highly valued commodities for humans (Molina et al. 1993; O'Dell et al. 1996).

Available Methods for Ecological Studies of Mycorrhizae

Sampling Strategies

An understanding of the spatial distribution of mycorrhizae is necessary to design effective sampling strategies. Mycorrhizae tend to be highly aggregated in natural systems because the distribution of mycorrhizal fungi is usually closely linked to the distribution of host plants (e.g., Allen and MacMahon 1985). Ideally, preliminary sampling should be performed to assess the spatial variability of each experimental system. Anderson et al. (1983) found that the variance of AM fungal spore counts was positively correlated with the size of the area sampled, and only by intensive sampling of a 0.5 × 0.5 m plot (combining 75 cores into 25 samples) could they achieve spore counts with a standard error less than 10% of the mean. Using species-effort curves, Tews and Koske (1986) determined that it was necessary to analyze about 30 samples (each 10 mL) to find nearly all the AM fungal species present in some sand dune systems in Rhode Island and Virginia (9 species and 18 species, respectively). Using a standard formula to estimate the number of samples required to achieve a certain degree of accuracy (Southwood 1978), Tews and Koske (1986) found that the required number of samples was often outside the range of practical sampling feasibility. Although an ideal sampling strategy can easily be designed on paper based on estimates of means and variances, the number of samples that a researcher can actually process is limited by time and expense. St. John and Koske (1988) recommended combining cores into composite samples (each composite representing the same replicate of a treatment) to reduce the variance, and consequently the number of samples necessary.

Assessment of Mycorrhizae in Field Samples

Two general approaches are used to assess mycorrhizae in field samples: direct observation and bait-plant bioassays. The two approaches provide different and complementary information, and for many studies it is advisable to use both methods. To estimate mycorrhizal biomass (including both host and fungal components), it is necessary to estimate root and fungal biomass concurrently. See Chapter 19, this volume, for protocols to estimate root biomass.

Direct Observation

The mycorrhizal status of a plant community can be assessed by directly examining root samples from the dominant plant species. Arbuscular mycorrhizae can be microscopically quantified in roots that have been cleared and selectively stained to accentuate the intraradical AM fungal structures (Kormanik and McGraw 1982; Brundrett et al. 1994, 1996). Spores of AM fungi can be quantified after they are extracted from soils using a wet-sieving method (McKenney and Lindsey 1987; Brundrett et al. 1994) or a flotation-adhesion method (Allen et al. 1979). Extraradical hyphae can also be extracted and quantified with or without vital stains to distinguish between living and dead fungal tissues (An and Hendrix 1988; Miller and Jastrow 1992; Sylvia 1992; Schaffer and Peterson 1993). Quantitative data on mycorrhizal fungal structures are useful for comparison purposes; however, there is no simple linear relationship between mycorrhizal functioning and the quantity of mycorrhizal fungi in a system. In fact, levels of root colonization and spore populations are often poorly correlated with the beneficial effects of mycorrhizal associations (e.g., Daniels et al. 1981; McGonigle 1988).

Ectomycorrhizae can usually be directly observed in field-collected roots with no special preparation, although some types of ectomycorrhizae in *Populus* and *Salix* are cryptic. When studying mycorrhizae in these taxa, it may be necessary to use the staining methodology recommended later for AM colonization and examine the roots with a compound microscope to determine if they have EM associations. Sporocarps of EM fungi can also be directly observed in the field (Bills et al. 1986; Arnolds 1992; Luoma et al. 1991; O'Dell et al. 1992). Occasionally, EM fungi form distinctive mats of extraradical hyphae (e.g., *Hysterangium* sp., and *Tricholoma* sp.) that also can be directly observed and quantified. Guides to the identification of EM fungi based on the morphology of root tips, sporocarps, and mycelial mats are available (Agerer 1987–1991, 1991; Brundrett et al. 1994; Goodman et al. 1996).

Bait-Plant Bioassays

The mycorrhizal colonization potential of soils can be estimated indirectly using a "bait-plant bioassay." These bioassays are designed to detect all viable propagules of mycorrhizal fungi, including spores, fragments of mycorrhizal roots, and extraradical hyphae, and may better reflect total activity of mycorrhizal fungi than direct counts of sporocarps, spores, or colonized root lengths (Brundrett et al. 1994; Abbott et al. 1995). Several methods have been developed that involve growing "bait-plants" in diluted field soils and assessing mycorrhizal colonization in the plants after a short time (e.g., Moorman and Reeves 1979; Porter 1979; Franson and Bethlenfalvay 1989). All these methods involve mixing field soils with sterilized media (usually a sand-soil mixture) according to a specific dilution series so that actual densities of propagules can be calculated. However, the process of mixing the field soil with the sterilized media has been shown to destroy hyphal networks that may be an important source of AM and EM inoculum (e.g., Jasper et al. 1989; Perry et al. 1989). An alternative approach is to sow seeds or seedlings of bait-plants directly into undisturbed soil cores. This method was successfully used to measure relative propagule densities of both AM and EM fungi in a *Eucalyptus* forest in Western Australia (Brundrett and Abbott 1994). This technique works well with most soil types except for extremely rocky and dry soils when it is difficult to acquire intact soil cores.

Species Composition

It is currently difficult to accurately assess the abundance of individual taxa of mycorrhizal fungi because the functionally important parts of mycorrhizae (the arbuscules, extraradical hyphae, and vesicles of arbuscular mycorrhizae, and the mycelium, Hartig net, or mantle of ectomycorrhizae) are often morphologically indistinguishable. However, using visual morphotyping techniques, EM structures can often be identified to the genus level, and the spores of AM fungi and the sporocarps of EM fungi can usually be differentiated to the species level. Molecular and biochemical methods currently being developed and refined will greatly facilitate the analysis of mycorrhizal fungal communities.

Direct Analysis of Spores or Sporocarps

Probably the cheapest and easiest method for examining the species composition of AM fungal communities is to collect, count, and identify spores. Populations of EM sporocarps can also be analyzed. Many studies have reported patterns in the composition of AM and EM fungal communities based on relative abundances of spores or sporocarps (e.g., Koske 1987; Johnson et al. 1991, 1992; Bills et al. 1986; Villeneuve et al. 1989; Luoma et al. 1991; O'Dell et al. 1992). There are strong limitations to spore and sporocarp analysis because the degree to which populations of mycorrhizal fungi sporulate varies between taxa, as well as seasonally or annually within taxa. There may be little relation between population densities of spores or sporocarps and levels of root colonization. Some fungi produce copious spores or sporocarps, while others rarely or never sporulate. At least 5 years of repeated observation during the fruiting season is required to detect most of the EM fungal species in a system (Arnolds 1992). Furthermore, identification of AM spores and EM sporocarps requires taxonomic expertise, and specimens collected from field samples are often degraded or parasitized, making their identification difficult or impossible. Reconstructing the species composition of mycorrhizal fungal communities from spore or sporocarp populations is analogous to reconstructing plant communities from pollen data. Populations of spores and sporocarps are useful for making relative comparisons across treatments but not for estimating the absolute abundance of individual species of mycorrhizal fungi. In the worst-case scenario, the mycorrhizal fungus with the most biomass in a system may not even be present in the spore or sporocarp community (e.g., Clapp et al. 1995; Gardes and Bruns 1996).

Trap Cultures

For arbuscular mycorrhizae, trap cultures are commonly established to overcome some of the problems associated with direct spore analysis. Trap cultures stimulate sporulation in the greenhouse and are useful for diversity estimates of AM fungi (Morton et al. 1993). They are established by mixing field soils or roots with sterile sand in pots and seeding them with an appropriate host plant. After 3–4 months the pots are checked for sporulation. Several "culture cycles" may be required before spores are formed. If no spores are present after 4 months, plant shoots are removed and pots are reseeded for a second or even a third trap culture cycle. The number of sporulating species can double or triple each culture cycle (Bever et al. 1996; Morton et al. 1995). Although trap cultures are time-consuming (requiring a minimum of 3 months), the fresh spores produced in the cultures are generally much easier to identify, and taxa not present as spores in field samples are often revealed (e.g., Johnson 1993).

Identification of EM Morphotypes

Because EM roots generally have distinguishable macroscopic features, probably the cheapest and easiest method to assess the species composition of EM fungal communities is to identify EM morphotypes of field-collected material. For this technique, EM fungi are identified from a series of morphological and anatomical variables as explained in Agerer (1991), Agerer (1987–1991), and Goodman et al. (1996). These descriptions are usually sufficient to identify EM fungi to genus and often to species.

Molecular and Biochemical Approaches

Recent advances in molecular techniques allow researchers to more accurately distinguish the fungal taxa in field-collected roots and soil. Nucleic acid–based identification methods are undergoing rapid change and development. Nonetheless, there are standard methods for routine identification of EM fungi within roots. The

DNA profiles of sporocarps can be matched with those of mycorrhizal roots to most accurately identify individual taxa of fungi forming ectomycorrhizae (Gardes et al. 1991). Molecular typing of EM fungi is advantageous because the plant and fungal species forming a single mycorrhiza can be identified from the same DNA extract by separately amplifying plant and fungal tissues using polymerase chain reaction (PCR) primers with appropriate specificity. Furthermore, fungal species not distinguished by EM morphology can be distinguished by their DNA. Comparisons of mycorrhizal diversity between samples may be more precise than those based exclusively on morphology. Methods for using specific DNA primers in conjunction with PCR to identify taxa of AM fungi are in a rapid stage of development (Clapp et al. 1995; Di Bonito et al. 1995; Lloyd-MacGilp et al. 1996; Dodd et al. 1996; Sanders et al. 1996). Biochemical techniques such as fatty acid profile analysis of fungal lipids (Graham et al. 1995) and immunofluorescence (Friese and Allen 1991b; Hahn et al. 1994; Thingstrup et al. 1995) are also being developed for identification and quantification of AM fungi.

Resources for General Information

Several excellent reference books thoroughly review mycorrhizal research methods (notably Schenck 1982; Norris et al. 1991, 1992; Brundrett et al. 1994, 1996). Brundrett et al. (1994) is a particularly useful and inexpensive general "how-to" manual, as is Goodman et al. (1996) for the identification of EM. An important resource for identification of AM fungi is the International Collection of Arbuscular and Vesicular-Arbuscular Mycorrhizal Fungi (INVAM 1997). BEG (1996) provides electronic taxonomic aids or computerized "expert systems" on CD-ROM for AM fungi. A delta-based system for characterization and determination of ectomycorrhizae (DEEMY) is available from R. Agerer (personal communication; rambold@botanik.biologie.uni-muenchen.de).

Recommended Protocols

Table 18.2 provides guidance for which of our seven recommended protocols to use for a given study objective. The estimated amount of time required to perform each protocol is provided in the pages that follow, and Table 18.2 classifies their difficulty from Level 1 (least technical) to Level 3 (most technical). The intensity of the sampling scheme will ultimately determine the total amount of time required to complete the analysis.

Sampling Design

The first step in any research project is to develop a sampling strategy that adequately assesses the range and variability in populations of plants and mycorrhizal fungi in the system. Although there are no standard sampling designs for mycorrhizae, these guidelines may be helpful. See also the discussion of sampling design for root biomass (see Chapter 19, this volume).

If feasible, use individual plants of a particular species as the experimental unit (instead of a plot), and use a stratified-random method to locate the plants to be sampled. Within-treatment variability can be reduced by combining several small samples into a single composite sample (e.g., a soil probe can be used to collect four or five cores a standard distance from the stem of an individual plant, with the cores thoroughly mixed before a subsample is taken for analysis). Sample as many treatment replicates as feasible (excess replicates can usually be stored) and use factorial experimental designs that can be collapsed during analysis if various factors are found to be insignificant.

Root and Soil Collection

Standard soil probes (e.g., Oakfield samplers or AMS soil core extractors) are commonly used to collect both soil and root samples for mycorrhizal analysis. Soil samplers are available in a variety of diameters and with a variety of tips (i.e., for dry, wet, or rocky soils). Generally, EM studies require larger soil volumes than AM studies. Root fragments in the cored soils can be collected on a coarse sieve. Alternatively, if it is necessary to know the exact identity of a host root, then the root must be excavated with a spade and followed back to the plant shoot or vice versa. Ziplock freezer bags are convenient for storing soil and root samples. Samples should be kept in a cooler, out of direct sunlight, until they are processed or stored (see guidelines later).

Several factors need to be considered in developing a sampling protocol. First, mycorrhizal activity varies greatly with depth. In agricultural systems most mycorrhizal activity occurs in the top 20 cm, the area of greatest root density, but rooting depth differs in natural ecosystems, and analyzing the mycorrhizal profile in some preliminary soil cores is useful to determine the appropriate sampling depth.

Second, the development of mycorrhizal colonization and the production of spores and sporocarps is influenced by season and the developmental stage of the hosts. Depending on the research goals, it may be necessary to collect samples at several times throughout the growing season. Alternatively, a single (standardized) time of sampling may be sufficient. Ideally, preliminary root and soil samples should be examined to determine the most appropriate sampling seasons.

Third, when samples are collected it is important to gather as much supporting information (metadata) as possible (see Chapter 1, this volume). Soil properties (e.g., pH, organic matter [OM] content, texture, and nutrient content) and biological properties (e.g., total plant cover, amount of shading, species composition of the plant community, and plant phenology) can be important predictors of mycorrhizal activity. Typically, well-established sites such as the U.S. LTER sites collect much of these supporting data (LTERNET 1996).

Sample Storage

If viability of field-collected mycorrhizae, hyphae, spores, or sporocarps is of interest, or if one is planning to conduct a bioassay, then samples should be processed immediately following collection. However, if viability is not of interest, then root and soil samples can be refrigerated for several weeks or frozen for long-term storage until they are processed. If trap cultures are to be established, then the timing of sampling within the growing season may be of concern because many temperate AM fungi have dormancy requirements (i.e., 4–8 weeks at 4–5 °C), and spores of some taxa actually require a cold treatment before they will germinate (Tommerup 1983; Safir et al. 1990). In contrast, tropical isolates of AM fungi may be very sensitive to cold temperatures, and many do not survive storage at even 4 °C (Morton et al. 1993).

Quantification of AM Colonization

We recommend a staining technique based on methods developed by Kormanik and McGraw (1982) and Koske and Gemma (1989). The recommended stain, trypan blue, works well on most roots; however, chlorazol black E or acid fuchsin may be better for some roots (Brundrett et al. 1994). The recommended methods for quantifying intraradical colonization using a dissecting microscope were developed by Giovannetti and Mosse (1980); McGonigle et al. (1990) and Brundrett et al. (1994, 1996) developed the recommended method for using a compound microscope.

Materials

1. Simport biopsy cassettes (Fisher Scientific Co., Pittsburgh, PA, USA, catalog no. 15-182-700)
2. Large beakers to hold cassettes
3. Heating element and water bath
4. 2.5% or 5% KOH. Add water to 25 g or 50 g of KOH pellets to bring the volume to 1 L.
5. 1% HCl. Add 1 mL concentrated HCl to 99 mL distilled water.
6. Optional: alkaline H_2O_2. Add 3 mL 20% NH_4OH (household ammonia) to 30 mL 3% H_2O_2.
7. Trypan blue stain. Combine 200 mL lactic acid, 200 mL glycerol, 200 mL distilled water, and 0.3 g trypan blue (Sigma Chemical Co., St. Louis, MO). Trypan blue is suspected to be carcinogenic; use gloves to protect hands when using stain solutions.
8. For evaluation using a dissecting microscope: 8.5 cm diameter plastic petri dishes etched in the bottom with grid lines spaced every 13 cm starting 6.5 cm from the edge of the dish. Hold the dish open side down over a piece of grid paper and use a dissecting needle and a straightedge to scratch the grid on the outside of the dish.
9. For evaluation using a compound microscope: microscope slides and 22 × 60 mm coverslips.
10. Polyvinyl alcohol (PVA) mountant. Mix 100 mL distilled water, 100 mL lactic acid, and 10 mL glycerol with 16 g PVA powder and place in a hot water bath (70–80 °C) to dissolve for about 4 hours.
11. Microscope slide weights made by filling small vials with BBs (used in shotgun cartridges, which can be purchased at sporting goods stores)

12. Dissecting microscope or compound microscope (fitted with a hairline graticule in the eyepiece)
13. Channel counter

Procedure

1. Collect a representative sample of roots from a known volume and mass of soil (using guidelines discussed earlier).
2. Wash roots and cut them into 2.5 cm pieces.
3. Randomly select a subsample of roots of known mass (between 0.25 and 0.5 g, fresh weight) and place in the Simport cassettes. Very fine roots may need to be placed between two layers of plastic-coated window screen inside the cassettes.
4. Place cassettes in a beaker containing 2.5% or 5% KOH and cover with aluminum foil. Warm the KOH in a water bath (70–90 °C) for 1 hour to 4 days inside a fume hood. This procedure clears the cytoplasm out of the cortical cells. Note: Clearing times and the optimum concentration of KOH (i.e., 2.5% or 5%) vary with root thickness and age. A long time and 5% KOH may be required to clear old, thick, lignified roots. A short time and 2.5% KOH may sufficiently clear young or fine roots. Insufficiently cleared roots retain cell contents and fail to absorb stain adequately; overcleared roots will lose their cortex. It is better to err on the side of insufficient clearing because roots can be returned to KOH for additional clearing if needed, whereas there is no way to recover the cortex if it is lost through overexposure to KOH. For very thick and/or pigmented roots it may be necessary to replace the KOH solution with fresh solution midway through clearing.
5. When roots are adequately cleared (with the appearance of cooked onions that have become translucent), decant KOH and gently rinse cassettes with tap water (change water at least six times).
6. If roots are still pigmented, soak them for 10–45 minutes in alkaline H_2O_2 at room temperature, then rinse the cassettes thoroughly with water. If the roots become darker in H_2O_2, repeat the KOH clearing procedure and the alkaline H_2O_2 clearing again.
7. Soak roots for 1 hour in 1% HCl. This acidifies the roots so that the trypan blue stain binds well to the AM fungal structures.
8. Decant HCl, add trypan blue stain to the beaker of cassettes, and warm in a water bath (90 °C) for 1 hour.
9. Decant and save stain (usable at least four times); rinse cassettes thoroughly with water. Trypan blue is a hazardous waste; contact occupational health and safety personnel to arrange for proper disposal.
10. Cassettes with stained roots can be stored in a 1:1:1 solution of lactic acid, glycerin, and water. Alternatively, they may be stored in distilled water in the refrigerator for up to 2 weeks or in a sealed container (dry) in the freezer for long periods of time.
11. Colonization of stained roots can be measured by the grid-line intersection method with either a dissecting or a compound microscope. Using a dissect-

ing microscope is faster and provides a direct estimate of specific root length; however, it provides less detailed information about the AM fungal structures in the roots. A compound microscope must be used to accurately quantify the relative abundance of vesicles, arbuscules, coils, and hyphae.

Grid-Line Intersection Method with a Dissecting Microscope

1. Evenly spread stained roots (of a known mass) across the bottom of an 8.5 cm diameter gridded petri dish containing about 10 mL of water.
2. Using a dissecting microscope at approximately 40× magnification, follow the lines with your eyes and evaluate each intersection between a line and a root for the presence or absence of AM fungal structures. For the assumptions of this method to be valid, it is important to consider the mycorrhizal status of the root only at the exact point of intersection between the root and the line.
3. Use a channel counter, or simply two hand-held counters, to count the total intersections and the AM intersections.
4. The percent AM root length is then calculated as:

$$\%\text{AM root length} = 100 \times [(\text{number of AM intersections})/(\text{number of total intersections})]$$

5. The dimensions of the grid in the petri dish are designed so that if all the lines are examined (both vertical and horizontal), the total number of intersections provides a direct estimate of root length in centimeters. This value can then be combined with root mass to calculate specific root length (cm/g root).

Grid-Line Intersection Method Using a Compound Microscope

1. Cut stained roots into 1 cm segments and use a fine forceps to arrange the segments lengthwise on a thin layer of PVA mountant on a microscope slide. Gently press a 22 × 60 mm coverslip onto the roots, starting at one side and working across to the other side to minimize air bubbles.
2. Flatten the coverslips (overnight or longer) using weights made by filling small vials with BBs.
3. Use a hairline graticule inserted into the eyepiece of a compound microscope to act as a line of intersection with the roots. At 200× magnification evaluate each intersection for AM fungal structures. There are seven possible, and mutually exclusive, categories of intersections (Fig. 18.3):

   p = no fungal structures
   q = arbuscules
   r = arbuscules and vesicles
   s = vesicles
   t = coils
   u = mycorrhizal hyphae (near but not at arbuscules or vesicles)
   v = hyphae not seen to be connected to arbuscules or vesicles (they may or may not belong to AM fungi)

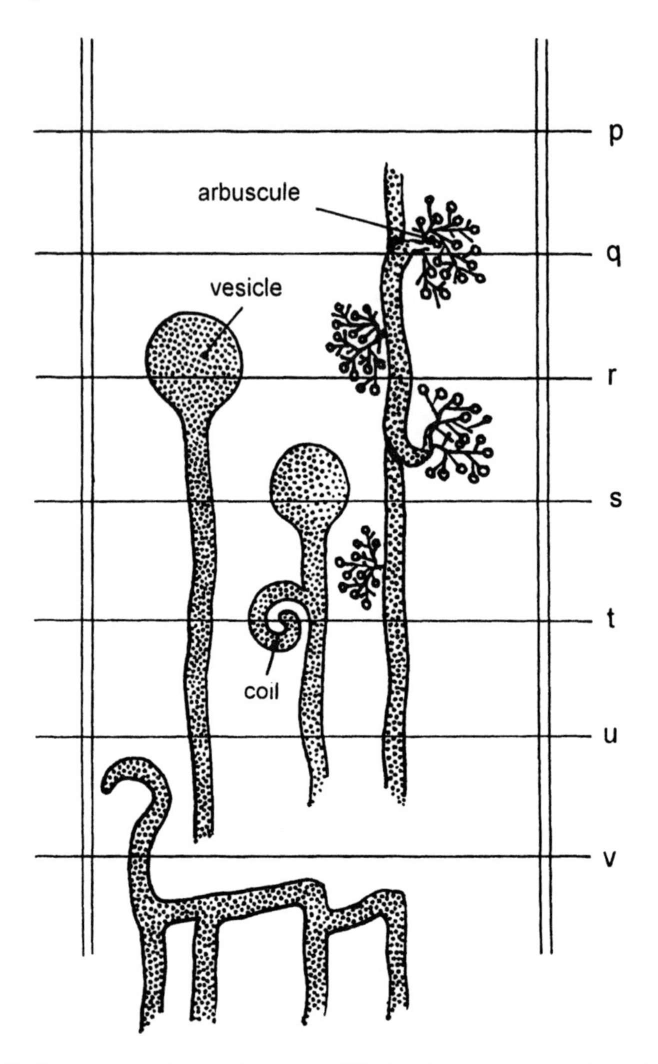

Figure 18.3. Roots mounted on a microscope slide showing seven categories of hairline root intersections (p–v). Modified from Brundrett et al. (1994).

Note that mycorrhizal hyphae are always intersected in *q, r, s* and *t,* and that mycorrhizal hyphae in *u* are known to be mycorrhizal because they are seen to be attached to arbuscules, vesicles, or both.

4. Examine 150 intersections for each root sample, scoring each intersection in one of the seven categories.
5. Where a total of $G = (p + q + r + s + t + u + v)$ intersections were inspected, the percentage of root length colonized by arbuscules and vesicles is calculated as

$$\text{arbuscular colonization} = 100 \times (q + r/G)$$

$$\text{vesicular colonization} = 100 \times (r + s/G)$$

6. The percentage of root length infected, which is equivalent to the percentage of root length colonized by mycorrhizal hyphae, can be calculated if the intersections of type v can be further subdivided into those with hyphae that are *not* mycorrhizal (v_1), and those that are mycorrhizal (v_2). In this case the percentage of root length colonized by mycorrhizal hyphae $= 100 \times [(q + r + s + t + u + v_2)/G]$.

Special Considerations

Approximately 5–30 minutes are necessary to prepare a root sample for staining. Once the roots are washed, cut, and placed into cassettes, they must soak in chemicals for up to 4 days. The type of microscope used (dissecting or compound) and the experience of the observer determine the amount of time required to assess colonization, ranging from 5 to 30 minutes per sample.

Quantification of EM Colonization

Ectomycorrhizal colonization of roots can be assessed by direct examination. Individual plants or soil cores may serve as the sampling units. Sampling individual plants is beneficial because the host can be identified to species, and other information about the host plant (age, height, crown diameter, diameter at breast height [DBH], etc.) can be collected. Unfortunately, sampling individual plants is not feasible when roots from many different plants are highly intermingled. For this reason, soil + root cores are often used. However, extracting roots from soil and organic matter in cores can be tedious if the core contains much organic matter (e.g., Vogt et al. 1982). In such cases, it may be advantageous to homogenize and subsample the core before washing.

Materials

1. Soil coring devices (e.g., AMS soil core sampler with hammer attachment, available from Forestry Suppliers, Jackson, MS)
2. 0.1 mol/L sodium pyrophosphate
3. Elutriator (available from Bel Art Inc., Pecquannock NJ); optional but optimal
4. Standard soil sieves with various mesh sizes
5. Forceps
6. Razor blades (double edged are sharpest and can be broken in half lengthwise for ease of use)
7. Plastic petri dishes
8. Small trays (approx. 10 × 20 × 3 cm)
9. Microscope slides and coverslips
10. Compound and dissecting microscopes (the latter preferably equipped with fiber-optic illumination)
11. Cotton blue in lactic acid. Combine 0.05 g cotton blue with 30 mL lactic acid (85–90%).

12. Formalin acetic acid (FAA). Combine 5 mL formalin, 90 mL 70% ethanol, and 5 mL acetic acid.

Procedure

1. Use a soil coring device to collect uniform volumes of soil. Samples can be taken systematically or randomly (we are not aware of any efforts to test different sampling approaches or minimum effort required for different levels of precision; see Chapter 19, this volume).
2. Place soil cores in Ziplock plastic bags and keep cool until processed. Soil samples can be stored at 2 °C for several weeks or at −20 °C for several months without obvious deterioration of mycorrhizae.
3. Gently break up the soil core and soak it in 0.1 mol/L sodium pyrophosphate for approximately 1 hour. Transfer core plus pyrophosphate solution to an elutriator, add approximately 500 mL water, and agitate with compressed air for about 1 hour. Introduce a slow flow of water to wash dissolved organic matter and fine soil particles from the sample. Allow the overflow to run through a sieve (0.5 mm mesh) to catch small root fragments and mycorrhizae. If an elutriator is not available, roots and mycorrhizae can be extracted from soils using a bucket, water, and a sieve. Place the soil core in a sieve and use a jet of water to remove as much of the mineral material as possible. Check to be certain that roots are not also being lost. It is best to initially compare sieves with various mesh sizes to find the mesh size that is most effective for a particular system. After the initial washing against the sieve, vigorous stirring of the soil sample in a bucket of water should cause the mineral material to sink to the bottom and the roots and organic debris to float to the surface, where they can be poured off and collected on a sieve. A second volume of water can be added and the process repeated until no further root material is acquired.
4. Place the material retained in the elutriator and/or on the sieve in a shallow tray of water and examine with a dissecting microscope. Use a forceps to separate EM root tips and secondary EM roots from humic materials and place them in a dish of clean tap water. Keep the extracted roots in water to prevent drying and loss of diagnostic features. Samples with a large amount of humus or organic matter may need to be examined a portion at a time.
5. Estimate total root length and numbers of EM root tips using the grid-line intersect method with a dissecting microscope as described earlier for AM colonization.
6. Confirm that root tips are EM using a compound microscope. Prepare slides by squashing individual tips between a coverslip and a microscope slide with a drop of cotton blue in lactic acid. Alternatively, use a razor blade or a hypodermic needle to cut radial or tangential sections that include a small amount of root cortex and mount in cotton blue in lactic acid. Examine the slides at 400× to observe mantle and Hartig net features.

For Estimating EM Fungal Diversity Based on EM Morphotypes

1. Use dissecting and compound microscopes to sort ectomycorrhizae with different morphological characters. Place morphotypes into small petri plates

with water. Clean ectomycorrhizae can be stored in water in the refrigerator for about a week before they start to decompose.

2. Characterize each morphotype based on external features such as color, texture, rhizomorph structure, and hyphal diameter, and internal features such as mantle thickness, mantle pattern, Hartig net development, specialized cells, hyphal junctions, and emanating hyphae. Refer to Agerer (1991), Agerer (1987–1991), and Goodman et al. (1996) for detailed descriptions of how to characterize ectomycorrhizae.
3. If care is taken to standardize the sampling and sorting process, it is possible to quantify dried morphotypes using a good balance.
4. Samples of morphotypes should be photographed, and samples of EM root tips can be frozen or lyophilized for DNA fingerprinting (PCR followed by restriction fragment length polymorphism (RFLP) analysis) or preserved in FAA for future morphological comparisons.

Special Considerations

It requires approximately 2–4 days for an experienced person to extract, sort, and describe mycorrhizae from a 5 × 15 cm core containing significant amounts of organic matter.

The Quantification and Identification of AM Fungal Spores

The spore extraction technique described here was developed by McKenney and Lindsey (1987). This technique is effective for recovery of high densities and diversities of spores from a wide range of soil types (note modifications for clayey and sandy soils); however, it is rather time-consuming. The flotation-adhesion spore extraction technique (Allen et al. 1979; Ianson and Allen 1986) is as effective in many (but not all) soil types and may save time. The described spore identification method is based on techniques developed by Schenck and Pérez (1990). Extracting and counting total spores is relatively easy (Level 1); identifying the spores is considerably more difficult (Level 3).

Materials

1. 250 mL Erlenmeyer flasks for clayey soils or a large pitcher or beaker (plastic or stainless steel is best) for sandy soils
2. Standard soil sieves with 250 μm, 90 μm, and 25 μm size mesh
3. Mechanical shaker for clayey soils
4. Sink with a small rubber hose attachment
5. 50 mL centrifuge tubes
6. Funnel
7. Tabletop centrifuge with swinging-basket attachment
8. 0.08 mol/L sodium hexametaphosphate. Combine 83.4 g sodium hexametaphosphate with 1 L water; bring pH to 8.5 using 0.1 N NaOH.
9. 2 mol/L sucrose. Combine 825 mL distilled water (100 °C) with 454 g table sugar; store in refrigerator and discard when moldy.

10. Filtration apparatus with a vacuum source
11. 0.45 μm membrane filters (47 mm diameter with 3.1 mm square grids)
12. Analyslide filter storage containers (Gelman Sciences, Ann Arbor, MI)
13. Compound microscope and coverslips (20 × 20 mm)
14. PVA mountant. Mix 100 mL distilled water, 100 mL lactic acid, and 10 mL glycerol with 16 g PVA powder, place in a hot water bath (70–80 °C) to dissolve for about 4 hours.
15. Microscope slide weights made by filling small vials with BBs (used in shotgun cartridges, which can be purchased at sporting goods stores)
16. Taxonomic references, e.g., Schenck and Pérez (1990), or keys from INVAM (1997) or BEG (1996)

Procedure

1. Air-dry and thoroughly homogenize soil samples.
2. If soils have a high clay content, place 25 g soil in a 250 mL Erlenmeyer flask, add 10 mL sodium hexametaphosphate solution and 90 mL deionized water. Shake 15 minutes with a mechanical shaker and go to step 4.
3. If soils are sandy, omit step 2. Place 25 g soil in a large pitcher or beaker. Use hose attachment to spray soil with a strong stream of water (approx. 400 mL). This suspends the soil and spores in the water.
4. Decant the suspension through a stack of sieves with progressively smaller mesh sizes: 250 μm on top, 90 μm in the middle, and 25 μm on the bottom. Do not pour the precipitate onto the sieves.
5. Use hose attachment to resuspend the precipitate in the flask, pitcher, or beaker. Vigorously fill with water, settle for approximately 10 seconds, and decant suspension onto sieves, leaving the precipitate behind.
6. Repeat step 5.
7. Use hose attachment to rinse sieves gently with water. Regularly check the 25 μm screen for clogging and if necessary clear it with a forceful spray of water, being careful not to lose any sample.
8. Rinse material on the 250 μm sieve directly into a petri dish and examine this faction with a dissecting microscope for spore clusters and large spores (*Gigaspora* and *Scutellospora* spp.).
9. Combine and rinse material from the 90 μm and 25 μm sieves into a 50 mL centrifuge tube.
10. Centrifuge at approximately 3000 rpm for 3 minutes.
11. Decant the supernatant onto the 25 μm screen and rinse into a beaker. Check for spores in the supernatant. If spores are present, save the supernatant and recombine it with the sievings in step 15. If no spores are present, discard.
12. Add 30 mL of 2 mol/L sucrose into centrifuge tubes and stir to resuspend the pellet.
13. Centrifuge at approximately 3000 rpm for 1.5 minutes.
14. Pour sucrose supernatant onto the 25 μm screen and *immediately* rinse with water (to minimize osmotic damage to the spores).
15. Rinse the sievings from the 25 μm sieve into a beaker and then into a filter apparatus fitted with a gridded membrane filter.

16. Evacuate water from the filter apparatus using a vacuum, being careful to disperse the spores uniformly over the filter. Store filters in Analyslide storage containers in the refrigerator.
17. If only total spore counts are necessary, use a dissecting and/or compound microscope to count the spores on the filterpaper (spores 25–250 μm diameter) and in the petri dish of water (spores >250 μm diameter). All of the spores on the filter paper can be counted (if numbers are low), or else a standard fraction of the paper can be examined and the total count estimated from that fraction. A total spore count per gram of dry soil can be calculated by adding the 25–250 μm count to the >250 μm count.

For Identification of AM Fungal Species

1. Label two microscope slides with the sample number and either ">250 μm" or "25–250 μm."
2. Place a small drop of water on the slide labeled ">250 μm" and use a jeweler's forceps to move all of the spores from the petri dish to the drop of water.
3. Slightly moisten the membrane filter containing the 25–250 μm fraction of spores. Use a forceps to pick up spores and place them in a drop of water on the slide labeled "25–250 μm." Spores on the surface of the damp membrane filter will adhere to anything damp. If there are many spores on the membrane filter, remove the spores only from a standard area (e.g., from 24 contiguous squares in the middle of the paper).
4. After all the spores have been transferred to the slide, allow the drop of water to dry; then carefully place a drop of PVA mounting media over the spores and cover with a coverslip.
5. Flatten the coverslips using weights made by filling vials with BBs and allow the slides to set overnight or longer. Important: PVA mountant never really dries, so the slides should always be stored flat or the PVA and spores will gradually slump out of the sides of the coverslips.
6. Because various mountants (PVA included) affect the thickness and structure of spore walls (Morton 1986), it is advisable to make initial microscopic observations of spores mounted in plain water. After making notes about the hyphal attachment, wall structure, color, and size, spore types can generally be grouped into categories. Then refer to references such as Schenck and Pérez (1990) or INVAM (1997) to identify the species. Researchers should expect that some of the spores will be undescribed species. Unknown spores can usually be identified to genus and assigned a number or a nickname (e.g., "small, yellow *Glomus*," or "textured, red *Scutellospora*").
7. Systematically examine the slides with a compound microscope and use a channel counter to count the number of spores of each species or spore type.

Special Considerations

Spore extraction requires approximately 15 minutes per sample, and estimation of the total number of spores in the sample requires approximately 5–15 minutes. Usually a centrifuge load of four or six samples is processed at one time.

Identification of spores can be time-consuming. Mounting extracted spores onto a microscope slide requires 15–45 minutes. Once spore types at a site are distinguished, counting and identification of all of the spores on a slide requires 15–60 minutes. These estimates vary with the experience of the observer, the number of spores, and taxa on the slide.

AM Trap Cultures

This protocol, developed by Morton et al. (1993), is the standard technique used at the International Culture Collection of arbuscular and VA Mycorrhizal fungi (INVAM 1997).

Materials

1. Sterilized pots (1 L is best)
2. Steamed no. 3 quartzite sand
3. Surface sterilized seed of sudangrass (*Sorghum sudanense*) or other host plant
4. Cork borer

Procedure

1. Collect at least 250 cm^3 of field soil plus roots and chop the roots into small pieces.
2. Thoroughly mix the soil in a proportion of 1:1 with steamed and moistened no. 3 quartzite sand.
3. Place soils into a sterilized pot and sow with 60–80 seeds of the host plant. The standard host used at INVAM (1997) is Sudangrass, although big bluestem (*Andropogon gerardii*) and Bahia grass (*Paspalum notatum*) have been successfully used in Minnesota and Florida, respectively.
4. After 3–4 months use a large cork borer to take a sample core from the pot. Extract and examine the spores from the core using the AM fungal spore protocol described earlier. If many spores are present, cut off the grass shoots and allow the pots to dry out in the greenhouse (approx. 2 weeks). Remove the soil plus roots from the pot and store in a Ziplock bag at room temperature or at 4 °C. If few or no spores are present, start a second culture cycle by gently removing the grass shoots and reseeding the pots. After 3–4 months repeat step 4.

Special Considerations

Setting up a culture requires only a few minutes; however, each culture must be maintained in a greenhouse for at least 3 months, and then spores must be extracted from the trap culture and examined using the protocol described earlier.

The Quantification and Identification of EM Sporocarps

If EM fungal diversity data are required, identifying sporocarps supplements the identification of EM morphotypes. This is generally a "Level 2 or 3" analysis because it takes considerable training to distinguish sporocarps produced by EM fungi from those produced by saprophytic fungi. Furthermore, sporocarps are ephemeral, so surveys are limited to the main fruiting seasons.

Materials

1. Nimrod (a device for determining circular plots, consisting of a measured length of chain attached to a stake or dowel)
2. Pocket knife or other digging and cutting tool
3. Truffle fork (four-tine garden cultivator)
4. Aluminum foil
5. Labels
6. Mushroom basket, frame pack, or 5 gallon plastic bucket
7. KOH. Make a 3–5% w/v aqueous solution.
8. Melzer's solution. Combine 100 g chloral hydrate, 100 mL distilled water, 1.5 g iodine, and 5.0 g potassium iodide; let mixture stand at least 48 hours before using. Note: chloral hydrate is toxic.
9. Combine 2 g p-dimethylaminobenzaldehyde (PDAB), 76 mL 95% ethanol, and 24 mL concentrated HCl.

Procedure

1. Carefully select the sampling location and plan the sampling design. Plan to study multiple stands of the EM host plants of interest. Because of extreme year-to-year variability in sporocarp production it is recommended that a site be studied a minimum of 5 years with observations every 1–2 weeks during peak fruiting times. One of two basic sampling designs can be used depending on study objectives: (1) Convenience sampling. If one desires a list of species, or a rough estimate of total richness in a habitat, then repeated scrutiny of sites during the fruiting seasons is adequate. Comparisons between sites are facilitated by standardizing the area examined for sporocarps. (2) Quantitative sampling. If variation in species richness between sites is to be measured, the sample area must be standardized to prevent bias due to unequal sampling effort. Total area examined should be approximately 1000 m^2 per stand. For long-term observations, rectangular plots divided into contiguous subplots are easiest to relocate. However, significantly more species may be detected by dispersing small (e.g., 4 m^2) plots along transects. Also, some plots must be raked with a truffle fork to detect sporocarps of hypogeous (belowground) species (see Gardner and Malajczuk 1988). Plots disturbed in this way are suspect for future sampling. We recommend a combination of permanent plots plus temporary transects where some plots are raked for hypogeous fungi.

2. To collect a mushroom or other sporocarp, carefully detach the entire base from the substrate. This may require digging to obtain entire specimens of taxa with deeply rooting bases (e.g., volva in *Amanita*).
3. It is important to collect specimens in all developmental stages. Specimens should be given a field label, wrapped in aluminum foil, and placed in a rigid container like a basket, frame pack, or plastic bucket.
4. Morphological features of sporocarp specimens must be documented soon after collection. Some taxa must be examined the same day that they are collected, while other taxa can be stored in a refrigerator for 2 or 3 days.
5. Group sporocarps into types and identify, when possible, using field guides to macrofungi (e.g., Arora 1986; Smith et al. 1981). Precise identification must reference technical monographs. The reaction of sporocarps to certain chemicals may be useful for identification. The most common reagents used for macrochemical tests are KOH, Melzer's solution, and PDAB.
6. After describing and photographing, sporocarps should be dried at 40–50 °C for two days and stored in a rigid container. Most herbaria use small cardboard boxes for specimen storage. Voucher collections must be deposited in a herbarium for all taxa encountered in a study. Dried specimens also serve as a source of DNA for molecular identification of mycorrhizae.

Special Considerations

It is easy to collect more sporocarp specimens than can be well documented and identified. Approximately 0.5–1.5 hours are required to describe an unknown collection.

Bait-Plant Bioassay

A bait-plant bioassay is a proxy measure of mycorrhizal propagule densities. This protocol was developed by Brundrett and Abbott (1994) to assess *total viable propagule densities* of either AM or EM fungi, and for this purpose it is simpler, and probably more accurate, than direct analysis of mycorrhizal structures from field samples. Methods to predict mycorrhizal colonization in the field based on bioassay results are discussed in Abbott et al. (1995). It is crucial that cultural conditions of bioassays are carefully designed to mimic conditions in the natural field environment. Several studies have shown that bioassay artifacts result from inappropriate host plants, soil media, or greenhouse environment (Wilson and Trinick 1982; Adelman and Morton 1986; An et al. 1990). For each study system, it is best to conduct a small pilot study to determine the optimum duration (ranging from 2 to 12 weeks) for the bioassay.

Materials

1. Soil corer
2. Pots
3. Greenhouse space

Procedure

1. Collect relatively undisturbed soil cores using a large corer. See Brundrett et al. (1994) for construction of a 12 cm diameter × 14 cm deep steel sampler. Alternatively, a tulip bulb planter or golf course cup cutter can be used in soils that are not hard or rocky.
2. Place the undisturbed core directly into a pot that has the same dimensions as the core (Brundrett and Abbott [1994] recommend using 1 L pots).
3. Establish control pots with pasteurized soil to quantify any greenhouse contamination.
4. Plant pregerminated seeds or seedlings into the pots. For arbuscular mycorrhizae, use either a bait plant that is native to the ecosystem or a standard host like *Zea mays, Sorghum vulgare,* or *Trifolium repens.* The advantage of using standard hosts is that they grow rapidly and uniformly; the disadvantage is that they may give erroneous results if the AM fungi in the system exhibit significant host specificity. Our experience indicates that the advantages of using a standard host plant (*Zea mays*) outweigh the disadvantages. Furthermore, Brundrett and Abbott (1994) did not find a significant difference in results using native versus nonnative AM hosts. However, in EM bioassays it is important to use bait plants from the same genera as the dominant plants in the study system because EM fungi often form mycorrhizae exclusively with particular plant taxa (Molina et al. 1992).
5. Harvest the plants after 2–12 weeks (time predetermined in a pilot study). Generally a shorter time is sufficient for AM hosts and a longer time for EM hosts. If desired, shoots and roots can be dried, weighed, and further analyzed for nutrient content or other properties.
6. Quantify AM colonization using the methods described earlier. For ectomycorrhizae, total root length and EM root tips can be counted on each bioassay seedling to give the proportion of root tips that are mycorrhizal and the relative frequency of each EM type.

Special Considerations

Approximately 10 minutes are necessary to collect the soil core, put it into a pot, and plant seeds. Then the pots must be maintained daily in the greenhouse for 2–6 weeks for arbuscular mycorrhizae or 4–12 weeks for ectomycorrhizae. Harvesting plants requires approximately 10–30 minutes per plant, depending on the type of data collected.

DNA Fingerprinting

The general approach for PCR is to extract DNA from tissues (mycorrhizae, sporocarps, leaves, etc.), add this to a buffer with dNTPs, Mg, Taq DNA polymerase and (usually 20-mer) oligonucleotide primers coding for sequences that flank the DNA sequence to be amplified. The mixture is placed in a thermal control device (thermocycler), and a program of heating and cooling cycles, in conjunction with the activity of DNA polymerase, should generate a several million–fold increase in the targeted DNA sequence. Primers that preferentially bind to fungal, basidiomycota, or plant DNA are available for some genes and intergenic regions (Egger 1995). The PCR product is then digested with restriction enzymes to give RFLPs. Restriction patterns from mycorrhizae are compared with those from fungal or plant tissue to identify symbionts. General references for these techniques include Sambrook et al. (1989), Ausubel et al. (1992), and Innis et al. (1989).

The area most often targeted for distinguishing species of fungi is the internally transcribed spacer (ITS) region of the nuclear encoded ribosomal genes. Targeting the ITS region has three advantages:

1. Genes are present in several hundred copies per cell which facilitates successful amplification from minute samples.
2. The ITS region is flanked by highly conserved 18S and 28S ribosome subunit genes, so primers that amplify all fungi or all basidiomycota are easily developed.
3. Within the ITS region, two highly variable regions (ITS-1 and 2) increase the chances of discriminating between closely related taxa. Other regions often targeted for molecular identification of fungi include the intergenic region of the nuclear-encoded ribosomal genes (IGS) and the chitinase synthase genes (Erland et al. 1994; Henrion et al. 1994; Mehmann et al. 1994).

Materials

1. 2.0% hexadecyltrimethylammonium bromide (CTAB) lysis buffer. Combine 100 mM tris HCl (pH 8.0), 1.4 mol/L NaCl, 20 mmol/L EDTA, 2.0 % w/v CTAB
2. Chloroform-isoamyl alcohol (24 parts chloroform:1 part isoamyl alcohol)
3. 2-propanol (chilled)
4. Wash solution (76% v/v ethanol, 10 mmol/L NH_4 acetate)
5. TE, pH 8.0 (10 mmol/L tris HCl [pH 8.0], 1 mmol/L EDTA [pH 8.0]).
6. 95% ethanol (chilled)
7. PCR buffer (supplied with Taq polymerase)
8. dNTPs: 2.0 mmol/L total final concentration (0.5 mmol/L each of G, A, C, T dTP in H_2O)
9. 25 mmol/L $MgCl_2$
10. 1.0% bovine serum albumin (BSA)
11. Oligonucleotide primers obtained from a commercial supplier. Dilutions are empirically determined and vary by source.
12. Taq DNA polymerase
13. Mineral oil
14. Sterile distilled H_2O. This must be free of DNA. Place double-distilled H_2O into acid-washed glassware and autoclave. Store in presterilized plastic tubes to minimize contamination.
15. Restriction enzymes and buffers, e.g., CFO1, HPA, RSA etc. Four base cutters are recommended; follow manufacturer's instructions.
16. Agarose: gel electrophoresis grade

17. Gel buffer (TAE) 0.04 mol/L tris-acetate, 1 mmol/L EDTA. A concentrated solution can be used as a stock solution: combine 242 g tris base, 57.1 mL glacial acetic acid, 100 mL 0.5 mol/L EDTA; dilute to 50 mL/L to use.
18. Ethidium bromide. Prepare stock solution by adding 10 mg per mL H_2O. To stain a gel add 10 μl to 200 mL H_2O.
19. 1.5 mL microfuge tubes
20. 0.5 mL microfuge tubes
21. Micropestles
22. Liquid nitrogen
23. Benchtop centrifuge
24. Water bath
25. Thermal control device (thermocycler)
26. Gel electrophoresis apparatus and power supply
27. UV transilluminator with Polaroid camera or digital imaging system

Procedure

DNA Extraction

1. For effective DNA extraction, EM root tips must be stored in water in the refrigerator and should be extracted within 1–2 weeks of collection. Alternatively, they can be stored frozen or lyophilized. Place 10–50 mg tissue (or one to three root tips) into a 1.5 mL eppendorf tube and freeze in liquid nitrogen.
2. Grind tissues, freeze in liquid nitrogen, and grind again.
3. Add CTAB solution 1:1 with tissue volume (30–50 μL) and mix.
4. Freeze and thaw three times.
5. Place in 65 °C water bath for 1 hour. Add chloroform-isoamyl alcohol 1:1 to the solution and vortex to emulsify.
6. Centrifuge at 12,500 rpm for 15 minutes. Precipitation may be facilitated by placing the material in the freezer overnight.
7. Carefully draw off the clean supernatant, place in a clean 0.5 mL tube, and add a volume of cold 2-propanol that is approximately two-thirds the volume of supernatant.
8. Centrifuge at 6000 rpm for 5 minutes. Pour off and discard supernatant very carefully to avoid losing the pellet.
9. Add wash solution and centrifuge at 10,000 rpm for 15 minutes. Pour off and discard supernatant vary carefully to avoid losing the pellet.
10. Add cold 95% ethanol to wash the pellet; air dry.
11. Suspend in 30 μL TE, and make a 1:200 dilution in TE.

DNA Amplification

1. Choose primers with varying specificity based on the sample to be amplified. There are fairly reliable primers with specificity to plants, fungi, and basidiomycota (Egger 1995). For example, combining a fungal-specific primer

and a basidiomycota-specific primer to amplify DNA extracted from EM will yield only basidiomycota (not plant or ascomycota) DNA.

2. Combine in a 0.5 mL microfuge tube: 12.2 μL sterile distilled H_2O, 2.5 μL 10× PCR buffer (supplied with Taq polymerase), 2.5 μL dNTPs, 1.5 μL 25 mmol/L $MgCl_2$, 0.25 μL 1.0% BSA, 0.125 μL Taq polymerase, and 0.5 μL each of "chosen" two primers (there should be a total of 20.075 μL of solution per tube).
3. Add 5.0 μL dilute genomic DNA to the solution prepared above and add a drop of oil on top of the reaction mixture to prevent evaporation.
4. Place tubes into the thermal control device ("thermocycler") and run a temperature cycle program (e.g., Gardes and Bruns 1993).
5. After the amplification period, remove the product from under the oil using a micropipetter and place it in a clean 0.5 mL tube (typically 15–20 μL of product can be obtained, which is enough material to check that amplification was successful and to run two or three restriction digests). If amplification was not successful, return to step 11 and try again using dilutions of 1:100 or 1:50 in 30 μL.

Restriction Digest

1. In a 0.5 mL tube combine 5.0 μL DNA (from PCR, earlier), 0.5 μL enzyme, 1.5 μL buffer (supplied with enzyme), and 8.0 μL sterile distilled H_2O.
2. Place in 37° C water bath for 1 hour.
3. Run digestion products through 1.8% agarose gel electrophoresis for 60–70 minutes at 100 volts, 50 mA, stain with ethidium bromide, and visualize with UV. Ethidium bromide is a carcinogen; wear gloves and follow the appropriate disposal guidelines. Also, be careful to protect your eyes from the UV light source.
4. Record gels with video imaging systems for easy analysis by importing the digital image into an image analysis program such as the public domain software, NIH Image (NIH 1997). Alternatively, gels can be photographed with a Polaroid camera using a UV filter.

Special Considerations

Usually 14 to 28 samples are prepared at one time, depending on how many can be conveniently run on one or two gels. To extract, amplify, and digest one set of 28 samples takes about 20 hours over 2–4 days.

Acknowledgments We thank Lyn Abbott, Louise Egerton-Warburton, Paul Grogan, Jean Lodge, Terence McGonigle, and an anonymous reviewer for their helpful suggestions. Financial support was provided to N. C. J by the National Science Foundation.

References

Abbott, L. K. 1982. Comparative anatomy of vesicular-arbuscular mycorrhizae formed on subterranean clover. *Australian Journal of Botany* 30:485–499.

Abbott, L. K., A. D. Robson, and M. A. Scheltema. 1995. Managing soils to enhance mycorrhizal benefits in Mediterranean agriculture. *Critical Reviews in Biotechnology* 15:213–228.

Adelman, M. J., and J. B. Morton. 1986. Infectivity of vesicular-arbuscular mycorrhizal fungi: influence of host-soil diluent combinations on MPN estimates and percentage colonization. *Soil Biology and Biochemistry* 18:77–83.

Agerer, R. 1987–1991. Colour Atlas of Ectomycorrhizae. 1st–5th delivery. Einhorn-Verlag, Schwäbisch-Gmünd, Germany.

Agerer, R. 1991. Characterization of ectomycorrhiza. Pages 25–27 *in* J. R. Norris, D. J. Read, and A. K. Varma, editors, *Methods in Microbiology. Volume 23, Techniques for the Study of Mycorrhiza.* Academic Press, London, UK.

Allen, E. B., M. F. Allen, D. T. Helm, J. M. Trappe, R. Molina, and E. Rincon. 1995. Patterns and regulation of mycorrhizal plant and fungal diversity. Pages 47–62 *in* H. P. Collins, G. P. Robertson, and M. J. Klug, editors, *The Significance and Regulation of Soil Diversity.* Kluwer Academic Press, Dordrecht, Netherlands.

Allen, M. F. 1991. *The Ecology of Mycorrhizae.* Cambridge University Press, New York, New York, USA.

Allen, M. F., and J. A. MacMahon. 1985. Impact of disturbance on cold desert fungi: comparative microscale dispersion patterns. *Pedobiologia* 28:215–224.

Allen, M. F., T. S. Moore Jr., M. Christensen, and N. Stanton. 1979. Growth of vesicular-arbuscular mycorrhizal and non-mycorrhizal *Bouteloua gracilis* in a defined medium. *Mycologia* 71:666–669.

An, Z. Q., and J. W. Hendrix. 1988. Determining viability of Endogonaceous spores with a vital stain. *Mycologia* 80:259–261.

An, Z. Q., J. W. Hendrix, D. E. Hershman, and G. T. Henson. 1990. Evaluation of the "most probable number" (MPN) and wet-sieving methods for determining soil-borne populations of Endogonaceous mycorrhizal fungi. *Mycologia* 82:576–581.

Anderson, R. C., A. E. Liberta, L. A. Dickman, and A. J. Katz. 1983. Spatial variation in vesicular-arbuscular mycorrhiza spore density. *Bulletin of the Torrey Botanical Club* 110:519–525.

Arnolds, E. 1992. The analysis and classification of fungal communities with special reference to macrofungi. *Handbook of Vegetation Science* 19:7–48.

Arora, D. 1986. *Mushrooms Demystified. 2nd edition.* Ten Speed Press, Berkeley, California, USA.

Ausubel, F. M., K. Struhl, J. A. Smith, J. Seidmen, D. Moore, R. Kingston, and R. Brent, editors. 1992. *Current Protocols in Molecular Biology.* Green, New York, New York, USA.

BEG. 1996. The Bank of European Glomales (BEG) expert system—a multimedia identification system for arbuscular mycorrhizal fungi. Available: http://wwwbio.ukc.ac.uk/beg/ (1999 March).

Bever, J. D., J. B. Morton, J. Antonovics, and P. A. Schultz. 1996. Host specificity and diversity of glomalean fungi: an experimental approach in an old field community. *Journal of Ecology* 84:71–82.

Bills, G. F., G. I. Holtzmann, and O. K. Miller Jr. 1986. Comparison of ectomycorrhizal Basidiomycete communities in red spruce versus northern hardwood forests of West Virginia. *Canadian Journal of Botany* 64:760–768.

Brown, M. F., and E. J. King. 1984. Morphology and histology of vesicular-arbuscular mycorrhizae. Pages 15–21 *in* N. C. Schenck, editor, *Methods and Principles of Mycorrhizal Research.* American Phytopathological Society, St. Paul, Minnesota, USA.

Brundrett, M. C. 1991. Mycorrhizas in natural ecosystems. *Advances in Ecological Research* 21:171–313.

Brundrett, M. C., and L. K. Abbott. 1994. Mycorrhizal fungus propagules in the jarrah forest. I. Seasonal study of inoculum levels. *New Phytologist* 127:539–546.

Brundrett, M. C., N. Bougher, B. Dell, T. Grove, and N. Malajczuk. 1996. *Working with Mycorrhizal Fungi in Forestry and Agriculture.* Monograph no. 32. Australian Center for International Agricultural Research, Canberra, Australia.

Brundrett, M., L. Melville, and L. Peterson, editors. 1994. *Practical Methods in Mycorrhiza Research.* Mycologue Publications, Sidney, British Columbia, Canada.

Clapp, J. P., J. P. W. Young, J. W. Merryweather, and A. H. Fitter. 1995. Diversity of fungal symbionts in arbuscular mycorrhizas from a natural community. *New Phytologist* 130:259–265.

Cox, G., and P. B. Tinker. 1976. Translocation and transfer of nutrients in vesicular-arbuscular mycorrhizas. I. The arbuscule and phosphorus transfer: a quantitative ultrastructural study. *New Phytologist* 77:371–378.

Daniels, B. A., P. M. McCool, and J. A. Menge. 1981. Comparative inoculum potential of spores of six VAM fungi. *New Phytologist* 89:385–391.

Di Bonito, R., M. L. Elliott, and E. A. Des Jardin. 1995. Detection of an arbuscular mycorrhizal fungus in roots of different plant species with the PCR. *Applied and Environmental Microbiology* 61:2809–2810.

Dodd, J. C., S. Rosendahl, M. Giovannetti, A. Broome, L. Lanfranco, and C. Walker. 1996. Inter- and intraspecific variation within the morphologically-similar arbuscular mycorrhizal fungi *Glomus mosseae* and *Glomus coronatum. New Phytologist* 133:113–122.

Egger, K. 1995. Molecular analysis of ectomycorrhizal fungal communities. *Canadian Journal of Botany* 73 (Suppl. 1):S1415–1422.

Erland, S., B. Henrion, F. Martin, L. A. Glover, and I. J. Alexander. 1994. Identification of the ectomycorrhizal basidiomycete *Tylospora fibrillosa* Donk by RFLP analysis of the PCR-amplified ITS and IGS regions of ribosomal DNA. *New Phytologist* 126:525–532.

Fitter, A. H. 1985. Functional significance of root morphology and root system architecture. Pages 87–106 *in* A. H. Fitter, D. Atkinson, D. J. Read, and M. B. Usher, editors, *Ecological Interactions in Soil. British Ecological Society Special Publication No. 4.* Blackwell Scientific, Oxford, UK.

Franson, R. L., and G. J. Bethlenfalvay. 1989. Infection unit method of vesicular-arbuscular mycorrhizal propagule determination. *Soil Science Society of America Journal* 53:754–756.

Friese, C. F., and M. F. Allen. 1991a. The spread of VA mycorrhizal fungal hyphae in the soil: inoculum types and external hyphal architecture. *Mycologia* 83:409–418.

Friese, C. F., and M. F. Allen. 1991b. Tracking the fates of exotic and local VA mycorrhizal fungi: methods and patterns. *Agriculture, Ecosystems and Environment* 34:87–96.

Gardes, M. and T. D. Bruns. 1993. ITS primers with enhanced specificity for basidiomycetes application to the identification of mycorrhizae and rusts. *Molecular Ecology* 2:113–118.

Gardes, M., and T. D. Bruns. 1996. Community structure of ectomycorrhizal fungi in a *Pinus muricata* forest: above- and below-ground views. *Canadian Journal of Botany* 74: 1572–1583.

Gardes, M., G. M. Mueller, J. A. Fortin, and B. R. Kropp. 1991. Mitochondrial DNA polymorphisms in *Laccaria bicolor, L. laccata, L. proxima,* and *L. amethystina. Mycological Research* 95:206–216.

Gardner, J. H., and N. Malajczuk. 1988. Recolonization of rehabilitated bauxite mine sites in Western Australia by mycorrhizal fungi. *Forest Ecology and Management* 24:27–42.

Giovannetti, M., and B. Mosse. 1980. An evaluation of techniques for measuring vesicular-arbuscular mycorrhiza infection in roots. *New Phytologist* 84:489–500.

Goodman, D. M., D. M. Durall, J. A. Trofymow, and S. M. Berch, editors. 1996. *A Manual of Concise Descriptions of North American Ectomycorrhizae.* Mycologue Publications, Sidney, British Columbia, Canada.

Graham, J. H., N. C. Hodge, and J. B. Morton. 1995. Fatty acid methyl ester profiles for characterization of Glomalean fungi and their endomycorrhizae. *Applied and Environmental Microbiology* 61:58–64.

Hahn, A., V. Gianinazzi-Pearson, and B. Hock. 1994. Characterization of arbuscular fungi by immunochemical techniques. Pages 25–39 *in* S. Gianinazzi and H. Schuepp, editors, *Impact of Arbuscular Fungi on Sustainable Agriculture and Natural Ecosystems.* Birkhaeuser, Basel, Switzerland.

Harley, J. L. 1989. The significance of mycorrhiza. *Mycological Research* 92:129–139.

Harley, J. L., and S. E. Smith. 1983. *Mycorrhizal Symbiosis.* Academic Press, London, UK.

Henrion, B., C. di Battista, D. Bouchard, D. Vairelles, B. D. Thompson, F. le Tacon, and F. Martin. 1994. Monitoring the persistence of *Laccaria bicolor* as an ectomycorrhizal symbiont of nursery-grown Douglas fir by PCR of the rDNA intergenic spacer. *Molecular Ecology* 3:571–580.

Hetrick, B. A. D. 1991. Mycorrhizas and root architecture. *Experimentia* 47:355–362.

Ianson, D. C., and M. F. Allen. 1986. The effects of soil texture on extraction of vesicular-arbuscular mycorrhizal fungal spores from arid sites. *Mycologia* 78:164–168.

Innis, M. A., D. H. Gelfand, J. J. Sninsky, and T. J. White. 1989. *PCR Protocols: A Guide to Methods and Applications.* Academic Press, London, UK.

INVAM. 1997. The international culture collection of arbuscular and VA mycorrhizal fungi (INVAM) [on-line]. Available: http://invam.caf.wvu.edu/ (1999 March).

Jasper, D. A., L. K. Abbott, and A. D. Robson. 1989. Soil disturbance reduces the infectivity of external hyphae of vesicular-arbuscular mycorrhizal fungi. *New Phytologist* 112:93–99.

Johnson, N. C. 1993. Can fertilization of soil select less mutualistic mycorrhizae? *Ecological Applications* 3:749–757.

Johnson, N. C., J. H. Graham, and F. A. Smith. 1997. Functioning of mycorrhizal associations along the mutualism-parasitism continuum. *New Phytologist* 135:575–585.

Johnson, N. C., D. Tilman, and D. Wedin. 1992. Plant and soil controls on mycorrhizal fungal communities. *Ecology* 73:2034–2042.

Johnson, N. C., D. R. Zak, D. Tilman, and F. L. Pfleger. 1991. Dynamics of vesicular-arbuscular mycorrhizae during old field succession. *Oecologia* 86:349–358.

Kormanik, P. P., and A.-C. McGraw. 1982. Quantification of vesicular-arbuscular mycorrhizae in plant roots. Pages 37–45 *in* N. C. Schenck, editor, *Methods and Principles of Mycorrhizal Research.* American Phytopathological Society, St. Paul, Minnesota, USA.

Koske, R. E. 1987. Distribution of vesicular-arbuscular mycorrhizal fungi along a latitudinal temperature gradient. *Mycologia* 79:55–68.

Koske, R. E., and J. N. Gemma. 1989. A modified procedure for staining roots to detect mycorrhizas. *Mycological Research* 92:486–488.

Leake, J. R. 1994. The biology of myco-heterotrophic ("saprophytic") plants. *New Phytologist* 127:171–216.

Lloyd-MacGilp, S. A., S. M. Chambers, J. C. Dodd, A. H. Fitter, C. Walker, and J. P. W. Young. 1996. Diversity of the ribosomal internal transcribed spacers within and among isolates of *Glomus mosseae* and related mycorrhizal fungi. *New Phytologist* 133:103–111.

LTERNET. 1996. National research sites with a common commitment. Data management. In U.S. Long Term Ecological Research (LTER) Program [on-line]. Available: http://www.lternet.edu/research/data/ (1999 March).

Luoma, D. L., R. E. Frenkel, and J. M. Trappe. 1991. Fruiting of hypogeous sporocarps in Oregon Douglas-fir forests: seasonal and habitat variation. *Mycologia* 83:335–353.

Massicotte, H. B., C. A. Ackerley, and R. L. Peterson. 1987. The root-fungus interface as an indicator of symbiont interaction in ectomycorrhizae. *Canadian Journal of Forest Research* 17:846–854.

McGonigle, T. P. 1988. A numerical analysis of published field trials with vesicular-arbuscular mycorrhizal fungi. *Functional Ecology* 2:473–478.

McGonigle, T. P., M. H. Miller, D. G. Evans, G. L. Fairchild, and J. A. Swan. 1990. A new method which gives an objective measure of colonization of roots by vesicular-arbuscular mycorrhizal fungi. *New Phytologist* 115:495–501.

McKenney, M. C., and D. L. Lindsey. 1987. Improved method for quantifying endomycorrhizal fungi spores from soil. *Mycologia* 79:779–782.

Mehmann, B., I. Brunner, and G. H. Braus. 1994. Nucleotide sequence variation of chitin synthase genes among ectomycorrhizal fungi and its potential use in taxonomy. *Applied and Environmental Microbiology* 60:3105–3111.

Miller, R. M., and J. D. Jastrow. 1992. Extraradical hyphal development of vesicular-arbuscular mycorrhizal fungi in a chronosequence of prairie restorations. Pages 171–176 *in* D. J. Read, D. H. Lewis, A. H. Fitter, and I. J. Alexander, editors, *Mycorrhizas in Ecosystems.* CAB International, Wallingford, UK.

Molina, R., H. B. Massicotte, and J. M. Trappe. 1992. Specificity phenomena in mycorrhizal symbioses: community-ecological consequences and practical implications. Pages 357–437 *in* M. F. Allen, editor, *Mycorrhizal Functioning.* Chapman and Hall, New York, New York, USA.

Molina, R., T. O'Dell, D. Luoma, M. Amaranthus, M. Castellano, and K. Russell. 1993. Biology, ecology, and social aspects of wild edible mushrooms in the forests of the Pacific Northwest: a preface to managing commercial harvest. USDA Forest Service, General Technical Report PNW-GTR-309. Portland, Oregon, USA.

Moorman, T., and F. B. Reeves. 1979. The role of endomycorrhizae in revegetation practices in the semi-arid West. II. A bioassay to determine the effect of land disturbance on endomycorrhizal populations. *American Journal of Botany* 66:14–18.

Morton, J. B. 1986. Effects of mountants and fixatives on wall structure and Melzer's reaction in spores of two *Acaulospora* species (Endogonaceae). *Mycologia* 78:787–794.

Morton, J. B., and G. L. Benny. 1990. Revised classification of arbuscular mycorrhizal fungi (Zygomycetes): a new order, Glomales, two new suborders, Glomineae and Gigasporineae, and two new families, Acaulosporaceae and Gigasporaceae, with an emendation of Glomaceae. *Mycotaxon* 37:471–491.

Morton, J. B., S. P. Bentivenga, and J. D. Bever. 1995. Discovery, measurement, and interpretation of diversity in arbuscular endomycorrhizal fungi (Glomales, Zygomycetes). *Canadian Journal of Botany* 73 (Suppl. 1):S25–S32.

Morton, J. B., S. P. Bentivenga, and W. W. Wheeler. 1993. Germ plasm in the international collection of arbuscular and vesicular-arbuscular mycorrhizal fungi (INVAM) and procedures for culture development, documentation and storage. *Mycotaxon* 48:491–528.

Mycolog. 1996. Mycolog publications [on-line]. Available: http://www.pacificcoast.net/mycolog/index.html (October 1998).

Newman, E. I., and P. Reddell. 1987. The distribution of mycorrhizas among families of vascular plants. *New Phytologist* 106:745–751.

NIH, 1997. NIH Image. National Institutes of Health [Online]. Available: http://rsb.info.nih.gov/nih-image/ (1999 March).

Norris, J. R., D. J. Read, and A. K. Varma, editors. 1991. *Methods in Microbiology. Volume 23,* Techniques for the Study of Mycorrhiza. Academic Press, London, UK.

Norris, J. R., D. J. Read, and A. K. Varma, editors. 1992. *Methods in Microbiology. Volume 24,* Techniques for the Study of Mycorrhiza. Academic Press, London, UK.

O'Dell, T. E., D. L. Luoma, and R. J. Molina. 1992. Ectomycorrhizal fungal communities in young, managed and old growth Douglas-fir stands. *Northwest Environmental Journal* 8:166–168.

O'Dell, T. E., J. E. Smith, M. A. Castellano, and D. L. Luoma. 1996. Diversity and conservation of forest fungi. Pages 5–18 *in* R. Molina and D. Pilze, editors, *Managing Forest Ecosystems to Conserve Fungus Diversity and Sustain Wild Mushroom Harvests.* USDA Forest Service, General Technical Report PNW-GTR-371. Portland, Oregon, USA.

Perry, D. A., M. P. Amaranthus, J. G. Borchers, S. L. Bouchers, and R. E. Brainerd. 1989. Bootstrapping in ecosystems. *BioScience* 39:230–237.

Porter, W. M. 1979. The "most probable number" method for enumerating infective propagules of VAM fungi in soil. *Australian Journal of Soil Research* 17:515–519.

Read, D. J. 1983. The biology of mycorrhiza in the Ericales. *Canadian Journal of Botany* 61:985–1004.

Read, D. J. 1984. The structure and function of the vegetative mycelium of mycorrhizal roots. Pages 215–240 *in* D. H. Jennings and A. D. M. Rayner, editors, *The Ecology and Physiology of the Fungal Mycelium: Symposium of the British Mycological Society.* Cambridge University Press, Cambridge, UK.

Rousseau, J. V. D., D. M. Sylvia, and A. J. Fox. 1994. Contribution of ectomycorrhiza to the potential nutrient-absorbing surface of pine. *New Phytologist* 128:639–644.

Safir, G. R., S. C. Coley, J. O. Siqueira, and P. S. Carlson. 1990. Improvement and synchronization of VA mycorrhizal fungal spore germination by short-term cold storage. *Soil Biology and Biochemistry* 22:109–111.

Sambrook, J., E. F. Fritsch, and T. Maniatis. 1989. *Molecular Cloning: A Laboratory Manual. 2d edition.* Cold Spring Harbor Laboratory Press, Cold Spring Harbor, New York, USA.

Sanders, I. R., J. P. Clapp, and A. Wiemken. 1996. The genetic diversity of arbuscular mycorrhizal fungi in natural ecosystems: a key to understanding the ecology and functioning of the mycorrhizal symbiosis. *New Phytologist* 133:123–134.

Schaffer, G. F., and R. L. Peterson. 1993. Modifications to clearing methods used in combination with vital staining of roots colonized with vesicular-arbuscular mycorrhizal fungi. *Mycorrhiza* 4:29–35.

Schenck, N. C., editor. 1982. *Methods and Principles of Mycorrhizal Research.* American Phytopathological Society, St. Paul, Minnesota, USA.

Schenck, N. C., and Y. Pérez. 1990. *Manual for the Identification of VA Mycorrhizal Fungi. 3d edition.* Synergistic Publications, Gainesville, Florida, USA.

Smith, A. H., H. V. Smith, and N. S. Weber. 1981. *How to Know the Non-Gilled Fleshy Fungi. 2d edition.* William C. Brown, Dubuque, Iowa, USA.

Smith, D. C., and A. E. Douglas. 1987. *The Biology of Symbiosis.* Edward Arnold, New York, New York, USA.

Smith, F. A., and S E. Smith. 1996. Mutualism and parasitism: diversity in function and structure in the "arbuscular" (VA) mycorrhizal symbiosis. *Advances in Botanical Research* 22:1–43.

Smith, S. E., and V. Gianinazzi-Pearson. 1988. Physiological interactions between symbionts in vesicular-arbuscular mycorrhizal plants. *Annual Review of Plant Physiology and Plant Molecular Biology* 39:221–244.

Smith, S. E., and D. J. Read. 1997. *Mycorrhizal Symbiosis. 2nd edition.* Academic Press, New York, New York, USA.

Southwood, T. R. E. 1978. *Ecological Methods with Particular Reference to the Study of Insect Populations. 2d edition.* Wiley, New York, New York, USA.

St. John, T. V., and R. E. Koske. 1988. Statistical treatment of endogonaceous spore counts. *Transactions of the British Mycological Society* 91:117–121.

Sylvia, D. M. 1992. Quantification of external hyphae of vesicular-arbuscular mycorrhizal fungi. Pages 53–65 *in* J. R. Norris, D. J. Read, and A. K. Varma, editors, *Methods in Microbiology. Volume 24, Techniques for the Study of Mycorrhiza.* Academic Press, London, UK.

Tews, L. L., and R. E. Koske. 1986. Toward a sampling strategy for vesicular-arbuscular mycorrhizas. *Transactions of the British Mycological Society* 87:353–358.

Thingstrup, I., M. Rozycka, P. Jeffries, S. Rosendahl, and J. C. Dodd. 1995. Detection of the arbuscular mycorrhizal fungus *Scutellospora heterogama* within roots using polyclonal antisera. *Mycological Research* 99:1225–1232.

Tommerup, I. C. 1983. Spore dormancy in vesicular-arbuscular mycorrhizal fungi. *Transactions of the British Mycological Society* 81:381–387.

Toth, R., and R. M. Miller. 1984. Dynamics of arbuscule development and degeneration in *Zea mays* mycorrhiza. *American Journal of Botany* 71:449–460.

Trappe, J. M. 1987. Phylogenetic and ecological aspects of mycotrophy in the angiosperms from an evolutionary standpoint. Pages 5–25 *in* G. R. Safir, editor, *Ecophysiology of VA Mycorrhizal Plants.* CRC Press, Boca Raton, Florida, USA.

Villeneuve, N., M. M. Grandtner, and J. A. Fortin, 1989. Frequency and diversity of ecto-mycorrhizal and saprophytic macrofungi in the Laurentide Mountains of Quebec. *Canadian Journal of Botany* 67:2616–2629.

Vogt, K. A., C. C. Grier, C. E. Meier, and R. L. Edmonds. 1982. Mycorrhizal role in net primary production and nutrient cycling in *Abies amabilis* ecosystems in western Washington. *Ecology* 63:370–380.

Widden, P. 1996. The morphology of vesicular-arbuscular mycorrhizae in *Clintonia borealis* and *Medeola virginiana. Canadian Journal of Botany* 74:679–685.

Wilcox, H. E. 1991. Mycorrhizae. Pages 731–765 *in* Y. Waisel, A. Eshel, and U. Kafkafi, editors, *Plant Roots, the Hidden Half.* Marcel Dekker, New York, New York, USA.

Wilson, J. M., and M. J. Trinick. 1982. Factors affecting the estimation of numbers of infective propagules of vesicular-arbuscular mycorrhizal fungi by the most probable number method. *Australian Journal of Soil Research* 21:73–81.

19

Measurement of Static Root Parameters

Biomass, Length, and Distribution in the Soil Profile

Caroline S. Bledsoe
Timothy J. Fahey
Frank P. Day
Roger W. Ruess

At all levels of ecological organization in terrestrial ecosystems, the structure and function of plant root systems are crucial. The physiological, architectural, and morphological features of individual root systems affect many processes—the development of soils; the composition of soil organisms; the turnover of soil organic matter; the acquisition of water and nutrients for plants; and the establishment, growth, reproduction, and mortality of plant populations. Competition for soil resources among individuals often determines survival and dominance of plants in the community. Moreover, the allocation of energy to belowground tissue fuels the complex soil community that regulates the cycling of nutrients and flow of energy through soils. Thus, our understanding of ecosystem structure and function and our ability to predict ecological responses to environmental change depend in large part on the availability of accurate information on plant root systems.

Unfortunately, root systems are difficult to study, primarily because they are spatially and temporally complex and because available methods for their study are labor-intensive and destructive. Nevertheless, currently available methods can tell us a great deal about root systems. In this chapter we briefly survey methods to measure various root parameters and recommend specific methods for each. We attempt to describe these methods in sufficient detail for use by those with little previous experience in root measurements. Our objective is to facilitate the use of standard methods and the production of comparable root data.

Focus and Scope

In comparative studies, root biomass data are often used primarily to understand factors affecting belowground root activities, particularly carbon distribution (Raich and Nadelhoffer 1989) and the acquisition of nutrients (Bledsoe and Atkinson 1991). Our emphasis is on comparisons among ecosystems, and we describe simple, widely applicable methods for the measurement of root biomass and root length for plants growing in natural conditions in field soils where root growth is not constrained by pots, walls, trenches, or other artificial barriers. While these methods may be applicable to plants growing in more artificial situations (e.g., pot studies, greenhouse work, nonsoil media, solution culture, nursery beds), they are not designed for these purposes.

In this chapter we cover selected static measurements of roots (biomass, length, and distribution within the soil profile); for a discussion of dynamic methods such as fine root production and root phenology, see Chapter 20, this volume, and Taylor (1987). For a more detailed discussion of root architecture, see Fitter (1994). Other methods particularly relevant to this chapter include mycorrhizal techniques (see Chapter; 18, this volume) and soil (including root) respiration (see Chapters 10 and 13, this volume).

Important Considerations

For comparative purposes, we believe that measurements of root biomass by depth, root length, and fine root production (including turnover and phenology) are the most important parameters for which standard methods are needed. Although other parameters could be measured, we argue that this minimum set should provide the best common denominator of standard techniques from which to build long-term data sets that describe and compare root dynamics for different sites. Further elaborations of physiological, architectural, and behavioral parameters can be developed to accommodate particular soil conditions, vegetation structure, or ecological interactions.

Root distribution varies significantly with soil depth; thus, roots must be sampled carefully by depth. We recommend that samples be taken at standard intervals (organic litter layer, surface-to-5 cm, 5–10 cm, 10–20 cm, etc. (see Chapter 1, this volume). Although we do not recommend sampling by horizon, due to difficulties in comparing data from soils with different depths of horizons, as well as exact determination of the depth of each horizon, we do recommend that depths of each horizon be noted because horizons strongly influence soils and root biomass. At some sites it may be necessary to subdivide the standard intervals (see earlier) to avoid inclusion of two horizons within one interval. We also recommend that samples be taken to the complete depth of the rooting zone. Most studies do not sample to a sufficient depth.

Since soils are spatially very heterogeneous, sampling must be designed carefully in order to sample adequately this heterogeneity. This often results in larger numbers of samples, with increased expense and time. When possible, live roots should be separated from organic matter and dead roots. These separations are time-consuming and inexact, but they are essential to the collection of high-quality root data. It is desirable to separate roots by plant species, but this is seldom possible in natural systems, where roots intermingle. If one is dealing with a small number of plant species, however, one may be able to develop site-specific morphological distinctions among species.

Finally, roots should be sampled by diameter classes (see recommended classes in the section "Soil Cores," below), since these classes are often strongly linked to root function. In our experience in a variety of forest ecosystems, as much as 75% of root length may be found in fine roots less than 5 mm in diameter (K. Pregitzer, personal communication). Fine roots are believed to be the most dynamic and active in water and nutrient uptake.

Definitions and Standard Methods

Roots, broadly defined, include roots, rhizomes, tubers, and other significant underground plant parts. Specific terms used in root studies follow; for all of the following definitions, depth of sampling must always be specified:

- Root biomass (g dry wt/cm^2) is the dry weight of roots per unit area of the core, since biomass is usually obtained from soil cores. Although it is desirable to report biomass of live roots only, it is difficult and time consuming to separate live from dead roots (see the section "Separation of Live and Dead Roots," below). Occasionally, one might use data from root windows (or minirhizotrons) to calculate biomass; then biomass is expressed as weight/unit window area (g dry wt/cm^2) as standards for cross-site comparison, we recommend presenting biomass as mass per unit ground surface area).
- Root length density (RLD; cm/cm^3) is total root length per unit soil volume. Occasionally length may be expressed as length/unit area (cm/cm^2), particularly when data are obtained from root windows (or minirhizotrons). Root length density and simple root length are sometimes used interchangeably. However, length (cm) is the less desirable and less comparable term, since length can be expressed in numerous ways (per plant, per sample, etc.). We recommend reporting root length density as length per unit soil volume.
- Root surface area (cm^2/g dry wt) is usually measured on harvested roots.
- Root diameter is expressed in mm (fine roots) or cm (larger roots).
- Annual fine root production or productivity ($g \cdot m^{-2} \cdot yr^{-1}$) estimates root growth over 1 year (see Chapter 20, this volume).
- Root initiation or birth is appearance of new roots, expressed as numbers of new roots/unit area (no./cm^2).
- Root senescence and death (mortality) is the disappearance of roots. Operationally this may be difficult to determine, since roots may become dark and moribund, then revitalize and grow again (Hendrick and Pregitzer 1996). Dead roots are those that totally disappear from root windows. In soil cores, dead roots are often very difficult to distinguish from inactive roots. Most researchers use visual criteria, since there are few satisfactory chemical or biological methods.

A standard set of root methods provides high-quality information and is sufficiently straightforward to be used by many researchers. Methods that require expensive equipment or extensive labor should be minimized when possible. For root studies, much of the work can be performed with simple equipment; minirhizotrons are an exception (see Chapter 20, this volume). We have recommended methods that are most likely to produce comparative information from different sites and to be reliable indices of long-term carbon storage in roots. Methodological details and information about these details (metadata; Michener et al. 1996; Federal Geographic Data Committee 1994) are important. For example, it is difficult to compare root biomass data if data are collected from different soil depths, since root biomass generally decreases significantly with depth. Standard units, while not essential, are also very helpful for assembling root data from different studies. Thus, complete standardization includes standard methods, protocols, units, descriptive (meta)data, and common formats and contents for data storage, access, and sharing (see Chinn and Bledsoe 1997).

Modeling

Root data are used in simulation modeling at many different scales. At larger temporal and spatial scales (years to decades, countries to continents), root data are used in models of global carbon storage that predict effects of climate change and global warming on atmospheric carbon dioxide concentrations (e.g., Raich and Nadelhoffer 1989). At smaller temporal and spatial scales (months to years, watersheds to regions), landscape models use root data to estimate net ecosystem productivity (both above- and belowground; e.g., Parton et al. 1987). At very small scales (days to months, centimeter to meter plots), root data can be used to predict effects of plant growth on soil processes, primarily through effects on soil carbon quality and quantity, and availability of water and nutrients. Although models at different scales require different root data, all models generally utilize root biomass data expressed as weight per unit area or length per unit volume. Thus, it is highly desirable for root researchers to produce data expressed in one or both of these formats, thus allowing their data to be used in model development, validation, and sensitivity analyses. Root data may also be used in conceptual models, such as the compartment-flow model of Santantonio and Grace (1987) and the nitrogen availability model of Nadelhoffer et al. (1985).

Indirect Estimates of Root Parameters

Since direct measurement of root parameters can be difficult, time-consuming, and expensive, alternatives have been developed to estimate parameters. The most common method to estimate root biomass is the use of allometric equations, which are often linear equations that regress root biomass (either fine root or coarse root biomass) against more easily measured aboveground plant parameters—shoot height, shoot diameter or circumference, and shoot (or aboveground) biomass are the most useful indices for estimates of fine root biomass (Kira and Ogawa 1968; Hermann 1977). Other independent variables include diameter at breast height (Baskerville 1966), bole cross-sectional area (Ovington et al. 1967), and length of bole (Young and Carpenter 1967). Vogt et al. (1985) reported a positive correlation ($r^2 = 0.85$) between fine root biomass (kg/ha) and starch content of living bark at breast height.

Allometric equations for coarse woody root biomass have been developed for many tree species, including eucalyptus (Westman and Rogers 1977); mixed deciduous species, and maple, beech, and birch (Whittaker et al. 1974); Mediterranean oaks in Spain (Canadell and Roda 1991); California blue oak (Millikin and Bledsoe 1997); Douglas fir (Santantonio et al. 1977); and miscellaneous other species (Kira and Ogawa 1968). Allometric equations are particularly useful for estimation of coarse woody root biomass because it is particularly difficult and expensive to obtain large woody root data from large trees. In many natural ecosystems, fine roots from different species tend to be intermingled, making it very difficult to determine fine root biomass for a given species using allometric equations.

Ancillary Measurements and Data Storage

Site descriptions should follow Chapter 2, this volume. Soil characteristics should include soil type, classification, texture (percent sand, silt, and clay), pH, bulk density, depth to bedrock or root restrictive layer (RRL), and descriptions and depths of major soil horizons. Other desirable parameters are percent soil organic matter (SOM), cation-exchange capacity (CEC), total carbon, nitrogen, and phosphorus, available phosphate, nitrate, and ammonium, and particle size. Vegetation at the site should be described to species with an estimate of percent cover or leaf area index (forests). It is highly desirable to include estimates of above- and below-ground net primary productivity when possible. For climate data, the minimum information should include location of the weather station, total monthly precipitation (cm/month), and air temperature (°C; mean, mean max, and mean min) based on multiyear averages (10+ years if possible).

Data content should include name of the site and location within the site, and units for all of the variables (e.g., root biomass in gm/cm^2). Data will ideally be stored in a "standard" spreadsheet format (e.g., Excel) or database (e.g., Microsoft Access, FoxPro). Several research groups are developing extensive root databases, which may soon lead to adoption of a de facto standard format for distribution of ecological data (see Bledsoe and Hastings 1996; Gross et al. 1995; Jackson et al. 1996; Vogt et al. 1996).

Available Methods

Several excellent reviews exist in the literature. Böhm (1979) reports in detail methods for measurement of static root parameters, including number, biomass, length, surface area, and volume; excavation, monolith, and wall methods are particularly well described. Static (biomass and length) methods are included in Vogt and Persson (1991), although their principal focus is on dynamic methods to measure growth and development of forest tree roots. Mackie-Dawson and Atkinson (1991) describe field methods, organized by type (extraction, observation, or indirect methods). The Tropical Soil Biology and Fertility Programme has published standard methods for determining root patterns (Anderson and Ingram 1993) and root length, biomass, production, and mortality (van Noordwijk 1993).

Available Methods for Coarse Woody Root Biomass, Length, and Diameter

Woody root dynamics change very slowly except during development of vegetation following large-scale disturbances; otherwise static, onetime measurements are generally sufficient. Few papers report changes in woody roots over time (but see Deans 1981; Kira and Ogawa 1968). Static measurements, on the other hand, have been made for a variety of species, including old-growth Douglas fir (Santantonio et al. 1977), Sitka spruce (Henderson et al. 1983), eastern red oak (Lyford 1980), eastern deciduous forest tree species (Stout 1956), woody Californian chaparral shrub species (Kummerow and Mangan 1981), northeastern mixed conifer-deciduous forest species (eight species; Whittaker et al. 1974), African woody savanna species (Rutherford 1983), montane Mediterranean oak forest in Spain (Canadell and Roda 1991), Sitka spruce in Scotland (Deans 1981), and Australian eucalyptus forest (Westman and Rogers 1977). Table 19.1 compares several of these papers, noting the diversity of methods, units, tree species, and ages and number of trees sampled.

Coarse woody root biomass sampling typically involves either tree excavation, quantitative soil pits, wall or trench profiles, or exploiting serendipitous events such as wind throws.

Excavation

Excavation typically involves using heavy equipment (e.g., backhoe or tractor) in conjunction with other methods (dynamite [e.g., Whitaker et al. 1974], water or air pressure, and/or hand tools) to loosen soil and remove an entire root system within a designated volume, often a cylinder of radius 1–2 m. Many researchers use a backhoe or tractor to trench around the tree, then use a water hose to wash away the soil. Drainage must be designed properly to avoid creating a mud pit. After removal from the excavation, the root ball and attached roots are cleaned. The root ball may be exposed by using air pressure to blow away the soil; this method works well only in very sandy or pumice soils. In heavier soils the root system may be exposed using a high-pressure water hose. See extensive discussions by Vogt and Persson (1991, pp. 488–490) and Böhm (1979, pp. 5–38). The choice of any one of these methods is based on characteristics of the particular site and soils, as well as cost and time considerations.

If the diversity of trees or shrubs is low, then it may be reasonable to excavate individuals of different sizes and to develop species-based regressions of root crown biomass versus basal area or height. In more diverse communities, species must be grouped into a few classes and allometric equations developed for each class. In any case, the excavations should obtain as large a proportion of the root crown as possible, and all lateral roots within a carefully measured distance of the center of the stem should be removed. This distance represents the radius of a circle around each

Table 19.1. Summary of Selected Coarse Woody Root Biomass Data for Field-Grown Tree Species

Species, Study Area	Trees/Sample	Tree Age	Units for Root Data	Methods	Citation
Red oak, USA	NG	40–70 y	# roots/tree	Excavation by shovel	Lyford (1980)
Oak, maple, beech, birch, USA	25	16–104 y	Ft2/tree	Excavation with fire hose	Stout (1956)
Eucalyptus, Queensland, Australia	31	NG	t/ha	Excavation	Westman and Rogers (1977)
Maple, beech, New Hampshire, USA	93	60–106	kg/tree	Excavation with dynamite	Whittaker et al. (1974)
Holm oak, Castanya, Spain	32	60–90 y	kg/tree	Excavation, hand tools	Canadell and Roda (1991)
Douglas-fir, Oregon, USA	3	94–135 y	kg/tree	Windfalls, hand tools	Sanantonio et al. (1977)
Sitka spruce, Greskine, Scotland	8	16 y	# roots/tree, m/tree	Excavation with hand tools	Henderson et al. (1983)

Notes: References are selected to emphasize the diversity of tree species, methods, and units of expression of root data.
NG = not given.

stem, the areas of which must be subtracted from the unit area of the stand from which the random pit samples were obtained; that is, any random point falling within these circles is not sampled by pits. Although the radial distance from stem centers beyond which lateral roots are removed need not be constant (i.e., smaller distance for smaller stems), this distance should increase as a linear function of stem diameter (or other easily measured plant trait) so that calculation of the excluded area is possible from stand survey data.

Quantitative Soil Pits

Quantitative soil pits, in conjunction with excavation and intensive evaluations of soil profiles and chemical constituents (e.g., Fahey et al. 1988), are very useful to determine lateral root distribution. The ideal dimensions of the quantitative pits will depend on the sizes and variation in distribution of woody roots. For deeply rooted plants such as desert shrubs, some method of extrapolating to greater depths will be necessary. The quantitative pits should be deep enough to recover over 90% of lateral woody root biomass. Root biomass usually declines markedly with depth, so the required depth often can be estimated by extrapolating the depth distribution of woody roots.

During quantitative pit sampling, large roots can be retained on coarse field screens (e.g., 1 cm mesh) after cutting at the pit face. To avoid double counting of small woody roots measured with cores, roots smaller than 1.0 cm are measured. The large roots are usually sorted into diameter classes (1–2 cm, 2–5 cm, 5–10 cm, >10 cm), and live and dead roots should be separated while roots are fresh (see section "Separation of Live and Dead Roots, below). Ordinarily it is not necessary to subsample in the field, but rather the dry weight of each size class in each depth class of each soil pit can be measured directly in the laboratory.

Subsampling involves digging by heavy equipment of "quantitative pits" with precise dimensions, usually 2–4 m on a side. Soil and roots are removed from this pit by backhoe or shovel and sieved through coarse sieves (1+ cm), although use of a backhoe may be limited by terrain or access in many natural systems. For example, Millikin and Bledsoe (1997) sampled quantitative pits of 0.5 m × 2 m × 1 m deep in an oak woodland in northern California. Fahey et al. (1988) sampled coarse woody roots in a northern hardwood forest at Hubbard Brook using 58 0.5 m^2 quantitative soil pits excavated to the depth of obstruction (fragipan or bedrock), or 1 m. The standard error was about 20% of the mean. Total woody root biomass (including root crowns) was estimated by adding root crown biomass estimates based on excavations and allometric equations (Whittaker et al. 1974). Sample design can be a problem, since random location of pits may include the root bole; thus random sampling must be modified to eliminate sampling boles of trees.

Wall Profile or Trench

The wall profile or trench method involves digging a long, deep trench at some distance (ideally random) from the tree and then carefully mapping the roots exposed along the surface on a two-dimensional graph, noting root diameters for each root. This method is very useful for mapping root distribution with depth but is not advised for biomass. Since no roots are harvested, biomass is only estimated, based on the number of root interceptions per unit area. In theory, this method is adequate. Woody root length is obtained for each size class, and then the length-to-weight ratio is measured for each class. Length estimate may be biased depending on the angles of intersection of roots with the wall. See Böhm (1979) for descriptions of this method.

Serendipitous Events

Since one may wish to avoid destroying large trees (for ecological, aesthetic, or financial reasons), accidental events in which nature or humans expose the root systems of trees may provide valuable opportunities for root study. For example, windthrow may uproot trees, allowing a researcher to measure the root system without having to excavate the tree. Humans also harvest trees during construction projects, providing an opportunity for sampling root biomass.

Coarse Woody Root Spatial Distribution

Several parameters—root length, diameter, and branching angle (Fitter 1994)—are measured in root spatial distribution studies. In many studies, only root length is measured. The three-dimensional distribution and architecture may be studied by observation of either excavated roots or relatively undisturbed roots in situ. The excavation method may be employed, although additional care must be taken to leave roots in place as the soil is removed. Typically, roots are "stabilized or anchored" by strings or wires to maintain their original orientation. For observations of relatively undisturbed roots, one may carefully place long metal rods throughout a soil monolith to stabilize it, then carefully brush or wash soil away to reveal the root architecture. Both methods are tedious and painstaking, although both are probably equally well suited for studying distributions.

Fine Root Biomass, Length, Diameter, and Spatial Distribution

Methods for biomass, length, diameter, and the spatial distribution of fine roots are discussed together, since most techniques allow simultaneous measurement of several of these four parameters. Some techniques are better adapted for one parameter than another. Generally, soil cores are well adapted to measurement of biomass, root length, and surface area. However, for diameter determinations, the wall profile method is often used, since diameters can be measured on relatively undisturbed fresh roots. Root windows or trenches can provide more detailed data on spatial distribution than can data on roots harvested from cores. The root ingrowth method is not discussed here, since it is not applicable to biomass; see Chapter 20, this volume, for a discussion of root ingrowth cores for measurement of fine root production.

Soil Cores

The classic method for measuring fine root biomass is to use soil cores to extract soils and roots, followed by washing, cleaning, and weighing of fine roots. These methods are described in detail by Vogt and Persson (1991) and Anderson and Ingram (1993). The soil core method has great value due to its simplicity and extensive use by many researchers at numerous sites over many years. In a comparison of three methods, Heeraman and Juma (1993) recommended destructive sampling by cores over monolith or minirhizotron methods, although results were species-specific.

Several reviews of root biomass have compiled extensive collections of data from the literature (Vogt et al. 1996; Jackson et al. 1996, 1997, Canadell et al. 1996). Root and soil data for LTER sites are accessible from http:\\lternet.edu(LTERnet.a,b,c).

Root Windows, Walls, and Minirhizotrons

Root windows, walls, trenches, and minirhizotrons can produce similar information. Minirhizotrons have the advantage of less disturbance to the rooting environment (see Chapter 20, this volume; Hendrick and Pregitzer 1993). Root windows (including larger root walls and trenches) have several advantages. Windows are inexpensive to install, easy to observe, and particularly useful for determination of spatial distributions. However, root windows are not ideal for measurements of length or biomass because one must assume that the window surface does not create unusual growing conditions, perhaps an unwarranted assumption. Root length is measured directly against the glass plate with a metric ruler and recorded. A dissecting scope may be used to view very fine roots (B. Haines, personal communication). Phenological observations can be performed, and root growth rates can be calculated from sequential measurements of root length. One advantage of this method is that root length is measured directly; however, root growth against the glass may be significantly different from growth in undisturbed soil. Unfortunately, there appears to be little or no empirical evidence to support or refute a "window" effect on root growth.

In the window technique a soil pit or trench is lined with transparent plastic or glass windows. After a period of time, roots growing along the face are measured and mapped. The size of the trench may vary from small (30–50 cm × 30–50 cm) to large (several meters). Larger trenches may be enclosed in a building and soil access maintained by removable windows. Examples of long-term rhizotrons include one at the East Malling Research Laboratory, UK, and another at the University of Michigan Biological Station in Pellston, MI, USA. Smaller trenches have been used at many sites, including several LTER sites (Coweeta, Hubbard Brook, Konza Prairie). An advantage of this method is the direct measurement of roots in relatively undisturbed soil. A disadvantage is the differential root growth across the glass plate. The two-dimensional distribution and architecture of large woody roots can be studied by digging a soil pit to expose roots, which are examined along a vertical plane (Vogt and Persson 1991; Bledsoe and Atkinson 1991). These methods are referred to as the "trench profile" method, the "profile wall" method, or the "glass wall" method (see Böhm 1979). All methods involve mapping roots on an exposed wall that can be covered with transparent glass or plastic sheets. These techniques are time-consuming and expensive in labor costs, but they produce extensive data on root distribution.

Root Screens

In many soils, the root coring technique is so time-consuming that it becomes prohibitively expensive for purposes of quantifying treatment effects in highly replicated experiments or spatial patterns of root distribution and abundance. A simple and inexpensive alternative approach for quantifying relative root growth and abundance is an in situ screen technique (Fahey and Hughes 1994). The number of fine roots growing through mesh screens inserted into the soil profile is assayed to provide an index of root biomass. The low cost and ease of sampling permit very large sample sizes. However, the process of placing a screen into rooted soil results in severing roots so that many roots growing through screens are sprouting from cut ends. In addition, the angle at which a screen intersects the soil will affect the calculated production or biomass. In summary, the in situ screen technique has not been evaluated widely enough to understand all its limitations or potential applications, and we do not recommend its use at this time.

Root Length Scanners

Most studies of fine root length measure length of fine roots harvested from soil cores. More elegant (and the most time-consuming) approaches rely on direct observations of intact, relatively undisturbed roots in soils (see earlier description of the wall profile method or a description of the minirhizotron method in Chapter 20). In general, measurements on harvested roots produce very acceptable estimates of root length. After soil cores have been processed to separate soil and organic matter from live roots, roots are washed and cut into short, uniform lengths (approx. 5–10 cm). The roots are then analyzed by one of several methods. The "scanner method" and the "line intercept method" are widely used; the "Xerox method" and "leaf-area-meter method" are less common. For all methods, roots must be displayed in a manner that minimizes overlap.

In the scanner method, fresh roots are placed in water in a tray; roots are scanned (video or optical scanner), and length and surface area are calculated from algorithms. For example, the Decagon image analysis system (Decagon Devices, Inc., Pullman, Washington, USA) assesses root lengths by the number of intersections between a series of horizontal scan lines and the root image on the monitor (Burke and LeBlanc 1988; Harris and Campbell 1989). There are several manufacturers of scanners, including Decagon Devices Inc., Ag-Vision's Monochrome Image Analysis System, and Regent Instrument's "WinRHIZO" System.

The line intercept method relies on counting the number of intersections of root segments with grid lines in a container, generally placed on a light table (Newman 1966). Then regressions are developed between the number of intersections and root length. This method can be easily calibrated using monofilament fishing line cut into segments of known length (K. Pregitzer, personal communication). In the Xerox method, roots are placed on a white background, minimizing overlap, and Xeroxed (C. Bledsoe, personal communication). The resulting black and white copy is scanned into a computer file, followed by use of a computer program to assay the image and calculate root length. Finally , a leaf area meter, commonly used to measure surface area of leaves, can be used to measure root surface area, followed by calculation of root length (C. Bledsoe, personal communication). The unit must be calibrated with "standard" wires of different diameters to determine the relationship between diameter and surface area. In calculating root length, one assumes that roots are a perfect cylinder.

Separation of Live and Dead Roots

The proportion of live to dead fine root tissues varies markedly with environmental conditions, time (season), and space (within plots, between watersheds, among biomes, etc.). Separation of live and dead roots is obviously important from a functional point of view, but the difficulty of sorting roots quantitatively into live and dead cannot be underestimated. Although some standardized criteria have been developed to distinguish live and dead roots on the basis of gross morphology, both the reproducibility and the absolute accuracy of these criteria are suspect. In addition, most soil cores contain large quantities of SOM fragments, which are difficult to separate from fine roots. Most root researchers work with fresh samples and hand-sort SOM from roots, relying on visual clues. After separation from the soil, roots should be separated from debris and organic matter. This is often performed with dissecting microscopes to remove any nonroot materials. It is a time-consuming process, and we suggest that a standard site-specific time period be allocated for each sample (e.g., 10–20 minutes). While this time will not be sufficient to remove all organic debris, it will allow removal of a majority. All SOM that is identifiable as roots should be so classified. Live and dead separations are then made on the fraction classified as roots. We recognize that often it is difficult to decide when a dead root is not a root but becomes part of the SOM fraction. When samples are fresh, separation of SOM from roots sometimes can be made on the basis of color, texture, bark, and sapwood characteristics.

We recommend that the proportion of live and dead tissues be quantified on carefully chosen subsamples and that this proportion be applied to total root biomass and length to estimate live root biomass. The subsample must be representative of the roots that are sorted from the cores, but of course it is important that they be handled so as to minimize death during the sorting procedure. We recommend that live and dead roots be separated based on visual clues, where live roots are generally lighter in color (not necessarily white), turgid, and not easily broken, and the cortex and periderm are not easily separated. Dead roots are generally brittle, dark brown or black, and shriveled or wrinkled.

If additional criteria are needed, vital stains may help distinguish between live and dead, although many researchers report mixed success and the use of vital stains does not appear to be widespread. Vital stains include tetrazolium chloride (Smith 1951), toluidine blue O, Congo red (Ward et al. 1978), and Evan's blue (Busso et al. 1989; Gaff and Okong' O-Ogola 1971). Joslin and Henderson (1984) used triphenyl-tetrazolium chloride with white oak (*Quercus alba*); both live and dead root extracts absorbed at 480 nm, although the slopes of the regressions were different. Shrub (*Artemisia*) and grass (*Agropyron*) roots could be distinguished by color and intensity of fluorescence after roots were extracted with 2 N NaOH (Caldwell et al. 1987). Use of molecular techniques has recently allowed the identification of genus and species for individual roots. Cullings (1992) developed primers to distinguish among roots of different plant species (ericaceous species, conifers, etc.) and mycorrhizal fungal species. Similarly, Rogers et al. (1989) used restriction enzyme ribosomal DNA hybridizations to identify conifer roots and their mycorrhizal fungi.

Chemistry of Coarse and Fine Roots

The analysis of the chemical composition of plant root material is critical for studies of nutrient turnover in ecosystems, the biotic pools of important biogeochemical elements, as well as for understanding nutrient and chemical limits to plant growth. The chemical analysis of plant materials is fairly straightforward, yet the analysis of plant roots presents several problems. Roots removed from soil need to be cleaned of soil contamination without unduly fragmenting roots or leaching water-soluble compounds. In addition, Fe and various other metals may form insoluble coatings on root surfaces, particularly in soils with fluctuating water saturation. These coatings are virtually impossible to remove, and thus interpretation of the chemical analysis of these elements from roots must include this pool of surface-bound, nonorganic material.

There are several published methods for organic matter digestion and chemical analysis, including both dry ashing and various wet digestion/ashing techniques. We do not discuss dry ashing techniques because, although ashing is quite fast and simple to perform, many researchers have noted the loss of elements from ashing due to volatilization, sorption on crucible surfaces, or actual loss of material during the ashing. Furthermore, ashing cannot be used for the analysis of volatile elements such as nitrogen or sulfur. The analysis of nitrogen and sulfur is most easily performed on a CHN or CNS analyzer (see Chapter 8, this volume). Wet ashing and analysis can be used if access to these analyzers is not possible. Wet digestion techniques vary in the acids and oxidants used to oxidize carbon and mineralize chemical constituents. We avoid perchloric acid procedures, since there are much safer alternatives to perchloric acid, which is extremely dangerous to use and requires specialized hoods. We recommend a wet digestion method where root materials are washed, preferably while still field-moist, dried, homogenized, digested, and then analyzed.

Coarse Woody Root Biomass, Length, and Diameter—Excavation

We recommend excavation of the tree root ball and of quantitative pits. Woody roots are difficult to study due to their large size and extreme spatial heterogeneity. High spatial variation in woody root biomass results from the high concentration of woody roots around the base of individual plant stems in forests and shrub lands; hence, random sampling alone is not an efficient method for measuring woody root biomass. Alternatively, root system excavation can provide direct estimates of woody root biomass on an individual plant basis, and allometric equations relating biomass to easily measured plant dimensions can be developed to convert to a stand or unit area basis. However, complete recovery of root systems is very difficult in many soils, especially for tree and shrub species with extensive root systems. Any protocol should combine two aspects—quantitative soil pits for random sampling of lateral root biomass, and the excavation of individuals, i.e., stem-based sampling of the entire plant (tree), including the root "crown."

By excluding a specified area around each stem from the area to be sampled with pits (the area associated with the allometrically defined root crown), one can obtain an accurate unbiased estimate of the total woody root biomass. In ecosystems without large trees or shrubs, it may be possible to sample all areas with pits and avoid excavations. In either case a direct estimate of wood root biomass by species will not be possible. However, an adequate degree of accuracy may be obtained on the basis of assumptions about the ratios of aboveground to belowground woody biomass. For most sites we recommend excavation of the tree root crown or ball followed by washing; quantitative pits are then dug to sample lateral coarse root distribution. For sites with high spatial heterogeneity, excavation is preferred due to its lower cost and shorter time per sample. For sites with low spatial heterogeneity, the "quantitative pit" method may be sufficient; this method requires less time and cost but still gives a reasonable estimate of biomass. If root crowns cannot be collected in the pits, however, some separate estimate of crowns is still needed.

Sampling design is very site-specific. Sample trees should be located on a slight slope, if possible, to facilitate drainage of water if used. Quantitative pits should be located randomly in an area adjacent to, but outside of, the excavated area. Each tree should be excavated from a cylindrical volume with a radius proportional to the diameter of the tree (Hanson 1981). Most large trees should be excavated from a volume with a radius of approximately 1–3 m. Skillful use of the backhoe to excavate and loosen the tree root crown will permit recovery of many lateral roots. In ecosystems with drought periods, it is useful to excavate during season(s) when soil is loose, moist, and easy to excavate.

Materials

1. Small backhoe
2. Hand tools (shovels, ax, digging trowel, etc.)
3. Metric measuring tape
4. Flagging tape
5. Large, coarse screen (preferably approx. 2 × 2 m screen with 1 cm diameter mesh)
6. Sampling bags, labels, marking pens, etc.

Procedure

1. Select a "typical" tree, estimate extent of canopy, and mark with flagging the circle to be excavated. Edge of canopy is a useful limit, unless tree canopy extent is quite large.

2. Measure tree height, and sever top from base, leaving a stump of approximately 0.5–1 m.
3. Using a backhoe, excavate around the tree base, following flagged circle. Dig a trench around the tree approximately 1–2 m deep. Use the backhoe bucket to loosen soil under tree stump.
4. Attach a chain to the stump, and pull stump with attached coarse roots from the ground with the backhoe.
5. Wash the stump and roots with a fire hose attached to a fire truck if available.
6. Subdivide root system into component parts. We recommend taking "slices" of the root system by depth (approx. 50 cm increments), then sorting roots into diameter classes of 1–2 cm, 2–5 cm, 5–10 cm, and >10 cm.
7. The length of larger roots may be measured directly, if desirable.
8. Dry the roots (approx. 55–70 °C for 24 hours or longer if necessary) and weigh. For large root systems, subsampling may be necessary.
9. Dig quantitative pits (approx. 1–3 m wide and 1–2 m deep) to sample lateral root biomass outside the excavated area. The size of pits depends on the estimated spatial heterogeneity of roots at a particular site.
10. Use a backhoe to scoop soil from quantitative pits and place soil on a large screen.
11. Shake the screen, allowing soil to fall through and leaving roots. Remove the roots, and bag and label for subsequent drying and weighing as described earlier.

Spatial Distribution of Coarse Woody Roots and Fine Roots—Trenches

The spatial distribution of larger coarse woody roots requires the use of large trenches, while the spatial distribution of fine roots can be studied with small windows. The methods are combined here because both can be accomplished with the same procedure, although one or the other can be applied alone where appropriate. Sampling design is site-specific and depends on the experimental treatments to be sampled. For example, in a vineyard, trenches were placed between the rows of grapes, one trench per variety of grape root stock (Moreno 1995). It is most efficient to have two persons—one in the pit to locate roots along the wall and to measure root diameters, and a second aboveground to record data. Root windows can be constructed of glass or Plexiglass, whose approximate size is 0.5 m × 0.5 m, or deeper if local soils have extensive roots extending below 0.5 m. It is generally sufficient to place four to eight root windows per site or per treatment.

Materials

1. Clear Plexiglass or glass for fine root windows (approx. 0.5 m × 0.5 m)
2. Lumber to provide support in trench or to support sides of small soil pits
3. Backhoe for deep trenches
4. Hand tools (shovels, trowels, etc.)
5. Sampling frame (approx. 1 × 1 m frame with cross wires) for trench
6. Metric ruler, notebook or computer for recording data

Procedure

1. Dig a trench (at least 1–1.5 m deep) or a smaller soil pit (fine roots). Carefully cut sides of trench or pit vertically. It is useful to avoid steep slopes in either case.
2. In some loose sandy soils, the sides of the trench may require support to avoid slumping. Care must be taken to protect the person in the trench from collapse of the trench walls.
3. A rigid grid system is placed against the trench wall, and locations and diameters of roots are mapped by the person in the trench and recorded by a second person.
4. For smaller pits, build a wooden support in the pit, leaving one or more sides open for viewing roots.
5. Insert clear window(s) into sides of box and against soil surface; secure window with wood trim or other devices.
6. It is often necessary to backfill between the window and the soil surface with additional soil, to avoid air gaps. Be certain to use soil from the same horizon if backfilling is required.
7. Place black plastic and insulated foam against the pit side of the window to eliminate light and reduce temperature extremes.
8. After a period when roots have grown along the window, map the location of roots by tracing roots with a marking pen. A grid (approx. 1 cm × 1 cm) on a piece of clear plastic can be placed on the window as a location aid. Alternatively, data may be recorded with a camera. For very fine roots, it is possible to mount a dissecting scope with light source against the glass.

Fine Root Biomass, Length, and Diameter—Soil Cores

Soil cores should be at least 5–7 cm in diameter. Smaller cores are not recommended because they are more subject to compaction, result in extensive severing of roots, and are less able to sample the spatial heterogeneity of roots in the soil. Heavy-duty cores for use in heavy or rocky soils may be purchased from the Giddings Machine Co. (Ft. Collins, CO, USA). Lightweight cores can be purchased from various sources (e.g., Forestry Supply Inc.) or constructed of metal or heavy plastic pipe. In heavy clay soils where extraction of the soil core is difficult, the core may be split, then clamped together. This will facilitate later removal of soil (Berman and Bledsoe 1998).

The number and placement of cores is site- and treatment-dependent. For example, sites with high spatial variability may require 10–15 cores per treatment or site to adequately sample the variability. Cores are driven into the soil to a depth sufficient to sample approximately 90% of the fine roots. For many species a depth of 50 cm is sufficient. Root biomass and length are generally measured directly on roots that have been separated by diameter class: tiny (<0.5 mm), fine (0.5–2 mm), small (2–5 mm), medium (5–10 mm), large (1–2 cm), and, if there are larger roots, very large (2–5 cm), and huge (>5 cm; modified from Böhm 1979, p. 131).

Materials

1. Soil corer (typically 5–7 cm diameter)
2. Devices (automobile jack, sleeve with handles, etc.) to remove cores from soil if working in heavy clay soils
3. In rocky soils, a stiff, small-diameter metal rod with a pointed end may be useful for locating a rock-free zone for inserting the corer.
4. Metric measuring tape
5. Sampling bags, labels, marking pens, etc.
6. Root hydropneumatic elutriation system (Gilson Variety Fabrication, Inc., 3033 Benzie Hwy., Benzonia, MI 49616; Smucker et al. 1982)

Procedure

1. Select a site for the core, using stiff rod (see earlier) if working in rocky soils.
2. Push or drive corer into soil, remove the corer, push soil core from corer, place in labeled bags in a cooler for transport to laboratory.
3. To avoid compaction of cores, it may be necessary to remove cores in increments (e.g., 20 cm each).
4. Store cores in a cold room (approx. 5–10 °C) until analysis. Most soil cores can be stored either frozen or cold for 1–6 months before analysis, unless mycorrhizal determinations are to be made. Then roots should be sampled within 2–6 weeks (see Chapter 18, this volume).
5. To separate roots from soil, rocks, and organic matter, gently wash soil cores either in an automated manner (e.g., root hydropneumatic elutriator; Smucker et al. 1982) or by hand using a series of screens and a gentle stream of water. The root elutriator works well for many types of soils, although with soils of high organic matter content, separation of roots from SOM can be a problem.
6. Working with a subsample of fresh roots (see the section "Separation of Live and Dead Roots," above), separate roots from organic matter, using visual characteristics. Further separate live from dead roots based on visual observations; live roots are generally lighter in color, turgid, and not friable. Dead roots are dark brown or black and easily broken; the root cortex can easily be separated from the conducting stele by pulling the root against a fingernail.
7. Roots may be sorted by diameter class most easily when fresh; see earlier for classes.
8. Root length is best determined on fresh roots, but dried material can be used. Root length is determined using one of several types of root length machines (see the section "Root Length," above).
9. After separation of SOM, live, and dead roots, dry roots for 24 hours at 55–70 °C and weigh.
10. A subsample should be ashed to determine ash-free dry weight. Then all samples should be corrected to eliminate weight due to contamination by soil particles.
11. Root data should be reported as g/m^2 of core area for each given depth.

Root Chemistry—Digestion for Macroelement Analysis

Block digestion using this Kjeldahl technique is appropriate for nonsulphur macronutrient analyses and necessary for nitrogen analyses; the microwave or hot plate technique (described later) is more appropriate for trace-element analysis. Roots may be digested fresh or dried. However, dried roots must be carefully cleaned before drying (see later). As with any sulfuric acid digest, $CaSO_4$ may precipitate; this digest should not be used to analyze for Ca. Since insoluble $CaSO_4$ precipitates may form in the glass tubes after repeated use, these tubes should never be used for Ca analysis. See also Chapter 8, this volume, for an equivalent technique for other woody tissue.

Materials

1. Block digester (20 or 40 positions) with tube rack
2. Volumetric glass digestion tubes calibrated to 75 mL (other volumes may be appropriate depending on digestor manufacturer)
3. Concentrated reagent-grade sulfuric acid
4. Reagent-grade (30%) hydrogen peroxide

Procedure

1. Wash roots prior to digestion by gently swirling in a phosphate-free dilute detergent solution in a clean (acid-washed) glass or polyethylene container.
2. Swirl roots in a 0.01 mol/L NaEDTA solution for 5 minutes to complex surface-bound cations, including metals.
3. Rinse roots in distilled deionized water and place immediately in paper bags in oven; dry at 70 °C for 24 hours.
4. Grind dry roots sufficiently to pass a 1 mm sieve, either by hand grinding with a mortar and pestle (critical if trace element analysis is performed) or with a Wiley mill to pass a 20-mesh screen.
5. Store dried and ground root material in a desiccator or a warm oven prior to analysis to avoid problems with weight change due to rehydration.
6. Weigh 250 mg dried, ground roots into acid-washed digestion tubes, placing material at bottom of tubes. Mass deos not need to be exact ±50 mg) so long as it is recorded and a sample-specific mass is used in the following calculation.
7. Add 5 mL concentrated H_2SO_4 to each tube, swirling the wet root material to wash down any powders from the sides.
8. Add 2 mL H_2O_2 very carefully to each tube, swirling constantly to reduce the vigorous boiling that will ensue.
9. When all tubes have finished boiling (1–5 minutes), place rack in the block preheated to no more than 170 °C; digest for 1 hour. Temperature must reach (but not exceed) 230 °C; if this temperature is reached before the hour, turn off the heater.
10. Remove the rack from the block, turn off heater, and allow tubes to cool.

11. When tubes are cool to touch, add 2 mL H_2O_2 to each tube again.
12. When the block has cooled to 175 °C or below, turn on heater, place rack with tubes in the block, and digest for an additional 2 hours, taking care that the final temperature does not exceed 350 °C. At least 1 hour of this final digest should be at 330–350 °C to ensure complete removal of the H_2O_2, which, if present, interferes with phosphorus analysis.
13. After tubes are cool, volumes are made up to mark with reagent-grade deionized water.
14. Analysis of nitrogen, phosphorus, potassium, magnesium, and some trace elements can be performed from this digest (see Chapters 5 and 8, this volume).

Calculations

$$\text{mg element/g root} = C \times F_{digestion}$$

where

C = concentration of element in the digestion solution as mg element/L solution.
$F_{digestion}$ = digestion fraction for tissue, as L of digestion solution/g dry root, e.g., 0.075 L solution/0.25 g root

Special Considerations

A standard plant sample (e.g., from the National Institute of Standards and Technology [NIST]) and at least two sample replicates should be brought through the digestion procedure with every batch of 40 samples. In addition, blank matrix material for the automated analysis should be made up by following the preceding procedure but without adding root tissue to the digestion tubes.

Root Chemistry—Digestion for Trace Element Analysis

Microwave or hot plate digestion of plant materials with HNO_3 is the most commonly used procedure for trace element analysis. Nitric acid is a more powerful oxidant than sulfuric acid, and thus peroxide is rarely needed. Obviously, nitrogen cannot be determined in these digests. This technique is particularly useful if automated microwave digestion equipment is available. If not, commercially available microwave ovens may be used with Parr bombs that are available from Cole-Parmer. The microwave procedure has the advantage of being the easiest digestion procedure, although the resulting digestate is highly concentrated in acid, which may cause problems for analysis if dilution is not possible. Hot plate digestion requires little technology but more operator time. In all cases, samples can be brought to dryness and redissolved in weak HNO_3 to avoid excess acid problems.

If using an automated microwave digestion system, follow the instructions included with the machine. If microwave digestion is done manually, the technique is equally simple. Specific methods for Parr bomb digestion depend on the size of bomb purchased; directions are included with the specific bombs.

Materials for Hot Plate Digestion

1. Closed Teflon vials, 20 mL size (available from Cole-Parmer)
2. Standard hot plate; with large hot plates, 20 samples may be digested at one time
3. Reagent-grade HNO_3

Procedure for Hot Plate Digestion

1. Add 200 mg root tissue to vials, 2 mL HNO_3, and close vials with lids. Root mass does not need to be exact ($\pm$50 mg) so long as it is recorded and a sample-specific mass is used in the following calculation.
2. Under a fume hood place vials on hot plate at approximately 120 °C for 1 hour or until HNO_3 reflux (condensation on lids) appears.
3. Remove lids and lower hot plate temperature to 75–100 °C until samples are evaporated to dryness. Take care to avoid charring the dried residue.
4. After drying, bring residues to volume (20 mL) with 3% HNO_3.
5. Analysis of trace elements and nonnitrogen macroelements can be performed from this digest (see Chapters 5 and 8, this volume).

Calculations

Calculations are the same as for macroelement analysis (see section "Root Chemistry—Digestion for Macroelement Analysis, above).

Special Considerations

If trace elements are to be determined, ultrapure HNO_3 must be used, and Teflon vials should be cleaned with boiling aqua regia between uses. Standard plant materials (e.g., from NIST) should be brought through the entire digestion procedure with every batch of 20 samples, along with one replicate and at least one blank. Although some researchers have added several drops of H_2O_2 to the digestion initially, this is generally more useful for recalcitrant materials (e.g., animal tissues with high lipid contents) than for plant materials. The analysis of standard materials should indicate whether recalcitrance is a problem.

Acknowledgments Work on this chapter was supported, in part, by the LTER Network Office Coordination Grant and other NSF Grants. We thank Robert Aiken, David Coleman, Kurt Pregitzer, Phillip Sollins, David Wedin, and two anonymous reviewers for their comments on this manuscript.

References

Anderson, J. M., and J. S. I. Ingram. 1993. Field procedures—Roots. Pages 22–46 *in* J. M. Anderson and J. S. I. Ingram, editors, *Tropical Soil Biology and Fertility: A Handbook of Methods.* CAB International, Oxford, UK.

Baskerville, G. L. 1966. Dry matter production in mature balsam fir stands: roots, lesser vegetation and total stand. *Forest Science* 12:49–53.

Berman, J. T., and Bledsoe, C. S. 1998. Soil transfers from valley oak (*Quercus lobata* Nee) stands increase ectomycorrhizal diversity and alter root and shoot growth on valley oak seedlings. *Mycorrhiza.* 7:223–235.

Bledsoe, C. S., and D. Atkinson. 1991. Measuring nutrient uptake by tree roots. Pages 207–224 *in* J. P. Lassoie and T. M. Hinckley, editors, *Techniques and Approaches in Forest Tree Ecophysiology.* CRC Press, Boston, Massachusetts, USA.

Bledsoe, C. S., and J. T. Hastings. 1996. Pages 1–16 *in US Long-Term Ecological Research Program, Report of the X-Roots Climate Mini-Workshop,* May 1996 [on-line]. Available: http:\\lternet.edu/research/im/xroots/aclim.htm(1999 March).

Böhm, W. 1979. *Methods of Studying Root Systems.* Springer-Verlag, Berlin, Germany.

Burke, M. K., and D. C. LeBlanc. 1988. Rapid measurement of fine root length using photo electronic image analysis. *Ecology* 69:1286–1289.

Busso, C. A., R. J. Mueller, and J. H. Richards. 1989. Effects of drought and defoliation on bud viability in two caespitose grasses. *Annals of Botany* 63:477–485.

Caldwell, M. M., J. H. Richards, J. H. Manwaring, and D. M. Eissenstat. 1987. Rapid shifts in phosphate acquisition show direct competition between neighboring plants. *Nature* 327:615–616.

Canadell, J., R. B. Jackson, J. R. Ehleringer, H. A. Mooney, O. E. Sala, and E. D. Schulze. 1996. Maximum rooting depth of vegetation types at the global scale. *Oecologia* 108:583–595.

Canadell, J., and F. Roda. 1991. Root biomass of *Quercus ilex* in a montane Mediterranean forest. *Canadian Journal of Forest Research* 21:1771–1778.

Chinn, H., and C. Bledsoe. 1997. Internet access to ecological information—the US LTER All-Site Bibliography Project. *BioScience* 47:50–57.

Cullings, K. W. 1992. Design and testing of a plant-specific PCR primer for ecological and evolutionary studies. *Molecular Ecology* 1:233–240.

Deans, J. D. 1981. Dynamics of coarse root production in a young plantation of *Picea sitchensis. Forestry* 54:139–155.

Fahey, T. J., and J. W. Hughes. 1994. Fine root dynamics in a northern hardwood forest ecosystem, Hubbard Brook Experimental Forest, New Hampshire. *Journal of Ecology* 82:533–548.

Fahey, T. J., J. W. Hughes, P. Mou, and M. A. Arthur. 1988. Root decomposition and nutrient flux following whole-tree harvest of northern hardwood forest. *Forest Science* 34:744–768.

Federal Geographic Data Committee. 1994. Content standard for non-geospatial meta data. *National Biological Survey Report,* November 1994.

Fitter, A. H. 1994. Architecture and biomass allocation as components of the plastic response of root systems to soil heterogeneity. Pages 305–323 *in* M. M. Caldwell and R. W. Pearcy, editors, *Exploitation of Environmental Heterogeneity by Plants.* Academic Press, New York, USA.

Gaff, D. F., and O. Okong'O-Ogola. 1971. The use of non-permeating pigments for testing the survival of cells. *Journal of Experimental Botany* 22:756–758.

Gross, K. L., C. E. Pake, E. Allen, C. Bledsoe, R. Colwell, P. Dayton, M. Dethier, J. Helly, R. Holt, N. Morin, W. Michener, S. T. A. Pickett, and S. Stafford. 1995. *FLED, Final Report of the Ecological Society of America Committee on the Future of Long-Term Ecological Data.* Ecological Society of America, Washington, DC, USA.

Hanson, E. A. 1981. Root length in young hybrid *Populus* plantations: its implication for border width of research plots. *Forest Science* 27:808–814.

Harris, G. A., and G. S. Campbell. 1989. Automated quantification of roots using a simple image analyzer. *Agronomy Journal* 81:935–938.

Heeraman, D. A., and N. G. Juma. 1993. A comparison of minirhizotron, core and monolith methods for quantifying barley (*Hordeum vulgare* L.) and fababean (*Vicia faba* L.) root distribution. *Plant and Soil* 148:29–41.

Henderson, R., E. D. Ford, E. Renshaw, and J. D. Deans. 1983. Morphology of the structural root system of Sitka spruce. 1. Analysis and quantitative description. *Forestry* 56:121–135.

Hendrick, R. L., and K. S. Pregitzer. 1993. The dynamics of fine root length, biomass and nitrogen content in two northern hardwood ecosystems. *Canadian Journal of Forest Research* 12:2507–2520.

Hendrick, R. L., and K. S. Pregitzer. 1996. Temporal and depth-related patterns of fine root dynamics in northern hardwood forests. *Journal of Ecology* 84:167–176.

Hermann, R. K. 1977. Growth and production of tree roots: a review. Pages 7–28 *in* J. K. Marshall, editor, *The Belowground Ecosystem: A Synthesis of Plant-Associated Processes.* Range Science Department Science Series, no. 26. Colorado State University, Fort Collins, Colorado, USA.

Jackson, R. B., J. Canadell, J. R. Ehleringer, H. A. Mooney, O. E. Sala, and E. D. Schulze. 1996. A global analysis of root distributions for terrestrial biomes. *Oecologia* 108:389–411.

Jackson, R. B., H. A. Mooney, and E. D. Schulze. 1997. A global budget for fine root biomass, surface area, and nutrient contents. *Proceedings of the National Academy of Sciences USA* 94:7362–7366.

Joslin, J. D., and G. S. Henderson. 1984. The determination of percentages of living tissue in woody fine root samples using triphenyltetrazolium chloride. *Forest Science* 30:965–970.

Kira, T., and H. Ogawa. 1968. Indirect estimation of root biomass increment in trees. Pages 96–101 *in* M. S. Ghilarov, V. A. Kovda, L. N. Novichkova-Ivanova, L. E. Rodin, and V. M. Sveshnikova, editors, *Methods of Productivity Studies in Root Systems and Rhizosphere Organisms.* USSR Academy of Sciences. Leningrad Publishing House, Nauka, USSR.

Kummerow, J., and R. Mangan. 1981. Root systems in *Quercus dumosa* Nutt. dominated chaparral in southern California. *Acta Oecologia* 2:177–188.

LTERnet. No date a. Gopher menu for the Core Dataset Catalog. *In U.S Long-Term Ecological Research Program (LTER)* [Online]. Available: gopher://lternet.washington.edu:70/11/catalog(1999 March).

LTERnet. No date b. Guide to current availability, through Gopher and the World Wide Web, of documented datasets (i.e., data + metadata) across the LTER Network. *In U.S Long-Term Ecological Research Program (LTER)* [Online]. Available: http://lternet.edu/about/research/data/(1999 March).

LTERnet. No date c. Guide to the LTER Network. *In U.S Long-Term Ecological Research Program (LTER)* [Online]. Available: http://lternet.edu/(1999 March).

Lyford, W. H. 1980. *Development of the Root System of Northern Red Oak* (Quercus rubra

L.). Harvard Forest Paper, no. 21. Harvard University, Harvard Forest, Petersham, Massachusetts, USA.

Mackie-Dawson, L.A., and D. Atkinson. 1991. Methodology for the study of roots in field experiments and the interpretation of results. Pages 25–47 *in* D. Atkinson, editor, *Plant Root Growth: An Ecological Perspective.* Special Publication No. 10. British Ecological Society, Blackwell Scientific Publishers, London, UK.

Michener, W. K, J. W. Brunt, J. J. Helly, T. B. Kirchner, and S. G. Stafford. 1997. Non-geospatial metadata for the ecological sciences. *Ecological Applications* 7:330–342.

Millikin, C. S., and C. S. Bledsoe. 1997. Woody root biomass of 40- to 90-year-old blue oaks (*Quercus douglasii*) in western Sierra Nevada foothills. Pages 83–89 *in* N. H. Pillsbury, J. Verner, and W. D. Tietje, editors, *Proceedings of a Symposium on Oak Woodlands: Ecology, Management, and Urban Interface Issues.* USDA Forest Service, General Technical Report PSW-GTR-160, Berkeley, California, USA.

Moreno, Lisa Dean. 1995. An analysis of *Vitus* species in rootstock crosses: vineyard root distribution patterns, root growth and metabolic responses to flooding in the greenhouse and associated soil and plant communities of wild vines. Ph.D dissertation. University of California, Davis.

Nadelhoffer, K. J., J. D. Aber, and J. M. Melillo. 1985. Fine roots, net primary production, and nitrogen availability: a new hypothesis. *Ecology* 66:1377–1390.

Newman, E. I. 1966. A method of estimating the total length of root in a sample. *Journal of Applied Ecology* 3:139–145.

Ovington, J. D., W. G. Forrest, and J. S. Armstrong. 1967. Tree biomass estimation. Pages 4–31 *in Symposium on Primary Productivity and Mineral Cycling in Natural Ecosystems.* University of Maine Press, Orono, Maine, USA.

Parton, W. J., D. S. Schimel, C. V. Cole, and D. S. Ojima. 1987. Analysis of factors controlling soil organic matter levels in Great Plains grasslands. *Soil Science Society of America Journal* 51:1173–1179.

Raich, J. W., and K. J. Nadelhoffer. 1989. Belowground carbon allocation in forest ecosystems: global trends. *Ecology* 70:1346–1354.

Rogers, S. O., S. Rehner, C. Bledsoe, G. Mueller, and J. F. Ammirati. 1989. Extraction of DNA from basidiomycetes for ribosomal DNA hybridizations. *Canadian Journal of Botany* 67:1235–1243.

Rutherford, M. C. 1983. Growth rates, biomass and distribution of selected woody plant roots in *Burkea africana–Ochna pulchra* savannah. *Vegetation* 54:45–63.

Santantonio, D., and J. C. Grace. 1987. Estimating fine root production and turnover from biomass and decomposition data: a compartment flow model. *Canadian Journal of Forest Research* 17:900908.

Santantonio, D., R. K. Hermann, and W. S. Overton. 1977. Root biomass studies in forest ecosystems. *Pedobiologia* 17:1–31.

Smith, F. E. 1951. Tetrazolium salt. *Science* 113:751–754.

Smucker, A. J. M., S. L. McBurney, and A. K. Srivastava. 1982. Quantitative separation of roots from compacted soil profiles by the hydropneumatic elutriation system. *Agronomy Journal* 74:500–503.

Stout, B. B. 1956. *Studies of the Root Systems of Deciduous Trees.* Black Rock Forest Bulletin *No. 15.* Harvard University, Cambridge, Massachusetts, USA.

Taylor, H. M. 1987. *Minirhizotron Observation Tubes: Methods and Applications for Measuring Rhizosphere Dynamics. ASA Special Publication, 50.* Madison, Wisconsin, USA.

van Noordwijk, M. 1993. Roots: length, biomass, production and mortality. Pages 132–144 *in* J. M. Anderson and J. S. I. Ingram, editors, *Tropical Soil Biology and Fertility: A Handbook of Methods.* CAB International, Oxford, UK.

Vogt, K. A., and H. Persson. 1991. Measuring growth and development of roots. Pages 477–501 *in* J. P. Lassoie and T. M. Hinckley, editors, *Techniques and Approaches in Forest Tree Ecophysiology.* CRC Press, Boca Raton, Florida, USA.

Vogt, K. A., D. J. Vogt, E. E. Moore, W. Littke, C. C. Grier, and C. Leney. 1985. Estimating Douglas fir fine root biomass and production from living bark and starch. *Canadian Journal of Forest Research* 15:177–179.

Vogt, K. A., D. J. Vogt, P. A. Palmiotto, P. Boon, J. O'Hara, and H. Asbjornsen. 1996. Review of root dynamics in forest ecosystems grouped by climate, climatic forest type and species. *Plant and Soil* 187:159–219.

Ward, K. J., B. Klepper, R. W. Rickman, and R. R. Allmaras. 1978. Quantitative estimation of living wheat-root lengths in soil cores. *Agronomy Journal* 70:675–677.

Westman, W. E., and R. W. Rogers 1977. Biomass and structure of a subtropical eucalypt forest, North Stradbroke Island. *Australian Journal of Botany* 25:171–191.

Whittaker, R. H., F. H. Bormann, G. E. Likens, and T. G. Siccama. 1974. The Hubbard Brook ecosystem study: forest biomass and production. *Ecological Monographs* 44:233–254.

Young, H. E., and P. M Carpenter. 1967. Weight, nutrient element and productivity studies of seedlings and saplings of eight tree species in natural ecosystems. *University of Maine Agricultural Experiment Station Technical Bulletin 28.* University of Maine, Orono, Maine, USA.

20

Fine Root Production and Demography

Timothy J. Fahey
Caroline S. Bledsoe
Frank P. Day
Roger W. Ruess
Alvin J. M. Smucker

Attributes of root systems appear to have resulted from pressures on plants to maximize the efficiency of soil resource acquisition and to minimize the risk of injury or death at environmental or competitive extremes (Yanai et al. 1994). These attributes include the morphology, physiology, demographics, distribution, timing, and the pattern of growth and longevity of both fine and coarse roots. In turn, these attributes of root systems strongly influence emergent properties of terrestrial ecosystems, including energy flow and biogeochemical cycles. Standard techniques to study the dynamics of fine roots should facilitate comparisons across sites and through time that could provide valuable insights into controls on the composition, structure, and function of terrestrial ecosystems.

Unfortunately, severe difficulties are encountered in studying roots in the soil environment. Despite a recent proliferation of efforts to quantify fine root dynamics in natural and agroecosystems, including a number of methodological comparisons, a solid objective basis for recommending a standard approach does not at present exist. Common techniques suffer from insufficient testing, unproven assumptions, or methodological difficulties in standardization. Nevertheless, recent research has considerably improved our basis for judging the adequacy of existing approaches in different ecosystem settings. In this chapter we briefly review five widely used protocols for studying fine root production (FRP); we summarize comparisons among these protocols and their strengths and limitations; and we describe in detail the applications and recent experience using the minirhizotron technique (Taylor 1987), with recommendations for further research to improve the basis for validating this promising approach.

Available Protocols

Except in the case where vegetation is reorganizing following severe disturbance, to estimate root production researchers must now assume that fine root systems are near steady state, so that on an annual basis the production of new roots is nearly balanced by the mortality of older roots. Annual FRP ($g/m^2/yr$) is the sum of the growth of existing roots and the initiation and extension of new roots, and under the steady-state assumption the median longevity of root tissues is the inverse of the turnover rate of roots (yr^{-1}). The dynamics of live fine root biomass depend on short-term (within-year) deviations from steady state, and many attempts have been made to use these deviations to assess fine root production (Vogt and Persson 1991). In the longer term (i.e., beyond a few years), departures from steady state may be detected as changes in fine root biomass, but the problems of spatial and temporal variation (see Chapter 19, this volume) will usually preclude detection of any such short-term departures. Hence, net fine root biomass production usually is assumed to be equivalent to fine root mortality, and either of these could be estimated as the ratio between fine root biomass (temporally averaged) and the median longevity of all fine roots.

For both conceptual and practical purposes it is important to distinguish among total carbon allocation to roots by plants, total belowground net primary production (BNPP), and gross or net changes in root biomass and length. Many methods are available to measure only the latter; however, to the extent that actual belowground production may "disappear" as root exudates and rhizodeposits, sloughed tissue, flux to extraradical mycorrhizal hyphae, and herbivory, methods in which only roots are directly measured will underestimate BNPP. Unfortunately, the magnitudes of fluxes associated with these disappearances of root C are notoriously difficult to quantify. Total root allocation (*TRA*) of carbon also includes respiratory losses associated with ion uptake and cellular maintenance (R_r), so that

$$\text{BNPP} = TRA - R_r$$

Finally, each of these terms also must take into account carbon allocation to woody roots or long-lived perennial parts of the root system. Woody root production is usually assumed to be proportional to aboveground production of woody tissues (Whittaker et al. 1974); thus, the standard approach for measuring woody root biomass (see Chapter 19, this volume) serves as the basis for estimating woody root production. The remainder of this chapter is concerned only with fine root production.

On shorter time scales (i.e., within-year), two additional aspects of fine root dynamics are so important to root system structure and function that standardized methodology would be valuable. First, root growth and death typically occur in pulses in many ecosystems, and consequent seasonal changes in root biomass contribute both to root system performance and to carbon and nutrient fluxes. Thus, standard techniques to quantify root phenology are needed. Second, the length, surface area, and physiological attributes of fine roots vary markedly with root diameter (Yanai et al. 1994) and with other morphological traits (e.g., suberized versus nonsuberized roots; mycorrhizal versus nonmycorrhizal), so that seasonal patterns in the demographics of fine root systems (Hendrick and Pregitzer 1992) must be quantified if an improved understanding of root function is to be realized. Hence, we emphasize the advantage of a technique for fine root measurement that concurrently quantifies production and demography.

Because it is essential for the calculation of energy and nutrient budgets, FRP has long been recognized as a crucial (but hard to measure) parameter, and many approaches have been devised and employed. In the past, the two most common techniques have been sequential coring and root ingrowth cores. As we describe later, despite their widespread usage, both of these techniques suffer from serious drawbacks in terms of standardization, error, bias, and labor requirements. Also, neither of these methods allows concurrent quantification of fine root demography. Their principal advantage is that they do not require any expensive instrumentation.

Other means for measuring FRP include direct measurements of major ecosystem carbon or nitrogen fluxes and the back-calculation of root production to balance C or N budgets. In herbaceous vegetation (e.g., grasslands, agroecosystems), isotopic labeling approaches using ^{14}C may be capable of providing accurate root production estimates. Finally, root observation windows (rhizotrons) have been used in concert with static measures of fine root length and biomass to calculate root production.

Sequential Coring Methods

Detailed descriptions of sequential coring methods for estimating FRP are available (Vogt and Persson 1991). In principle, these techniques rely on periodic changes in live and dead fine root biomass to calculate root production. For example, if root growth occurs in a distinct and brief seasonal pulse, then root production can be estimated from the difference between annual minimum and maximum biomass (Aber et al. 1985). An important assumption of such a simple sequential coring approach is that coincident growth and death/decay/herbivory during the interval between core samplings are minor, an assumption that usually is violated (Burke and Raynal 1994).

An improved approach to the calculation of FRP using sequential coring relies on a compartmental flow model that accounts for coincident growth and decomposition (Santantonio and Grace 1987). In addition to precise estimates of periodic changes in live and dead root biomass, the compartment flow model requires information on the in situ decay rate of roots (and its dependence on temperature in temperate environments). Publicover and Vogt (1993) analyzed the sensitivity of FRP estimates obtained with this model to a variety of measurement errors and concluded that a principal problem was obtaining reliable information on in situ fine root decomposition. Direct measurements of fine root decay using trenched plots and litter bag incubations appear not to provide reliable information relevant to the in situ decay of individual roots in the mycorrhizosphere (Gholz et al. 1986; Fahey and Arthur 1994; Hendrick and Pregitzer 1996); hence, until improved methods for fine root decay are available, the accuracy of the sequential coring/compartment flow model approach will be questionable.

A variety of other problems and limitations of sequential coring methods have also been noted. First, in some soils and soil horizons, especially organic ones, the separation of roots from soil can be very difficult. For most mineral soil horizons, however, automated methods (e.g., hydropneumatic elutriation; Smucker et al. 1982) have greatly increased the efficiency of root-soil separation (Box 1996). Second, depending on the statistical criteria for judging the significance of temporal changes in fine root biomass, estimates of root production by sequential coring can vary markedly for the same data set (Singh et al. 1984; Kurz and Kimmins 1987). This problem will prove particularly vexing for any attempt at standardization across ecosystems because of marked differences in spatial and temporal variability of root biomass and production among ecosystems. Third, in most cases sequential coring techniques require the separation of live and dead roots (Vogt and Persson 1991). Although vital stains may be capable of providing objective information on fine root vitality, these are impractical for routine separations, and normally other visual criteria are required (Publicover and Vogt 1993). Unfortunately, these criteria are not always reliable or repeatable. For example, when Gholz et al. (1986) sorted roots of *Pinus elliottii* into three classes (live, dead, and unknown), the unknown category was sometimes as large a proportion of total roots as was the live roots category. Moreover, observations with minirhizotrons have revealed resumption of growth even of extremely moribund roots (Hendrick and Pregitzer 1996).

Ingrowth Cores

Soil ingrowth cores have been widely used to estimate FRP in perennial vegetation, especially in shallow or stony soils (Vogt and Persson 1991). The great advantage of the ingrowth core technique is its simplicity and low cost. However, there are at least four reasons to expect artifacts when the technique is applied in intact natural ecosystems: (1) all the roots in the plane of the core are cut so that many of the ingrowing roots are growing from damaged tips; (2) the preparation of soil for placement into the cores usually will alter resource availability and soil structure; (3) the method essentially measures the rate of root colonization of a large soil volume, much different from root growth in a fully colonized soil profile; and (4) as with sequential cores, concurrent growth and mortality during the recolonization interval cannot be measured directly. Also, ingrowth cores do not readily provide desired information on fine root demography.

From the point of view of standardization, there may be important differences among sites in the magnitude and direction that each of these potential sources of error influences estimates of root production, but no quantitative basis currently exists to judge this problem. Among the most prominent considerations that must be confronted are decisions on the duration of incubation and timing of ingrowth core placement in the field. In principle, a combination of long-term (seasonal to annual) and overlapping short-term (e.g., monthly) incubations could allow estimation of root turnover during sampling intervals (Steen 1985). However, Neill (1992) indicated that short-term bags greatly underestimated production in a prairie marsh, presumably because of the disturbance to root systems associated with placement of in-growth cores.

Carbon and Nitrogen Budgeting Approaches

Many of the major carbon and nitrogen fluxes in ecosystems can be measured accurately using standardized approaches, including fluxes in soils. If the assumption that the large soil pools of C or N are near steady state can be made, then in many ecosystems it may be possible to obtain accurate estimates of FRP and total C and N allocation to roots by closing the ecosystem budgets for these elements (Nadelhoffer et al. 1985; Raich and Nadelhoffer 1989). For example, Haynes and Gower (1995) compared FRP between control and fertilized pine plantations by measuring aboveground detritus production (P_a), total C emission from soil (R_s), and root respiration (R_r), under the assumption that the soil organic matter pool was constant, i.e.,

$$R_s - P_a = \text{FRP} + R_r$$

Possibly the most vexing problem with this promising approach is obtaining accurate measurement of in situ root respiration (Vogt et al. 1989). Root respiration has been measured both directly by gas exchange methods on excised roots (Bloom and Caldwell 1988; Fahey and Hughes 1994; Kelting et al. 1995; Burton et al. 1996) and as the difference in total soil respiration between control and trenched plots (Ewel et al. 1987; Bowden et al. 1993). The reliability of these estimates of R_r has not been adequately evaluated.

A conceptually similar approach that relies on closing the ecosystem N budget yielded reasonable values for root production in several temperate forests (Aber et al. 1985; Fahey et al. 1985; Nadelhoffer et al. 1985). If gaseous and leaching losses of N are small or can be measured accurately, and reliable estimates of soil N mineralization are available, then the N budget approach may provide good FRP values (Nadelhoffer and Raich 1992). In most cases the crucial measurement here is net mineralization of soil N (see Chapter 13, this volume). Although further testing and methods comparison are warranted, these budgetary approaches appear to provide some of the most reliable estimates of BNPP, and they may be particularly useful to verify estimates obtained by more direct methods and to evaluate treatment responses in field experiments (Haynes and Gower 1995; Kelting et al. 1995).

Isotopic Labeling Methods

A variety of approaches that rely on measurements of the fate of the radioisotope ^{14}C have been devised to estimate FRP and turnover (Caldwell and Camp 1974; Milchunas et al. 1985; Milchunas and Lauenroth 1992; Swimmen et al. 1994, 1995). Because it is necessary to distribute the ^{14}C label relatively uniformly through the plant root system, these techniques have been employed only in herbaceous vegetation. In principle, the simplest of these techniques for estimating only FRP is the

^{14}C turnover method, in which estimates of the turnover of carbon in the root system are obtained by regressing the ^{14}C content of the root system over time; FRP is then obtained as the quotient of mean root biomass and the time for 100% turnover of root carbon (Milchunas and Lauenroth 1992). Although a ^{14}C dilution approach could in principle provide estimates of FRP, Milchunas and Lauenroth (1992) questioned its efficacy in the complex soils that characterize most natural ecosystems. Other approaches that examine the fate of a short-term, early season ^{14}C label (e.g., Swimmen et al. 1995) may provide more detailed information on energy flow in the soil food web, but have not yet been adapted for perennial vegetation. At present, a standardized, cross-site approach to root production utilizing isotopic labeling cannot be recommended because of difficulties in application to forest and shrubland ecosystems; however, additional research toward the development of novel approaches utilizing radioactive or stable isotopes is warranted.

Minirhizotrons

Root observation windows or rhizotrons have long been used to observe the in situ growth of plant roots (Head 1966). Rhizotrons permit repeated, nondestructive observation of individual roots—their growth, morphological changes, and longevity. However, profile-wall rhizotrons so markedly influence the soil profile and especially the patterns of root growth (e.g., the measured roots grow mostly along the window surface) that Vogt and Persson (1991) concluded rhizotrons cannot be used to measure root production or growth on a stand level. The development of minirhizotrons for viewing roots through small-diameter soil access tubes may largely correct this problem of profile-wall rhizotrons because only a small fraction of measured root length is in contact with the rhizotron surface, and minimal disturbance to the soil profile occurs during rhizotron installation.

The minirhizotron technique provides opportunities for repeated, nondestructive observation of individual roots, soil pores, mesofauna, nodules, and other associated environments. Transparent cylindrical access tubes are installed in the soil (Box 1996; Taylor 1987), and a miniature video camera (Upchurch and Ritchie 1983) is used to photograph permanent quadrats positioned along the surface of each access tube. Various characteristics of each of the roots captured in the photographic images are recorded over a sequence of sample dates, so that the "birth," subsequent growth, morphological changes, and ultimately the death of individual roots can be observed and quantified manually or by image analysis (Smucker et al. 1987; Smucker 1993). As explained later, this information can be used to calculate FRP, phenology, and demography.

The most serious limitation to the adoption of the minirhizotron technique is the high initial cost of hardware and software. Although labor costs for field installation and sampling are moderate, information management can be time-consuming. Efficient systems for developing a database and automated digitization and image analysis are essential to avoid data overload.

In principle, this technique should prove a reliable and standardized approach for measuring FRP, but only limited comparative testing of the minirhizotron approach against other methods has been conducted (see Aerts et al. 1989; Majdi et al. 1992; Hansson et al. 1995; Majdi 1996; Swimmen et al. 1995). The most crucial assumption that must be tested when estimating FRP via minirhizotrons is that root longevity (or root length growth and mortality) is not affected by the minirhizotron access tube. If this assumption is violated, then the accuracy of the minirhizotron approach is questionable. In comparison with profile wall rhizotrons, the minirhizotron should have much less influence on root behavior because of the minimal extent of root contact with the tubes.

Summary of Comparisons of Root Production Methodologies: Recommended Procedures

Although several comparisons of root production methodologies have been conducted in a variety of natural and managed ecosystems, these do not provide a sufficient quantitative basis for concluding that one or another method is the most reliable under the broadest suite of soil and vegetation types. This situation is partly a result of the numerous uncertainties and assumptions that are associated with each method and with the high cost of comprehensive methodological comparisons. However, the results to date do provide some indications of known limitations of some of the techniques in some situations and point towards fruitful areas for future FRP comparisons.

Nadelhoffer and Raich (1992) evaluated literature estimates of aboveground net primary production (ANPP), BNPP, and TRA for a variety of forests to provide insights into the probable errors and biases of several commonly used methods (sequential and maximum-minimum coring, ingrowth cores, and N budgeting). They concluded that sequential coring has often overestimated FRP and that errors associated with the technique are often large. Also, they contended that maximum-minimum coring usually underestimates FRP, often by a large amount. These conclusions echo previous analyses (Singh et al. 1984; Kurz and Kimmins 1987) and raise serious questions about any methodological conclusions that can be derived from comparisons among soil coring methods and the other approaches being evaluated.

Nadelhoffer and Raich (1992) regarded ingrowth core estimates as too few to provide conclusive evidence about performance of the method. Although some studies have noted reasonably good agreement between ingrowth core and sequential core approaches (Persson 1983; Symbula and Day 1988), the previous contention should be kept in mind. Fahey and Hughes (1994) applied a correction for measured within-season root turnover (obtained using in situ screens) to long-term ingrowth core data and found reasonable agreement with a C budget estimate of BNPP. However, Hansson et al. (1995) concluded that both ingrowth cores and sequential coring seriously underestimated FRP in a semiarid shrub land, despite the combination of short-term and long-term ingrowth cores designed to partly correct for within-season turnover. Despite the clear advantages of ingrowth cores in terms of costs, it seems likely that FRP comparisons across ecosystems using this technique are unlikely to be reliable, and for purposes of standardized methodology for FRP, ingrowth cores cannot be recommended.

Nadelhoffer and Raich (1992) were impressed by the parsimonious results obtained by the N budget approach in several temperate forests. Both this method and the conceptually similar C budget approach appear promising, especially for cross-site comparisons, and it is surprising that more widespread use of these methods has not occurred to date. The most important uncertainty in these budget approaches is the accuracy of root respiration and soil N mineralization estimates (see Chapter 13, this volume) because the FRP measurements are highly sensitive to error or bias in these estimates.

Most recently a few comparisons of the minirhizotron technique against other methods of measuring FRP have been conducted. Aerts et al. (1989, 1992) compared sequential coring and minirhizotrons in wet and dry heathland in the Netherlands. The two methods gave reasonably similar values in dry heathland but much different results in wet heathlands. They regarded the results from minirhizotrons as more reliable, largely because of the usual uncertainties associated with sequential coring. The comparison of Hansson et al. (1995) for *Artemisia* shrubland in China indicated that FRP values derived from both ingrowth cores and sequential coring were much lower than for minirhizotrons. They, too, considered the minirhizotron estimates more reliable, but again the basis of this conclusion was largely distrust of the coring methods. Finally, Swimmen et al. (1995) compared ^{14}C-based estimates of fine root dynamics in agroecosystems with those derived from minirhizotron observations on the same fields (Van Noordwijk et al. 1994). The minirhizotron estimates of root turnover were 14–37% lower than for the isotope pulse label technique, a difference they ascribed in part to the root length basis of the minirhizotron estimate contrasted with the carbon-based isotope estimate. Although many comparative evaluations of minirhizotron estimates of root length and biomass have been reported (see later), to our knowledge these are the only comparisons for FRP. Clearly, the accuracy of FRP estimates by the minirhizotron technique requires additional testing, and for natural perennial vegetation, comparisons with C and N budget estimates are a very high research priority.

Although there is not sufficient information to conclude that one standard method for measuring FRP should be adopted (see earlier), the minirhizotron method shows high promise of providing both accurate FRP estimates and concurrent information on fine root demography in most ecosystems. For this reason we recommend the minirhizotron technique as the method of choice for LTER but at the same time caution that its full interpretation awaits further work in many habitats.

Minirhizotrons for Estimating Fine Root Production and Demography

Taylor (1987) presents a comprehensive overview of minirhizotron technologies. As described previously, the basic minirhizotron system (Fig. 20.1) is composed of a clear acrylic tube about 5 cm in diameter that is placed into the soil at an angle. At intervals of a week or more a specialized video camera is inserted into the tube to record changes in root number and size at specific locations along the tube. These changes are used to infer root turnover and demographics.

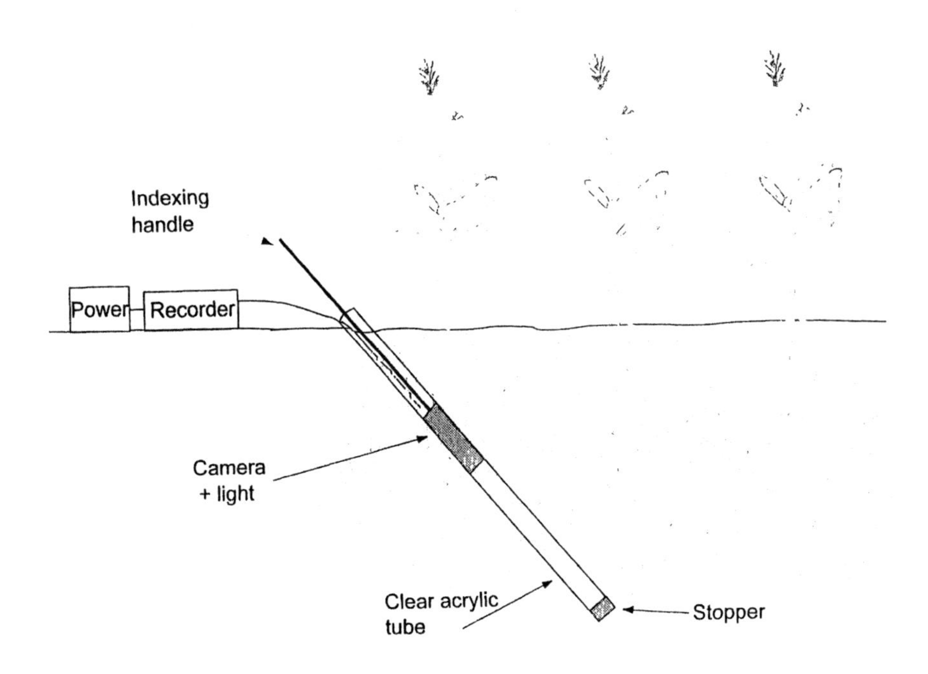

Materials

1. Minirhizotron video system. At the time of this writing the only U.S. source of minirhizotron camera systems is Bartz Technology Co. (Santa Barbara, CA); however, some researchers utilize endoscope or boroscope systems available from many medical supply distributors to obtain images, particularly when inflatable minirhizotron access systems are necessary (see later; Lopez et al. 1996). The video-based minirhizotron systems consist of a color charge-coupled device (CCD) video camera and indexing handle (Ferguson and Smucker 1989); a power supply and system control box; various attachments to remotely control lighting, focus, and the position of the video sensor; and a SVHS or Hi-8 video recorder and monitor to view images. As in the section "Data Collection, Image Analysis, Data Management" (below), computer hardware and software also are needed to automate the digitization and analysis of images.
2. Minirhizotron access tubes, of a diameter appropriate for the minirhizotron camera (usually 50 or 32 mm i.d.). Clear acrylic or polybutyrate tubes available from major plastic distributors should be cut to an appropriate length and marked to provide permanent reference points. To this end a series of sampling frames slightly smaller than the camera lens (e.g., 15 × 10 mm) should be positioned vertically along the tube. Craine and Tremmel (1995) describe a low-cost approach using transparency film, while Kloeppel and Gower (1995) describe an automated tube etching approach. To maximize the amount of information available from each tube, they should be close together along

the tubes. Ferguson and Smucker (1989) designed an index handle and associated computer software that enable users to identify each video frame along the minirhizotron tube, and manually count and record root numbers within each video frame without etching the minirhizotron tube.

Procedures

Access Tube Installation

Insert access tubes into soil core holes extracted to the maximum depth (e.g., to obstruction) or to a depth sufficient to encompass most of the plant rooting zone using an auger or Giddings probe. Access tubes should be installed at an angle to minimize artifacts owing to the tube effects on soil resources. The optimal angle of placement will vary with soil properties, and it is only essential that the angle be measured so that the actual depth of each sample can be calculated. Sampling frames should be oriented upward toward the soil surface. Prior to installation, the bottom of the tubes should be permanently stoppered to prevent seepage of water from below. To assure that soil profile temperature is not greatly affected by the access tube, it may be important to insulate the tubes against daily and seasonal fluctuations in air temperature; this is easily accomplished using foam pipe insulation inserted into the tubes.

Several comparisons of root length density between minirhizotron and soil core sampling have indicated that both depth distributions and absolute amounts of root length density may not be accurately measured by minirhizotron (Parker et al. 1991; Heeraman and Juma 1993; Volkmar 1993; Wiesler and Horst 1994). These artifacts have been ascribed to the influence of tubes on root growth (Brown et al. 1991; DeRuijter et al. 1996), effects of tube placement on soil properties adjacent to the tubes (Hummel et al. 1989), and poor contact or voids between tubes and soil (Parker et al. 1991; Gijsman et al. 1991; Volkmar 1993). Although these influences on root length density might not adversely affect FRP estimates based on the root longevity approach (described later), it is nonetheless clear that great care is needed in field installation of tubes. Especially in very stony or swelling soils, the problems of poor soil-tube contact and installation effects on soil properties have led to recent efforts to perfect a pressurized wall or inflatable minirhizotron access system (Gijsman et al. 1991; Lopez et al. 1996; Merrill 1992). However, at present these systems have not been adapted to allow the repeated measurement of indexed roots necessary for estimating FRP.

Soil coring techniques for minirhizotron tube placement will vary with soil properties, and we recommend consulting with regional soil scientists. In general, core holes must be straight and provide a very snug fit for the access tubes (Box et al. 1989). In some soils, tubes are best installed when the profile is moist to minimize clay smearing of the tube surface, which can be a serious problem (Meyer and Barrs 1991). The portion of the access tube protruding from the soil surface should be darkened to prohibit entry of solar radiation, and the top should be capped or stoppered to prevent rainfall entry. In many soils it may be necessary to anchor the access tubes to prevent movement that would cause shifting of the position of the sample frames. Such shifts are usually obvious but can prove very difficult to correct.

Sampling Design

The sampling design for minirhizotron tube placement will depend on vegetation structure and composition and the specific aims of the root measurement program. Similarly, the optimum number of access tubes will depend mostly on variation in root dynamics across the site. At this point there is little quantitative basis on which to judge the optimum sample size for most natural ecosystems; however, comparative studies of root coring and minirhizotrons for root length density measurements in agroecosystems may provide a limited basis for judging sample size requirements for croplands (Box et al. 1989; Majdi et al. 1992; Heeraman and Juma 1993; Murphy and Smucker 1995). The limiting step in most minirhizotron studies is likely to be information management (digitization and analysis), the demands of which depend mostly on the total number of roots captured in the images. Hence, in cases where root length density or FRP is expected to be low, more access tubes will be feasible than where root length density or FRP is high. Studies to date indicate that in most ecosystems it will be necessary to pool the data from several tubes to obtain sufficient information to calculate FRP (Hendrick and Pregitzer 1996), but replication may be obtained by stratified groups of tubes.

Some observations indicate large differences in fine root dynamics among species in natural vegetation (Aerts et al. 1989, 1992; Hansson et al. 1995). If such differences are known or anticipated, then direct measurements of the root systems of individual plants may provide more efficient measurement of FRP than random sampling. However, in species-rich vegetation and in most forests where the degree of overlap in the distribution of individual plant roots is great, randomized placement of tubes is usually indicated. Although high spatial variation in fine root biomass significantly afflicts the accuracy and precision of biomass estimates, it may not represent a serious problem for FRP estimates because the principal objective is an accurate and unbiased estimate of root longevity. Other than the aforementioned species effects, there is little basis to judge possible spatial variations in root longevity within ecosystems.

The length of the conditioning period prior to the initiation of detailed sampling has not yet been categorically investigated. Depending on root growth characteristics in a particular ecosystem, during a period lasting from several months to several years after installation the root length density on the minirhizotron tubes will increase as roots colonize the soil adjacent to tubes. For example, Hendrick and Pregitzer (1996) indicated full colonization after about 1 year in several temperate forests, whereas even after 3 years root length density was still increasing in a dry heathland ecosystem (Aerts et al. 1992). Although there is currently no quantitative basis with which to judge, we recommend that a conditioning period of one growing season be employed as a standard practice. As described later, especially when roots of natural perennial vegetation are measured in cohorts, it is necessary to obtain images of all the roots that are present on the tubes prior to the first date of quantitative sampling so that new root segments can be distinguished.

For the purpose of quantifying FRP and describing root phenology and demography, sampling at a frequency of 20–30 days should be adequate in most temperate zone forest ecosystems. Sampling frequencies of 10–14 days are recommended for the accurate evaluations of root turnover in grassland and agricultural ecosystems (Pietola and Smucker 1995; Box et al. 1989; Majdi et al. 1992). Again, limitations of data management may be encountered with more frequent sampling, particularly if the annual interval of root growth is long, such as in tropical and warm temperate climates. Conversely, where root growth activity is concentrated into brief, favorable intervals, more frequent sampling may be advisable. It is not necessary that the interval between sampling dates be constant throughout the year.

Data Collection, Image Analysis, and Data Management

Photographic images from the minirhizotron camera are collected on videotape. Each sampling frame is taped for a few seconds and can be subsequently reviewed for digitizing. Several choices of hardware and software are available to digitize and analyze the images. In terms of standardization, the digitization and data management systems should permit each root to be indexed, measured, and classified according to rules that allow standard calculations. Typically, digitization programs include a video capture board, a data compression algorithm, and an image processing program that accurately analyzes and records appropriate root morphologies in each video image (Smucker 1993; Smucker and Aiken 1992).

For example, ROOTS, an interactive PC-based software program developed at Michigan State University, is one system that has been designed to digitize minirhizotron images stored on SVHS tape (Hendrick and Pregitzer 1992). The program is designed to be used with a TARGA (Truevision, Indianapolis, IN) video board, which temporarily captures images on the monitor for digitizing. The length and diameter of each root are traced using a mouse, and ROOTS calculates and saves these measurements to a dBASE file (Ashton-Tate, Torrance, CA). Each individual root within a frame is assigned a unique number and morphological category, and digitized tracings of each root are stored (Hendrick and Pregitzer 1992). When the same frame from the subsequent sampling date is digitized, individual roots are reidentified, and ROOTS recalls each numbered tracing. Thus, the growth, morphological changes, and death/disappearance of individual roots are readily quantified. A conceptually similar software program for MacIntosh computers also is available (Craine and Tremmel 1995).

One possible problem for purposes of standardization is that the most appropriate root categories, in terms either of diameter, morphology, or hierarchy classes, are likely to differ among vegetation types. However, if the necessary information is available for each root category, then differences among sites and studies in the categories employed will not preclude standardized calculations and comparisons of root production.

Root image processing is becoming more rapid and quantitative. Recent developments in the Root Image Processing Laboratory at Michigan State University have made significant improvements in the digitalization, processing, and databasing of minirhizotron video images. The computer program MR-RIPL measures numerous root parameters from video-recorded images by the minirhizotron microvideo camera. Using multiple filters, MR-RIPL algorithms measure root length, surface area, volume, and total length for five separate width classes of roots. The program processes each image or frame, 320 × 240 pixels, from 1 to 16 seconds, on a SunSparc Ultra 1 and generates files containing the preceding root parameters. A tangential program processes these video images into one image and generates a compressed statistical database for several minirhizotron tubes for purposes of cross-comparisons. These programs are implemented in the C language and tested on Sun workstations and PC computers, using the LINUX operating system. A complete operation manual for the MR-RIPL is available for review on the RIPL Home Page of the World Wide Web at the location http://rootdata.css.msu.edu. Using a combination of ridge detection and mole filters, centerlines of video-recorded roots, observed at the surfaces of minirhizotron tubes, are measured for length, width, parallelism, intensity, and angle changes. This approach, combined with some artificial intelligence, provides opportunities to quantify 20 morphological root parameters.

Recent studies of woody vegetation (Hendrick and Pregitzer 1992) indicate that four categories of roots can usually be distinguished readily with the current technology: white roots, tan roots, brown roots, and black roots, the latter category often being assumed to be dead. Observation of black, apparently dead roots should continue until they disappear because these may occasionally show continued growth or branching.

Calculations

Roots of most perennials and some annuals should be analyzed as cohorts. That is, all segments of roots that appear in the interval between two sampling dates should be assigned to a cohort, with the known or assumed date of "birth" being the midpoint between sampling dates. Subsequently, these cohorts must be analyzed for changes in size, morphology, and/or hierarchical status through their entire life span, i.e., until they disappear from the images. By following cohorts in this way, the median longevity of each cohort can be measured and fine root length production calculated as the ratio of root length density to longevity (Hendrick and Pregitzer 1992). In practice, these calculations become much more complex because the longevity and length associated with the various cohorts will differ. For energy and nutrient budget purposes, root length production estimates must be augmented with information on specific root length (SRL; root length/mass) to permit calculation of biomass production. Moreover, because individual root segments often undergo diameter changes during their lifetime (and hence specific root length changes), calculation of biomass production and nutrient fluxes becomes still more complex.

As a starting point we recommend that the simplest estimates of FRP may be the most appropriate standardized protocol. This estimation technique depends only on the calculation of median longevity and root length for each cohort. Additional morphological evaluations (e.g., root branching frequencies, diameters, calculated surface areas and volumes, numbers of root tips, root parameters versus soil depth) will be based on particular experimental objectives.

The information captured by minirhizotrons allows two somewhat independent calculations of FRP and mortality, and we recommend that both calculations be carried out to provide a comparative basis for better evaluating minirhizotron production estimates. Both protocols rely on an independent destructive measurement of fine root biomass, obtained at the beginning of the growing season or at the time of expected annual average biomass. For the first approach, minirhizotrons are used to obtain estimates of root longevity for cohorts produced during each month of the growing season, and the overall median fine root longevity is calculated by weighting the cohort longevities according to their relative contribution to total root length production for the year. FRP (or mortality) under the steady-state assumption is calculated as the ratio between average fine root biomass (g/m^2) and weighted median cohort longevity (yr). This approach requires only that minirhizotrons provide accurate information on fine root longevity (Hendrick and Pregitzer 1992).

The second approach is to calculate the ratio of the annual production of root length to the initial root length in the minirhizotron images. This ratio is multiplied by the initial fine root biomass obtained with soil cores at the start of the growing season to calculate production (Hendrick and Pregitzer 1993). Similarly, annual root mortality can be estimated using the ratio of annual length mortality to initial length. This method does not assume long-term steady state, but it requires that the minirhizotron surface does not significantly affect the length or growth patterns of fine roots. Based on the aforementioned observations of tube influences on root growth and distribution, there is reason to expect the former method to yield more reliable estimates of FRP. For both methods and in many ecosystems, the accuracy and applications of these estimates might be significantly increased by separating both the biomass and longevity estimates for different soil horizons or depth increments. For example, in many soils root longevity and root length densities vary markedly with depth (Persson 1983; Gholz et al. 1986; Fahey and Hughes 1994; Hendrick and Pregitzer 1996). As a standard procedure we recommend calculation by three depth categories: organic horizons (where present), A and B horizons, and below the B horizon.

The survivorship data generated using minirhizotrons are analogous in many ways to survival data obtained in other fields of inquiry, and sophisticated approaches to analyzing factors contributing to differential survivorship have been developed (Cox and Oakes 1984). No standardized technique can be recommended at this time for either within-site or cross-ecosystem comparisons because the particular forms of auxiliary data will likely vary depending on the nature of the comparisons being made.

Fine Root Demography

Comparative estimates of FRP and its response to environmental variation and change will greatly improve our understanding of patterns of energy flow and nutrient cycling in terrestrial ecosystems. However, a mechanistic picture of community and ecosystem dynamics would require more detailed information; for example, a mechanistically based nutrient cycling model might depend on daily estimates of fine root length and root nutrient demands. The information collected with minirhizotron systems permits the quantification of fine root demography from which periodic estimates of the distribution of fine root length, among the various other functional categories, can be obtained. A standard procedure for calculating and expressing patterns of fine root demography would permit cross-ecosystem comparisons of root function and its responses to environmental and biotic cues.

Hendrick and Pregitzer (1992) described a simple demographic model in which the transition probabilities between various root morphology classes were calculated from minirhizotron data. If standardized root morphology categories could be devised, then root phenology and demography could be compared across ecosystems and incorporated into models. The recognition of morphology classes depends on the quality of minirhizotron images as well as on the root characteristics themselves. For example, current technology does not always permit easy recognition of mycorrhizal versus nonmycorrhizal roots, nor can the fine roots of different species of the same life form be readily distinguished. In fact, the morphological criteria that have been applied to date (e.g., color) are not clearly associated with any well-characterized functional differences. Thus, although the minirhizotron technique eventually should be capable of providing standardized demographic parameters that are connected with root system function, a standardized protocol for fine root demography must await technological development to the point that functionally defined categories can be distinguished and quantified.

A Final Consideration

A lingering question must be addressed before full confidence can be afforded to minirhizotron estimates of FRP and demography: do roots growing adjacent to minirhizotron tubes exhibit the same dynamics as those growing in bulk soil? As noted earlier, comparisons with root cores indicate that root length density may not be accurately represented by minirhizotrons, but this deficiency would not affect our proposed protocol for root production as long as the tubes do not alter the longevity of fine roots or the *relative* production and mortality of root length. Conclusive tests of these assumptions will be difficult because alternative approaches to FRP are problematic. However, comparisons with the C and N budgeting approaches would seem prudent.

One partial test has been reported that compared the morphological characteristics of roots obtained with cores against those from minirhizotron images (Majdi et al. 1992). The frequency distributions of total root lengths were similar. This test could be improved, however, by comparing frequency distributions of root length and duration by size classes and hierarchical (i.e., branching) classes. A second partial test could be obtained using in situ root screens (Fahey and Hughes 1994), which generate the same information as minirhizotrons but under conditions where the measured roots do not grow along an access tube. This technique is feasible only in the organic horizons of acid forest soils that undergo little faunal mixing; however, if results from this method were very similar to those obtained with minirhizotrons, we could be more confident in the efficacy of the latter under other soil and vegetation conditions.

References

Aber, J. D., J. M. Melillo, K. J. Nadelhoffer, C. A. McClaugherty, and J. Pastor. 1985. Fine root turnover in forest ecosystems in relation to quantity and form of nitrogen availability: a comparison of two methods. *Oecologia* 66:317–321.

Aerts, R., C. Bakker, and H. DeCaluwe. 1992. Root turnover as a determinant of the cycling of carbon, nitrogen and phosphorus in a dry heathland ecosystem. *Biogeochemistry* 15:175–190.

Aerts, R., F. Berendse, N. M. Klerk, and C. Bakker. 1989. Root production and root turnover in two dominant species of wet heathlands. *Oecologia* 81:374–378.

Bloom, A. J., and R. M. Caldwell. 1988. Root excision decreases nutrient absorption and gas fluxes. *Plant Physiology* 87:794–796.

Bowden, R. D., K. J. Nadelhoffer, R. D. Boone, J. M. Melillo, and J. B. Garrison. 1993. Contributions of aboveground litter, belowground litter and root respiration to total soil respiration in a temperate mixed hardwood forest. *Canadian Journal of Forest Research* 23:1402–1407.

Box, J. E., Jr. 1996. Modern methods for root investigations. Pages 193–237 *in* Y. Waisel, A. Eshel, and U. Kafkafi, editors, *Plant Roots: The Hidden Half.* 2d edition. Marcel Dekker, New York, New York, USA.

Box, J. E., A. J. M. Smucker, and J. T. Ritchie. 1989. Minirhizotron installation techniques for investigating root responses to drought and oxygen stress. *Soil Science Society of America Journal* 53:115–118.

Brown, D. P., T. K. Pratum, C. Bledsoe, E. D. Ford, J. S. Cothern, and D. Perry. 1991. Non-invasive studies of conifer roots: nuclear magnetic resonance (NMR) imaging of Douglas-fir seedlings. *Canadian Journal of Forest Research* 21:1559–1566.

Burke, M. K., and D. J. Raynal. 1994. Fine root growth phenology, production and turnover in a northern hardwood forest ecosystem. *Plant and Soil* 162:135–146.

Burton, A. J., K. S. Pregitzer, G. P. Zogg, and D. R. Zak. 1996. Latitudinal variation in sugar maple fine root respiration. *Canadian Journal of Forest Research* 26:1761–1768.

Caldwell, M. M., and L. B. Camp. 1974. Belowground productivity of two cool desert communities. *Oecologia* 17:123–130.

Cox, D. R., and D. Oakes. 1984. *Analysis of Survival Data.* Chapman and Hall, New York, New York, USA.

Craine, J., and D. Tremmel. 1995. Improvements to the minirhizotron system. *Bulletin of the Ecological Society of America* 76:234–235.

DeRuijter, F. J., B. W. Veen, and M. VanOijen. 1996. Comparison of soil core sampling and minirhizotrons to quantify root development of field grown potatoes. *Plant and Soil* 182:301–312.

Ewel, K. C., W. P. Cropper Jr., and H. L. Gholz. 1987. Soil CO_2 evolution in Florida slash pine plantations. II. Importance of root respiration. *Canadian Journal of Forest Research* 17:330–333.

Fahey, T. J., and M. A. Arthur. 1994. Further studies of root decomposition following harvest of a northern hardwood forest. *Forest Science* 40:618–629.

Fahey, T. J., and J. W. Hughes. 1994. Fine root dynamics in a northern hardwood forest ecosystem, Hubbard Brook Experimental Forest, NH. *Journal of Ecology* 82:533–548.

Fahey, T. J., J. B. Yavitt, J. A. Pearson, and D. H. Knight. 1985. The nitrogen cycle in lodgepole pine forests, southeastern Wyoming. *Biogeochemistry* 1:257–276.

Ferguson, J. C., and A. J. M. Smucker. 1989. Modifications of the minirhizotron video camera system for measuring spatial and temporal root dynamics. *Soil Science Society of America Journal* 53:1601–1605.

Gholz, H., L. C. Hensley, and W. P. Cropper Jr. 1986. Organic matter dynamics of fine roots in plantations of slash pine (*Pinus elliottii*) in north Florida. *Canadian Journal of Forest Research* 16:529–538.

Gijsman, A. J., J. Floris, M. van Noordwijk, and G. Brouwer. 1991. An inflatable minirhizotron system for root observation with improved soil-tube contact. *Plant and Soil* 134:261–270.

Hansson, A. C., Z. Aiferi, and O. Andren. 1995. Fine-root production and mortality in degraded vegetation in Horqin sandy rangeland in Inner Mongolia, China. *Arid Soil Research and Rehabilitation* 9:1–13.

Haynes, B. E., and S. T. Gower. 1995. Belowground carbon allocation in unfertilized and fertilized red pine plantations in northern Wisconsin. *Tree Physiology* 15:317–325.

Head, G. C. 1966. Estimating seasonal changes in the quantity of white unsuberized roots on fruit trees. *Journal of Horticultural Science* 41:197–206.

Heeraman, D. A., and N. G. Juma. 1993. A comparison of minirhizotron, core and monolith methods for quantifying barley (*Hordeum vulgare* L.) and fababean (*Vicia faba* L.) root distribution. *Plant and Soil* 148:29–41.

Hendrick, R. L., and K. S. Pregitzer. 1992. The demography of fine roots in a northern hardwood forest. *Ecology* 73:1094–1104.

Hendrick, R. L., and K. S. Pregitzer. 1993. The dynamics of fine root length, biomass and nitrogen content in two northern hardwood ecosystems. *Canadian Journal of Forest Research* 23:2507–2520.

Hendrick, R. L., and K. S. Pregitzer. 1996. Applications of minirhizotrons to understand root function in forests and other natural ecosystems. *Plant and Soil* 185:293–304.

Hummel, J. W., M. A. Levan, and K. A. Sadduth. 1989. Minirhizotron installation in heavy soil. *Transactions of the American Society of Agricultural Engineers* 32:770–776.

Kelting, D. L., J. A. Burger, and G. S. Edwards. 1995. The effects of ozone on the root dynamics of seedlings and mature red oak (*Quercus rubra* L.). *Forest Ecology and Management* 79:197–206.

Kloeppel, B. D., and S. T. Gower. 1995. Construction and installation of acrylic minirhizotron tubes in forest ecosystems. *Soil Science Society of America Journal* 59:241–243.

Kurz, W. A., and J. P. Kimmins. 1987. Analysis of some sources of error in methods used to determine fine root production in forest ecosystems: a simulation approach. *Canadian Journal of Forest Research* 17:909–912.

Lopez, B., S. Sabate, and C. Garcia. 1996. An inflatable minirhizotron system for stony soils. *Plant and Soil* 179:255–260.

Majdi, H. 1996. Root sampling methods—applications and limitations of the minirhizotron technique. *Plant and Soil* 185:255–258.

Majdi, H., A. J. M. Smucker, and H. Persson. 1992. A comparison between minirhizotron and monolith sampling methods for measuring root growth of maize (*Zea mays,* L.). *Plant and Soil* 104:127–134.

Merrill, S. D. 1992. Pressurized-wall minirhizotron for field observation of root growth dynamics. *Agronomy Journal* 84:755–758.

Meyer, W. S., and H. D. Barrs. 1991. Roots in irrigated clay soils: measurement techniques and responses to root zone condition. *Irrigation Science* 12:125–134.

Milchunas, D. G., and W. K. Lauenroth. 1992. Carbon dynamics and estimates of primary production by harvest, ^{14}C dilution, and ^{14}C turnover. *Ecology* 73:593–607.

Milchunas, D. G., W. K. Laurenroth, J. S. Singh, C. V. Cole, and H. W. Hunt. 1985. Root turnover and production by ^{14}C dilution: implication of carbon partitioning in plants. *Plant and Soil* 88:353–365.

Murphy, S. L., and A. J. M. Smucker. 1995. Evaluation of video image analysis and line-intercept methods for measuring root systems of alfalfa and ryegrass. *Agronomy Journal* 87:865–868.

Nadelhoffer, K. J., J. D. Aber, and J. M. Melillo. 1985. Fine roots, net primary production, and nitrogen availability: a new hypothesis. *Ecology* 66:1377–1390.

Nadelhoffer, K. J., and J. W. Raich. 1992. Fine root production estimates and belowground carbon allocation in forest ecosystems. *Ecology* 73:1139–1147.

Neill, C. 1992. Comparison of soil coring and in-growth methods for measuring belowground production. *Ecology* 73:1918–1921.

Parker, C. J., M. K. V. Carr, N. J. Jarvis, B. O. Puplampu, and V. H. Lee. 1991. An evaluation of the minirhizotron technique for estimating root distribution in potatoes. *Journal of Agricultural Science* 116:341–350.

Persson, H. 1983. The distribution and production of fine roots in boreal forests. *Plant and Soil* 71:87–101.

Pietola, L., and A. J. M. Smucker. 1995. Fine root dynamics for alfalfa and following a late invasion by weeds. *Agronomy Journal* 87:1161–1169.

Publicover, D. A., and K. A. Vogt. 1993. A comparison of methods for estimating forest fine root production with respect to sources of error. *Canadian Journal of Forest Research* 23:1179–1186.

Raich, J. W., and K. J. Nadelhoffer. 1989. Belowground carbon allocation in forest ecosystems: global trends. *Ecology* 70:1346–1354.

Santantonio, D., and J. C. Grace. 1987. Estimating fine root production and turnover from biomass and decomposition data: a compartment flow model. *Canadian Journal of Forest Research* 17:900–908.

Singh, J. S., W. K. Laurenroth, H. W. Hunt, and M. D. Swift. 1984. Bias and random error in estimators of net root production: a simulation approach. *Ecology* 65:1760–1764.

Smucker, A. J. M. 1993. Soil environmental modifications of root dynamics and measurement. *Annual Review of Phytopathology* 31:191–216.

Smucker, A. J. M., and R. M. Aiken. 1992. Dynamic root responses to soil water deficits. *Soil Science* 154:281–289.

Smucker, A. J. M., J. C. Ferguson, W. P. DeBruyan, R. K. Belford, and J. T. Ritchie. 1987. Image analysis of video recorded plant root systems. Pages 67–80 *in* H. M. Taylor, edior, *Minirhizotron Observation Tubes: Methods and Applications for Measuring Rhizosphere Dynamics. ASA Special Publication 50.* American Society of Agronomy, Madison, Wisconsin, USA.

Smucker, A. J. M., S. L. McBurney, and A. K. Srivastava. 1982. Quantitative separation of roots from compacted soil profiles by hydropneumatic elutriation systems. *Agronomy Journal* 74:500–503.

Steen, E. 1985. Root and rhizome dynamics in a perennial grasscrop during an annual growth cycle. *Swedish Journal of Agricultural Research* 156:25–30.

Swimmen, J., J. A. van Veen, and R. Merckx. 1994. Rhizosphere carbon fluxes in field-grown spring wheat: model calculations based on ^{14}C partitioning after pulse-labelling. *Soil Biology and Biochemistry* 26:171–182.

Swimmen, J., J. A van Veen, and R. Merckx. 1995. Root decay and turnover of rhizodeposits in field-grown winter wheat and spring barley estimated by ^{14}C pulse-labelling. *Soil Biology and Biochemistry* 27:211–217.

Symbula, M., and F. P. Day Jr. 1988. Evaluation of two methods for estimating belowground production in a freshwater swamp forest. *American Midland Naturalist* 120:405–415.

Taylor, H. M., editor. 1987. *Minirhizotron Observation Tubes: Methods and Applications for Measuring Rhizosphere Dynamics.* ASA Special Publication 50. American Society of Agronomy, Madison, Wisconsin, USA.

Upchurch, D. R., and J. T. Ritchie. 1983. Root observations using a video recording system in minirhizotrons. *Agronomy Journal* 75:1009–1015.

van Noordwijk, M., G. Brower, H. Koning, F. W. Meijboom, and W. Grzebisz. 1994. Production and decay of structural root material of winter wheat and sugar beet in conventional and integrated cropping systems. *Agriculture, Ecosystems and Environment* 51:99–113.

Vogt, K. A., and H. Persson. 1991. Measuring growth and development of roots. Pages 477–501 *in* J. P. Lassoie and T. M. Hinckley, editors, *Techniques and Approaches in Forest Tree Ecophysiology.* CRC Press, Boca Raton, Florida, USA.

Vogt, K. A., D. J. Vogt, E. E. Moore, and D. G. Sprugel. 1989. Methodological considerations in measuring biomass, production, respiration and nutrient resorption for tree roots in natural ecosystems. Pages 217–232 *in* J. G. Torrey and L. J. Winship, editors, *Applications of Continuous and Steady-State Methods to Root Biology.* Kluwer Academic Publications Dordrecht, Netherlands.

Volkmar, K. M. 1993. A comparison of minirhizotron techniques for estimating root length density in soil of different bulk density. *Plant and Soil* 157:239–245.

Whittaker, R. H., F. H. Bormann, G. E. Likens, and T. G. Siccama. 1974. The Hubbard Brook Ecosystem Study: forest biomass and production. *Ecological Monographs* 44:233–254.

Wiesler, F., and W. J. Horst. 1994. Root growth of maize cultivars under field conditions as studied by the core and minirhizotron method and relationships to shoot growth. *Zeitschrift fuer Pflanzenernaehrung und Bodenkunde* 157:351–358.

Yanai, R. D., T. J. Fahey, and S. L. Miller. 1994. Efficiency of nutrient acquisition by fine roots and mycorrhizae. Pages 75–103 *in* W. K. Smith and T. M. Hinckley, editors, *Physiological Ecology of Coniferous Forest: A Contemporary Synthesis.* Academic Press, San Diego, California, USA.

Index

LaVergne, TN USA
14 July 2010

189414LV00001B/8/A